T0140080

Lecture Notes in Networks and Systems

Volume 47

Series editor

Janusz Kacprzyk, Polish Academy of Sciences, Warsaw, Poland
e-mail: kacprzyk@ibspan.waw.pl

The series "Lecture Notes in Networks and Systems" publishes the latest developments in Networks and Systems—quickly, informally and with high quality. Original research reported in proceedings and post-proceedings represents the core of LNNS.

Volumes published in LNNS embrace all aspects and subfields of, as well as new challenges in, Networks and Systems.

The series contains proceedings and edited volumes in systems and networks, spanning the areas of Cyber-Physical Systems, Autonomous Systems, Sensor Networks, Control Systems, Energy Systems, Automotive Systems, Biological Systems, Vehicular Networking and Connected Vehicles, Aerospace Systems, Automation, Manufacturing, Smart Grids, Nonlinear Systems, Power Systems, Robotics, Social Systems, Economic Systems and other. Of particular value to both the contributors and the readership are the short publication timeframe and the world-wide distribution and exposure which enable both a wide and rapid dissemination of research output.

The series covers the theory, applications, and perspectives on the state of the art and future developments relevant to systems and networks, decision making, control, complex processes and related areas, as embedded in the fields of interdisciplinary and applied sciences, engineering, computer science, physics, economics, social, and life sciences, as well as the paradigms and methodologies behind them.

More information about this series at http://www.springer.com/series/15179

Michael E. Auer · Reinhard Langmann
Editors

Smart Industry & Smart Education

Proceedings of the 15th International
Conference on Remote Engineering
and Virtual Instrumentation

 Springer

Editors
Michael E. Auer
Carinthia University of Applied Sciences
Villach, Austria

Reinhard Langmann
Competence Center Automation Duesseldorf
(CCAD)
Duesseldorf University of Applied Science
Duesseldorf, Germany

ISSN 2367-3370   ISSN 2367-3389 (electronic)
Lecture Notes in Networks and Systems
ISBN 978-3-319-95677-0   ISBN 978-3-319-95678-7 (eBook)
https://doi.org/10.1007/978-3-319-95678-7

Library of Congress Control Number: 2018947462

This Springer imprint is published by the registered company Springer Nature Switzerland AG
The registered company address is: Gewerbestrasse 11, 6330 Cham, Switzerland

Preface

The REV conference is the annual conference of the International Association of Online Engineering (IAOE) and the Global Online Laboratory Consortium (GOLC).

REV2018 was 15th in a series of annual events concerning the area of remote engineering and virtual instrumentation. The general objective of this conference is to contribute and discuss fundamentals, applications, and experiences in the field of Remote Engineering, Virtual Instrumentation, and related new technologies like Internet of things, Industry 4.0, cyber security, M2M, and smart objects. Another objective of the conference is to discuss guidelines and new concepts for education at different levels for the above-mentioned topics including emerging technologies in learning, MOOCs & MOOLs, open resources, and STEM pre-university education.

REV2018 has been organized in cooperation with Hochschule Duesseldorf University of Applied Sciences and the International E-Learning Association (IELA) from March 21 to 23, 2018, in Duesseldorf, Germany.

REV2018 offered again an exciting technical program as well as networking opportunities. Outstanding scientists accepted the invitation for keynote speeches:

- Andreas Schreiber
  Phoenix Contact, Corporate Technology & Value Chain, Director Business & Product Innovation, Blomberg, Germany
- Ishwar K. Puri
  AAAS Fellow, ASME Fellow, CAE Fellow, Dean and Professor, Faculty of Engineering, McMaster University, Hamilton, Ontario, Canada
- Lee Chee Hoe
  Yokogawa Electric International, Global IA Cyber Security Evangelist, Security Consultant for Europe
- Mario Trapp
  Acting Director, Fraunhofer ESK Institute, Munich, Germany
- Rahman Jamal
  Global Technology and Marketing Director at NI worldwide

- Seeram Ramakrishna
  Fellow of UK Royal Academy of Engineering and Professor of Mechanical Engineering at the National University of Singapore.

It was in 2004, when we started this conference series in Villach, Austria, together with some visionary colleagues and friends from around the world. When we started our REV endeavor, the Internet was just 10 years old! Since then, the situation regarding online engineering and virtual instrumentation has radically changed. Both are today typical working areas of most of the engineers and are inseparably connected with

- Internet of things
- Cyber-physical systems
- Collaborative networks and grids
- Cyber cloud technologies
- Service architectures

to name only a few.

With our conference in 2004 (fifteen years ago), we tried to focus on the upcoming use of the Internet for engineering tasks and the problems around it. And as we can see, it was very successful.

The REV2018 conference takes up the following topics in its variety and discusses the state of the art and future trends under the global theme "Smart Industry & Smart Education":

- Online engineering
- Cyber-physical systems
- Internet of things and industrial internet of things
- Industry 4.0
- Cyber security
- M2M concepts and smart objects
- Virtual and remote laboratories
- Remote process visualization and virtual instrumentation
- Remote control and measurement technologies
- Networking, grid, and cloud technologies
- Mixed-reality environments
- Telerobotics and telepresence
- Collaborative work in virtual environments
- Smart city, smart energy, smart buildings, and smart homes
- New concepts for engineering education in higher and vocational education
- Augmented reality and human–machine interaction
- Standards and standardization proposals
- Applications and experiences.

All submission types have been accepted:

- Full paper and short paper
- Work in progress and poster
- Special sessions
- Interactive demonstrations, workshops, and tutorials.

All contributions were subject to a double-blind review. The review process was very competitive. We had to review nearly 270 submissions. A team of 200 reviewers did this terrific job. Our special thanks go to all of them.

Due to the time and conference schedule restrictions, we could finally accept only the best 110 submissions for presentation. The conference had again about 140 participants from 39 countries from all the continents.

Michael E. Auer
REV General Chair
Reinhard Langmann
REV2018 Chair

Committees

General Chair

Michael E. Auer   IAOE President & CEO

REV 2018 Chair

Reinhard Langmann   Duesseldorf University of Applied Sciences

International Advisory Board

Abul Azad   President Global Online Laboratory Consortium,
                                   USA
Philip Bailey   MIT, Cambridge MA, USA
Denis Gillet   EPFL Lausanne, Switzerland
Bert Hesselink   Stanford University, USA
Zorica Nedic   University of South Australia
Cornel Samoila   University of Brasov, Romania
Franz Schauer   Tomas Bata University, Czech Republic
Tarek Sobh   University of Bridgeport, USA

Program Co-chairs

Harald Jacques   Duesseldorf University of Applied Sciences
Doru Ursutiu   IAOE President

Technical Program Chair

Danilo G. Zutin   IAOE Vice president

IEEE Liaison

Manuel Castro, Spain

Workshop and Tutorial Chair

Andreas Pester, Austria

Special Session Chair

Teresa Restivo, Portugal

Demonstration and Poster Chair

Christian Geiger   Duesseldorf University of Applied Sciences, Germany

Publication Chair and Web Master

Sebastian Schreiter, France

International Program Committee

Akram Abu-Aisheh	Hartford University, USA
Laiali Almazaydeh	Al-Hussein Bin Talal University, Jordan
Yacob Astatke	Morgan State University, USA
Gustavo Alves	ISEP Porto, Portugal
Chris Bach	University of Bridgeport, USA
Nael Bakarad	Grand Valley State University, USA
David Boehringer	University of Stuttgart, Germany
Michael Callaghan	University of Ulster, Northern Ireland
Manuel Castro	UNED Madrid, Spain

Arthur Edwards	University of Colima, Mexico
Torsten Fransson	KTH Stockholm, Sweden
Javier Garcia-Zubia	University of Deusto, Spain
Denis Gillet	EPFL Lausanne, Switzerland
Olaf Graven	Buskerud University College, Norway
Ian Grout	University of Limerick, Ireland
Christian Guetl	Graz University of Technology, Austria
M. Carmen Juan Lizandra	Universidad Politécnica de Valencia, Spain
Alexander Kist	University of Southern Queensland, Australia
Vinod Kumar Lohani	Virginia Tech, VA, USA
Petros Lameras	Coventry University, UK
Sergio Cano Ortiz	Universidad de Oriente, Cuba
Carlos Alberto Reyes Garcia	INAOE Puebla, Mexico
Joerg Stefan-Reiff	Wildau Technical University of Applied Sciences, Germany
Jörg Stöcklein	University of Paderborn, Frauenhofer IEM, Germany
Ananda Maiti	University of Southern Queensland, Australia
Dominik May	TU Dortmund, Germany
Zorica Nedic	University of South Australia, Australia
Kalyan B. Ram	Electrono Solutions Pvt Ltd, India
Ingmar Riedel-Kruse	Stanford University, USA
Franz Schauer	Tomas Bata University, Czech Republic
Juarez Silva	University of Santa Catarina, Brazil
Vladimir Uskov	Bradley University, USA
Matthias Christoph Utesch	Technical University of Munich, Germany
Igor Verner	Technion Haifa, Israel
Dieter Wuttke	TU Ilmenau, Germany
Katarina Zakova	Slovak University of Technology, Slovakia
Holger Reckter	Mainz University of Applied Sciences, Germany
Volker Paelke	Bremen Univ. of Applied Sciences, Germany
Shen Yan	SIPAI Shanghei, China
Stefan Marks	Auckland University of Technology, New Zealand
Prabhu Vinayak Ashok	Nanyang Polytechnic, Singapore

Contents

Internet of Things and Industry 4.0

A Cloud-Based Blended Learning Lab for PLC Education   3
Reinhard Langmann and Matthias Coppenrath

Environmental Sound Recognition with Classical Machine
Learning Algorithms   14
Nikolina Jekic and Andreas Pester

Early Signs of Diabetes Explored from an Engineering Perspective   22
Jenny Lundberg and Lena Claesson

Digitalization of Engineering Education: From E-Learning
to Smart Education   32
Irina Makarova, Ksenia Shubenkova, Dago Antov,
and Anton Pashkevich

Investigating Rate Increase in Aerospace Factory By Simulation
of Material Flow Operations   42
Laura Lopez-Davalos, Amer Liaqat, Windo Hutabarat, Divya Tiwari,
and Ashutosh Tiwari

Demonstration: Cloud-Based Industrial Control Services   50
Reinhard Langmann and Leandro Rojas-Peña

Poster: Teaching Automation and Logistics with Virtual
Industrial Process   57
Florence Lecroq, Jean Grieu, and Hadhoum Boukachour

Remote Control and Measurements

Use of VISIR Remote Lab in Secondary School: Didactic
Experience and Outcomes   69
Manuel Blazquez-Merino, Alejandro Macho-Aroca, Pablo Baizán-Álvarez,
Félix Garcia-Loro, Elio San Cristobal, Gabriel Diez, and Manuel Castro

On Effective Maintenance of Distributed Remote Laboratories   80
Tobias Fäth, Karsten Henke, René Hutschenreuter, Felix Seidel,
and Heinz-Dietrich Wuttke

A Multi-agent System for Supervisory Temperature Control Using
Fuzzy Logic and Open Platform Communication Data Access   90
Martha Kafuko and Tom Wanyama

Combining Virtual and Remote Interactive Labs and Visual/Textual
Programming: The Furuta Pendulum Experience   100
Daniel Galan, Luis de la Torre, Dictino Chaos, and Ernesto Aranda

TRIANGLE Portal: An User-Friendly Web Interface
for Remote Experimentation   110
Almudena Díaz-Zayas, Alberto Salmerón Moreno,
Gustavo García Pascual, and Pedro Merino Gómez

"Hands-on-Remote" Laboratories   118
Frantisek Lustig, Pavel Brom, Pavel Kuriscak, and Jiri Dvorak

Virtual Power-Line Communications Laboratory
for Technology Development and Research   128
Asier Llano Palacios, Xabier Osorio Barañano,
David de la Vega Moreno, Itziar Angulo Pita, and Txetxu Arzuaga Canals

Development of an Automatic Assessment in Remote
Experimentation Over Remote Laboratory   136
Abderrahmane Adda Benattia, Abdelhalim Benachenhou,
and Mohammed Moussa

Demonstration: Using Remotely Controlled One-Way Flow
Control Valve for Speed Regulation of Pneumatic Cylinder   144
Brajan Bajči, Slobodan Dudić, Jovan Šulc, Vule Reljić, Dragan Šešlija,
and Ivana Milenković

Demonstration: Virtual Lab for Analog Electronic Circuits   153
K. C. Narasimhamurthy, Ankit Sharma, Shorya Shubham,
and H. R. Chandan

Poster: Wireless Sensor Network to Predict Black Sigatoka
in Banana Cultivations   159
Andrés Subert-Semanat

Poster: Influence of the Direction of Movement of Earth-Moving
and Construction Machines on the Stability of Remote Control Data
Transmission via Mobile Communication Channels   165
Tatyana Golubeva, Sergey Konshin, Sergey Leshchev,
Natalia Mironova, and Boris Tshukin

Virtual and Remote Laboratories

Development and Implementation of Remote Laboratory as an
Innovative Tool for Practicing Low-Power Digital Design
Concepts and Its Impact on Student Learning 175
Shatha AbuShanab, Marco Winzker, and Rainer Brück

Remote Labs for Electrical Power Transmission Lines
Simulation Unit 186
Kalyan B. Ram, Panchaksharayya S. Hiremath, M. S. Prajval, B. Karthick,
Prasanth Sai Meda, M. B. Vijayalakshmi, and Priyanka Paliwal

Sustainability of the Remote Laboratories Based on Systems
with Limited Resources 197
Galyna Tabunshchyk, Tetiana Kapliienko, and Peter Arras

What Are Teachers' Requirements for Remote Learning Formats?
Data Analysis of an E-Learning Recommendation System 207
Thorsten Sommer, Valerie Stehling, Max Haberstroh, and Frank Hees

Evaluating Remote Experiment from a Divergent Thinking Point
of View 217
Cornel Samoila, Doru Ursutiu, and C. A. Neagu

"Electromagnetic Remote Laboratory" with Embedded
Simulation and Diagnostics 226
Franz Schauer, Michal Gerza, Michal Krbecek, Das Sayan,
Mbuotidem Ime Archibong, and Miroslava Ozvoldova

Smart Grid Remote Laboratory 234
Kalyan B. Ram, S. Arun Kumar, Manish Ahlawat,
Sanjoy Kumar Parida, S. Prathap, Preeti S. Biradar, and Vishnu Das

e-LIVES – Extending e-Engineering Along the South
and Eastern Mediterranean Basin 244
Manuel Gericota, Paulo Ferreira, André Fidalgo, Guillaume Andrieu,
Abdallah Al-Zoubi, Majd Batarseh, and Danilo Garbi-Zutin

A Reliability Assessment Model for Online Laboratories Systems 252
Luis Felipe Zapata-Rivera and Maria M. Larrondo-Petrie

Digital Remote Labs Built by the Students and for the Students 261
J. Nikhil, J. Pavan, H. O. Darshan, G. Anand Kumar, J. Gaurav,
and C. R. Yamuna Devi

Virtual Learning Environment for Digital Signal Processing 269
Yadisbel Martinez-Cañete, Sergio Daniel Cano-Ortiz,
Frank Sanabria-Macias, Reinhard Langmann, Harald Jacques,
and Pedro Efrain Diaz-Labiste

Online Experimenting with 3D LED Cube 277
Katarína Žáková

Management of Control Algorithms for Remote Experiments 283
Matej Rábek and Katarína Žáková

**Demonstration: Web Tool for Designing and Testing of Digital
Circuits Within a Remote Laboratory** 290
Javier Garcia-Zubia, Eneko Cruz, Luis Rodriguez-Gil, Ignacio Angulo,
Pablo Orduña, and Unai Hernandez

**Poster: An Experience API Framework to Describe Learning
Interactions from On-line Laboratories** 298
Pedro Paredes Barragán, Miguel Rodriguez-Artacho, Elio San Cristobal,
Manuel Castro, and Hamadou Saliah-Hassane

**Poster: Remote Engineering Education Set-Up of Hydraulic Pump
and System** 304
Milos Srecko Nedeljkovic, Novica Jankovic, Djordje Cantrak, Dejan Ilic,
and Milan Matijevic

Poster: "Radiation Remote Laboratory" with Two Level Diagnostics . . . 312
Michal Krbecek, Sayan Das, Franz Schauer, Miroslava Ozvoldova,
and Frantisek Lustig

Cyber Physical Systems and Cyber Security

Enabling Remote PLC Training Using Hardware Models 323
Alexander A. Kist, Ananda Maiti, Catherine Hills, Andrew D. Maxwell,
Karsten Henke, Heinz-Dietrich Wuttke, and Tobias Fäth

**Towards Data-Driven Cyber Attack Damage and Vulnerability
Estimation for Manufacturing Enterprises** 333
Vinayak Prabhu, John Oyekan, Simon Eng, Lim Eng Woei,
and Ashutosh Tiwari

**Practical Security Education on Combination of OT and ICT
Using Gamification Method at KOSEN** 344
Keiichi Yonemura, Ryotaro Komura, Jun Sato, and Masato Matsuoka

SEPT Learning Factory Framework 354
Dan Centea, Mo Elbestawi, Ishwar Singh, and Tom Wanyama

Remote Structural Health Monitoring for Bridges 363
Mohammed Misbah Uddin, Nithin Devang, Abul K. M. Azad,
and Veysel Demir

XOR 378
Christoph Vorhauer and Klaus Gebeshuber

Development Models and Intelligent Algorithms for Improving the Quality of Service and Security of Multi-cloud Platforms 386
Irina Bolodurina and Denis Parfenov

The Application of the Remote Lab for Studying the Issues of Smart House Systems Power Efficiency, Safety and Cybersecurity 395
Anzhelika Parkhomenko, Artem Tulenkov, Aleksandr Sokolyanskii, Yaroslav Zalyubovskiy, Andriy Parkhomenko, and Aleksandr Stepanenko

Human Machine Interaction and Usability

Human-Computer Interaction in Remote Laboratories with the Leap Motion Controller 405
Ian Grout

Visual Tools for Aiding Remote Control Systems Experiments with Embedded Controllers 415
Ananda Maiti, Andrew D. Maxwell, and Alexander A. Kist

Process Mining Applied to Player Interaction and Decision Taking Analysis in Educational Remote Games 425
Thiago Schaedler Uhlmann, Eduardo Alves Portela Santos, and Luciano Antonio Mendes

An Approach to Teaching Blind Children of Geographic Topics Through Applying a Combined Multimodal User Interfaces 435
Dariusz Mikulowski

Development of a Virtual Environment for Environmental Monitoring Education 443
Jeremy Dylan Smith and Vinod K. Lohani

School Without Walls - An Open Environment for the Achievement of Innovative Learning Loop 451
Carole Salis, Marie Florence Wilson, Franco Atzori, Stefano Leone Monni, Fabrizio Murgia, and Giuliana Brunetti

Low-Cost, Open-Source Automation System for Education, with Node-RED and Raspberry Pi 458
Phaedra Degreef, Dirk Van Merode, and Galyna Tabunshchyk

Demonstration: Face Emotion Recognition (FER) with Deep Learning – Web Based Interface 466
Andreas Pester and Kevin Galler

Poster: An Approach for Supporting of Navigation of Blind People in Public Building Based on Hierarchical Map Ontology 471
Dariusz Mikulowski and Marek Pilski

**Poster: A Mobile Application for Voice and Remote Control
of Programmable Instruments** 479
Burak Ece, Ayse Yayla, and Hayriye Korkmaz

Biomedical Engineering

**Organic Compounds Integrated on Nanostructured Materials
for Biomedical Applications** 489
Cristian Ravariu, Elena Manea, Florin Babarada, Doru Ursutiu,
Dan Mihaiescu, and Alina Popescu

**Towards an Automated Analysis of Forearm Thermal Images
During Handgrip Exercise** 498
Pedro Silva, Ricardo Vardasca, Joaquim Mendes,
and Maria Teresa Restivo

Handgrip Evaluation: Endurance and Handedness Dominance 507
Ricardo Vardasca, Paulo Abreu, Joaquim Mendes,
and Maria Teresa Restivo

Digital Health for Computer Engineering Classes: An Experience 517
Lucia Vaira and Mario A. Bochicchio

**A Support System for Information Management Oriented
for the Infant Neurodevelopment Study** 528
Sergio Daniel Cano-Ortiz, Yadisbel Martinez-Cañete,
Lienys Lombardía-Legrá, Reinhard Langmann, and Harald Jacques

**Demonstration: Online Detection of Abnormalities in Blood
Pressure Waveform: Bisfiriens and Alternans Pulse** 536
Daniel Nogueira, Rafael Tavares, Paulo Abreu, and Maria Teresa Restivo

Augmented and Mixed Reality

**The Effect of Augmented Reality in Solid Geometry Class on Students'
Learning Performance and Attitudes** 549
Enrui Liu, Yutan Li, Su Cai, and Xiaowen Li

**Multimodal Data Representation Models for Virtual, Remote,
and Mixed Laboratories Development** 559
Yevgeniya Sulema, Ivan Dychka, and Olga Sulema

**Voice Driven Virtual Assistant Tutor in Virtual Reality
for Electronic Engineering Remote Laboratories** 570
Michael James Callaghan, Gildas Bengloan, Julien Ferrer, Léo Cherel,
Mohamed Ali El Mostadi, Augusto Gomez Eguíluz, and Niall McShane

**Using Unity 3D as the Augmented Reality Framework
for Remote Access Laboratories** 581
Mark Smith, Ananda Maiti, Andrew D. Maxwell,
and Alexander A. Kist

A Literature Review on Collaboration in Mixed Reality 591
Philipp Ladwig and Christian Geiger

REMLABNET and Virtual Reality 601
Tomas Komenda and Franz Schauer

**Exposing Robot Learning to Students in Augmented
Reality Experience** 610
Igor Verner, Michael Reitman, Dan Cuperman, Toria Yan,
Eldad Finkelstein, and Tal Romm

**Framework for Augmented Reality Scenarios
in Engineering Education** 620
Matthias Neges, Mario Wolf, Robert Kuska, and Sulamith Frerich

**Poster: SIMNET: Simulation-Based Exercises for Computer
Network Curriculum Through Gamification and Augmented Reality** . . . 627
Alvaro Luis Fraga, María Guadalupe Gramajo, Federico Trejo,
Selena Garcia, Gustavo Juarez, and Leonardo Franco

Applications and Experiences

**Using Learning Theory for Assessing Effectiveness of Laboratory
Education Delivered via a Web-Based Platform** 639
Shyam Diwakar, Rakhi Radhamani, Nijin Nizar, Dhanush Kumar,
Bipin Nair, and Krishnashree Achuthan

Vocational Education for the Industrial Revolution 649
Enrique Blanco, Fernando Schirmbeck, and Claiton Costa

Students' Perception of E-library System at Fujairah University 659
Ahmad Qasim Mohammad AlHamad
and Roqayah Abdulraheim AlHammadi

**Virtual Working Environment Scheduling of the Cloud System
for Collective Access to Educational Resources** 671
Irina Bolodurina, Leonid Legashev, Petr Polezhaev, Alexander Shukhman,
and Yuri Ushakov

Activities of Euro-CASE Engineering Education Platform 678
Petar Bogoljub Petrovic and Milos Srecko Nedeljkovic

**Virtual Instrumentation Used in Engineering Education Set-Up
of Hydraulic Pump and System** 686
Milos Srecko Nedeljkovic, Djordje Cantrak, Novica Jankovic,
Dejan Ilic, and Milan Matijevic

**Study of Remote Lab Growth to Facilitate Smart Education
in Indian Academia** 694
Venkata Vivek Gowripeddi, Kalyan Ram Bhimavaram, J. Pavan,
Nithin Janardhan, Amrutha Desai, Shubham Mohapatra, Apurva Shrikhar,
and C. R. Yamuna Devi

**Ant Colony Algorithm for Building of Virtual Machine Disk
Images Within Cloud Systems** 701
Irina Bolodurina, Leonid Legashev, Petr Polezhaev,
Alexander Shukhman, and Yuri Ushakov

Work in Progress: Pocket Labs in IoT Education 707
Christian Madritsch, Thomas Klinger, and Andreas Pester

**Demonstration: Using IPython to Demonstrate the Usage of Remote
Labs in Engineering Courses – A Case Study Using a Remote
Rain Gauge** 714
Alberto Cardoso, Joaquim Leitão, Paulo Gil, Alfeu S. Marques,
and Nuno E. Simões

**Poster: LabSocket-E, LabVIEW and myRIO in Real-Time/Embedded
Systems Student Teaching and Training** 721
Doru Ursutiu, Andrei Neagu, and Cornel Samoila

**Poster: Smart Applications for Raising Awareness of Young Citizens
Towards Using Renewable Energy Sources and Increasing
Energy Efficiency in the Local Community** 728
Radojka Krneta, Snežana Dragićević, Andreas Pester, and Andreja Rojko

Author Index 737

Internet of Things and Industry 4.0

A Cloud-Based Blended Learning Lab for PLC Education

Reinhard Langmann$^{(\boxtimes)}$ and Matthias Coppenrath

Hochschule Duesseldorf University of Applied Sciences, Duesseldorf, Germany
langmann@ccad.eu

Abstract. This paper presents a concept, its implementation, and evaluation for a cloud-based, blended learning lab (CBLL) for PLC education. The CBLL combines training on a real on-site PLC with technology models from the cloud. The technology models are loaded into the learner's web browser from the cloud where they are run. The models are linked directly with the inputs and outputs of the PLC via a CloudIO adapter to enable a PLC program to control the technology models. Management of the technology models is performed via an IIoT platform where model use is offered as a service. The learner can use the models on-site without any licenses or software being installed.

Keywords: Blended learning lab · Cloud-based learning · PLC programming
PLC education · Virtual system model

1 Introduction

Industrial controls, and particularly PLC controls, form a key technological basis for automating technical processes today and will likely continue to in the future. Learning how to program PLC controls and their applications is therefore a central educational task in all courses of study for automation technology in universities and colleges, as well as in training for specialist workers, e.g. mechatronics and process control engineering at vocational schools.

Typically, PLC training takes place in laboratory practical sessions where the learners work on physically present controls and can control various technological model systems linked to them. Due to the high costs and, partly, due to non-feasible operational and maintenance expense in the training field, real systems typical to industry are rarely used.

An alternative to the necessary on-site device technology for PLC training would be the use of remote labs, such as those described in the current state of technology (e.g. in [1, 2]), in which the learners can be educated on a PLC programming system with corresponding training facilities via web browser, webcam and remote access. Until now, this type of PLC training could not be offered on a large scale. As in the past, PLC training is still carried out on real-world controls in all training facilities.

With the promotion of the Industry 4.0 concept, digitalisation of production and the Industrial Internet of Things (IIoT), as well as secure cloud computing, however, new opportunities are arising for training in core areas of automation technology which have hitherto hardly been used. The present paper describes an option for modernising PLC

© Springer International Publishing AG, part of Springer Nature 2019
M. E. Auer and R. Langmann (Eds.): REV 2018, LNNS 47, pp. 3–13, 2019.
https://doi.org/10.1007/978-3-319-95678-7_1

training within this context using cloud computing and IIoT in the form of a blended learning lab.

2 State-of-the-Art

There are three fundamental aspects required for training on a PLC:

- A PLC control (as a device or soft PLC in a PC);
- A PLC programming system;
- A model of the technological process to be controlled (technology model).

PLC control and programming systems are generally industrial products, such as S7-1500 with TIA-Portal (Siemens) or ILC 350 with PC Worx (Phoenic Contact). Special didactic modifications for these are neither desired nor available. Learners should be trained on a practical basis using industrial controls.

There is a different situation regarding the third above-mentioned part, the technology model:

Technology models are not offered by control manufacturers as an industrial product, which means that each training facility needs to procure such models from providers of didactic learning systems or self-develop them in order for them to correspond to their requirements and the options they wish to provide. As a result, there are various PLC training systems in training facilities whose efficiency for training can vary greatly.

When using technology models for PLC training, there are basically four types of situation:

- *Case 1*: Use of a real, on-site industrial system as a technology model.
- *Case 2*: Combinations of simple switches/buttons, LEDs, potentiometers etc. are used as a technology model (simulation boards).
- *Case 3*: Technology models are designed as special device-specific system models and process simulators.
- *Case 4*: Within a computer, simulated models of various real systems are generated and connected to the periphery (inputs/outputs) of the real PLC as a technology model.

Case 1 would be the best solution for practical training. In most cases, though, this alternative is not feasible, due to reasons associated with cost.

From the authors' point of view, most PLC training systems use simple simulation boards pursuant to Case 2 due to reasons associated with cost. Many providers of didactic learning systems use simulation boards as technology models. However, only simple technological processes without internal process behaviour can be simulated. The models are therefore not realistic, and learners can hardly understand the process which is to be controlled.

Case 3 offers the possibility for more complex technology models with a much greater degree of realism. Examples of these are technology models based on Lego [3] or the system models and process simulations pursuant to [4]. Even with these models, industry-typical applications can only be reproduced to a very limited extent. In addition, these technology models are less flexible in use and usually exceed the financial resources that are available per learner in the training institutions.

With Case 4, powerful and flexible technology models can be set up and used which have internal process behaviour and also give the learner a realistic idea of the process which is to be controlled. In addition, 3D models are often used. The process environment for the technology models is a PC. In essence, two methods are used to connect these models to the PLC controller:

- Direct connection of the model sensors/actuators to the PLC input/output modules via special hardware adapters [5–8]. This option does not place special demands on the PLC, is usually applicable to controllers of any manufacturer and corresponds to an industry-typical design, since the PLC process signals can be connected directly with the technology model in terms of hardware.
- The technology model is connected via a communication interface (e.g. Ethernet, MPI) of the PLC, i.e. the PLC also has to have one [7–9]. In addition, this also has to be taken into account when configuring and programming the PLC, since the PLC input/output signals are not connected and they can only be connected indirectly to the technology model via a network and special drivers.

With Case 4, very powerful and complex technology models for education and training can be created with a high degree of reality. A disadvantage of these systems, however, is that it is classic stand-alone software based on the licence model. Neither cloud computing nor the use of the software as a service is provided for. The installation of the systems also requires a number of conditions, is currently rather complex, and the use is not very flexible. In part, the use is also only limited to selected PLC controllers, such as those from Siemens (e.g. SIMIT [7]).

In summary, it can be estimated that the technology models for PLC training which are available according to the state of the art so far do not use either principles and methods from Industry 4.0 or solutions from the IIoT area. In scientific publications addressing the use of virtual technology models in connection with the evaluation/commissioning of PLC programs, as well, only the above-mentioned systems of Case 4 are used. Global networking, service principles and the Internet of Things so far do not play a role in this regard [10–13].

3 Concept

The basic idea of the present concept consists of technology models being saved in a cloud-based platform, which can be loaded into a learner's web browser as required and run from their location. The connection of the PLC input/output signals to the sensors and actuators of the model is performed via a special cloud adapter (CloudIO), which transmits the input/output signals via appropriate web protocols to the technology model. This results in a *Cloud-based Blended Learning Lab (CBLL)*, in which the user works directly via on-site controls with the various technology models being sourced from an online cloud. Figure 1 presents the principal diagram.

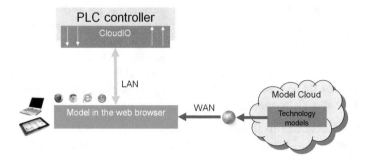

Fig. 1. Principle diagram for a cloud-based blended learning lab (CBLL) for PLC training

The technology models use web technologies, such as SVG and/or X3D for their generation. JavaScript can be embedded into the models when programming model behaviour.

Since the Internet can be accessed via the PLC's input/output signals via the CloudIO, all appropriate systems available online can generally be controlled using the on-site PLC. This results in a variety of options for potential technology models. For instance, real systems (including their sensors and actuators) are accessible via a web connector on the Internet and thus used for PLC training. Figure 2 presents such a variant.

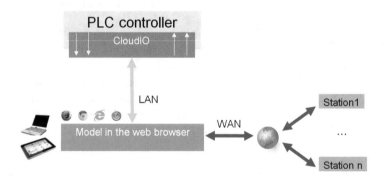

Fig. 2. Real systems in a CBLL as a technology model for PLC training

Another benefit of the CBLL concept for PLC training is that the technology models can be used in line with the service principle known within cloud computing, i.e. the use of technology models can, for example, be charged in a time-dependent manner ("Rent a model"). This only requires suitable IoT/IIoT platforms via which the models (and stations also) can be controlled/managed.

On-site, learners no longer require stand-alone installations of software or licenses. Execution of the technology models can be performed by the learner on every PC or post-PC (smartphone, tablet, etc.).

4 Implementation

Realisation and implementation for the CBLL concept is divided into three components.

4.1 CloudIO

Prototypical realisation on the basis of PiXtend from the company *qube solution* was performed for the cloud adapter (CloudIO). The device contains a Raspberry Pi and expands upon this with various analogue and digital connections according to PLC standard IEC 61131. Of the available digital and analogue inputs and outputs, 8 digital inputs and outputs were used for the prototype as well as two analogue inputs and outputs. A Linux OS operates on the device. Proprietary application programmes can be implemented on the basis of the C-library provided.

Three variants were implemented and tested in the prototype for the connection of the process inputs and outputs to the Internet:

- *Variant 1*: Process data communication is performed by using the Node-RED [14] that is preinstalled on the Raspberry Pi and available Linux test programmes.
- *Variant 2*: Preparation of the process data for transmission to an MQTT broker is performed via a C-program. The Open-Source-Broker EMQTT [15] is used as an MQTT broker, which is also implemented on the CloudIO in the Raspberry Pi. This allows the technology models to be linked to the CloudIO using MQTT protocol.
- *Variant 3*: A third variant involves using the C++-Library Websocket++ for a websocket connection [16] instead of MQTT, using which the process data of the PiXtend communicates with the technology model via a "lean" application protocol (WOAS protocol [17]).

Figure 3 depicts the basic structure of the CloudIO for the MQTT and websocket variants.

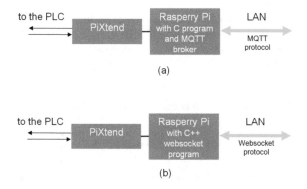

(a)

(b)

Fig. 3. Basic structure of the CloudIO for variants 2 (a) and 3 (b)

The CloudIO is built onto a DIN A4 board within the didactic learning system Eduline created by Phoenix Contact, which ensures that the cloud adapter can easily replace the simulation board in this PLC training system. The connection to the inputs and outputs of the Eduline-PLC ILC 191 ME/AN or ILC 131 ETH is performed via two 24-pole D-SUB cables (see Fig. 4).

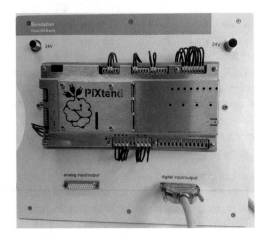

Fig. 4. Structure of the CloudIO on an eduline board from phoenix contact

The CloudIO can also be linked to any other digital/analogue process signals (e.g. Siemens PLC) for the purposes for connecting these to the Internet/Intranet via a web protocol.

4.2   Technology Models

The IIoT platform WOAS (WOAS portal – http://woas.ccad.eu) [18] is used to create the technology models.

The platform provides two services with which technology models based on SVG graphics and based on 3D modelling with X3D can be realised. Table 1 illustrates essential features of these services. If required, other specific model services can also be integrated into the WOAS portal.

Table 1. Model services in the WOAS IIoT platform

Service name	Graphic	Number of channels
service4svgModel	SVG (preferred for 2D models)	12 input channels (actuators)
		12 output channels (sensors)
service4x3dModel	X3D (preferred for 3D models)	12 input channels (actuators)
		12 output channels (sensors)

The internal model behaviour (such as the rise of a liquid in the tank when filled) is realised by a JavaScript file linked to the respective service instance.

To create a technology model, the specific model services pursuant to Table 1 may be combined with other services of the WOAS portal (e.g. with real-time plotters, controls) so that powerful technology models can be created. The connection to the CloudIO occurs via Virtual Devices [19] in the WOAS portal, which is available for MQTT and Websocket in the portal

4.3   Management of the Models

Management of the technology models can be performed in a multi-user-capable IIoT platform (e.g. ThingWorx [20]). Currently, however, the free-to-use IIoT platform 'WOAS' is deployed. The greatest advantage of the WOAS portal is, amongst others, that once a technology model has been downloaded to the learner's web browser, process data communication only takes place within the local network (LAN) between the CloudIO and the technology model.

Via the IIoT platform, the user (learner) receives a login to the models they want and they can use it for training on their PLC on-site. If required, through this platform the use of model services can be charged via a clearing system.

5   Evaluation and Application

5.1   Time Characteristics

Using special measurement scenarios, the response times for the three above-mentioned implementation variants of the CloudIO can be determined during the course of long-term tests (20 h).

The basic measurement setup is shown in Fig. 5.

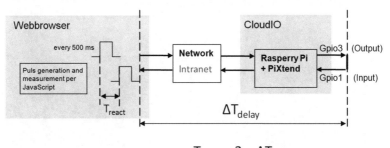

$$T_{react} = 2 \times \Delta T_{delay}$$

Fig. 5. General measuring structure for the measurement of the reaction time using the CloudIO

Every 500 ms, a pulse is sent from the web browser to a digital output of the CloudIO (Gpio3), which is directly connected to a digital input (Gpio1). The resulting response time in the web browser is recorded as a histogram over a longer period of time. For the three implementation variants, the average reaction times listed in Table 2 were measured over a period of 20 h.

Table 2. Time behaviour of the CloudIO implementation variants

Variant	Web protocol	Response time (ms)	Comment
1	Websocket	150	via Node-RED
2	MQTT	50	Standard deviation <5 ms
3	Websocket	45 ms	Standard deviation ca. 10 ms

The results show that with the proposed CBLL concept, technology models with process times of >45 ms in variant 3 can be successfully used for PLC training. As a result, a large number of different production processes can be realistically modelled.

5.2 Application

Two applications were created for evaluating the CBLL concept.

Controlling a Heater Boiler: The technology model features a heater boiler that can be filled with liquid via an inlet. After heating to a pre-selected temperature, the heater boiler can be emptied via an outlet. The heater boiler has been created as a 2D-SVG graphic. Within the technology model, this SVG graphic is combined with further services, such as a real-time plotter for the liquid filling level and temperature progression. Figure 6 depicts the technology model in the web browser of a Galaxy 10 tablet using Android OS.

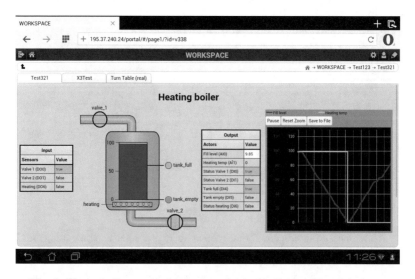

Fig. 6. Heater boiler as a technology model in the Galaxy 10 web browser

Controlling a Processing Station: The 3D model of a FESTO processing station (Fig. 7a) was converted into the web world via X3DOM for use as a technology model (Fig. 7b) and equipped with behaviour by a JavaScript program. After integration as a service into the WOAS portal, this model can be linked with the training PLC and controlled via CloudIO.

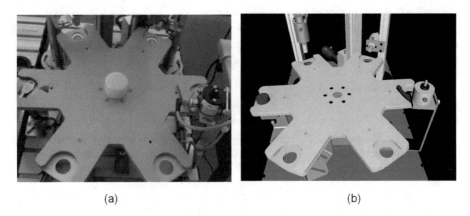

(a)   (b)

Fig. 7. Processing station as a real station (a) and as a technology model (b) in a web browser

Both technology models were tested successfully using the Eduline system PLC trainer (ILC 191 ME/AN control). The PLC programs required can thus be developed and tested by a learner in a realistic model. The learner only requires a WiFi-capable tablet/smartphone/PC, as well as a login to the IIoT platform WOAS in order to be able to use the models in their device's browser. Installations of software or licenses are no longer required on-location with the learner.

6   Summary and Future Works

The article describes the structure, application and evaluation of a special Cloud Connector (CloudIO) for use in PLC training. The CloudIO can be used to connect any realistic models of technological processes to the on-site training PLCs as well as to visualise and operate them via a web browser. Stand-alone simulation systems are no longer needed for the technology models. As a model, virtual 2D/3D models as well as real equipment can be used remotely. This results in a cloud-based blended learning lab (CBBL) for PLC training.

An evaluation showed that technology models with process times of ca. 50 ms can be used via this method. This makes the models applicable for most training tasks in PLC programming.

The proposed solution also make possible a completely new business model in the field of training: technology models are no longer installed stand-alone with a fixed licence; instead, they can be realised using a suitable IIoT platform (e.g. the WOAS portal) pursuant to the SaaS principle (software as a service) rented and billed based upon time.

The other R&D work focuses in particular on an expansion of the number of technology models, on a clearing platform for the settlement of model use as well as the remote use of real, existing equipment in the CCAD through any local PLC training system.

In 2018, a comprehensive evaluation with trainees in four vocational colleges is planned for the use of the CBBL for PLC training. For this purpose, further 2D and 3D models of technological processes are currently being developed and a further four CloudIO modules are being assembled.

References

1. Avila, J.L., Amaya, D.: Using remote laboratories for education in industrial processes and automation. In: Proceedings of the 2014 IAJC-ISAM International Conference (2014)
2. Tasneem, H.R.A., et al.: Introduction of PLC-based remote laboratory for modular mechatronics system (MMS). Int. J. Emerging Technol. Adv. Eng. 2(7) (2012)
3. Maloney, D.: Desktop Factory Teaches PLC Programming. http://hackaday.com/2017/04/16/desktop-factory-teaches-plc-programming/. Accessed Apr 2017
4. Lucas-Nülle: Product Catalogue. www.lucas-nuelle.com (2017)
5. Pittschellis, R.: 3D-Simulation im Ausbildungsbereich. In: Virtual Efficiency Congress (2011)
6. FESTO: Learning Systems Software (2017). http://www.festo-didactic.com/ov3/media/customers/1100/software_2.pdf
7. Siemens: SIMIT (V9.0). User Manual. Accessed May 2016
8. WinMOD: Mewes & Partner (2017). http://winmod.de/en/
9. FACTORY-I/O: Realgames (2017). https://factoryio.com/
10. Salazar, E.A., Macías, M.E.: Virtual 3D controllable machine models for implementation of automations laboratories, In: Proceedings of 39th ASEE/IEEE Frontiers in Education Conference, 18–21 Oct 2009
11. Dzinic, J., Yao, C.: Simulation-based verification of PLC programs. Master of Science thesis in production engineering, Chalmers University of Technology, Sweden (2013)
12. Bhowmik, P.K., et al.: PLC based operation of a process control model - a learning aid for undergraduate students. In: Proceedings of the International Conference on Mechanical Engineering and Renewable Energy 2011 (ICMERE 2011), 22–24 Dec 2011
13. ISG: Simulation based engineering and virtual commissioning. Presentation, ISG Steuerungstechnik. http://www.isg-stuttgart.de/fileadmin/user_upload/virtuos/2017-01-18_ISG-virtuos-Praesentation-2017engl.pdf. Accessed 18 Jan 2017
14. Node-RED: https://nodered.org/. JS Foundation (2017)
15. EMQTT: http://emqtt.io/. EMQ Enterprise (2017)
16. Websocket++: https://www.zaphoyd.com/websocketpp. Zaphoyd Studios (2017)
17. Competence Center Automation Düsseldorf (CCAD): WOAS Protocol Gateway API. R&D Document (2014)

18. Langmann, R., Meyer, L.: Architecture of a web-oriented automation system. In: 18th IEEE International Conference on Emerging Technologies and Factory Automation (ETFA 2013), 10–13 Sept 2013
19. Competence Center Automation Düsseldorf (CCAD): Description of a Virtual Device (in German). R&D Document (2014)
20. ThingWorx: https://www.thingworx.com/, PTC (2017)

Environmental Sound Recognition
with Classical Machine Learning Algorithms

Nikolina Jekic[✉] and Andreas Pester

Carinthia University of Applied Sciences, Villach, Austria
Nikolina.Jekic@edu.fh-kaernten.ac.at,
A.Pester@fh-kaernten.at

Abstract. The field of study interested in the development of computer algorithm for transforming data into intelligent actions is known as machine learning. The paper investigates different machine learning classification algorithms and their effectiveness in environmental sound recognition. Efforts are made in selecting the suitable audio feature extraction technique and finding a direct connection between audio feature extraction technique and the quality of the algorithm performance. These techniques are compared to determine the most suitable for solving the problem of environmental sound recognition.

Keywords: Machine learning · ESR · Python · Audio feature extraction
MFFC · k-Nearest Neighbors · Logistic regression · Support Vector Machines

1 Introduction

Machine learning has been running in the background for years, powering mobile applications and search engines. Recently, it has become a more widely used in engineering, medicine, business etc., due to all latest technological advancements. An impressive rise in data and computing capabilities has made this exponential progress possible. The ongoing connection of the IoT and the Machine Learning world will give an impulse for the development of remote engineering too in the next years.

In this work, research in audio recognition in the field of Environmental Sound Recognition (ESR) is presented. Environmental sound recognition is still in its development and one of the main problems is the lack of a universal database. The motivation for the methods presented in this work was established from fields of Automatic Speech Recognition (ASR) and Music Information Retrieval (MIR), for which extensive bibliography exists.

All presented examples are implemented in Python and Jupyter Notebook is used as the computational environment for the development of training examples. The results are evaluated using an evaluation metric.

© Springer International Publishing AG, part of Springer Nature 2019
M. E. Auer and R. Langmann (Eds.): REV 2018, LNNS 47, pp. 14–21, 2019.
https://doi.org/10.1007/978-3-319-95678-7_2

2 Environmental Sound Recognition

Hearing is one of the human species most important senses. After sight, it is the sense that is most used to gather information about the surrounding environment. By environmental sounds, we refer to different everyday sounds, both natural and artificial, appearing in daily life other than speech and music. ESR can be significant in audio search applications, surveillance applications, home-monitoring environments, smart houses, speaker identification and language recognition with environmental sounds in the background, robot navigation, etc.

Training of data for machine learning algorithms can be described as the vector value of the functions, but in audio data it is not that simple. In an attempt to classify environmental sounds, the main question is "How is a sound represented?". A sound needs to be converted into a series of values that describe it adequately. Extraction of audio features is needed in order to use extracted data to train and test our machine learning model that will classify environmental sound.

In Fig. 1, the diagram of environmental sound recognition is represented. In order to classify environmental sound, two steps must be followed. The first step is to extract audio features from environmental sounds and the second step is to feed machine learning classification algorithms with the extracted data, in order to obtain classified environmental sound. This approach was established from fields of Music Information Retrieval (MIR) and Automatic Speech Recognition (ASR), based on the results from these fields.

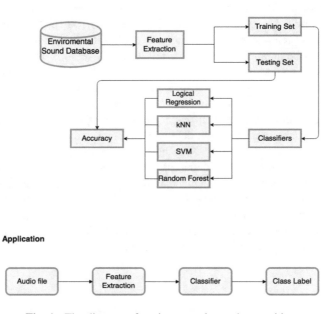

Fig. 1. The diagram of environmental sound recognition

2.1 Audio Feature Extraction

The feature analysis is considered the most crucial and important part in environmental sound recognition [1, 2]. The aim of feature extraction is to convert the sound signal into a sequence of feature vectors in order to produce a set of characteristic features that describe the sound signal [2–4].

Signals are usually analyzed in time or frequency domain. Therefore, acoustic features can be grouped into two groups, temporal and spectral features that can be extracted using time and frequency domain respectively [2, 4]. By nature, the environmental sound is considered as unstructured data and many features are needed to describe the audio signals. In this work, several analysis methods were considered, inspired by the MIR and ASR fields. Sound signals were detected using the standard deviation of normalized power sequences. Multiple feature extraction techniques like Zero Crossing Rate, Spectral Centroid, Spectral Contrast, Spectral Bandwidth, Spectral Rolloff, Mel Frequency Cepstral Coefficient (MFCC) and Fast Fourier Transform (FFT) were applied on the raw sound signal.

3 Database and Methodology

After review of already existing and online available databases, the "BDLib" library of environmental sounds has been chosen. [5] The following twelve classes are used as a training and testing examples: airplanes, alarms, applause, birds, dogs, footsteps, motorcycles, rain, rivers, sea waves, thunder and wind. Each class contains 10 audio samples, representing real life situations. The field of ESR is still in development and this number of audio samples will be enough for the research in this thesis. The recordings are in WAV format with a sampling rate of 44,100 Hz and 16-bit resolution. WAV format is preferred when working with audio files in Python.

In all tests, the number of classes per test dataset is set to 6. The reason for this is that after careful review of studies done on this topic, it can be noticed that in the multi-classification problems with more than 5 classes, the accuracy of recognition is very low. Classes are chosen in the way that some of them have similar spectrogram (e.g. rain and rivers) and some of them have greater diversity between each other (e.g. airplane and birds). After obtaining data with different audio feature techniques, four classifiers are trained: k-Nearest Neighbors, Logistic Regression, Support Vector Machine and Random Forest. The training and test datasets contain 60 files and their labels. The classifier should be universal for a problem of environmental sound recognition. This means that method developed in this work should be able to classify different environmental sounds, not only sounds in the examples of this work. The idea behind is to test the method on different datasets in order to determine the best approach for the problem of environmental sound recognition.

4   Classification and Results

Three examples of multi-classification are presented in this work. In binary classification, the accuracy of 50% is a worst case possibility, as it would also be achieve by random guessing. In a multi-classification problem with 6 different classes, by random guessing the accuracy would result in only 16.7%.

All tests were done in Python environment and Jupyter Notebook (with IPython kernel). Different Python libraries NumPy, SciPy, Scikit-learn, Talkbox SciKit, IPython, Pandas, Librosa and Matplotlib, are used [14–18].

In the first test, audio feature extraction was done with Fast Fourier Transformation (FFT) and Mel Frequency Cepstral Coefficients (MFCC). The first attempt in training extracted data was done with logical regression classifier. 70% of the total data was used for training and the remaining 30% for testing.

The frequencies from reading a raw sample were extracted by applying the Fast Fourier Transformation (FFT) and used to feed a classifier used for classification of environmental sounds. The goal of this task was to create a fingerprint of a sound using FFT. Assigning sounds to corresponding classes as labels, training data was collected. Not having a deep understanding of data, the attempt to train a logistic regression classifier using extracted FFT features for classifying 6 different sound classes with reasonable accuracy did not give good enough results. Only one class out of total 6 classes has a recognition accuracy of more than 50%. Second attempt to train a logistic regression classifier using MFFC features showed better performance. However, 4 classes out of total 6 classes have a percentage of recognition less than 50% and this points that the model needs to be improved. For this reason, investigation of audio feature extraction techniques is continued with the focus on MFCC features and their combination with other features. In the next section of this work, examples contain different audio extraction technics which are used for training classifiers. Usage, comparison and efficiency of different classifiers are demonstrated.

As already seen, audio features can be grouped into two groups: temporal and spectral features. Focus in the second test is an investigation in the area of temporal (Zero Crossing Rate) and MFCC features. Obtaining temporal and MFCC features was done in Python environment version 3.5 with package Librosa. Librosa is a python package structured as collection of submodules for music and audio analysis. With help of package Librosa, Zero Crossing Rate (ZCR) and MFCC features can be easily extracted and manipulated. For every feature the mean value and the standard deviation were computed. Table 1 lists the mean value and the standard deviation of four features and target variable for first four audio files from existing database. Training set had 60 training examples. Each training example had 28 features (the mean value and the standard deviation for 1 temporal feature and 13 MFCC coefficients) and one target variable (class).

In this example four classifiers are trained: k-Nearest Neighbors, Logistic Regression, Support Vector Machine and Random Forest. 70% of the total data was used for training and the remaining 30% for testing.

Table 1. Extracted features of environmental sound files

meanZCR	stdZCR	meanMfcc1	stdMfcc1	class
−0.78397	−0.91778	1	−0.9994	Airplane
0.80451	−0.77247	−0.26708	−0.97995	Rain
−0.75783	−0.09163	−0.14732	−0.87365	Birds
−0.87369	−0.8231	0.25731	−0.42407	Dogs

All the undergoing tests are multi-class problems. A confusion matrix was used for the purpose of distinguishing between different classes and directly to see which classes actually cause confusion, as well as to measure accuracy in the multi class problems (Fig. 2).

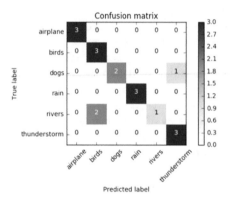

Fig. 2. Confusion matrix of classification with kNN algorithm

The process of classification with kNN algorithm involves identification of k training examples that are nearest in the distance to the test data features. The k-Nearest Neighbors algorithm was optimized in way that k takes all values in range from 1 to 11. This was done in order to determine the maximum accuracy with this method. Maximum Accuracy is 83.3% on the test dataset is with 4 neighbors. Most of the classes have the percentage of recognition of 100% and this points that the model is good. In classification with SVM, the obtained accuracy is 77.8% on the test dataset. Similar like in classification with kNN, most of the classes have the percentage of recognition 100%. Achieved accuracy with Random Forest and Logistic Regression algorithm is 0.833%. Following general approach of every machine learning algorithm, it can be noticed that preparing and analyzing data, is very important in the problem of environmental sound recognition.

The data gathered from previous example indicates extracting data with different audio feature technique and combining them will improve the performance of a classifier. The more carefully data is extracted, the better accuracy rate will be achieved. In next test, the model performance is improved by combining temporal, spectral and MFCC features. For the purposes of examples comparison, datasets used for classification in Test 2 will remain the same. The spectral features that are used: Spectral

Centroid, Spectral Contrast, Spectral Bandwidth, Spectral Roll-Off. The mean value and standard deviation of spectral features will be calculated and added to the temporal and MFCC features obtained in Test 2. The calculation is similar to the calculation that has been done for pervious features. The final training set has 60 training examples. Each training example has 36 features (the mean value and the standard deviation for 1 temporal feature, 4 spectral features, and 13 MFCC coefficients) and one target variable (class). Following general approach of every machine learning algorithm, all steps: collecting data, preparation, and analysis of the input data, training the algorithm, testing the algorithm and using algorithm will be similar to the previous example. The same four classifiers are trained: k-Nearest Neighbors, Support Vector Machine, Random Forest and Logistic Regression. 70% of the total data was used for training and the remaining 30% for testing. Recognition rate are as follow for the corresponding algorithms k-Nearest Neighbors-88.9%, Support Vector Machine-83.3%, Random Forest-83.3%, Logistic Regression-83.3%.

Having a deep understanding of data, training the classifier with good accuracy is achieved. Applying different audio extraction technics performance of classifiers is improved in comparison with Test 2. The highest accuracy in classification is achieved with kNN algorithm in Test 3.

Table 2. Accuracy of machine learning algorithms in environmental sound recognition

Algorithm	Accuracy of the test with 28 features	Accuracy of the test with 36 features
k-nearest neighbors	83.3%	88.9%
Support vector machine	77.8%	83.3%
Random forest	83.3%	83.3%
Logistic regression	83.3%	83.3%

5 Conclusion

In this work, a comparison between the most commonly used methods in the ESR field is presented. Four machine learning algorithms were tested in combination with different feature sets, which resulted from a feature selection process. In the first example, not having a deep understanding of data, the attempt to train a logistic regression classifier using extracted FFT and MFFC features for classifying 6 different sound classes with reasonable accuracy did not give good enough results.

Improved model performance was achieved in the second and the third example using different audio extraction techniques for training classifiers. From the extracted audio features, the best results are achieved in the third example using following features: MFCC, ZCR, Spectral Centroid, Spectral Roll-off, Spectral Contrast and Spectral Bandwidth. In the both examples, four machine learning classification algorithms are trained: k-Nearest Neighbors, Logistic Regression, Support Vector Machine and Random Forest.

The experimental results showed promising performance in classifying 6 different sound classes. Regarding the classifiers, k-NN provided the highest recognition rates among the algorithms used in this work. However, remaining three algorithms showed comparable performance. Table 2 lists achieved accuracy.

5.1 Outlook

The future work will first focus on extending the proposed environmental sound library, in order to incorporate more sound classes and investigate the method that would show satisfactory performance in the problem of recognition more than 6 classes.

Furthermore, since most audio features were originally designed for the domains ASR and MIR domains, the design of a new set of features targeted at the domain of ESR keeps the potential for further improvement of the results and, eventually, the elimination of confusion between classes sharing similar characteristics.

Deep Learning with different kind of Neural Networks like deeper Networks, Recurrent Neural Network, Convolutional Neural Network, which gave good results in the fields of ASR and MIR.

Finally, as mentioned above, this thesis focused only on training four machine learning classifiers. Future research will direct towards Deep Learning with different kind of Neural Networks like deeper Networks, Recurrent Neural Network, Convolutional Neural Network, which gave good results in the fields of ASR and MIR.

References

1. Dufaux, A.: Detection and Recognition of Impulsive Sounds. University of Neuchtel, Switzerland (2001)
2. Cowling, M., Sitte, R.: Comparison of techniques for environmental sound recognition (2003)
3. Peeters, G.: A large set of audio features for sound description (similarity and classification) in the CUIDADO project (2004)
4. Pillos, A., Alghamidi, K., Alzamel, N., Pavlov, V., Machanavajhala, S.: A real-time environmental sound recognition system. In: Detection and Classification of Acoustic Scenes and Events (2016)
5. Bountourakis, V., Vrysis, L., Papanikolaou, G.: Machine Learning Algorithms for Environmental Sound Recognition: Towards Soundscape Semantics. Thessaloniki, Greece (2015)
6. Coelho, L.P., Richert, W.: Building Machine Learning. Packt Publishing, Birmingham-Mumbai (2015)
7. Yang, J., Luo, F.-L., Nehorai, A.: Spectral contrast enhancement: algorithms and comparisons (2003)
8. Jiang, D.-N., Lu, L., Zhang, H.-J., Tao, J.-H., Cui, L.-H.: Music type classification by spectral contrast feature (2002)
9. Schindler, A.: Music information retrieval (2016)
10. Harrington, P.: Machine Learning in Action. Manning Publications Co., Shelter Island, NY 11964 (2012)

11. Lantz, B.: Machine Learning with R. Packt Publishing, Birmingham (2015)
12. Chu, S., Narayanan, S., Kuo, C.: Environmental sound recognition with time frequency audio features. IEEE Trans. Audio Speech Lang. Process. (2009)
13. https://www.sas.com/en_us/insights/analytics/machine-learning.html
14. http://matplotlib.org
15. https://www.python.org/
16. http://scipy.org
17. https://matplotlib.org/
18. http://jupyter.org/

Early Signs of Diabetes Explored from an Engineering Perspective

Jenny Lundberg[1] and Lena Claesson[2(✉)]

[1] Linnaeus University, Växjö, Sweden
jenny.lundberg@lnu.se
[2] Blekinge Institute of Technology, Karlskrona, Sweden
lena.claesson@bth.se

Abstract. Undetected diabetes is a global issue, estimated to over 200 million persons affected. Engineering opportunities in capturing early signs of diabetes has a potential due to the complexity to interpret early signs and link it to diabetes. Persons with untreated diabetes are doubled in risk of getting cardiovascular diseases and may also suffer other consequent diseases. In Sweden, approximately 450 thousand have diabetes where 80–90% are of type 2 with 1/4 unaware of it, i.e. approx. 100 thousand. Screening approaches, searching specifically for diabetes in persons not showing symptoms has been initiated with positive results. However, some general drawbacks of screening such as false sense of security are an issue. In this publication, we focus upon in home measurements and empowering of the individual in identifying early signs of diabetes. The methods in this publication are to gather data, evaluate and give suggestion if clinical test to confirm or reject diabetes. In home measurements, education process with companies for innovation possibilities.

Keywords: Engineering education · Diabetes · Internet-of-Things

1 Introduction

Societal challenges such as the global health epidemic of diabetes puts high demands upon the individual in understanding early signs of health risks. The occurrence of diabetes, a chronic disease, is of epidemic nature. 2015, 415 million worldwide have diabetes and it is estimated that in 2040, 642 million worldwide will have diabetes, see Fig. 1, [1]. Identifying signs of disease at an early stage, perhaps including behavior leading to unhealthy behavior can be of further interest. Exploring data using Artificial Intelligence (AI) and Big Data approaches can have the potential in empowering the individual in making informed decisions.

The common types of diabetes have different causes and affect different populations: Type I or Juvenile-Onset Diabetes, often diagnosed pre-adolescence, this autoimmune disease is characterized by the failure of the pancreas to produce the insulin needed to transform glucose into energy. Type II or Adult-Onset, although adult-onset diabetes can affect any age group, the disease often occurs in overweight adults. Although the pancreas produces insulin, excess weight interferes with the body's ability to use the

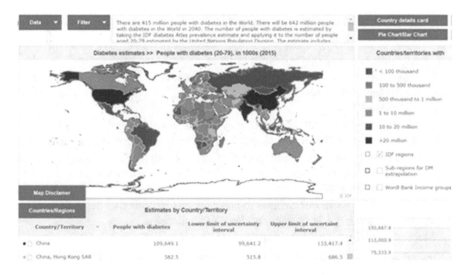

Fig. 1. Global spread of Diabetes in 2015 presentation made by the International Diabetes Federation (IDF), where dark green is "good" i.e. <100 thousand and dark purple <20 million is bad [5].

insulin to convert glucose into energy, and, over time, the pancreas may lose its ability to produce insulin. Treatment often involves weight-loss recommendation, medication to increase insulin production or slow blood glucose production and, in hard-to-treat cases, insulin injections. Gestational Diabetes this type of diabetes occurs during pregnancy and is almost always resolved after the woman gives birth. For further reading about diabetes and pregnancy, see [2, 3]. Although treatments for the three types differ, warning signs and long-term risks are similar [4].

Diabetes is a collection of heterogeneous diseases that are commonly characterized by the blood sugar being too high. The disease is affected by genetic risk factors, but scientists are still looking for answers to the biological mechanisms behind it. Lifestyle related issues, such as obesity, stress and smoking, are known risk factors for type 2 diabetes. The Swedish Cardio Pulmonary Bio Image Study, SCAPIS, is a world-unique research project within heart, vascular and lung. The project provides great opportunities for new research breakthroughs. With the collected material from 30,000 people, SCAPIS becomes a unique knowledge base for researchers around the world trying to find out why diseases such as stroke, COPD, sudden cardiac arrest and myocardial infarction occur, why some get them and others not, and how to prevent them from occurring [4].

Untreated diabetes can give consequent diseases, and early detection in screening approaches has been researched. According to [6]:

– diabetes cases diagnosed by screening were detected on average 4.6 years earlier than those detected by the healthcare.
– compared with the diabetes cases detected by screening, diabetes cases found in the hospital had twice as high a risk of dying during the follow-up period.

– patients diagnosed in health care also had 55% higher risk of cardiovascular disease, more than twice as high risk for kidney disease and 66% higher risk of eye disease.

This study aims to answer the question;

How can we use engineering for example sensor/actuator and data from a person to identify early signs of diabetes, if there are any signs (i.e. a person can have diabetes without any signs)?

In this project, the goal is to identify possibilities in linking early signs of diabetes related to the human and bodily/emotional/behavioral signs or other data to engineering solutions to identifying if the person has early stages of diabetes. Thus, collecting background data of relevance to diabetes, and human body, behavior data and more to identify early signs of diabetes into creating awareness about the signs enabled by engineering. Then to identify issues and/or changes in behavior that indicate that the individual has got diabetes or are in the early stages of developing diabetes.

More specifically we hold the purposes below:

1. To identify engineering requirements possibilities in implementing a lab platform identifying early signs of diabetes and to explore these.
2. To work with home & other environment to identify where sensor/actuators can be implemented in lab, smart home & mobile, wearable solutions.
3. To present an educational approach and platform in how to identify and connect engineering requirements in relation to early signs of diabetes.
4. Identify Artificial Intelligence and Big Data opportunities including visualization and prediction models to use for different types of data input.
5. Additional to the purposes above is to understand business possibilities in matching the cost - benefit from the industrial perspective. That is the approach is directed towards students from industry as a part of expert competence for innovation with possible business opportunities.

2   Early Signs and Health Monitoring

The approach is to consider early signs indicating if the person is having diabetes, see Fig. 2. This includes monitoring of behavior and behavior changes. Physiological changes making changes in the behavior.

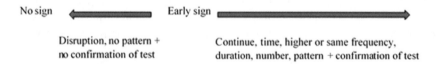

Fig. 2.   The approach to consider early signs indicating if the person is having diabetes

There are differences between early signs and signs. These different types of signs can be related in different ways. The confirmation of behavior being a sign of diabetes can only become a sign if it is clinically confirmed that it really is diabetes.

According to [7] a person can have diabetes type 2 without symptoms. According to [8] there are differences in how the signs appear given which type of diabetes. Type 2 diabetics are the main group since they can have undiscovered diabetes for a long time.

For type 1 they often have a rapid disease course. They have low c-peptide and blood glucose levels are often cloud with most often autoantibodies to insulin-producing cells. They are often below 25 years old, they can have normal/low body weight, severe weight loss, dehydration, significant diabetic symptoms and ketones in urine and blood.

For type 2, the heredity, such as obesity, high blood pressure and blood lipid disorders. The debut is sneaking with lighter symptoms and progress. Usually there are no ketones in blood and urine. The c-peptide value is high along with high blood sugar.

Active and passive health monitoring and analysis are part of this suggested approach. The active is the health form and the active note taking, and health monitoring using health equipment and the passive are the health monitoring where it is included in the everyday life, see Fig. 3. The symptoms for diabetes type 2 are the same as for type 1 however the debut for type 2 are more insidious.

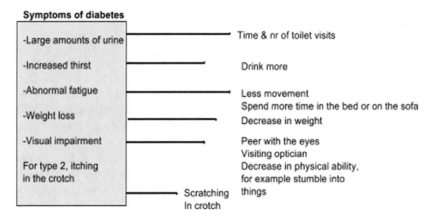

Fig. 3. Symptoms of diabetes 1 and 2 according to [4] Hjärt-Lung Fonden and 1177 National Swedish Health and Care Services [7]. For type 2 there can also be no symptoms, and for type 1, there can also be have a stomach ache and smell of acetone from the mouth as a sign.

Impairment of vision are one of the early signs connected to diabetes. One approach in considering this is to document eyes and movements. As shown in Fig. 4 the picture to the left, the person is reading a document in normal text size. The face is relaxed but focused on the reading task. The face on the right with the small sized text document, the person has moved closer to the screen, the eyes are a bit smaller, and in this case the light from the computer makes the face look lighter (dependent on lighting conditions in room and other aspects). How much time the person spends before the computer, how to assure it is true, i.e. the sensitivity, and differentiate this from general vision loss related to age are an issue. For how long and how often the monitoring should be performed, how to handle integrity issues and how to train data analytics are further issues.

Fig. 4. Left, the person is reading a document in normal text size. The person on the right reading a small sized text document has moved closer to the screen.

Brighter color in the face due to approaching the screen, smaller eyes and wrinkling of the forehead, small lines beside the eyes. Here AI and big data approaches can be one possible solution. There are on the market AI solution to this, according to [9] however not in the compilation of tracking early signs of diabetes to our best knowledge.

3   Method

The approach implemented is in cooperation academia - company, with student in the course acting company representative to identify innovation & business possibilities. The teacher sets up the course with a focus area, allowing the student to identify possible areas of innovation. The specific course on master level that has been set up i.e. ET2592, in springtime 2017, Analysis and modeling 2, Smart Home & Health system at Blekinge Institute of Technology (BTH).

Within the framework of this Knowledge Foundation (KKS) funded project, "Remote Diagnosis - online engineering, at master level" BTH students and employees of the companies participating in the project but also other professionals in the industry, could attend the newly designed distance learning courses [10]. KKS the research financier for universities with the task of strengthening Sweden's competitiveness and ability to create value [11].

The courses are divided into two blocks, one block that covers methods for diagnosis and one block concerning technology for remote diagnosis. The courses in the block concerning methods for diagnosis are based on each other while the courses in the block which covers technology can be studied independent of each other. Each course with green frame in the block diagram shown in Fig. 5 results in 3 European Credit Transfer and Accumulation System credits (ECTS-credits), except for the Ethics and Architecture's lecture series. One academic year corresponds to 60 ECTS-credits.

The course block that concerns methods for diagnosis is divided into three specializations; Mechanical systems, Electrical systems and Smart homes and health applications. Each course consists of six lecture sessions (2×45 min), six exercises (1×45 min) and one laboratory session (3×45 min). Each course is given during a period of eight weeks. Each course is given as a distance course and includes experimental work on remotely controlled experimental setups. Thus, the lab assignments in the courses are carried out remotely. Adjustments to the course schedule were made based on interest

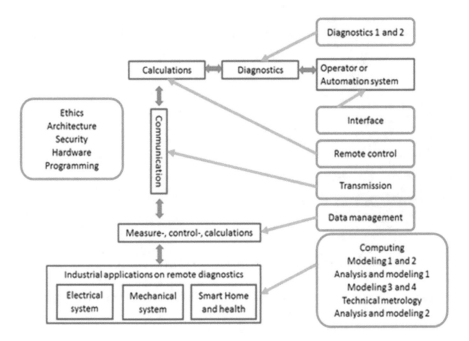

Fig. 5. Block diagram of the courses developed in the project and how different courses and areas of expertise support a remote diagnostic system. Real equipment, with control, calculations and automation inside blue frames and courses inside green frames.

from the companies involved. The courses include application modules and these can be adjusted according to the participants' interests.

Furthermore, the first time a course was given within the project only employees form the companies participating in the project attended the course. The second time a course was given both students from academia and industry may attend the course.

Research front, State of the art, and research and engineering opportunities are discussed. What is interesting within this focus area from the point of view of innovation. What could you possibly do a business model of? What type of expert competence are needed to build engineering solutions in relation to diabetes? What does the company think they want to focus on for innovation within the specific focus? The teacher and the student discuss innovation opportunities and the teacher adjusts the teaching according to the student's interest. The pedagogical approach has been that of blended learning, i.e. the students were provided with course material in digital resources before each lecture, see Fig. 6, and the lecture focused on deeper discussions and analysis concerning the course subject. Requirements for lab is set out together within the course.

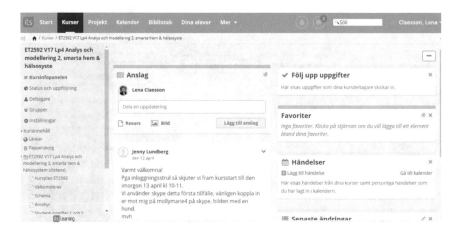

Fig. 6. Screenshot of the digital resources used in the course Analysis and modeling 2, Smart Home & Health system at Blekinge Institute of Technology (BTH).

We have used the public information from Swedish health authorities [12] to identify early signs of diabetes.

Furthermore, a suggested approach based on the following points aspects are considered.

1. Documentation in risc test [12] according to Diabetesförbundets 8 questions, with risc points according to the following scale and prediction, >7p small, 1 of 100, over 20p very big approx. 50%.
2. Identifying normative behavior, how long time, which parameters record behavior over time (glucose values, body weight), individual patterns in behavior.
3. How many individuals are needed to get statistically ensured basis.
4. Big data, methods for matching signs of diabetes, enquiry and statistics.
5. Prediction models in risks in developing diabetes type 2.

Typically, sensitivity is used in an ex test if you have diabetes or not, we can use this to combine different types of input, i.e. parameter 1, 2, 3, 4, 5 and then make a qualified guess and then send to the doctor for real investigation.

4   Results

The outcomes are as follows:

A course package on Master level focusing on specific societal challenges such as diabetes.

Figure 7 presents in and out flow of liquid of a human. The measurements have been done with several smartphone apps for support. The person that has been measured is a non-diabetic.

In Fig. 8 the glucose values for 20 h are displayed. The interstitial fluid is measured in mmol/L.

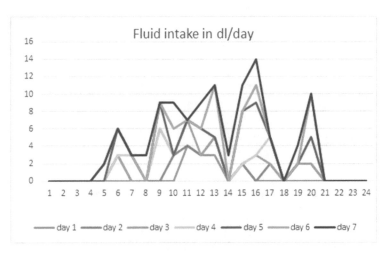

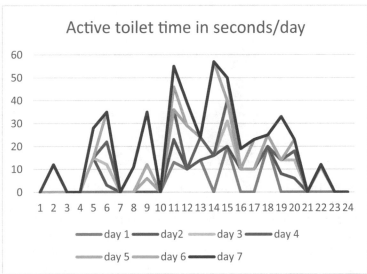

Fig. 7. Diagram at top, measured a week from Monday to Sunday, the stacked line diagrams present the fluid intake measured in dl per hour per day are displayed. Diagram at bottom, the active toilet time, i.e. the time for urinating measured in seconds per hour.

The sensor on the arm, see Fig. 9, communicates with a device, sending values in a preset interval, it is displayed in the log for documentation every 15 min. The values can be measured anytime with the device reader being swept over the arm.

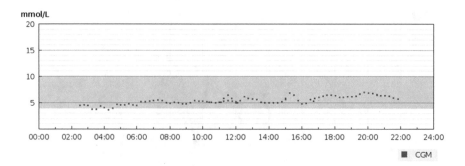

Fig. 8. 24 h glucose measurements sampled every 15 min in mmol/L.

Fig. 9. The glucose sensor Abbott Libre [13] cinched on the left upper arm.

The Abbott Libre sensor [13] are cinched on the left upper arm. The process of putting the sensor on the skin is quite easy in this case. It is done by the person him/herself with just one hand. Swimming, washing and sport performance, like swimming, is no limitation. When changing clothes without thinking about the sensor, some clothes can get stuck a bit. Other Continuous Glucose Monitoring sensors has been tested, one required support from another person, taking in mind that the subject measuring is a novice in CGM. The sensor is 35 mm in diameter and 5 mm high, weighing 5 g. A sliver oxide battery provides energy for up to 14 days. The sensor values are stored for 8 h and are measured every 15 min.

There are compatibility and integration issues between the different sensors and monitoring devices. These issues are partly solved by some companies holding platforms integrating different types of sensors/actuators. However, there are still issues and this affects the synchronization as well, making analysis of big data between several devices complex. Furthermore, some changing in numbers of the values, i.e. sometimes day and value are saved as one value, and if you want only the value and not the date displayed for every instant, a manual process of shaping the data to desired form are to be performed.

5   Conclusion

The conclusions of this paper are that the engineering of health systems is of interest from industrial perspective. More specifically, when it comes to diabetes, considering a specific focus catching early signs of diabetes are of interest from an industrial representative student. Holding competence in smart home & health applications holding a specific focus upon a specific disease and finding early signs of this are of specific interest.

In this publication, we have described remote and diagnosis possibilities in connecting early signs of diabetes to engineering opportunities sensing the signs. We present a conceptual framework and approach in combining sign – symptom - sensor – diagnosis. We use medical well-established symptoms of diabetes, discuss possible functions in monitoring normative behavior and present technical solutions to monitor over time.

As a part of the course development, big data analysis, prediction models etc. are going to be part of a new course within the project, "Remote Diagnosis - online engineering, at master level" at BTH.

References

1. International Diabetes Federation (IDF), 02 December 2017. http://www.idf.org/about-diabetes/facts-figures
2. Hod, M., et al.: Textbook of Diabetes and Pregnancy. CRC Press, London (2016)
3. Toledano, Y., et al.: Diabetes in Pregnancy, International Textbook of Diabetes Mellitus, 4th edn., pp. 823–835 (2015)
4. Hjärt och Lungfonden, 25 November 2017. https://www.hjart-lungfonden.se/Sjukdomar/Hjartsjukdomar/Diabetes/Symptom-diabetes/
5. U.S. Department of Health & Human Service, 01 December 2017. https://www.niddk.nih.gov/about-niddk/strategic-plans-reports/Documents/Diabetes%20in%20America%202nd%20Edition/chapter5.pdf
6. Feldman, A.L., et al.: Screening for type 2 diabetes: do screen-detected cases fare better? Diabetologia **60**, 2200–2209 (2017)
7. National Care Guide for Information and Services for Health and Care 1177, 29 November 2017. https://www.1177.se/Skane/
8. Living Healthy with Diabetes, 30 November 2017. http://lhwd.net/warning-signs-diabetes-2
9. Smart Eye, 02 December 2017. http://www.corp.smarteye.se/sv/press/pressmeddelanden/details?releaseId=2450302
10. BTH - Diagnos på distans – 'online engineering' på mastersnivå", 15 November 2017. www.bth.se/diagnospadistans (20171115)
11. KK-Stiftelsen, , 28 November 2017. http://www.kks.se (20171128)
12. Diabetesförbundet, 01 November 2017. https://www.diabetes.se/globalassets/forbundet/diabetes/journalistguide-visuell-identitet/risktest_final.pdf (20171101)
13. FreeStyle Libre, 02 December 2017. https://www.freestylelibre.se/ (20171203)

Digitalization of Engineering Education: From E-Learning to Smart Education

Irina Makarova[1(✉)], Ksenia Shubenkova[1], Dago Antov[2], and Anton Pashkevich[2]

[1] Kazan Federal University, Naberezhnye Chelny, Russian Federation
kamIVM@mail.ru, ksenia.shubenkova@gmail.com
[2] Tallinn University of Technology, Tallinn, Estonia

Abstract. Digitalization of all scopes of activities along with the rapid accumulation of information, development of technologies and the processes' intellectualization pose global challenges in both the economy and education. Information technologies become an integral part of the human living space, causing the emergence of a new digital (networking) generation of people, for whom a mobile phone, a computer, and the Internet are the natural elements of their life. A universal approach to Smart Education is needed. Business requires engineers who can design, create and operate complex technical systems. At the same time, principles of sustainable development and minimization of negative environmental impacts must be observed. Educational system should ensure the quality of training engineers, who are needed by business and society. To realize this goal, there are opportunities associated with the use of such educational technologies as modeling, simulators, augmented and virtual reality. Ways to improve educational process with use of simulators and a virtual reality, as well as examples of using such training technologies to increase students' motivation when training them for the automotive company are presented in the article.

Keywords: Engineering education · Smart Education · Students' motivation
Virtual reality · Digitalization · Industry 4.0

1 Introduction

The XXI century has exacerbated the problems facing the world community, that caused by the depletion of natural resources and critical conditions of ecosystems. Society development is associated with processes of urbanization, which are accompanied by the negative impact on environment and lead to climate change. These processes cause the changes in patterns of employment – old professions are disappearing and new ones are appearing. According to analysts' forecasts, the Fourth Industrial Revolution will force the mankind to reconsider the attitude to work and the working process. In the World Economic Forum's report [1] it was noted that in the 5 years the cardinal changes will affect more than 35% of the modern working skills. Such rates of technics and technologies development require shifting the paradigm of the fairly inert educational system. The employment of the population and the sets of necessary skills were analyzed in the report [1], as well as the strategy of labor resources' development in the future was presented.

World employers have expressed their opinion on the issue, what will be affected by such a rapid development of the technological process, and in what direction will the labor market develop. As a result of the survey [2], 10 skills that will be the most actual by 2020 were identified. One of the three most popular skills will be creativity, as it will be necessary to invent ways and places of application of new technologies, creation of new products and services. Human capital is a key factor for growth, development and competitiveness. The Human Capital Report calls for a human-centric vision of the future of work, one where people acquire and use their knowledge and creativity as key drivers of a prosperous and inclusive economy [3].

To provide the possibility of training engineers "for the future", conditions for changing people's consciousness should be created, because the need for engineers with creative thinking is becoming more acute. We need a transition to a society of creators, which, in the context of intellectualization of all spheres of activity, requires the creation of prerequisites for a conscious choice of engineering profession. The changes should affect not only the application of new tools and teaching methods, but also build a systematic strategy. This strategy should ensure sustainability of the educational system, the possibility of its continuous improvement and development. This is to be corresponded to the needs of the real sector of the economy.

2   Smart Society and Engineering Education

2.1   Smart Education and University 4.0: Paradigm Shift

Today we can confidently talk about technologies that will change the world in 5–10 years. Technologies, which experts consider the most promising, are associated with the concept of the Fourth Industrial Revolution. One of its reasons is the desire of a person to provide comfortable living conditions and daily activities through the expense of smart decisions. Smart technologies are aimed at finding the best solutions that will cause less harm to the environment.

With the notion of the Fourth Industrial Revolution is connected to the initiative of the German government, supported by manufacturers of various equipment, the so-called Industry 4.0. Industry 4.0 implies the use of Internet of Things (IoT) and Big Data in the manufacturing process, when any components of the system are linked together via a network and independently find ways to reduce costs and to fulfill orders. Industry 4.0 presupposes the rational use of natural and technical resources, the most effective energy saving, the secondary processing of all waste and reception new products, raw materials or energy from them. It is the fundamentally new paradigm: "Repair instead of a new purchase, the hire instead of property". Industry 4.0 implies cyber-mechanical systems that are designed to fulfill new tasks: by analyzing customer requirements, they themselves change the technological process, and also serve themselves. This will require a completely new engineering approach: digital enterprise models; engineers and programmers who will operate with them; smart devices that can exchange information with each other and with a digital model. The most important is the modeling of the management system. Industry 4.0 will allow reducing critically the dependence on the human factor due to the increasing percentage of intelligent labor. Besides that, Industry 4.0. involves the creation of a virtual process and its connection to the actual physical process that takes place in the enterprise. This allows debugging the process virtually, saving money for installation and commissioning. Implementation of a virtual simulator system will allow staff to practice real actions in virtual workstations.

SMART society poses a new global task for universities: training specialists with creative potential, who are able to think and work in a new world. Since, on the one hand, types of activity done by an engineer start to be more and more multivarious and, on the other hand, companies want to get a ready specialist for carrying-out of a certain work, a conflict between educational and business goals arises. One more problem is a decrease of young people's motivation to get engineering education, especially that one, which assumes to deal with highly intelligent activities in the future. Digitalization of manufacturing within the framework of transition to Industry 4.0 requires the engineer to have new competencies that involve the possession of digital technologies and such practical skills as communication in social networks, selecting useful information from large data sets, working with electronic sources, creating personal knowledge bases, which requires a change in the nature of the educational process. In the next twenty years, automation of industry will lead to a reduction of about 40% of jobs. Along with it, by 2020–2022, employers' demands on the labor market will shift to innovative professions in the fields of robotics, programming, IT infrastructure, artificial intelligence. This implies changes in the system of personnel training. The educational system is more inert, therefore, it is necessary to change the paradigm for its deep "reconfiguration" to meet the needs of the economy and society by training engineers taking this into account.

The new conditions require the education to train efficiently the creative self-motivated individuals, who are capable to solve complex problems using innovative and flexible tools. To realize this, first of all, the transfer from reproductive approach in the organization of educational system and educational process to creative one must be done. The engineer must be able to adapt quickly in a changing world, be able to apply digital technologies for the design, creation, management of technical systems throughout their life cycle. Therefore, when creating educational content, in our opinion, it is necessary to focus students on acquisition of such skills to use software packages and mathematical models that they could clash in professional life.

Universities and research centers play a key role in the innovation ecosystem. Universities generate knowledge for the enterprises of the city and serve as a meeting point for educational and scientific centers, students and the world of business. Therefore, it is important that cities strive to develop cooperation between enterprises and educational and scientific centers, thereby contributing to economic and social development. The World Bank has determined expected structure of national wealth for those countries, where an innovative SMART society, the main idea of which is development of human potential, will be formed. Natural resources will constitute only 5% of national wealth; material and production capital – 18%, while 77% of national wealth will consist of personal knowledge and skills that will determine the future of person.

2.2 Situation in Automobile Industry: Problems with Personnel and Possible Solutions

Requirements to the automotive industry have sharply increased recently: cheap vehicles, a wider range of models and higher quality are necessary to cover the global market. The intelligent functions of the vehicle's onboard systems are expanded through driver assistance systems and IT integration with various devices. Therefore, the issue of production digitalization for automotive producing is even more relevant than for the modern economy in general. To increase the industry efficiency, it is important to plan a virtual plant, because 3D modeling of the prototype and a virtual model of the technological process allows reducing the time spent from planning to production almost twice (approximately from 30 to 18 months). New technologies will provide opportunities for producing unique parts for individual orders. This new market of materials, equipment, software and services will require new specialists. For example, jobs, a third part of which has never existed before, were created in the United States last quarter of a century. They have appeared in new areas, such as information technology, the creation of new applications and production of equipment.

Thus, Siemens uses the elements of Industry 4.0 in business and takes the leading positions in the field of automation. Data analysis, artificial intelligence, robotics and autonomous security systems are the key technologies here. Education and training are the most important issues for Siemens. The company has invested 510 million euros in employees around the world: 240 million euros in education, and 270 million euros in training. The demand for creative and highly qualified specialists is growing, therefore Siemens supports new ideas and non-standard approaches, cooperates with start-ups and 25 partner universities. Training is a strategically important factor for maintaining competitiveness and ensuring work in the future. So long as, on the one hand, types of activity done by an engineer start to be more and more multivarious and, on the other hand, companies want to get a ready specialist for carrying-out of a certain work, a conflict between aims of education and business arises. Another issue is a decrease of young people's motivation to get engineering education, especially, which assumes to deal with highly intelligent activities in the future.

In the context of the above-mentioned challenges the use of game-based learning methods allows, on the one hand, to understand deeper the core of real processes and, on the other hand, to increase interest of students. An additional point is that detailed analysis of processes and risks, which can appear, allows students later to avoid occupational injuries. It will encourage a sustainability of both educational and manufacturing systems. Sustainability of the education system will be ensured by increasing its flexibility in educational resources (content) as well as by reducing costs of equipment and materials that are required for changing educational courses during the transition of the business to new technologies. Sustainability of the manufacturing system is provided by a quick adaptation of young specialists at their workplace as well as closer cooperation with educational institutions by using new educational technologies.

2.3   Simulators and Game Based Learning Methods in the Engineering Education

Organization of the game requires a computationally modeling complex that includes a control set and a set of models. The teacher develops the script of the game, inputs initial data and rules, determines the duration of the assignment. During the game students are analyzing the data and are making decisions, observing the restrictions. The teacher monitors the compliance with rules and evaluates students' actions (Fig. 1a). Computer models, as a rule, are not universal. Each of them is designed to solve its own special range of problems. As far as computer models are based on mathematical models (which contain the control parameters), they could be used not only to demonstrate manufacturing aspects that are difficult to reproduce in the educational process, but also to determine the influence of certain parameters on the studied processes. This allows them to be used as simulators of laboratory installations and for testing the managerial skills of simulated processes. Computer-aided technologies allow not only working with ready-made models of objects, but also making them out of individual elements.

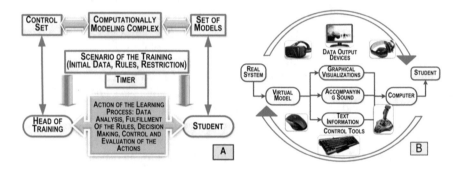

Fig. 1. The conceptual scheme of the computer simulator complex: (a) an interaction of teacher and the students during the game; (b) the architecture of virtual simulator complex

Electronic laboratory course allows simulating real processes or conducting an experiment that cannot be realized in the real world. In the process, simulator imitates not only the real installation, but also the study objects and the experimental conditions. Laboratory simulators allow students to determine the optimal experimental parameters, gain initial experience and skills, facilitate and accelerate the work with real experimental installations and objects.

Computer models, construction sets and simulators allow students to solidify knowledge and to gain skills of its practical implementation in situations that simulate reality. In addition to the professional skills formation, computer simulators successfully develop creativity, professional intuition and, what is the most important, teamwork skills. All of this allow significantly improving the quality of engineering education. The effectiveness of the educational process by using the simulator is largely determined by: (1) the kind of the subject area model, i.e. the quality of conformance of the visual interpretations of the designed models that are used in the simulator to the real analogue, (2) completeness of the simulator's script, i.e. quality of conformance of the scenario to the processes in the simulated objects of the physical world. Thus, the greater the quality of conformance of the virtual models to their real analogues, the better simulator promotes assimilation of knowledge. The architecture of any computer simulator complex is determined by its purpose, by the list of tasks, functions and by the type of the models used in the simulator complex (Fig. 1b). Five types of computer simulator complexes can be identified according to their intended purpose:

1. Simulators developing motor skills are widely spread to train driving of different types of vehicles, shooting, welding, sport games, etc.
2. Simulators that train pattern recognition skills are used in education of medical diagnosticians, operators of various specialties, complex systems managers, etc.
3. Simulators that train to work by algorithm are intended for training specialists to operate and maintenance the sophisticated equipment. These simulators are based on a static model of the world. They simulate operation only of the properly functioning equipment. The simulator usually has pretty inflexible educational script: student can feel free to act only in the intervals between control situations, and the right decision (situation) is always unique. Thus, student, as a final result, must reproduce exactly this unique decision (situation).
4. Simulators that train behavior in emergency situations are used to train personnel and operators of electric power stations, nuclear power stations, as well as to train driving skills in complicated situations when there is a risk of traffic accident.
5. Simulators that teach to solve problems with a branched decision tree. The main focus of these simulators is to check the decision that was proposed by student. This simulator is used to train skills of designing, assembly and installation of technical systems, troubleshooting and equipment repairing.

Virtual simulators have a number of weaknesses that significantly limit their use: (1) closeness – the inability to change the predetermined range of equipment and related training materials, (2) redundancy of the interface that significantly overloads the script of the laboratory work by operations of setting up the simulator's functional, and (3) in most cases, the behavior of the studied devices and measuring equipment is not simulated. It is calculated without taking into account the dynamics or, at the best case, it is calculated by the ready-made formulas of the appropriate differential equations with unchangeable parameters.

The last one causes a predetermined invariant assembly of virtual elements into the final scheme. At the same time, the real stands always allow some changes in configuration (e.g. to change places of the series circuit elements). In addition, the teacher does not have an access to the processing the script of laboratory work, he can't promptly pay student's attention to the necessary aspects of his actions and the teacher has to use only facilities that are inflexibly built into the purchased software product.

2.4   Directions to Apply Virtual Reality in Engineering Education

Virtual reality is currently an instrument whose possibilities are far from being exhausted. Development of computer equipment and technologies significantly expanded the possibilities of modeling and communications. It became possible to model not only physical objects, creating their computer analogue in the form of a 3D-model, but also processes, creating a virtual production environment. Modern means of modeling, virtual and augmented reality allow a person not only to see the object from different sides, to look inside it, to see what it consists of, but also become a participant of the process. This applies equally to the creation of a virtual production environment and a virtual educational environment. Analysis of numerous papers in the field of virtual reality application for engineering education showed that researches are usually done in the following directions:

- Complicated systems study models. This is such systems as atomic energy, space, construction, transport and logistics [4].
- Creating a virtual educational environment. Such developments allow the teacher to remotely control the learning process. In addition, it allows you to create "virtual classes", in which you can observe the actual production process with the help of a web camera [5, 6].
- Such technologies can be used for networks cooperation between universities for the purpose to organize teamwork of students [7].

3   Results and Discussions

3.1   Application of Interactive Educational Methods When Training Engineers for the Automotive Industry

Automotive industry is one of the locomotive in European countries. The concept of "Industry 4.0" itself, as well as some other advanced technologies, first arose and were used here. Project of autonomous cars is a world trend for the intellectualization of management in all spheres of human activity: ideas of "Smart City", "Smart Energy", "Smart Healthcare", "Smart Education" are intimately related to the concepts "Smart Mobility" and "Smart Transport". We have conducted a research on the tasks and types of professional activity of engineering students for the automotive industry. Questionnaires were developed and a survey among employers was conducted that allowed identifying problems for different areas of engineering that are experiencing young professionals. In addition, the survey among students was conducted that has helped to define the most challenging problems they faced during the training period. It was revealed that automotive industry faces an amount such challenges as an increase of energy and fuel efficiency of vehicles as well as their level of environmental friendliness; an intellectualization of transport system; a reduction of a negative environmental impact by manufacturing, maintenance, service and recycling of vehicles. Besides that, an engineer should have skills to create digital models of car itself, its production as well as systems of service supports and intelligent transport systems.

Practical approval of the proposed methodology was realized in several stages. At the first stage, the survey among representatives of the potential employees was carried out. The aim of this survey was to identify the most significant competences of university graduates, which are required for successful work and career progress. The second stage included the development, using the SMART principles, of curriculum and educational content directed to produce the required competences of the future engineers. At the third stage, test groups of students, whose study was carried out with the application of new learning methods, were formed. Depending on direction of educational program, each student group developed projects to solve the real manufacturing problems using appropriate technical and software tools.

3.2 Experience of Students' Participation in Real Projects

When teaching specialists in the field of logistics, students get a material to study the specialized transportation software (e.g. AnyLogic, PTV Vision), as well as models of problem areas of city road network. As a result, students come to a classroom lessons with suggestions for optimizing parameters of the street-road network, discuss them, check various options which can be used to obtain optimal solution to the problem. The purpose of optimization may be: reduction the possibility of road accidents in this area [8], increasing the road network's capacity [9], reduction of the environmental load [10], etc. As an example of student's project Fig. 2a presents a model of the existing intersection configuration, which is needed to be optimized and Fig. 2b presents a model proposed by students during the classes [11] (Fig. 3).

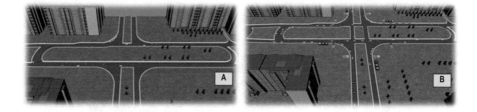

Fig. 2. Optimization of the road network's configuration

Proposed methodology was applied when training engineers for the scientific and technical center of automotive company. Students had obtained practical experience, after that they worked as engineers in scientific and technical center. At the same time, they were carrying out their graduation project. As part of an exploration, students resolved the real problems. Obtaining the skills of working with Siemens software while learning, allows students to carry out real projects on design, technological preparation of production, launching the vehicle in series and directly producing. Students participate in such projects as "Unmanned Vehicle", "Digital Production", "Smart Factory", etc. This increases their motivation and promotes the acquisition of new competencies.

To assess the effectiveness the comparison of academic performance and the quality of graduation projects of the last 5 years have been conducted. The analysis shows, that the quality of graduation projects in experimental groups is higher (Fig. 4a). This can be explained by the higher motivation of students, as well as by the focus on creativity development. Comparison of academic performance of master-students and experimental groups was performed. Groups were formed from students of experimental and traditional groups, who wanted to study master course. The quality of learning was assessed by the results of two examination periods for 2016/2017 academic year. The results are presented in Fig. 4b.

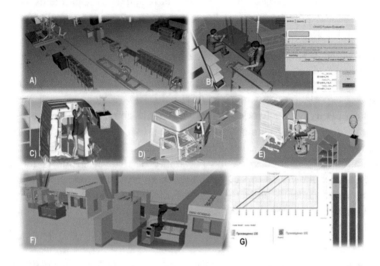

Fig. 3. Examples of students' graduation projects: (A) virtual production; (B) checking ergonomics of technological process; (C–E) simulation of the process of assembling the cabins; (F) simulation of the robotic section; (G) monitoring of technological process

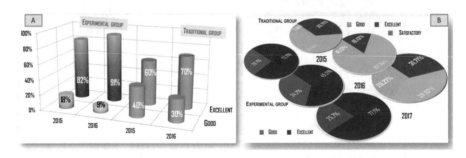

Fig. 4. Comparing of experimental and traditional groups: (a) quality of graduation projects; (b) academic performance of students

4   Conclusions

Smart Education concept implies, first of all, the creation of a such education system that will shape not only professional competencies of future engineers, but also will develop their environmental awareness and social responsibility for future generations. Implementation of progressive teaching methods will have a synergistic effect, increasing the quality of processes in industrial, transport, educational and other systems of Smart Cities. The gained experience shows that the use of new teaching methods in engineering education contributes to the development of engineering competencies that are required for high-technology industries. Besides that, cooperation with business allows teaching staff to study new technologies and to apply them for the training of students. Thus students will be ready to implement complex solutions when designing, modernizing and managing production systems, which will increase their efficiency, sustainability and will ensure the quality and competitiveness of the products. An integrated platform for implementing such solutions will improve the quality of both operational and strategic management as in both production and educational systems. Moreover, intellectualization of management processes will increase sustainability of these systems and will facilitate their adaptation to the changing needs of the market.

Acknowledgement. This paper would not have been possible without the financial support in the frame of the European Regional Development Fund's Doctoral Studies and Internationalisation Programme DoRa Plus which is carried out by Archimedes Foundation.

References

1. The Future of Jobs Employment, Skills and Workforce Strategy for the Fourth Industrial Revolution (2016). http://www3.weforum.org/docs/WEF_Future_of_Jobs.pdf
2. 10 skills that will be the most actual by 2020 (in Russian). http://womo.ua/10-professionalnyih-navyikov-kotoryie-budut-nuzhnyi-v-2020-godu/
3. Global Human Capital Report (2017). http://reports.weforum.org/global-human-capital-report-2017/human-capital-report-2017/
4. Behzadan, A.H., Kamat, V.R.: A framework for utilizing context-aware augmented reality visualization in engineering education. In: 12th International Conference on Construction Application of Virtual Reality, Taipei, Taiwan, 1–2 November 2012, pp. 292–299 (2012)
5. Araya, D.: Education in the Creative Economy: Knowledge and Learning in the Age of Innovation. Peter Lang, New York (2010)
6. Brown, J.S.: Learning, Working and Playing in the Digital Age (2014). http://serendip.brynmawr.edu/sci_edu/seelybrown/
7. Zaugg, H., Davies, R., Parkinson, A.R., et al.: Creation and implementation of a backpack course to teach cross-cultural and virtual communications skills to students in an international capstone experience. In: 2011 ASEE Annual Conference and Exposition, Vancouver, BC, June 2011. https://peer.asee.org/17679
8. Makarova, I., Khabibullin, R., Pashkevich, A., et al.: Modeling as a method to improve road safety during mass events. Transp. Res. Procedia **20**, 20430–20435 (2016)
9. Makarova, I., Pashkevich, A., Shubenkova, K.: Ensuring sustainability of public transport system through rational management. Procedia Eng. **178**, 137–146 (2016)
10. Tosa, C., et al.: A methodology for modelling traffic related carbon monoxide emissions in suburban areas. Transport **28**(2), 1–8 (2013)
11. Makarova, I., Shubenkova, K., Tikhonov, D., et al.: An integrated platform for blended learning in engineering education. In: CSEDU 2017 – Proceedings of the 9th International Conference on Computer Supported Education, vol. 2, pp. 171–176 (2017)

Investigating Rate Increase in Aerospace Factory By Simulation of Material Flow Operations

Laura Lopez-Davalos[1], Amer Liaqat[2], Windo Hutabarat[3],
Divya Tiwari[3(✉)], and Ashutosh Tiwari[3]

[1] Cranfield University, Bedfordshire, UK
Laura.Lopez-davalos@cranfield.ac.uk
[2] Airbus, Broughton, UK
amer.liaqat@airbus.com
[3] Automatic Control and Systems Engineering,
The University of Sheffield, Sheffield, UK
{w.hutabarat,d.tiwari,a.tiwari}@sheffield.ac.uk

Abstract. The main challenge aerospace industries are facing in recent times has been triggered by the remarkable increase in commercial aircraft demand. To address this challenge, aircraft manufacturers need to explore ways to increase capacity and workflow through process optimisation and automation. This study focusses on the optimisation of component flow and inventory during the assembly of the A320 Family wings' at Airbus (Broughton, UK) plant through Discrete Event Simulation (DES).

This research measured the likely impact of future changes in the wing assembly process, using simulation by: mapping of component flow from delivery to the point of use, simulation of current logistics scenario (AS-IS), simulation of future logistics scenarios (TO-BE) that include proposed changes for optimising flow and managing capacity surge, and testing and validation of mapping and simulation. The developed DES model demonstrated the impact of changes planned to be implemented by showing a considerable increase in production capacity growth, by achieving a target of 50% increase of aircraft rate/month within one year. It also highlighted the main problems causing blockages and other non-value activities in the process.

Keywords: Discrete Event Simulation · Aircraft manufacturing
Witness

1 Introduction

Airbus is a global leader in aeronautics, space and related services. According to Airbus' latest Global Market Forecast 2017, the world's passenger aircraft fleet is set to more than double in the next 20 years, and more than 10 years would be required to meet the current backlog [1]. In order to address this challenge, aircraft manufacturers need to explore ways to increase capacity and workflow through process optimisation and automation. This project utilised simulation for measuring the impact of future

© Springer International Publishing AG, part of Springer Nature 2019
M. E. Auer and R. Langmann (Eds.): REV 2018, LNNS 47, pp. 42–49, 2019.
https://doi.org/10.1007/978-3-319-95678-7_5

changes in the wing/spars' assembly process. The movement of spars, which is the main structural member of a wing, was chosen as the main object of study. Spars are the most structurally demanded components of the wing. They are located perpendicularly to the fuselage, attached to it and expanded towards the wingtip [2]. The single-aisle wings have two spars: the leading edge (LE) or front spar is located close to the front of the wing and the trailing edge (TE) or rear spar is located at two-thirds of the distance to the wing's rear.

Simulation models have the ability to predict behavior with a higher level of detail, making it possible to solve problems before they occur [3]. Discrete Event Simulation (DES), was utilized as a tool for system modelling, analysis and later lean implementation. It provides insight into the performance, capacity and constraints of a factory enabling manufacturers to perform changes to a factory model until the desired performance is achieved, with the expectation that the new-built factory or newly reconfigured production line is "right-first-time" thereby reducing the need to physically prototype the process. The research work involved the following steps:

i. Data gathering and material flow mapping by developing process chart of the LE/TE spar movements, starting from arrival of spar to Airbus plant and finishing at the final assembly points.

ii. Development of four simulation models for current and future scenarios:

- AS-IS 1: Current scenario (Month 1).
- AS-IS 2: Included changes suggested and represented shop-floor on Month +6.
- TO-BE 1: Presented the to-be shop-floor scenario of Month +12.
- TO-BE 2: Presented the to-be shop-floor scenario of Month +18.

iii. Analysis and evaluation of simulation models considering the Key Performance Indicators (KPIs) and the effect of adding extra shifts on Saturdays and Sundays.

iv. Validation and conclusions.

2 Methodology

2.1 Data Gathering and Material Flow Mapping

The spars LE and TE were the selected elements for study, which started with the unloading of the spars from the trailers, and finished when the assembled wings were taken out from the jigs. The on-site data was gathered by:

- Preparing flow-process charts for manufacturing activities, times and specifications.
- Informal interviews with operators and documents related to the process.

After on-site data gathering, the material flow was mapped and visualized using process chart. Visualising the work process provided enough information about activities, relationships, flows, and hidden waste. It can bring a deep understanding and major breakthroughs in productivity and other performance [4].

2.2 Simulation Models

The simulation models were built in WITNESS Horizon (Lanner Group, UK). The models were based on the data gathered on the floor shop and the information obtained from the logistic team including the tactical plan for the assignation of wings to jigs. The four simulation models are detailed as follows:

1. AS-IS 1 model: It represented the current scenario (Month 1) in the Airbus plant. It contained 212 elements including 130 variables to facilitate required performance of the model. This model was presented to Airbus and the suggested changes were incorporated in AS-IS 2 model.
2. AS-IS 2 model: This model included the suggested changes by Airbus and it presented the shop-floor scenario on July 2017. The main changes were:

   - Arrivals and launches: The number of trailers arriving per week was increased by 16% in one unit. An additional launcher was added in the storage area.
   - Incorporation of the Step Change line 1: Step Change is a fully automated assembly line that works as a jig. It consists of 7 stations, being able to hold 7 wings (14 spars). The number of paths increased by 21% in this model.
   - Shifts: Addition of shifts on Friday night, Saturday and Sunday mornings.

3. TO-BE 1 model: It included the changes that were planned to be implemented in the shop floor in January 2018. The main changes in this model were:

   - Arrivals and launches: The number of trailers arriving was increased by 33% per week. An additional launcher was added in the storage area.
   - Incorporation of the Step Change line 2: The Step Change line 2 is a replication of Step Change line 1. The number of paths increased by 35%.

4. TO-BE 2 model: Represented shop-floor scenario on July 2018. It contained 300 elements with 197 variables and 1164 lines of coding. The main changes were:

   - Arrivals and launches: The number of trailers arriving increased by 50% per week. A third launcher was added in the storage area.
   - Incorporation of Step Change line 3: The Step Change line 3 is a replication of Step Change line 1 and 2. The number of paths increased by 50%.

3 Results

The KPIs taken into account while extracting the results from the models were capacity (number of finished goods or throughput) and Work In Progress (WIP). For the four models, 8–9 different scenarios were tested and the top three displaying best results were chosen for further analysis. For each of these three scenarios, two additional variations were tested:

- Implementing the Saturday morning shift in the shed group (heavy gang, inspection and wing removal operators).
- Implementing the Saturday and Sunday morning shifts in the shed group.

The decision about best scenarios for further testing was a trade-off between the number of finished goods and the WIP. Thus, other parameters were taken into account, such as the percentage of idle time of the jigs and Step Change lines. The scenarios in which the highest idle time percentage was lower than 2% were considered. The maximum time spars spend in the storage area was also a measured index. The results were extracted by simulating for 30 weeks with a warm up period of 4 weeks.

3.1 AS-IS 1 Model

A set of variables was created to calculate the number of wings that went out of the jigs. These variables show the WIP and finished goods at any point while the simulation is running. The finished goods were classified by weeks and by north/south jigs. The results have been summarised in Table 1. The values corresponding to AS-IS 1 are presented in Table 1 as "100 units" and the values corresponding to other models (AS-IS 2, TO-BE 1, TO-BE 2) are presented relative to the AS-IS 1 values. As seen from Table 1, the number of spars waiting to be assembled in the jigs was more than 40% of the total WIP. The remaining 60% were being assembled in the jigs. Even if the percentage of idle time of the jigs was minimum (0.18%), it was noticed that the storage area was full. The average waiting time and maximum waiting time for spars to be transported to the jigs is represented as T and MT days. Upon presenting the model to Airbus for validation, it was suggested that an additional variable was created that would calculate the number of wings assembled per month.

Table 1. Summary of results from AS-IS 1, AS-IS 2 and AS-IS 2 + Saturday and Sunday shifts for 30 weeks. The values corresponding to AS-IS 1 are presented as "100 units" and the values corresponding to other models (AS-IS 2, TO-BE 1, TO-BE 2) are relative to the AS-IS 1 values

	AS-IS 1	AS-IS 2	AS-IS 2 + Saturday shift	AS-IS 2 + Saturday + Sunday shift
Finished wings	100	122	124	125
Lead time (minutes/wing)	100	81.6	80.8	80
Work in progress (spars)	100	106	94	91
WIP waiting (spars)	100	89	61	53
Jigs idle time (%)	[0–0.18]	[0–0.45]	[0–0.26]	[0–0.22]
Spars in storage area	100	100	50	21
Max spars in storage area	100	100	100	100
Average time in storage area (min)	T	0.69 T	0.59 T	0.56 T
Max time in storage area (min)	MT	0.95 MT	0.84 MT	0.83 MT

3.2   AS-IS 2 Model

As suggested by Airbus, this model included the variables that calculated the number of assembled wings per month. In this model, the average assembled wings per month increased by 22% in 30 weeks. Table 1 depicts the results obtained from AS-IS 2 model, AS-IS 2 with a Saturday morning shift, and AS-IS 2 with a Saturday and Sunday morning shift. In AS-IS 2 the lead-time decreased considerably compared to AS-IS 1. With the incorporation of the Step Change line 1, the WIP increased by 6%, and the number of spars waiting to be transported to the jigs decreased by 11%. Furthermore, the WIP decreased considerably when the Saturday morning shift was implemented. The percentages of idle time in both jigs and Step Change were less than 1.2%. With the implementation of shifts on Saturday and Sunday, the shed and storage were less congested, and the average waiting time of spars decreased.

3.3   TO-BE 1 Model

In this model, the average number of wings per month increased to 132, which stands for 66 set of wings per month. This throughput exceeds the 63 aircraft/month Airbus' goal. Table 2 depicts the results obtained from TO-BE 1 model, TO-BE 1 added with a Saturday morning shift, and TO-BE 1 added with a Saturday and Sunday morning shift. A comparison of Table 1 and 2 demonstrates an appreciable 15% increase in WIP. The reduction of WIP with the implementation of Saturday and Sunday morning shifts' was due to operators being able to transport spars to the jigs/Step Change and the wings out of them during the weekend. Hence, the storage areas were less congested, decreasing the average waiting time in the storage area by more than 50% and the maximum waiting time by approximately 16%. The implementation of the Saturday morning shift appeared to have a more significant influence on the results than the latter implementation of Sunday morning shift.

3.4   TO-BE 2 Model

In this model, the average number of wings assembled per month increased by 58%. The WIP rose up by 21%, higher than the data obtained from previous models. However, the number of spars waiting to be transported to the assembly points decreased by 25%. This was a result of the implementation of Step Change line 3, which increased the capacity of the assembly points in 14 spars. Table 2 depicts the results obtained from TO-BE 2 model, adding a Saturday morning shift to the model and adding both the Saturday and Sunday morning shift to the model. The lead-time showed a decrease compared to the previous models, but it did not variate significantly from one scenario to the other. The reason behind the decrease in WIP was that operators were able to transport the spars into the assembly points and the wings out of them during the weekend. In this model, the average time spars wait in the storage area decreased considerably, the storage area was less congested, and the buffer in the shed was not even used. Having more space on the shop floor makes the transport of spars between stations more fluent and blockages less likely to appear.

Table 2. Summary of results obtained from TO-BE 1, TO-BE 1 + Saturday and Sunday shifts, TO-BE 2, TO-BE 2 + Saturday and Sunday shifts. The values corresponding to AS-IS 1 presented in Table 1 as "100 units" and the values corresponding to other models (AS-IS 2, TO-BE 1, TO-BE 2) are presented relative to the AS-IS 1 values

	TO-BE 1	TO-BE 1 + Sat shift	TO-BE 1 + Sat + Sun shift	TO-BE 2	TO-BE 2 + Sat shift	TO-BE 2 + Sat + Sun shift
Finished goods (wings)	139	140	141	156	157	158
Lead time (min/wing)	71.7	71.1	70.9	63.9	63.5	63.3
Work in progress (spars)	115	106	101	121	113	112
WIP waiting (spars)	85	64	53	75	57	53
Jigs idle time (%)	[0–0.45]	[0–0.26]	[0–0.22]	[0–0.45]	[0–0.24]	[0–0.24]
Spars in storage area	79	50	20	58	41	20
Max spars in storage area	91	88	91	91	83	83
Average time in storage area	0.61 T	0.5 T	0.48 T	0.5 T	0.38 T	0.37 T
Max time in storage area	0.87 MT	0.84 MT	0.84 MT	0.83 MT	0.74 MT	0.74 MT

4   Discussion and Validation

The initial AS-IS 1 model was presented to the Airbus logistics' team for validation and the proposed suggestions and expectations for the future scenarios were used as a baseline for the subsequent models. Figure 1 depicts the relative quantities of finished goods after 30 week simulation time. The implementation of the Saturday morning shift appears to make a significant impact on the increase in the throughput. According to Fig. 1, by January 2018 (TO-BE 1), the production capacity is expected to increase by 41% as compared to the current scenario (AS-IS 1 model). The final model (TO-BE 2), shows that the production capacity would increase by 58% as compared to the current scenario. Figure 2 depicts the WIP records obtained after 30 week simulation time. The results are classified per model and scenario. The number of spars that can be assembled at the same time increases by 6% in AS-IS 2, 15% in TO-BE 1, and finally increasing up to 21% in TO-BE 2. Therefore, the number of spars in waiting decrease in line with the increase in WIP from one model to the other.

Fig. 1. Finished goods (wings) in 30 weeks per model and scenario

Analysis of WIP depicts an increase from AS-IS 1 to AS-IS 2, however, with the addition of additional shifts in ASIS 2 the WIP decreases. Addition of Saturday morning shift demonstrates a significant influence on WIP reduction, whereas a further addition of Sunday morning shift shows less significant influence on it. The WIP reduction leads to an increase in the idle time of the machines. Based on the results shown in Figs. 1 and 2, the implementation of the Step Changes is highly beneficial for the growth of the company. However, a problem that could arise with the capacity increase is a poor management of the WIP, leading to blockages and delays in the

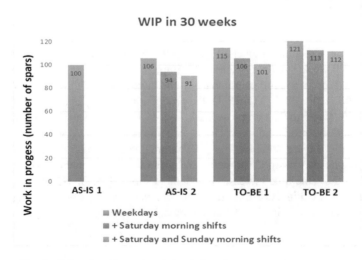

Fig. 2. WIP after 30 weeks of simulation time per model and scenario

production. Therefore, the WIP, more specifically, the amounts of spars waiting in the buffers and footprints should always be maintained to a minimum. The implementation of Saturday morning shift has proved to be the optimum solution as it offers a trade-off between low WIP and low idle time percentage of the jigs and Step Changes.

5 Conclusions

This research measured the likely impact of future changes in the wing assembly process using material flow mapping and simulation of current and future logistics scenarios. The developed DES model demonstrated an increase in production capacity growth by achieving a target 50% increase in aircraft rate/month within one year. It highlighted the main problems causing blockages and other non-value activities.

Acknowledgements. This project was supported by the Royal Academy of Engineering under the Research Chairs and Senior Research Fellowships scheme. The authors would also like to acknowledge Airbus for their support.

References

1. Airbus: Airbus Market Order - Deliveries (2017). http://www.airbus.com/company/market/orders-deliveries. Accessed 28 Nov 2017
2. Dababneha, O., Kipourosb, T.: A review of aircraft wing mass estimation methods. Aerosp. Sci. Technol. **72**, 256–266 (2018)
3. Cunha, P.F., Mesquita, R.M.: The Role of discrete event simulation in the improvement of manufacturing systems performance. In: Camarinha-Matos, L.M., Afsarmanesh, H. (eds.) Balanced Automation Systems II. IFIP Advances in Information and Communication Technology, pp. 137–146. Springer, Boston (1996)
4. Singh, B., et al.: Value stream mapping: literature review and implications for Indian industry. Int. J. Adv. Manuf. Technol. **53**, 799–809 (2011)

Demonstration: Cloud-Based Industrial Control Services

Reinhard Langmann[✉] and Leandro Rojas-Peña

Hochschule Duesseldorf University of Applied Sciences, Duesseldorf, Germany
langmann@ccad.eu

Abstract. The demonstration shows a cloud-based industrial controller by an example of an automated processing and testing station. The station is completely controlled by a Cloud-based Industrial Control Service (CICS). The control program is implemented as an IEC 61131-3 program by conventional industrial programming tools, for example, by PC WORX or CODESYS.

Keywords: Control service · Cloud-based controller · Web-based control system
Automation system

The demonstration is based on the paper from the REV2017 "Cloud-based Industrial Control Services - The next generation PLC". This paper describes the concept and a prototype implementation for the new type of a PLC controller in which the controller functions (programs) will be implemented as control services in a cloud. The programming of this new PLC occurs as is usual in industry pursuant to the standard IEC 61131-3. The above mentioned paper introduces a methodology for the definition and structural configurations of Cloud-based Industrial Control Services (CICS).

1 Concept

1.1 CICS Basis Model

A CICS base model must take into account both, the aspects of control engineering and the web technology features.

From a control engineering point of view, a CICS control based on traditional PLC consists of the following components (Fig. 1):

- *CISC program (CICS-P)*: IEC61131-3 control program in the PLCopen XML notation. It includes only the program and the variables, but no controller configuration.
- *CICS runtime (CICS-RT)*: Execution environment for the CICS program. It can be cycle controlled or event-based.
- *CICS router (CICS-R)*: Device configuration for a CICS controller, i.e. it is determined which CPS components (which automation devices) are connected to the controller.

M. E. Auer and R. Langmann (Eds.): REV 2018, LNNS 47, pp. 50–56, 2019.
https://doi.org/10.1007/978-3-319-95678-7_6

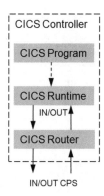

IN/OUT CPS

` **Fig. 1.** General structure of a CICS controller

By separating CICS-P, CICS-RT and CICS-R in a CICS controller and the principle arbitrary distribution of the individual components in an IP network utilising cloud technologies, both a change in the control program code (control algorithm) as well as a change in the device configuration with, for example, the replacement of modules and Plug & Work in real time are possible. CICS-R and CICS-P can be exchanged on-the-fly during a program cycle.

For the identification of a viable CICS basic model, it is also necessary to show possible solution variants for CICS, starting with the basic principles on the web, and then to reflect these in the available web technologies.

In principle, two types of network computers are available in the Internet (Web) as a world-wide computer network:

- Server computers that can provide and run IT entities (objects, services, programs).
- Client computers that can only run IT entities.

As a working principle on this server/client computer network, the client-server principle, i.e. a client must first submit a request to run an IT entity on the server. This means that application technology IT entities in the server cannot act on their own (self-acting). The client is usually a web browser on the client computer.

If one considers any application-specific functional system, which is implemented with web technology tools, the models for the execution (RUN) of this system are the result, as shown in Fig. 2:

(a) The functional system is only stored on the server and is only executed there. The execution of the system in the server is started by the client (server mode).
(b) The functional system is stored on the server and is loaded into the client via a request. The system is only executed in the client (client mode).
(c) The functional system is stored on the server and components of the functional system are also executed on the server. Other components are loaded into the client via a request and these components are also executed there. The execution of the system components in the server is started by the client (mixed mode).

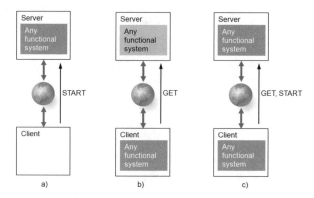

Fig. 2. Models for executing a web technology based functional system (block with black font = only saved, block with white font = is being executed)

In all three cases, the functional system can be distributed over several servers (cloud) or even several clients. If the control technology based CICS structure, shown in Fig. 1, is adapted to the web based functional systems, seen in Fig. 2, two server and client based CICS base models are obtained:

1. *Server Mode (SM)*: The CICS runtime is statically linked to a fixed CPS component in a configuration process. After the CICS control has been started via the client, the CICS controller automatically connects to the associated CPS component via the IP network and executes the control program.
2. *Server-based Mixed Mode (SMM)*: Before starting the CICS runtime, a CICS router is loaded from the server to the client. After the CICS runtime is started, this router dynamically connects the CPS component with the CICS runtime on the server. The process data from the automation device are now routed to the server via the client.
3. *Client Mode (CM)*: CICS runtime and CICS routers are executed as an instance on the client (Web browser). The client is an inherent part of the CICS control system and is necessarily required for executing the control program. The server is no longer required at the runtime.
4. *Client-based Mixed Mode (CMM)*: The control program runs in the CICS runtime on the client, but the communication to the CPS component runs over a dynamically reconfigurable CICS router in the server.

Figures 3 and 4 illustrate the four basic models of a CICS control.

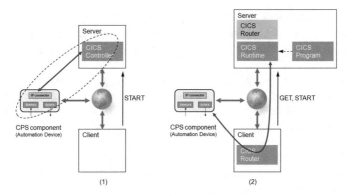

Fig. 3. Component structure and communication paths for server-based CICS solutions (1) – Server Mode (SM); (2) – Server-based Mixed Mode (SMM)

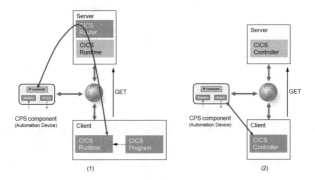

Fig. 4. Component structure and communication paths for client-based CICS solutions (1) – Client-base Mixed Mode (CMM), (2) – Client Mode (CM)

1.2   Control Services

The controller features of a CICS controller should no longer be available as classic control functions, but rather as control services according to the service paradigm. A CICS (literally: Cloud-based Industrial Control Service) can thereby use all the features of cloud computing and, thus, create new business models, such as the rental of control services.

In terms of information technology, CICS has to be produced, parametrised, distributed, stored and recalled as objects by means of methods. Since the components of a CICS control are no longer available as hardware, but only as software objects in the IT or Web world and generally also are stored there in databases, it makes sense to use data models for the modelling of the CICS service architecture.

A CICS controller is realised with two services: Runtime service and Router service. Both CICS services are built according to the principle of web-oriented automation services.

CICS-RT: Runtime Service

Corresponding to the state machine in a traditional PLC, the CICS-RT also has a defined sequence behaviour as the most important component in order to execute a control program.

A CICS-RT can be operated in cyclic mode and event-based mode. In cycle mode, the I/O image is updated, equivalent to a traditional PLC. In event-based operation, the control program is executed only when the value of an input variable changes or an internal event occurs (for example, the execution of a timer).

CICS-R: Router Service

The I/O configuration service of a CICS control system (CICS-R) is separated from the CICS runtime for the following reasons:

- Securing a dynamic reconfiguration, i.e. in the case of an identical control program, the I/O configuration can be changed within a program cycle.
- Identical machines/systems can be operated with the same control program despite different I/O modules.
- A distributed separate configuration service forms the basis for a future automatic IIoT[1]-based device configuration.

2 Implementation

The demonstration shows a prototype implementation for a CICS controller in the Server-based Mixed Mode (SMM – Figs. 2 and 3). The CICS runtime is executed in the server (cloud) as an server instance and the CICS router in the client (browser) as an client instance (Fig. 5).

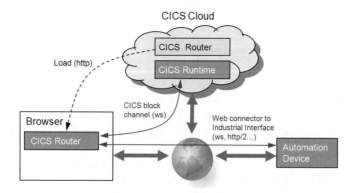

Fig. 5. CICS control in the server-based mixed mode (SMM)

Here, as well, a direct process data communication takes place only between the client and devices. Between CICS-R and CICS-RT, there is a special bidirectional CICS

---

[1] IIoT – Industrial Internet of Things.

block channel for the transmission of I/O images. The process data are transmitted by this channel as Strings over WebSocket. The CICS runtime is operated via an HMI proxy on the client. In terms of technical implementation, the CICS-SMM controller is a distributed elaborate solution. However, process data connection to the devices can also be performed locally and the CICS-RT can use the full performance of the server anyway. A dynamic re-configuration is easy and possible.

3   Application

The CICS components (CICS-RT and CICS-R) are constructed as services and are instantiated as JavaScript objects using uniform and consistent methods. They can therefore be used in any IIoT platform or as stand-alone web pages. The freely available IIoT platform WOAS is used for the demonstration (http://woas.ccad.eu).

The applications therefore did not have to be programmed, but instead were configured in the IIoT platform in a browser-based EDIT mode with little effort.

The CICS-SMM controller has been extensively tested in various application examples. In the demonstration, a CICS-SMM controller controls a processing and test stations. Figure 6 shows the technological structure for the demonstration.

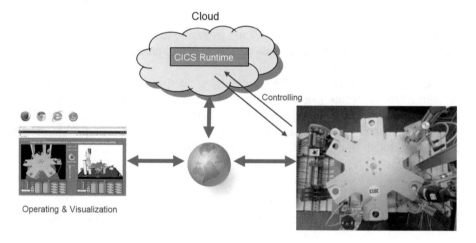

Fig. 6.   Technological structure of the CICS-SMM demonstration

4   Evaluation

A CICS control system uses IP networks for data transmission, regardless of the solution variation. From the perspective of an automation technician, these networks are a priori neither reliable nor deterministic and are not within the jurisdiction of the respective technical automation solution. Extensive time measurements for different communication structures were therefore performed for the CICS prototypes.

A practice-oriented method was chosen for the time measurements, which allows direct statements about the reaction time of a CICS controller. With a CICS SMM solution, the reaction times likewise with a probability of 95% are about 100 ms.

If the CICS controller is operated only in the Intranet, response times of under 40 ms can certainly be achieved. Altogether, the statement can be made that technical processes with process times of >150 ms (simple assembly process, temperature and mixing processes, climate and energy processes, etc.) can already be performed successfully from the cloud by means of a CICS.

5 Summary

The demonstration shows a prototypical application of a mixed client/server-based CICS controller in a scenario from production automation. An evaluation of this CICS applications showed that simple technical processes with process times of greater than 150 ms can already be controlled reliably over the standard Internet.

Using the CICS concept, a new type of industrial control was developed and tested that allows for the complete detachment of control function and associated equipment to globally distributed, cloud-based software control services. A CICS control is operated by a classic IEC61131-3 control program, thus ensuring the interoperability and industrial compatibility of the control system. The application of the service paradigm for industrial control functions significantly increases flexibility, meets industry 4.0 requirements such as changeability, reconfiguration and autonomy, and enables new business models to lease automation functions.

The CICS demonstration confirms that previous hardware-oriented and centralised procedures for the control of automated devices, machines and systems (e.g. PLC controllers) can be distributed and used transparently for uncritical real-time conditions (e.g. environmental processes, logistics processes, energy processes, simple assembly processes) through IP-network-distributed software services.

Reference

Langmann, R., Stiller, S.: Cloud-based Industrial Control Services - The next generation PLC

Poster: Teaching Automation and Logistics with Virtual Industrial Process

Florence Lecroq[✉], Jean Grieu, and Hadhoum Boukachour

PIL (Pôle Ingénieur et Logistique), LITIS (Laboratoire d'Informatique de Traitement de d'Information et des Systèmes), Normandy Le Havre University, Le Havre, France
florence.lecroq@univ-lehavre.fr

Abstract. In the 60's, the first Computer-Aided Design (CAD) Software appeared. Nowadays, it belongs to the most common tools used in industry. Nowadays, digital simulations of complete industrial processes are implemented in factories. This new concept is one of the main components of the Industry 4.0 paradigm. Indeed, thanks to the Industrial Internet of Things (IIoT), and the huge amount of data produced by the different systems of the production lines, it is now possible to simulate all these systems with virtual twins. These twins are used in order to validate real industrial processes before to be built. They are also efficient learning tools, allowing the students to be trained in absolute safe conditions. In this article, we present how students use virtual industrial processes driven by Programming Logic Controllers (PLCs) programmed in Sequential Function Chart (SFC) language.

Keywords: Industry 4.0 · Virtual process · Virtual Reality · PLCs

1 Introduction

Today we are living the fourth industrial revolution [1]. This fundamental industrial evolution is called Industry of the Future in France, Industry 4.0 in Germany and Smart manufacturing in the USA.

The first industrial revolution was the arrival of mechanical production. It is based on coal, metallurgy, textiles and the steam engine. It began in Britain at the end of the 18th century, then in France at the beginning of the 19th century. The second is mass production with the mechanical production of electricity at the end of the 19th century, and Taylorism, which is a scientific organization of labour which allows to increase the productivity of the employees. The third industrial revolution appeared in the middle of the 20th century with the automation of production. Period during which Internet appeared. At that time, the industrial programmable logic controllers were introduced in industries for controlling machines.

Today, the fourth industrial revolution is on its way. The plant becomes an interconnected global system. The industrial production centre is equipped with flexible, fully automated units. Machines communicate with one another without human intervention. Secure transport of simple or complex data must be allowed from sensors to advanced control systems, whether local or remote. The collection of the data

© Springer International Publishing AG, part of Springer Nature 2019
M. E. Auer and R. Langmann (Eds.): REV 2018, LNNS 47, pp. 57–65, 2019.
https://doi.org/10.1007/978-3-319-95678-7_7

produced by different components of the production line also makes it possible to produce a virtual replica of all or part of this chain. That way, we can generate process simulations or tests. Moreover, the numerical twin permits to enable future technicians to familiarize themselves with work tools and complex procedures or facilitate repairs and predictive maintenance. As much as the simulation product is today acquired with the CAD (Computer Aid Design), as much the simulation process, upstream of the production, is still to be implanted. The use of the Virtual Reality will therefore be a key point of Industry 4.0.

In this context of industrial process simulations in Industry 4.0, we propose to present our teaching approach and the students feedbacks. This work deals with the use by students of virtual operative parts simulations for learning industrial programmable logic controllers (PLCs).

The first part of this paper will present the material used; the second part will discuss the PLCs programming language; then, the next part will show the different virtual operative parts used. Finally, we present in the last part a survey in order to appreciate the students' reactions to this new pedagogical tool.

2   Equipment

Within the framework of practical work for the teaching of PLC, we have at our disposal a room containing eight similar workstations (Fig. 1).

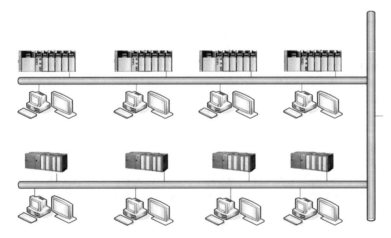

Fig. 1. The equipment available in the room for PLC teaching.

Each workstation contains (Fig. 2):

- A personal computer (PC) with two screens, with programming software of the PLC on one screen and the operative part simulator on the second screen;
- A PLC with 32 digital inputs, 32 digital outputs, and an Ethernet interface card to connect the PLCs to the network and to each other.

Fig. 2. The details of one workstation.

Our students learn to work with two different brands of PLCs. These are SIEMENS S7 315 2DP controllers and SCHNEIDER TELEMECANIQUE Premium TSX 57 [2, 3]. On SIEMENS we use the SIMATIC STEP 7 programming software and on SCHNEIDER we use the UNITY Pro XL programming software. Thus, turn by turn, our students can use the two different systems. It makes them more autonomous about PLC programming during their internship at the end of the diploma.

Inputs and Outputs of the PLC are connected to an interface box. Then, on the other side, this interface is linked to the PC via USB. This is where the PLC sends commands to actuators or receives data from sensors placed in the virtual operative parts visible on the second screen of the PC. The first screen of the PC allows to view the program carried out by the students in SFC (Sequential Function Chart) (Fig. 2).

In the next part, we will present the language used to program the PLCs.

3 PLC Programming

3.1 The Different Programming Languages

The programming languages of automated systems are described in IEC 610161-3 established by the International Electrotechnical Commission [4, 5]. The last edition dates from 2013. It specifies the syntax and semantics of a unified suite of 5 programming languages for industrial programmable logic controllers (PLCs). Two languages are textual and three are graphic. The five languages are:

1. FBD (Function Block Diagram): This graphic language allows to program using blocks, representing variables, operators or functions.
2. LD (Ladder Diagram): This graphic language is mainly dedicated to the programming of Boolean equations (TRUE/FALSE).
3. ST (Structured Text): This high-level text language allows the programming of any type of algorithm more or less complex.

4. IL (Instruction List): This low-level textual language allows one instruction per line. It can be compared to the assembler language.
5. SFC (Sequential Function Chart): Derived from the GRAFCET language [6], this high-level graphic language allows easy programming of all sequential processes. It is an object-oriented programming framework that organizes actions written in IEC programming languages (FBD, LD, ST or IL) into a unified sequential control program.

In Fig. 3, we present the same program written with different languages.

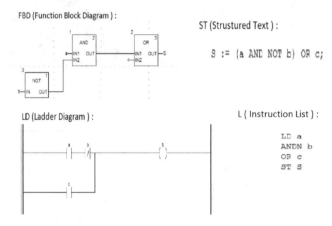

Fig. 3. Example of the same code written with four different languages.

In addition to the SFC language, students use the LD on SIEMENS and the ST on SCHNEIDER to complete their program,

In the next section, we present the SFC language.

3.2   The SFC Language

SFC stands for «Sequential function Chart» [7]. It is a sequential control system language that it is used for process flow control especially for devices like PLCs. SFC switches from one state of operation to the next subject to system conditions. SFC is almost entirely used in its graphical form which can be quick to visualize and easy to learn. The Fig. 4 presents the main elements of a SFC diagram.

Step

The most important element is an SFC 'Step' which represents an order or an action. A number is assigned to a Step. In order to associate the actions with SFC Steps, the following properties are available:

- **N** – Non-stored: the action is active when the step is active;
- **R** – Overriding Reset: the action is deactivated;
- **S** – Set (stored): the action is activated and remains active until a reset;

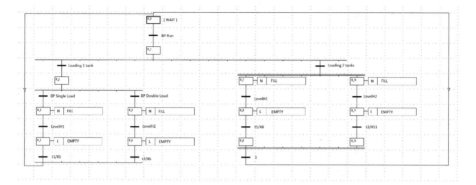

Fig. 4. The main elements of a SFC diagram

- **L** – Time Limited: the action is activated for a limited time;
- **D** – Time Delayed: the action becomes active after a certain time as long as the step is still active;
- **P** – Pulse: the action is executed just one time if the step is active;
- **DS** – Delayed and Stored: the action is activated after a certain time as long as the step is still active and remains active up to a reset.

In a Step, it is possible to have several Outputs of the PLC, or an action with a condition. That means the action is produced when the Step is active and the condition is true.

Transition
Between two Steps, there are so-called Transitions. A Transition condition must have the value TRUE or FALSE. Thus, it can consist of either a Boolean variable (input of PLC), a Boolean address or a Boolean constant. It can also contain a series of instructions having a Boolean result, either in Structured Text syntax or in Ladder language. A syntax rule imposes the alternation Step/Transition and only one Transition between two Steps.

Alternative branch
Two or more branches in SFC can be defined as alternative branches. The conditional branches can be added with 'Divergence and Convergence in OR or AND' (Fig. 4). When you start an OR divergence, each branch must begin and end with a Step. When you start a AND divergence, each branch must begin with a Step and end before the convergence with a Step.

Active Step
After calling the SFC POU (Program Organization Unit), the action (surrounding by a double border in Fig. 4) belonging to the initial Step is executed first. The initial Step is represented by the initial values of its internal and output variables. A Step, whose action is being executed, is called active. If the Step is active, then the appropriate action is executed once per cycle. The Step is active until the Boolean condition of the next Transition is TRUE. Then, the Step become inactive and the next Step will be executed in the next cycle of the PLC.

In the next part, we will present the virtual operative parts software named ITS PLC.

4   Virtual Operative Part Software

The ITS software was developed by Real Games Company [8, 9]. It offers five distinct operative parts containing sensors and actuators connected to the PLC's actual inputs and outputs via a USB box that returns to the PC. We use a screen to program the PLC and a screen to check the action of the PLC program on the virtual operative parts. The five distinct operative parts are (Fig. 5):

Fig. 5. The five virtual operative parts of ITS PLC.

- **Sorting:** Carrying boxes from the feeder to the elevators, and sorting them by size.
- **Batching:** Mixing primary colors (red, green and blue) to obtain the desired color.
- **Palletizer:** Palletizing boxes in three layers.
- **Pick and Place:** Putting parts inside boxes, using a three axis manipulator.
- **Automatic Warehouse:** Transporting, storing and retrieving boxes from racks.

One of the main features of the software is the ability to interact with movable objects in real time. By a simple click and drag, we can move the object where we want. That way, we can:

- interact with the system as we do with a real one;
- add and remove objects from the production line, at any time during the simulation;
- generate intentional failures during the process;
- test individual parts of the system like a conveyor table.

All these systems can be controlled automatically by the external PLC with the program of the students.

In the next part of these paper, we will describe the students' behaviour.

5   The Student's Activity

Every five years [10], we conduct a survey in order to estimate the students' interest in the use of virtual (3D) operative parts for teaching PLC programming. Here, we propose to present the results of the last survey among eighty students.

5.1   The Student's Profiling

Our students are young (20.5 years on average) in second year at University (80% of students surveyed) or third year at University (20%). They are mainly from scientific syllabus (95%). They are mainly boys (92.5%). Only 5% do not use social networks. The others have a FACEBOOK account (87.5%), YOUTUBE (70%) to post their videos, or 42.5% go to TWITTER. At home, 90% have a HD TV or 4 K HD (3D). The remaining 10% use only a PC screen. Most of them (90%) are gaming enthusiasts (only 10% do not play) on PC (67.5%) or Smartphone (67.5%) or home console (57.5%). They play mainly online and on multi-player platforms. 32.5% are playing every day during 3 to 4 h. Another question also allows us to say that 77.5% prefer to go out with their friends than to test the last game released. 25% recognize hacking games to get them, otherwise they buy them mostly by download or new DVDs.

We deduce from these first results that our students know very well the virtual worlds and know how to use them perfectly. We also discover that they are almost addict players who do not work after classes.

The remainder of the questionnaire deals with teaching PLC.

5.2   Teaching PLCs with 3D Virtual Worlds

Our students did not know the PLC (80%) before their training. They work on three-hour sessions, mainly in pairs (97.5%). Students who had previously worked on PLC did not use virtual operative part (85.5%). They observed the result of their programming by lighting LEDs on the outputs of the PLCs according to the switches placed on the inputs.

What they prefer in the features of the virtual operative parts is the realism of the simulation, followed by the reactivity of the system (real time), and then the physics engine that makes the environment credible. Finally, they appreciate being able to test an industrial process without risking injury, or damaging equipment. At 77.5%, they prefer to work with these virtual operative parts, rather than with simple switches or LEDs connected to the PLC to simulate an industrial process.

6 Conclusion

In this paper, all the advantages given by the learning process using virtual operative parts driven by PLC programming, offer good opportunities for students' skills according to the needs of the Industry 4.0. Indeed, the Industry 4.0 today proposes the creation of virtual twins for the validation of production lines before their real setting-up. These twins allow the future technicians to be trained without risking to be injured, but also to realize substantial savings for training centers. Due to the structures of the programs, the use of SFC language for programming PLCs' is easy to understand and to code by students. These students, also video gamers, have no problems to understand the use of these virtual worlds. They approved by a large majority the use of 3D environments to simulate industrial or logistics processes. They claim for even more integration of virtual equipment's in the syllabus.

Acknowledgments. This project has been supported by the European Commission under the ERDF (European Regional Development Fund) Programme through the 5.5 Action of the GRR CLASSE Programme.

References

1. GIMELEC: Industrie 4.0: L'Usine connectée. http://www.gimelec.fr/Publications-Outils/Industrie-4.0-l-usine-connectee-Publication
2. SIEMENS: Support Industry. https://support.industry.siemens.com
3. SCHNEIDER: Automation and Control for Industry. www.schneider-electric.fr/fr/work/products/automation-and-control.jsp
4. International Electrotechnical Commission IEC 610161-3 (2013). https://webstore.iec.ch/publication/4552
5. Jouvray, N.: Langages de programmation pour systems automatisés: Norme CEI 61 131-3. Revue Techniques de l'Igénieur, ref. S8030, pp. 1–31 (2008)
6. ADEPA (Agence nationale pour le DEveloppement de la Productique Appliquée à l'industrie): "Le Grafcet", AFCET (association française pour la cibernétique economique et technique), ed. Cépadues (2002). ISBN: 2854283805
7. John, K.-H., Tiegelkamp, M.: IEC 61131-3: Programming Industrial Automation Systems, Concept and Programming Languages, Requirements for programming Systems, Decision-Making Aids. Springer (2010)
8. ITS PLC Professional Edition. http://www.realgames.pt

9. Riera, B., Vigario, B.: Virtual systems to train and assist control applications in future factories. In: 12th IFAC Symposium on Analysis, Design, and Evaluation of Human-Machine Systems, 2013, Las Vegas, NV, USA, pp. 76–81 (2013)
10. Lecroq, F., Grieu, J., Boukachour, H., Person, P., Galinho, T.: Learning PLC's with GE3D: the students' feedbacks. In: IEEE Global Engineering Education Conference (EDUCON), Amman, 2011, pp. 319–323 (2011)

Remote Control and Measurements

Use of VISIR Remote Lab in Secondary School: Didactic Experience and Outcomes

Manuel Blazquez-Merino[✉], Alejandro Macho-Aroca, Pablo Baizán-Álvarez,
Félix Garcia-Loro, Elio San Cristobal, Gabriel Diez, and Manuel Castro

Electronics, Electric and Control Engineering Department,
UNED - Universidad Nacional de Educación a Distancia, Madrid, Spain
{mblazquez,amacho,pbaizan,fgarcialoro,elio,gdiaz,
mcastro}@ieec.uned.es

Abstract. VISIR remote lab, designed as a learning tool in subjects related with Electricity and Electronics in undergraduate and graduate courses, has unlikely been used in secondary school technology teaching up today. In the following document, a research work is presented in order to show the development, implementation and learning outcomes of an educational experience for secondary students, using VISIR remote lab as a means to carry out practices of measuring and introducing a methodology based on the features of Bloom-Anderson's taxonomy. The experience have been guided by a work document in which explanations and didactic activities have been included, specifically designed under a cognitive skills development focus.

Keywords: Remote laboratory · VISIR · Secondary education
Bloom-Anderson · Electric measures · Electricity · Teaching learning process

1 Introduction

VISIR is a remote lab focused to practices in Electricity and Electronics, well known in academic and engineering scope [1, 3]. Its design corresponds with a learning tool for practice in undergraduate and graduate course, but it has been used in very few occasions. As a resource to teach Technology related topics at Secondary school up to day. In this document, a research work is presented to show the design, development, implementation and learning outcomes from a educational experience with secondary students, whose purpose is the optimization of the learning of electric magnitudes and circuits. In fact, to promote the optimization of the learning of electricity in this type of so young students, technological tools are not enough. A specific methodology has also been designed to carry out the teaching-learning process. To do this, the methodology has been based on the characteristics of the Bloom-Anderson taxonomy [2], in order to organize the learning stages by activating a group of cognitive skills of secondary stage students. VISIR remote lab [6] has been the used resource in order to face the nucleus of the activities, by means of which, electric magnitudes measurements have been carried out in various parts and circuits. Both proposals have shaped the didactic experience from two levels: the methodological and the contextual. The methodological

© Springer International Publishing AG, part of Springer Nature 2019
M. E. Auer and R. Langmann (Eds.): REV 2018, LNNS 47, pp. 69–79, 2019.
https://doi.org/10.1007/978-3-319-95678-7_8

approach has been applied to all students since we are convinced that learning must be centered on the students and adapted to the needs of each one of them. But, to measure the differences in the contextual approach, that is, in the resources that are used to promote the learning of the contents of electricity, two working groups have been created among the participating students. A first group has used common instrumentation devices in a actual laboratory and has been formed as the reference group, while a second group of students has made the measurements with the remote VISIR laboratory, thus establishing the control group.

The experience has been guided by a working dossier in which explanations and didactic activities have been included, specifically designed by a constructivist approach [11] what aims to outstand the development of cognitive skills [2].

2   Methodological Justification

In the world of education and above all, in education in Engineering, teachers use to focusing learning towards memorization and repetition of processes as the fundamental didactic tool. Thus, one of the pillars of learning is based on a teacher-focused didactic approach, who transmits his/her knowledge to students, whose priority is memorizing and the making of exercise to understand the theory. Conceptual understanding, therefore, is relegated to the student's capabilities. This approach is based, therefore, on applying a purely behavioural methodology [12], in which the teacher don't pay attention to how student achieves meaningful learning by him/herself. The way to evaluate is simplified, in most cases, to the making of exams and tests, which measure in a timely manner the degree of acquisition of knowledge by the student. In a way, it is about evaluating a dynamic situation through a photo finish. The proposed methodology therefore seeks to evaluate the progress of students in each of the actions and activities carried out and precisely in this regard, we have found the approach proposed by Bloom-Anderson as a guide to track students' developments.

2.1   Bloom-Anderson's Taxonomy

Bloom, firstly in 1948, and Anderson, in a later review in 2001 [2], developed a classification of cognitive skills, from the simplest to the most complex in order to establish a rule to guide the learning evaluation processes. These skills were grouped into two levels, each of them representing those psychological abilities focused to the use of memory or creative development. Both levels, defined as thinking skills or cognitive abilities, are defined as Low Order Thinking Skills (LOTS) and High Order Thinking Skills (HOTS), respectively. To be more precise, and according the scheme presented in Fig. 1, LOTS correspond with the cognitive skills related with the memory, the understanding of theoretical concepts of knowledge and its application in problem solving. By other side, HOTS are those skills that contribute in the students to develop a certain ability to analyze a problem, evaluate different solutions or contexts and create new solving scenarios.

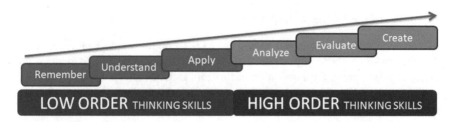

Fig. 1. Scheme of Bloom-Anderson's taxonomy stages according to cognitive levels

Therefore, one of the objectives of the research work here presented, is intending to assess the degree of consolidation of the LOTS and as a consequence, its relationship with the degree of HOTS achievement. We believe that this approach is correct from the constructivist standpoint and will help to high school students to optimize the learning of electrical phenomena and thus, improve their cognitive skills when applying a correct strategy to calculate electrical circuits.

2.2   VISIR Remote Lab

There are some references to electric circuits simulators and virtual laboratories [10, 13], resources that widely proliferated in the 1990s. However, with the turn of the millennium, many researchers thought about the possibility of using laboratories that could be handled remotely, using the web as a means of access. This is the case of VISIR (Virtual Instrument Systems In Reality), developed in the first instance by Ingvaar Gustavsson from Blekinge Institute of Technology in 2001 [8]. Although VISIR has already been extensively described in other publications, it's a system, as described by Tawfik et al. in 2011 [1], configured to offer a virtual environment to the user through web access, through which the connections of a system of electrical and electronic devices can be manipulated (PXI platform). The constitution of the devices available in the laboratory to carry out the measurements, is programmed through the Component List (CL) that will be loaded in the Equipment Server Software (ESS). The devices in CL are then combined with the Measurement Server to perform the measurements required by the user. Between both sides, User Interface and Server Equipment, some intermediate stages to automate the user's access and communication can be found. The process, from the user's standpoint, is simple, since it focuses on accessing the system through its credentials and operating in the user interface. The equipment counts with a Web server and a database manager in order to provide communication with the user through the interface. This interface has an operation similar to that which would be used in any classroom laboratory, using a breadboard to assemble the different electrical or electronic components and proceeding to incorporate the various connections that make up a given circuit. On the other hand, one of the utilities of VISIR is the possibility of having instrumentation to measure voltage, electrical intensity and electrical resistance, with which the student will be able to autonomously perform the measurements required in any practices protocol.

2.3   Design of the Learning Experience

Conceptual contents are progressive, and are focused on the study of electrical magnitudes, resistors and other components of electrical circuits, circuit layouts such as series, parallel and compound, the association of resistors and the concept of equivalent resistance. The practical part is containing the measurement procedures, instrumentation and measurement devices and the concept of error in the measurement. In order to be able to compare the impact on learning, the assimilation of contents and the development of skills related to the measurement of electrical magnitudes, the same practices protocol has been given to the whole of the students, containing both theoretical explanations and practical work proposals and which has been designed following a specific design according the particularities of Bloom's taxonomy.

Additionally, a Moodle developed MOOC (http://62.204.201.27/moodle/), has been created, in which documentation that compose practices protocols can be found, as well as other appropriate teaching materials such as tutorial videos for the use of the remote laboratory and to facilitate the understanding of the practices making. The videos included are of short duration (between 2 min and 8 min) and their objective is to guide the students in the sequencing of theoretical and practical activities. Students participating in this research work are attended in the firsts two years of secondary school. In total, they are 147 students (63 girls and 84 boys). The didactic experience has been carried out throughout 6 sessions, each of them 55 min long, where it is observed that, in addition to the activities of the experience, two questionnaires have been provided for the students to be filled out at the beginning and at the end of the experience. Both questionnaires contain questions about their knowledge and expectations regarding electrical measurements and the use of instrumentation. The questionnaires have the purpose of measuring the students' subjective perception in the didactic experience. In order to be able to contrast the learning outcomes, it has been thought to divide students into two groups, a first reference group and a second control group. Both groups have used the same time for activities and the same documentation of practices. The students of the reference group, which has been called Classroom Laboratory (74 students - 42 boys and 32 girls) have to handle measurement devices as well as actual electrical devices and therefore, they have carried out the practices by manually connecting various electrical resistances in a breadboard and measured with a multimetre. On the other hand, the control group (73 students - 42 boys and 31 girls), which has been called Remote Laboratory, has used the VISIR remote laboratory to make the connections in the circuits and connect the instrumentation.

3   Learning Outcomes and Results

The research work presented here has two clear educational objectives. The first one focuses on the evaluation of the impact that the making of the practices has had on the acquisition of theoretical knowledge in order to assess the suitability of the VISIR remote laboratory as a learning resource. The second will consist of the validation of the hypothesis that the degree of consolidation of low order thinking skills influences the acquisition of high order thinking skills. Each of the activities that makes up the student dossier has been corrected and valuated in a scale from 0 (minimal grade) to 10 (maximum grade). After correcting, an average grade has been calculated for each of students, and finally a final mean number has been obtained per Level, as shown in the Table 1. In view of these values, the grades obtained in Level 1 are quite mediocre and very similar between students of the Classroom Lab and VISIR remote lab. In this case, there are two circumstances that explain these low qualifications. On the one hand, students have never studied issues related to Electricity and the use of mathematical rules is difficult for them. On the other, it is also the first time that they use instrumentation. We think that it's possible the short duration of the activities to be an additional factor.

Table 1. Average grades of students depending the type of lab used and gender

	Type of Lab		Gender	
	Classroom	VISIR	Girls	Boys
Students	74	73	63	84
Level 1 avg.	4.71	4.66	5.43	4.17
Level 2 avg.	5.86	4.69	6.14	4.56
Level 3 avg.	4.33	5.43	6.26	3.41

These differences are broadened when it comes to comparing the results of Level 2 activities, improving in the case of the Classroom Lab. It is possible that the use of VISIR, requires some extra practice that young students have not done and this could be a reason explained by the grades obtained by students of second year in Level 3, the most complex, in which VISIR students obtain significantly better grades. With respect to the analysis of the data related to gender, the difference in the scores between boys and girls is very remarkable. The differences are greater as the difficulty to handling the laboratory increases. We believe that such a difference between the average number of girls (6.26/10) and that of boys (3.41/10) to be significant. Boys have got the lowest average in all the levels, while the girls have obtained the highest percentage.

3.1 Impact of the Practice Making to Learning

The students have used the practice protocol as a guide document in which they have answered the proposed activities. Some of them have been of a theoretical nature and in others, they had to write the results of their measurements as part of the practical activity. In the following Figs. 2, 3 and 4 we have represented a cloud of dots in which the X axis values correspond to the assessment of practical activities such as handling of instrumentation and making of measurements and in the Y axis to the theoretical activities assessment.

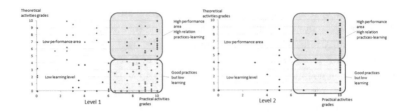

Fig. 2. Relation of theoretical and practical activities grades (All students)

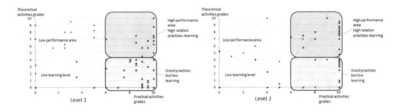

Fig. 3. Relation of theoretical and practical activities grades (Actual lab students)

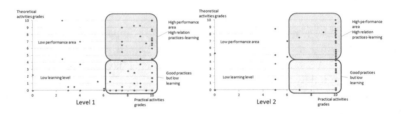

Fig. 4. Relation of theoretical and practical activities grades (VISIR remote lab students)

As can be seen in the series of Figs. 2, 3 and 4, the results are much better in Level 2 activities, despite being a more complex level. This phenomenon we understand that has much to do with the familiarity in the use of the laboratory in Level 2 and the difficulty of taking the first steps in Level 1.

In a more exhaustive examination of the evaluations of the students according to the type of laboratory used, a greater concentration of dots is observed in the area defined as "*High performance area - High relation practice-learning*" in the students who have used VISIR. On the other hand, we found it interesting to make a comparison between students according to gender. This analysis has been motivated by observing the answers to the questions of the questionnaires on the part of the girls, that we thought that they offered a perception of their learning lower to that of the boys. In the following Figs. 7 and 8, we have proceeded to represent the dots clouds by levels. Thus, Fig. 5 corresponds to the measurement values of theoretical-practical learning in the girls (left of the graph) and the boys (right of the graph) for the activities of Level 1. In Fig. 6 the valuations for level 2 of activities are represented. As it can be observed, a paradoxical situation is given since in both levels L1 and L2, more amount of dots are located in the "*High performance*" and "*Good practices but low learning*" areas in female's graphic, it's said, grades achieved by girls are confirmed as better than boys'. This situation is opposite to the girls' more modest opinion retrieved in the questionnaires about their own learning than boys.

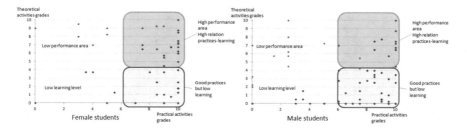

Fig. 5. Comparison of Level 1 activities grades by gender

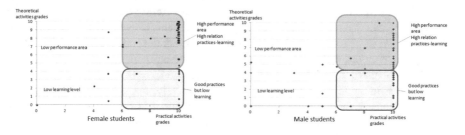

Fig. 6. Comparison of Level 2 activities grades by gender

3.2 LOTS-HOTS Mutual Relation

In a second type of analysis, the activities included in the students' dossiers have been identified according the incidence each one has got in the cognitive skills. Thus, two groups of activities have been formed, those that contribute in the development of Low Order Thinking Skills (LOTS) and those that allow students to strengthen their High Order Thinking Skills (HOTS). Our hypothesis indicates that to be able to develop HOTS is essential to have worked enough their lower skills (LOTS). In this way, we have been measuring each student's performance in both types of activities and we have considered the student's age factor, since it's very possible to be one of the more important factors in the cognitive development, mainly as teenagers. Thus, in Fig. 7, the graph represents the average grades of the students in the first course (12–13 years old) but grouping LOTS-type and HOTS-type activities.

Students' grades have been grouped according to the ranges indicated in X-axis. By other side, in the Y-axis the percentage of students, whose grades have been of each range, are determined. It can be observed that most of students' LOTS grades are concentrated in the range between 5 and 8 points out of 10, while most of students' HOTS grades are lower than 5 points out of 10. In certain way, this corroborates partially our hypothesis since it hasn't been possible to develop higher cognitive skills such as analysis and creation of situations while lower skills haven't been mostly assimilated.

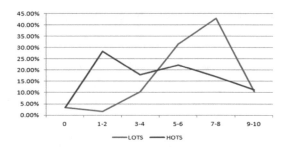

Fig. 7. Scores according the type of cognitive activity (LOTS/HOTS) in 1st year students

To complement these data, it is important to note what has occurred with the 2nd year students. In the Fig. 8 a graph with the same format as in Fig. 7 has been represented, so that they can be compared. In this case, the curve that represents HOTS skills is more similar to the one that represents LOTS. A parameter that is convenient to analyze, since LOTS and HOTS are two variables with a high degree of dependence, is the correlation factor (R). Using the values of the covariance and the particular variances for each of the frequency distributions, the values of $R1 = 0.49$ and $R2 = 0.86$ have been obtained, for the grades in the first and the second year respectively. In the Fig. 9 we have represented simultaneously clouds of points for the students of 1st year and 2nd year have been represented in which it's clearly observed why the correlation factor in the 2nd year students is higher than in the students of the first year. The coordinates of each point in the 1st year are more concentrated in the center of the graph, which reduces the mutual relationship between both variables, while in the case of 2nd year, the points are concentrated along a straight sloped line. This clearly identifies a strong and linear mutual relationship between both LOTS and HOTS variables, with which in this case we can affirm the validity of our hypothesis.

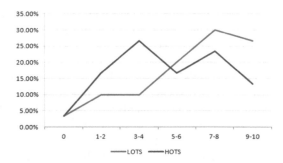

Fig. 8. Scores according the type of cognitive activity (LOTS/HOTS) in 2nd year students

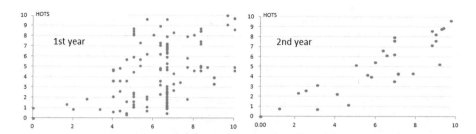

Fig. 9. Mutual relation (correlation) between LOTS and HOTS variables in 1st and 2nd course

4 Conclusions

At the beginning of the research work that is presented in this paper, several objectives were proposed, all framed into a hypothesis. The hypothesis has been formulated in the context of the use of unusual resources in the teaching of Electricity. Thus, given the availability of the VISIR remote laboratory and its adaptability to different educational levels, it has allowed us to think about the possibility of including it as an optimal learning resource. Determining the degree of learning optimization of VISIR is one of the objectives set in the hypothesis. Secondly, it was thought that the remote laboratory, widely used in the context of engineering education, could be complemented with a newly created methodology, although based on the indications given in the Bloom-Anderson taxonomy. Thus, from these two resources, technical and methodological, respectively, a group of high school students has been experimenting an innovative proposal to learning.

Students, whose ages are between 12 and 14 years old, have to learn in the context of Secondary School curricula, in which topics related to Electricity, electric phenomena, electric magnitudes and circuits are included. So far, most of teachers are teaching following a traditional behavioural model and we, authors, have thought whether this model could be suitable or even, if we will be able to maximize the students' learning. The didactic experience derived from this approach has been satisfactory, since some conclusions have been obtained that allow us to continue delving into this research.

As a first conclusion it has to be indicated that the use of the VISIR laboratory has an impact on student learning very similar to the use of a real laboratory for students of the first year of secondary education (12 years of age), although we have been able to verify that this impact is significantly higher for older students (14 years of age).

As a second conclusion it has been observed that the girls have been able to take more advantage of the laboratory and the methodology than the boys, having obtained higher grades than boys of up to 20% difference. In spite of everything and in a para-doxical way, the impression and perception that the girls have on their own learning is lower to that of the boys, who believe they have learned more than they have actually done. It is very possible that these conclusions can be justified from the level of Psychology, but in our case, we have only been able to verify this significant fact.

The age of the students is also an important factor when it comes to assuming the presented methodology. We have verified that the relationship between the consolidation of Low Order Thinking Skills and the development of High Order Thinking Skills, through which students put in practice the use of analytical cognitive tools and reflective and creative thinking, has a strong correlation when it comes to 14-year-old students, while this correlation is weaker in students in the first year of secondary education (12 years old).

These conclusions, together with the figures and statistics generated in the research work are neither conclusive nor definitive, although certainly significant, and the authors commit ourselves to continue studying and improving the resources used in two ways. A first way will be explored in the sense of improving the methodology in order to achieve a process that optimizes the learning of Electricity in a general way regardless of the age of the students. A second way will be the creation of new adaptations in the VISIR remote laboratory for the learning of electronic circuits.

Acknowledgements. The authors acknowledge the support of the "Escuela de Doctorado de la UNED", eMadrid project (Investigación y Desarrollo de Tecnologías Educativas en la Comunidad de Madrid) - S2013/ICE-2715, VISIR+ project (Educational Modules for Electric and Electronic Circuits Theory and Practice following an Enquiry-based Teaching and Learning Methodology supported by VISIR) Erasmus+ Capacity Building in Higher Education 2015 n° 561735-EPP-1-2015-1-PT-EPPKA2-CBHE-JP and PILAR project (Platform Integration of Laboratories based on the Architecture of visiR), Erasmus+ Strategic Partnership n° 2016-1-ES01-KA203-025327.

References

1. Tawfik, M., et al.: VISIR deployment in undergraduate engineering practices. In: GOLC Workshop in the 41st ASEE/IEEE Frontiers in Education Conference (FIE 2011), Consortium Remote Laboratories Workshop (GOLC). Global Online Laboratory (2011). 978-1-4577-1944-8
2. Anderson, L.: Taxonomy of educational objectives. In: Encyclopedia of Educational Theory and Philosophy Anonymous, 2001, pp. 789–791. Sage (2001)
3. García-Zubía, J., et al.: Easily integrable platform for the deployment of a remote laboratory for microcontrollers. In: Annual Global Engineering Education Conference, EDUCON 2010 Conference Book. IEEE Education Society (2010). ISBN: 978-84-96737-70-9
4. Chen, X., Zhang, Y., Kehinde, L., Olowokere, D.: Developing virtual and remote undergraduate laboratory for engineering students. In: 40th ASEE/IEEE Frontiers in Education Conference Proceedings, Arlington, VA, 13 January 2010
5. Angulo, I., et al.: El proyecto VISIR en la universidad de Deusto: laboratorio remoto para electrónica básica. In: Conference Proceedings TAEE 2010 (2010). ISBN: 978-84-96737-69-3
6. Gustavsson, I.: Laboratory experiments in distance learning. In: Proceedings of the ICEE 2001 Conference, Oslo/Bergen, Norway, 6–10 August 2001 (2001). http://www.ineer.org/. Accessed 27 Feb 2017
7. Gustavsson, I., et al.: The VISIR project - an open source software initiative for distributed online laboratories. In: Remote Engineering and Virtual Instrumentation (REV 2007), June 2007

8. Deniz, D.Z., Bulancak, A., Özcan, G.: A novel approach to remote laboratories. In: 32nd ASEE/IEEE Frontiers in Education Conference Proceedings, Boston, MA, 6–9 November 2002 (2003)
9. Pastor Vargas, R.: Especificación formal de laboratorios virtuales y remotos: Aplicación a la Ingeniería de Control. Doctoral Thesis presented at Escuela Técnica Superior de Ingeniería Informática of Universidad Nacional de Educación a Distancia, UNED (2006)
10. Bruner, J.: Studies in Cognitive Growth: A Collaboration at the Center for Cognitive Studies. Wiley, New York (1966)

On Effective Maintenance of Distributed Remote Laboratories

Tobias Fäth, Karsten Henke[✉], René Hutschenreuter, Felix Seidel,
and Heinz-Dietrich Wuttke

Ilmenau University of Technology, Ilmenau, Germany
{tobias.faeth,karsten.henke,rene.hutschenreuter,
felix.seidel,dieter.wuttke}@tu-ilmenau.de

Abstract. Within two TEMPUS projects, we implemented the GOLDi remote lab infrastructure at ten universities in Armenia, Georgia, Germany and Ukraine. Our interactive hybrid online lab – called GOLDi (Grid of Online Lab Devices Ilmenau) – stands for a grid concept to realize a universal remote lab platform. In addition to the actual use as a remote lab for educational purposes, maintenance is of crucial importance - especially if the lab is part of a cluster of distributed labs.

Until this day there are still major issues concerning the maintenance and reliability of the hardware setup at each GOLDi location. In this article, we propose a new internal Ethernet based communication implementation. We compare the new implementation with the existing CAN based implementation in regards to different characteristics such as cost of implementation, reliability and maintainability.

Keywords: Virtual and remote labs · Control engineering education
Web-based education · Web-based design tools · Distance learning

1 Introduction

Our hybrid online lab GOLDi (Grid of Online Lab Devices Ilmenau), described in several papers, supports the design process of digital control systems. The development process usually consists of the conceptual formulation and the design of the control algorithm to finally achieve a validated control [1–5]. For the functional description, we offer different specification techniques by using noncommercial development tools for various web-based control units in the remote lab:

- a Finite State Machine (FSM) based design to describe digital automata - executed within a client-side FSM interpreter,
- a software-oriented design in C or assembler executed on microcontrollers,
- a hardware-oriented design in hardware description languages or schematic block design by using FPGAs.

The ongoing development of the GOLDi platform focuses on to different aspects of the system, usability, reliability and maintainability. While the usability and reliability

© Springer International Publishing AG, part of Springer Nature 2019
M. E. Auer and R. Langmann (Eds.): REV 2018, LNNS 47, pp. 80–89, 2019.
https://doi.org/10.1007/978-3-319-95678-7_9

are crucial for a successful application for educational purposes, maintainability issues mainly impact future developments. As a distributed platform, GOLDi faces different challenges that are caused by the physical inaccessibility of the hardware for the core development team. In the following section, we will look at three different milestones that were taken to ensure a seamless operation of the platform.

2   GOLDi Remote Lab Development Milestones

In this section, these three milestones will be chronologically explained in detail.

2.1   Installation of the GOLDi Remote Labs at Each Partner Institution

Our first approach at the beginning of the projects in 2012 was to duplicate the entire remote lab setup from the Ilmenau University of Technology (remote lab server, control units, electromechanical hardware models, remote lab infrastructure) to each partner institution. While this step allowed a simple deployment without the necessity of changes to the system, we quickly noticed that maintenance of the whole system became a major issue. Following the idea of independent deployments at the different locations, the first concept was that each partner institution should maintain its laboratory independently. In this model, the Ilmenau team would only provide support, while specially trained staff at each GOLDi location was maintaining their local deployment. This approach turned out to be very ineffective.

After installing the first GOLDi remote labs at our partner universities, we had to realize the following [6]:

- Each institution has specific modifications according to their own requirements,
- Each university has different network architectures,
- Each university has different remote lab configurations,
- To have access to an experiment of the partner labs, a separate user account on the corresponding partner website is necessary.

We decided to reengineer the network of all distributed GOLDi remote labs to a GOLDi cloud system (see Fig. 1).

2.2   Reengineering of the Network of All Distributed GOLDi Remote Labs into the GOLDi Cloud System

Available GOLDi Servers (each corresponding to a partner remote lab) are registered in the GOLDi cloud. The new cloud structure has the following advantages [7]:

- Maintenance of the whole system on one central location: www.goldi-labs.net,
- All partner universities have the same GOLDi software version,
- New functionalities are immediately available for all partners,
- Usage of all partner labs with one central goldi-labs.net user account,
- The ability to quickly change the experiment location if local hardware is unavailable or blocked by another user,

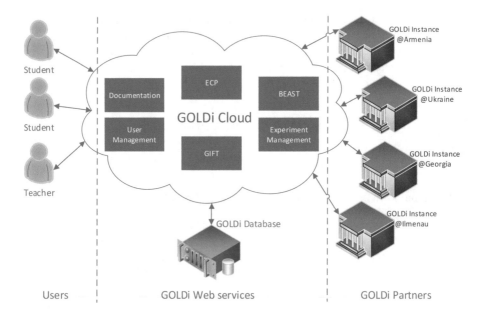

Fig. 1. GOLDi Cloud Architecture

- The user can determine which lab has the lowest delay (best reaction time) regarding his current geographic position and internet connection,
- Overview about all running experiments in all partner labs worldwide with the possibility to observe these experiments.

Each GOLDi user communicates with the GOLDi cloud to access the following GOLDi Web services:

- Experiment Control Panel (ECP): User interface to perform remote experiments and provide the communication with the GOLDi Partners,
- Experiment Management: Experiment booking and pre-planning,
- User Management,
- GOLDi Documentation for control units and electromechanical hardware models,
- GOLDi Design Tools:
  - Block diagram Editing and Simulation Tool (BEAST): User interface to develop block diagram based designs,
  - Graphical Interactive Finite State Machine Tool (GIFT): User interface to develop Finite State Machine based designs.

2.3   Revision of the Hardware Infrastructure and Maintenance Processes of the GOLDi Cloud System

For some time now, there have been increasing problems and delays when the laboratory was extended with new functionalities of the installed hardware setup. The firmware of the hardware components has to be updated every time a feature is added or a bug fix introduced. Up until now, each hardware component had to be manually disconnected from the running lab, connected via USB to a separate computer, running the GOLDi Firmware Updater software and reconfigured manually at each partner institution. This procedure caused considerable difficulties for some of our partners and slowed the overall process of the GOLDi Lab development.

That is why we decided once again to revise the GOLDi cloud system – but this time concerning the hardware components at each distributed partner remote lab and the communication between them. From now on, it is possible to update all components automatically from the distance (by the Ilmenau Team as GOLDi coordinator) without any manual intervention on site necessary.

As this milestone is still under active development, the results will be presented in a future work.

3   Analysis and Comparison of Internal Remote Lab Communication Systems

One benefit of the ongoing revision of the GOLDi infrastructure for our partner institutions is a significant reduction in the maintenance effort of their own remote labs. This can be attributed first and foremost to the conversion of the existing CAN Bus system for internal communication of all hardware components within the GOLDi infrastructure to an Ethernet based system presented in this article. For this purpose, we will be comparing the existing GOLDi CAN communication with our new proposal, an Ethernet based solution.

3.1   GOLDi CAN Communication System

Within the GOLDi CAN implementation the CAN Bus system is used, which has been standardized in [8]. This bus system has been developed for the use within the automotive industry and is widely used and tested there. We use this concept for interconnecting all the lab devices as shown in Fig. 2:

- Physical System Protection Unit (PSPU): Protects the electromechanical hardware models against malicious commands issued by a control unit. Performs the translation between the communication bus and 24 V I/O,
- Bus Protection Unit (BPU): Protects the communication bus against overload and blocking caused by a malicious user programs,
- Bus Control Unit (BCU): As the Goldi Lab Server is a standard PC, there has to be an interface between it and the CAN bus system. This interface is performed by the BCU. The BCU translates the CAN messages to a UART via USB protocol. In addition, the BCU is used as a switching device to interconnect up to 8 lab devices. BCUs can be chained together to achieve bigger networks. All the lab devices connected to each BCU are powered by a single 24 V power supply connected to the BCU. This results in each lab device only having one outgoing cable connection.

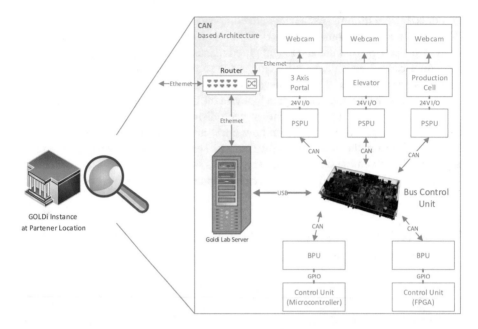

Fig. 2. GOLDi Hardware Setup (CAN implementation)

For the Lab Server to be able to connect to the GOLDi cloud and to access the webcams placed at each electromechanical hardware model a separate Ethernet router also containing a firewall is used.

3.2 GOLDi Ethernet Communication System

In this article, we propose the following communication layout displayed in Fig. 3. We replaced the different hardware communication layers used in the GOLDi CAN implementation (CAN, USB, Ethernet) with a single Ethernet based solution. This results in the Lab Server being able to communicate directly with all the connected lab devices. As Ethernet interfaces are available for standard PCs no USB connection is necessary. Therefore, the BCU which now would only perform the task of a switching device can be replaced by an off the shelf Ethernet switch. For cost effectiveness, we are not using Power over Ethernet (PoE), as these devices are still not widely available for the implementation in low cost solutions. This results in each lab device needing its own power supply. The advantages and disadvantages of this power distribution concept and several other characteristics of the described communication systems will be laid out in detail in the next section.

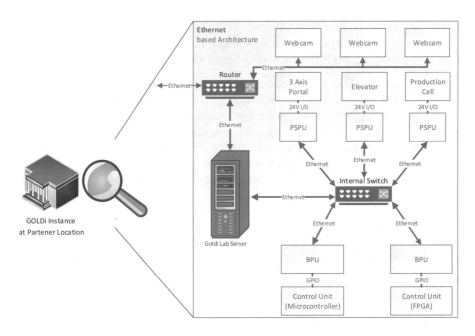

Fig. 3. GOLDi Hardware Setup (Ethernet implementation)

3.3 Comparison of Possible Internal Remote Lab Communication Systems

In this section, the characteristics displayed in Table 1 will be assessed in detail for both selected Communication methods: CAN and Ethernet.

Table 1. CAN vs. Ethernet internal remote lab communication

Characteristics	CAN	Ethernet
Cost	Cheap ICs available	ICs considerably more expensive
Power Management	Centralized	Distributed
Cable Types	Custom GOLDi cables	Off the shelf
Switching Solution	Custom GOLDi BPU	Off the shelf
Scalability	8 lab devices per BPU	Unlimited
Logical Topology	Shared medium	Isolated communication via TCP
Transfer rate	125 kBit/s	10 Mbit/s
Debugging Features	Separate debugging hardware and software necessary	Debugging in place e.g. by using Wireshark
Size of Datagrams	Limited to 8 Bytes	64 kByte
Error Behavior	Error detecting	Automatic TCP retransmit in case of errors

Cost

The cost characteristic is referring to the implementation cost for integrating the communication method into the PCB design of the lab devices.

- *CAN:* For the GOLDi CAN implementation, we selected the MCP2515 and MCP 2551 by Microchip for the implementation. The purchasing cost for both chips is <1€. Therefore, the CAN based communication is cost effective.
- *Ethernet:* For the GOLDi Ethernet implementation, we selected the W5300 chip by Wiznet packaged in a breakout board (WIZ830 MJ). The purchasing cost for these breakout boards has come down since the implementation of the GOLDi CAN communication but with approx. 20€ per board is still higher than the CAN implementation.

Power Management

The power management characteristic represents the way power is distributed to the lab devices within the laboratory.

- *CAN:* When using the GOLDi CAN implementation, power is distributed by each BCU. Therefore, a 90 W power supply is recommended to be connected to each BCU at a GOLDi instance. The result of this approach is fewer power supplies needed for each GOLDi instance. However, the power supplies most of the time do not operate under enough load which results in a worsened efficiency. This approach also introduces a single point of failure as a failing power supply or a short circuit in any lab device or hardware model can result in the inoperability of the entire GOLDi instance.
- *Ethernet:* With the WIZ830 MJ used for the Ethernet implementation not supporting power over Ethernet, the power supply within the GOLDi Ethernet implementation is distributed. Each lab device uses its own power supply. Therefore, the lab devices with low power consumption, such as the Microcontroller or FPGA Control Units, can be supplied with inexpensive low power supplies and the higher power PSPUs with connected hardware models can be powered with supplies for their specific power consumption. This approach results in more supplies installed at each partner location but overall is more cost effective and improves reliability, as a failing supply will only affect one lab device.

Cable Types

The cable types characteristic represents the physical cables needed to connect the different lab devices.

- *CAN:* Within the GOLDi CAN implementation, proprietary GOLDi cables are used to be able to provide the power described in the power management characteristic. These cables are custom built by the GOLDi maintenance staff in Ilmenau. This results in the inability to buy replacements locally at each GOLDi installation (especially in Eastern Europe) as well as an increased cost of production and shipping. In addition, the used cables have proven to be prone to short circuits making them unreliable.

- *Ethernet:* Within the GOLDi Ethernet implementation, standard Ethernet cables are used for device interconnections. These cables are proven to be reliable, cost effective and widely available.

Switching Solution

The switching solution characteristic describes how the physical connections between the devices are connected by a switch and what hardware is necessary for this purpose.

- *CAN:* In the GOLDi CAN implementation, the switching is done by a BCU. This is necessary as there are no off the shelf components for the GOLDi cables and power management. As the BCU is a custom-made component, the cost for each BCU exceed the cost of Ethernet switching solutions by a factor of approx. 30.
- *Ethernet:* In case of the GOLDi Ethernet implementation, standard 100Mbit Ethernet switches are used. These components are widely and inexpensively available in different port capacities.

Scalability

The scalability characteristic describes the ability for both methods to handle growing numbers of lab devices.

- *CAN:* For 8 lab devices each a BCU is needed. The BCUs can be interconnected, which means that only one BCU has to be connected to the Lab Server. Nevertheless, each BCU is cost expensive compared to an Ethernet switch limiting the CAN implementations scalability.
- *Ethernet:* As there are a wide variety of switches with different counts of Ethernet ports available which can also be connected can easily scale up to high lab device numbers. Compared to the BCU Ethernet switches are inexpensive.

Logical Topology

This characteristic describes if device communications influence each other.

- *CAN:* As the CAN bus is a fully shared medium, each device can influence the messages sent by other devices. This results in a defective device blocking the entire CAN bus.
- *Ethernet:* As the TCP connections are isolated, a device cannot influence any other connection. This improves the overall reliability of the system.

Transfer Rate

This characteristic describes the data transfer rate of the communication method.

- *CAN:* The GOLDi CAN communication supports transfer rates up to 125kB/s.
- *Ethernet:* The GOLDi Ethernet communication operates at 10Mbit/s.

Debugging Features

This characteristic describes the ease of access for debugging features.

- *CAN:* Special hardware is needed for debugging the CAN bus. For the GOLDi implementation, a hardware analyzer along with the required software "CANalyzer" from Vector Informatik [9] was chosen (see Fig. 4).

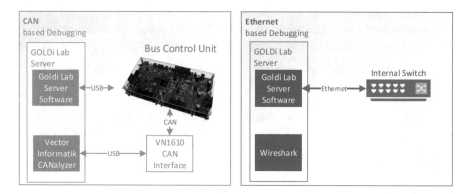

Fig. 4. GOLDi Debugging overview

- *Ethernet:* Standard Ethernet analyzers such as Wireshark [xx] can be used. There are no additional hardware requirements.

Size of Datagrams
This characteristic describes the maximum size of each message.

- *CAN:* Each message can be up to 8 Bytes long.
- *Ethernet:* Each message can be up to 64 kB long as the GOLDi Ethernet protocol does not use packet fragmentation.

Error Behavior
This characteristic is referring to the ability to detect and handle errors.

- *CAN:* The CAN bus protocol is designed to automatically detect errors.
- *Ethernet:* The TCP/IP based Ethernet solution will automatically detect errors and issue a retransmit of the faulty message.

3.4   Evaluation

The results of the prior analysis are that our proposed Ethernet based implementation achieves a great improvement in regards to the maintainability and reliability as well as lowering the cost of hardware for the setup of a GOLDi instance. The choice of implementing the GOLDi CAN architecture was based on historic development of hardware prices as well as the availability of such hardware components. The current pricing of the required Ethernet components led to the redesign of our network architecture.

4   Conclusion

The benefit of the ongoing third revision of the GOLDi infrastructure for our partner institutions is a significant reduction in the maintenance effort of their own remote labs. This can be attributed first and foremost to the conversion of the existing CAN bus system for internal communication of all hardware components within the GOLDi infrastructure to an Ethernet based system.

When the third milestone is fully implemented, we will consider further improvements such as the power supply of all GOLDi components via Power over Ethernet.

Nevertheless, all these improvements cannot replace the manual maintenance of the actual electromechanical hardware models at each partner institution in the future.

Acknowledgment. The authors would like to acknowledge the work of Tobias Vietzke and Stephen Ahmad for their work within the GOLDi framework and Felix Seidel in particular, as most parts of the implementation are based on his B.Sc. Thesis [10].

References

1. ICo-op project Website: http://www.ICo-op.eu
2. DesIRE project Website: http://tempus-desire.thomasmore.be
3. Henke, K., Vietzke, T., Wuttke, H.-D., Ostendorff, S.: GOLDi – Grid of Online Lab Devices Ilmenau. Int. J. Online Eng. (iJOE) **12**, 11-13 (2016). ISSN: 1861–2121
4. Henke, K., Ostendorff, S., Wuttke, H.-D., Vietzke, T., Lutze, C.: Fields of applications for hybrid online labs. Int. J. Online Eng. (iJOE) 9, 20–30 (2013)
5. Henke, K., Ostendorff, S., Vogel, St., Wuttke, H.-D.: A grid concept for reliable, flexible and robust remote engineering laboratories. Int. J. Online Eng. (iJOE), **8**, 42–49 (2012)
6. Henke, K., Vietzke, T., Hutschenreuter, R., Wuttke, H.-D.: The remote lab cloud goldilabs.net. In: 13th International Conference on Remote Engineering and Virtual Instrumentation, REV 2016, Madrid, February 2016
7. GOLDi-labs cloud Website: http://goldi-labs.net
8. The CAN wiki pages: http://www.can-wiki.info/doku.php
9. Vector Informatik Homepage: http://vector.com
10. Seidel, F.: Konzeption und prototypische Implementierung einer modularen Backend-Softwarearchitektur für das GOLDi-Remotelab, B.Sc. Thesis, TU Ilmenau (2017)

A Multi-agent System for Supervisory Temperature Control Using Fuzzy Logic and Open Platform Communication Data Access

Martha Kafuko[1] and Tom Wanyama[2(✉)]

[1] HATCH Ltd., 2800 Speakman Drive, Mississauga, ON, Canada
[2] W Booth School of Engineering Practice and Technology, Faculty of Engineering,
McMaster University, Hamilton, ON, Canada
wanyama@mcmaster.ca

Abstract. Independent temperature controllers are usually robust and energy efficient because they do not heat or cool the entire building, including spaces that are not occupied, to the same temperature. Moreover, they can use tested and true temperature control techniques such as Proportional-Integral-Derivative (PID). Unfortunately, independent temperature controllers usually do not coordinate temperature control of multiple spaces in buildings or multiple buildings maintained by the same organizations to optimize total energy usage. In this paper, we present a multi-agent system that uses fuzzy logic to supervise independent PID temperature controllers. The system has a supervisor agent that adjusts the setpoints of the PID controllers based on total energy cost and other parameters managed by specialized agents. The agents are built using a mix of commercial off-the-shelf software, including LabView, Microsoft Excel, KEPServer Ex5 OPC server, and custom written code. In addition, the multi-agent system uses Open Platform Connectivity (OPC) technology to share data. OPC technology enables the supervisor agent to communicate with the independent temperature controller agents over the Internet making the system scalable to multi-building or multi-unit supervisory control – even if the buildings or the units are not in the same geographical location.

1 Introduction

Increased awareness of the negative effects of energy use combined with increasing worldwide energy demand has resulted in energy saving policies that advocate and encourage rational energy consumption [6]. Since buildings account for a sizeable amount of the total energy use, energy saving in buildings is critical to attaining overall reduction in energy consumption. Worldwide, buildings account for 40% of the energy use. Energy use statistics from Canada, the United States of America (USA), and Europe show similar share of energy use, with buildings representing 31% of Canadian, 42% of USA, and 40% of European energy consumption [2, 7, 11]. Heating and cooling is one of the main uses of energy in buildings. The ratio of the amount of energy used for heating and cooling to the total building energy use depends heavily on the weather

© Springer International Publishing AG, part of Springer Nature 2019
M. E. Auer and R. Langmann (Eds.): REV 2018, LNNS 47, pp. 90–99, 2019.
https://doi.org/10.1007/978-3-319-95678-7_10

pattern of the country and it varies between 18% and 73% [10]. In Canada, this ratio is 53% while in USA it is 39% [7, 11]. Consequently, efforts to reduce energy consumption in buildings in countries such as Canada must focus on the heating and cooling energy use among other energy uses in buildings.

There are many approaches to saving building heating and cooling energy. Some examples of these approaches include:

- Keeping window coverings closed in the summer to block heat from the sun.
- Installing energy efficient heating and cooling equipment.
- Installing energy efficient windows and doors.
- Insulating the building and sealing all leaks.

Temperature control is one of the few heating and cooling energy saving approaches that can be automated, and can be done with little alteration to the building structure or architecture. Consequently, many energy saving temperature control schemes have been proposed in literature. These schemes fall into two main categories, namely: independent controllers and centralized controllers [6]. Generally, independent controllers offer better energy performance because they allow for different spaces to have different temperature and energy utilization profiles. Furthermore, they can use proven control methods such as hysteresis (on/off) and Proportional-Integral-Derivative (PID).

In this paper, we present an agent-based temperature control system that has multiple local temperature controllers and a supervisory controller. The local temperature controllers are associated with various spaces, units, or buildings whose temperature is being controlled; whereas the supervisory controller manages the setpoints of the local controllers. The system is implemented using multi-agent technology to make it flexible, easy to deploy and easy to manage. Under some predefined energy consumption, cost, and space usage conditions the supervisory controllers moves the setpoints of the local controllers to save energy. Since the system uses Open Platform Connectivity (OPC) technology as an agent information highway, the supervisory controllers can manage local temperature controllers of multiple buildings or units in the same or different geographical locations.

The rest of this paper is arranged as follows: Sect. 2 covers the literature review on energy saving temperature control systems. In Sect. 3 we present the proposed agent based heating and cooling system and Sect. 4 deals with the testing and results of the case study. Section 5 covers the conclusion.

2   Literature Review

Conventional temperature controllers are deduced from control theory techniques that focus on determining how systems work to establish the associated system control strategies. These control approaches work well for linear Single Input Single Output (SISO) systems. For such systems, PID is the most used control strategy in industry because it was primarily designed for them. Generally, temperature controllers are powerful process control tools that offer very simple operation. They take a signal from a temperature device, such as a thermocouple or RTD, and maintain a setpoint using an output signal. For example, AutomationDirect's SOLO digital temperature controller is a single loop dual output temperature limit controller that can control both heating and cooling simultaneously. The controller supports four types of control modes: PID, ON/OFF, Ramp/Soak and Manual.

There are temperature control strategies that are designed to control nonlinear Multiple Input Multiple Output (MIMO) systems. For example, fuzzy logic controllers support MIMO systems and focus on what the system does instead of how it works. Such controllers are required in supervisory temperature control that minimizes energy usage. Singhala et al. [9] propose a fuzzy logic temperature controller that takes the temperature error, the difference between the setpoint and the actual temperature, as the input. The output is a Pulse Width Modulation (PWM) signal that controls the heater temperature and the fan speed. This is implemented using a microcontroller. Basu [1] presents a fuzzy logic temperature controller that uses temperature error and rate of change of temperature error (derivative of temperature error) as inputs. The output of Basu's controller is an analog signal that controls the electrical power of the heater. The system is implemented using the fuzzy logic toolbox in Matlab; and having two inputs improves its performance. Kobersis et al. [5] presents a temperature control system that is similar to the one presented by Basu [1], that is implemented using LabView.

The fuzzy logic temperature control systems proposed by Singhala et al. [9], Basu [1], and Kobersis et al. [5] are not superior to conventional temperature controller such as those based on PID. The superiority of nonlinear MIMO systems control methods is realized when nonlinear parameters such as energy, cost, and time of use are inputs to the temperature control system. Martinčević et al. [6] presents an energy saving Model Predictive Controller (MPC). The MPC problem formulation for temperature control consists of a minimization of energy consumption with respect to temperature constraints that are set by the system user and the system's physical limitations. Martinčević et al.'s [6] temperature control system saves energy by letting the temperature to slide from the setpoint, within a predefined range. Huberman and Clearwater [3] propose a multi-agent based temperature control system. The system models the temperature control problem as an optimization of the building thermal resource. The resource is distributed to the agents that represent the individual room temperature controllers through an auction process. The auction process rations energy because it is designed to ensure that agents can consume extra portions of the thermal resource only if there is another agent that is willing to sell (give up) all or some of its share. The temperature in the individual rooms is controlled used PID. Paris et al. [8] propose a hybrid PID-fuzzy temperature control system. The focus of their controller is the management of the building's two energy sources, namely renewable energy and fossil energy. The PID controls the use of the renewable energy while fuzzy logic controls fossil energy use. In addition, fuzzy logic is used to supervise the PID so as to minimize total energy usage.

3   Multi-agent System for Supervisory Temperature Control

3.1   System Structure and Architecture

Our multi-agent system for supervisory temperature control has agents that control temperature in the different spaces in buildings (Fig. 1). These agents ensure that temperature in each of the spaces remain within an acceptable range with respect to the use of the space. Figure 1 shows that at the local level, our system has two agents for every space. These are the Temperature Agent responsible for controlling the space temperature, and the Energy Agent that monitors the energy consumption of the space. Moreover, the figure shows that at the supervisory level, the system has the following agents:

- Energy Price Agent – This agent is responsible for monitoring the price of energy for each of the temperature agents on an hourly basis.
- Weather Agent – The weather agent is responsible for collecting local weather information of the various temperature control agents. The supervisor agent uses this information to manage the total energy.
- Supervisor Agent – This agent supervises the temperature control agents by adjusting their temperature setpoints based on the allotted total power, the energy price, weather conditions, and the space's thermal needs.

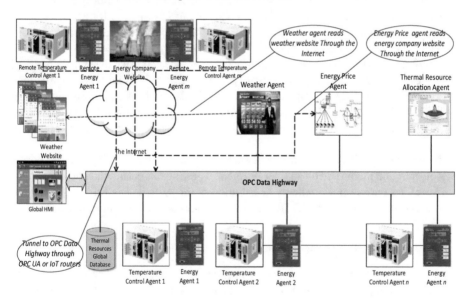

Fig. 1. Information sharing among agents of the temperature control system

Figure 1 shows that the agent of our temperature control system shares information using Open Platform Connectivity (OPC) technology. This technology allows the agents to communicate over the Internet making it possible for the supervisor agent to manage temperature control agents in buildings that are in different geographical locations. In addition, Fig. 1 shows that OPC technology enables the system to collect and store global data into a single database that is accessed by agents regardless of their physical location. This creates a single point of "truth", which improves the performance of the system.

3.2 Information Gathering for Total Energy Optimization

The multi-agent system for supervisory temperature control uses a series of Commercial Off-The-Shelf (COTS) hardware and software that are integrated using OPC technology and custom glue-ware applications. It achieves total energy optimization by using information from various sources as follows:

- The Temperature Control Agents store the temperature setpoints and the actual temperatures (Fig. 2) of spaces, units, or buildings involved in the energy optimization. This information is available to the Supervisor Agent by the Programmable Logic Controllers (PLCs) that are used to implement the Temperature Control Agents.
- The Weather Agent obtains weather information, such as temperature, from various local weather websites in the area where the buildings involved in total energy optimization are located. This agent is implemented using Microsoft Excel macroinstructions written in Visual Basic Applications (VBA) and updates its information every hour. At the local level, it is used as a source of weather information for the user, while at the global level it is used a source of information for energy optimization as shown in Fig. 2.
- The Energy Price Agents obtain local energy pricing for the locations where the buildings involved in total energy optimization are located. Like the Weather Agent, the Energy Price Agent is implemented using Microsoft Excel macroinstructions written in Visual Basic Applications (VBA). It also updates its information every hour. At the local level, it is used as a source of energy price information for the user, while at the global level its information is used for energy optimization as shown in Fig. 2.
- Every space, unit, or building involved in the energy optimization, has its power consumption monitored by an Energy Agent. These agents ensure that the Supervisor Agent has current power/energy distribution of the system. The Energy Agents are implemented using the SEL751A IEC61850 compliant intelligent power meters.

3.3 Supervisory Temperature Control

Figure 2 shows how the supervisory controller is integrated with distributed local temperature controllers. Note that the available thermal resource to heat/cool all the spaces, units, or buildings involved in the energy optimization is measured in terms of total allotted power. This information is integrated with other temperature control related information such as energy pricing, power consumption, and weather condition by the Supervisor Agent using fuzzy logic to determine the optimum (adjusted) temperature setpoints (SP′) of the distributed temperature controllers. In our energy efficiency and temperature control test platform, we implemented the Supervisor Agent using LabView.

The Supervisor Agent runs on a PC that acts as the system supervisory controller. In manual model, the supervisory controller lets the distributed local temperature controllers to decide whether to use the optimum temperature setpoints (SP′) or the user setpoints (SP); and in the automatic model, all distributed local temperature controllers use the optimum temperature setpoints (SP′) determined by the Supervisor Agent.

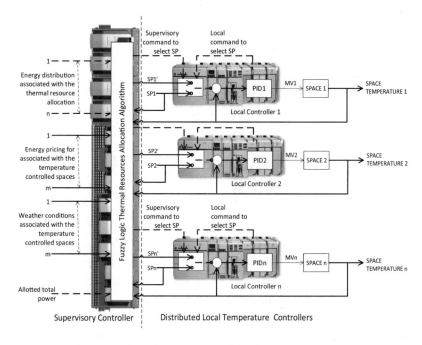

Fig. 2. Integration of supervisory and local temperature controllers

3.4 Temperature Control

The Temperature Agents control temperature through a PID process shown in Fig. 3. This process is described by Martha Kafuko and Tom Wanyama in their paper titled "Integrated Hands-On and Remote PID Tuning Laboratory" [4] as follows: "The set $T_s(t)$ point (SP) is the desired temperature of the house. Note that this parameter can be replaced by the optimum temperature setpoints (SP′). The manipulated variable (MV) $f(t)$ of the PID is the 0–5 V analog input to the KT-5194 DC Motor PID speed controller. This controller features two control modes, namely: open loop PWM control, and closed loop PID control. In our platform, the controller is used in the open loop PWM control mode were the 0–5 V input signal determines the value of the 0–24 V (10 A-maximum) PWM output. The 0–24 V power supply to the fan DC motor is the control variable (CV) $c(t)$ of our PID loop. The temperature of the fictitious house is measured using an RTD probe whose resistance varies from 100 Ω at 0 °C to 220 Ω at 300 °C. The RTD signal is fed in a signal conditioner that produces a proportional 0–10 V analog signal. The signal conditioner output is the input to a Micrologix 1400 PLC that has an ADC the converts the analog signal to a 0-4095 digital variable. This variable is scaled to produce the actual temperature of the house, which is the process variable (PV) $T_a(t)$ of the PID loop. Moreover, the PLC generates the hysteresis signal that control the blow heater respectively." Note that the PLC also stores settings that determine whether the user temperature setpoints (SP) or the optimum temperature setpoints (SP′) are $g(t)$ applied to the PID.

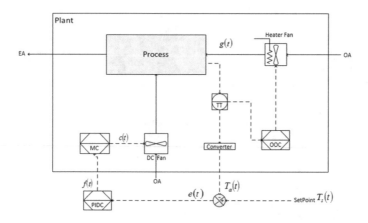

Fig. 3. Process diagram of the temperature agent PID controller

4   Supervisory Temperature Control Testing

Testing the operational effectiveness of our multi-agent system for temperature control was accomplished by controlling a cooling system of a fictitious house. That house is heated by a hair dryer, which is fitted with a 120 V heater and 120 V fan motor. The heater and fan motor are controlled by the PLC through 24 V relay. The cooling system in the house has a small fan ran by a 24 Volts DC motor. The Temperature Agent controls this motor using a PID controller described in Sect. 3.4. In addition, the house has two 40 W lamps. The following steps were taken to test the system:

i. The house temperature (user setpoint SP) was set to 30 °C; the supervisory controller was set to automatic, meaning the PID had to use the adjusted setpoint (SP′) as the input; and the heater was set to heat the house at a constant rate, with the test platform set to limit the maximum temperature to 48 °C.

ii. The Supervisor Agent was set to determine an optimum setpoint (SP′) for the PID based on the energy consumption of the house.

iii. The system allotted power was set to 0.4 kW, which is the power consumption of our test platform without the two lamps and the cooling fan. Note that the allotted power determines the classification of power consumption in the membership function of Fig. 3.

iv. With both lamps turned off, SP, SP′, actual house temperature, and cooling fan power were logged.

v. Step iv was repeated with one lamp turned on, and with two lamps turned on.

4.1   Test Results

Figure 4 shows the variation, with time, of the optimum (adjusted) setpoints (dotted graph lines) and the associated actual house temperature (solid graph lines). Note that the actual temperature is the output of the PID controller.

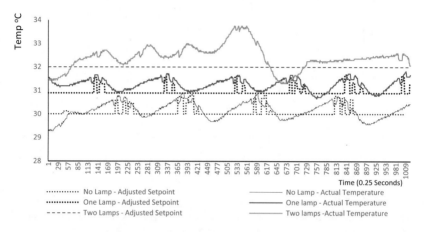

Fig. 4. Adjusted temperature setpoints and associated actual temperatures

Figure 5 shows the power consumed by the cooling fan when no lamp, one lamp, or two lamps are turned on.

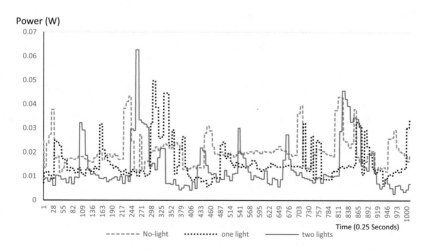

Fig. 5. Fan motor power consumption under different supervisory conditions

Figure 6 show the percentage energy saving versus the optimum (adjusted) setpoints.

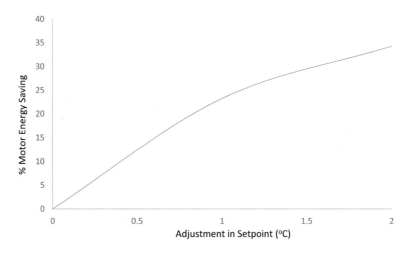

Fig. 6. Variation of fan motor energy saving with adjustment in temperature setpoint

4.2   Discussion of Results

The test results in Fig. 4 show that fuzzy logic works well as a supervisory controller of PID temperature controllers, since it can adjust the PID setpoints based on predefined criteria (in this test-case the criterion is power consumption of the building). Moreover, Fig. 5 shows that adjusting the PID setpoint leads to reduced power consumption of the cooling system, resulting into energy saving. In fact, Fig. 6 shows that 1 °C adjustment of the setpoint leads to about 25% saving in energy, while a 2 °C leads to about 35% energy saving.

The huge energy saving potential of our fuzzy logic thermal resource allocation systems based on the supervisory control test results come at the price of reduced stability of the PID temperature controllers. The instability is due to the following activities of the system:

i.   When temperature is high, PID increases the speed of the motor, which increases power consumption of the system.
ii. Fuzzy logic adjusts (increases) the setpoint of the PID in response to increased power consumption.
iii. Increasing the setpoint reduces the temperature error, thus the PID reduces the speed of the fan motor. This results in increased temperature since the heating is constant, which takes the system to step 1.

5   Conclusion and Future Work

This paper describes a system that achieves total energy saving through thermal resource distribution using fuzzy logic algorithms. The fuzzy logic algorithms is managed by the

Supervisor Agent which supervisors the temperature control agents that use PID controllers to maintain the temperature of the spaces at the setpoints. Test results show that adjusting the PID setpoints by the Supervisory Agent leads to huge energy saving. However, the energy saving comes at the price of reduced stability of temperature control.

In the future, we would like to stabilize the PID controllers. We believe that this will be achieved by filtering the power consumption of building using Kalman filters. In addition, we would like to develop an algorithm for distributing the allotted power to the Heating Air Conditioning and Ventilation (HVAC) system since this power represents the available thermal resources of buildings.

References

1. Basu, S.: Realization of fuzzy logic temperature controller. Int. J. Eng. Technol. Adv. Eng. **2**(6), 9 (2012)
2. European web portal for energy efficiency in buildings. www.buildup.eu. Accessed 2 Feb 2017
3. Huberman, B., Clearwater, S.H.: A multi-agent system for controlling building environments. In: Proceedings of the First International Conference on Multiagent Systems (1995)
4. Kafuko, M., Wanyama, T.: Integrated hands-on and remote PID tuning laboratory. In: Proceedings of the Canadian Engineering Education Association Conference, Hamilton – Canada, June 2015
5. Kobersi, I.S., Finaev, V.I.: Control of the heating system with fuzzy logic. World Appl. Sci. J. **23**(11), 1441–1447 (2013)
6. Martinčević, A., Lesic, V.: Model predictive control for energy-saving and comfortable temperature control in buildings. In: 24th Mediterranean Conference on Control and Automation (MED), Athens, Greece, 21–24 June 2016
7. Natural Resources Canada (NRCan): Energy Use Data Handbook, 1990–2009
8. Paris, B., Eynard, J., Grieu, S., Polit, M.: Hybrid PID-fuzzy control scheme for managing energy resources in buildings. Appl. Soft Comput. **11**(8), 5068–5080 (2011)
9. Singhala, P., Shah, D.N. Patel, B.: Temperature control using fuzzy logic. Int. J. Instrum. Control Syst. **4**(1) (2014)
10. Ürge-Vorsatz, D., Cabeza, L.F., Serrano, S., Barreneche, C., Petrichenko, K.: Heating and cooling energy trends and drivers in buildings. Renew. Sustain. Energy Rev. **41**, 85–98 (2015)
11. World Business Council for Sustainable Development: Energy efficiency in buildings - transforming the market, 2009. www.wbcsd.org/web/eeb. Accessed 2 Feb 2017

Combining Virtual and Remote Interactive Labs and Visual/Textual Programming: The Furuta Pendulum Experience

Daniel Galan, Luis de la Torre$^{(\boxtimes)}$, Dictino Chaos, and Ernesto Aranda

UNED, 28040 Madrid, Spain
{dgalan,ldelatorre,dchaos}@dia.uned.es, earanda@bec.uned.es

Abstract. This paper proposes a new way of experimenting with online (virtual or remote) interactive laboratories. Experimentation possibilities can be opened by allowing students to interact with the online laboratory using a visual and textual programming language that can communicate with the laboratory application and which includes tools to define and plot graphs. The combination of interactive laboratories with a visual and textual programming language, benefits both teachers and students: the former have a wider range of possibilities when considering the assignments that can be proposed and the latter acquire a greater knowledge of the plant under study by facing a more inquiry-based approach for online experimentation. To demonstrate the usefulness and possibilities of this environment, a Furuta pendulum system has been successfully used in both its remote and its virtual version.

Keywords: Virtual labs · Remote labs · Online experimentation

1 Introduction

Education in science, technology, engineering and mathematics (STEM) disciplines require practical experimentation. In traditional education, this experimentation has usually been performed in laboratories or the field. However, in online and blended education, computers constitute a very interesting tool for carrying out experiments. From the necessity of performing experiments under the online and blended education methodologies, simulations or virtual laboratories and remote laboratories were born.

This type of laboratories is useful for the two previous educational scenarios because of their advantages over hands-on laboratories [2,5,7,9,11].

The vast majority of nowadays virtual and remote laboratories implementations are based on interactive experimentation [8]. In such labs, students change parameters in the lab graphical user interface and see the evolution of the system in real-time. However, even with its many benefits, interactive experimentation has some important drawbacks. Perhaps, the most relevant one is that students

© Springer International Publishing AG, part of Springer Nature 2019
M. E. Auer and R. Langmann (Eds.): REV 2018, LNNS 47, pp. 100–109, 2019.
https://doi.org/10.1007/978-3-319-95678-7_11

sometimes try to solve problems in a trial and error process [6]. Instead of reflecting in advance on what result they want to obtain from the experiment/system and thinking on intelligent approaches for reaching it, they just manipulate the laboratory to see what happens, hoping that some combination of buttons and input parameters may result on a response that is close to the one they are looking for. Therefore, working with these labs does not necessarily require the knowledge of the underlying concepts. In contrast to this interactive experimentation, automated experiments do not have this lack. Students must think in advance how to face a problem and how to work with the lab in order to obtain the results they might be looking for.

This paper presents a visual and textual programming experimentation environment that mixes interactivity and automated experiments. On the one hand, interactive experimentation allows students to modify parameters to see immediately reflected the output changes and the evolution of the model behavior in the graphical user interface. On the other hand, when some (textual or visual) programming needs to be done by the students, the experimentation process requires more involvement from them than just pressing buttons, entering values in input fields, moving sliders and/or watching at data being automatically plotted in predefined graphs. We believe that, in such scenario, students would not only be more motivated to work, but they will also: (1) be required to demonstrate more knowledge about how to solve the problem and (2) they will better develop their technical/scientific skills.

2   The Visual/Textual Experimentation Environment

We propose an experimentation environment (EE) that supports: (1) using existing laboratories without requiring any extra adaptation; (2) writing experiments for those labs in a visual programming language; (3) writing experiments in a textual programming language; (4) running the experiments with an interpreter (so that experiments are executed on the fly); and (5) collecting and plotting data from the experiments.

The EE proposed in this work consists of three elements, which are clearly visible in Fig. 1: (1) the experiment editor (marked with a red rectangle at the bottom), (2) the custom charts panel (blue box at the top right) and (3) the online laboratory (green box at the top left).

Experiment Editor. The experiment editor is the component in which the programming of the automated experiments is done. It is formed by a script area and a menu, located on the left, with the different categories in which the students can do the programming. The programs created here must be automatically interpreted. This way, students do not need to recompile the experiments over and over with every small change they make in their code. Instead, they can just run it, see if the results are what they expected and continue creating the experiment.

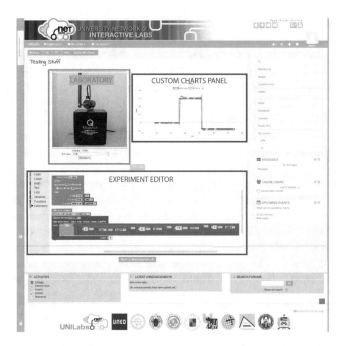

Fig. 1. General architecture of the experimentation environment

Custom Charts Panel. The EE proposed in this work allows the creation and customization of charts. Students first decide the data they want to visualize and then, they can define the title, the sampling period and the maximum number of points they want to plot. Once defined, the charts are automatically created, and the information is shown in real time. It must be also possible to create different charts for the same experiment or to confront data from different experiment executions.

Online Laboratory. Laboratories are automatically prepared to be used in the environment with no extra adaptation or implementation. The type of the laboratory, whether virtual or remote, is entirely transparent for the EE and its use is not influenced by the nature of the laboratory at all.

3   Implementation

This section details how the implementation of the three components (the experiment editor, the custom charts and the online labs) of the visual/textual experimentation environment is addressed.

3.1   Blockly for the Experiment Editor

Blockly, an open-source project developed by Google, is a JavaScript library for visually programming code with blocks. The project started in 2011 and nowadays it is used in a large list of web applications. This project aims to teach basic programming knowledge to new users by linking pieces of code like a jigsaw puzzle. The scripts created with Blockly can generate JavaScript, Python, Dart, and PHP. However, it can be customized with new blocks and with new functions to generate other computer languages [12]. In the presented EE, the generated language is JavaScript because it fits perfectly with the laboratories implementation. New blocks have been defined in Blockly to access the online lab variables and functions, allowing a complete control of the lab through programming.

3.2   Chart.js for the Custom Charts

Chart.js is an open-source JavaScript library that enables the creation of HTML5 charts. It is the library chosen by Moodle to create their charts, so it fits nice with the environment. It has several remarkable features:

- It includes eight different types of charts that can be mixed, customized or even animated.
- As it is built upon HTML5, it has a great performance in modern browsers, which is very important in real-time charts like the experimentation ones.
- The charts have a responsive design so the experimentation environment can be used with computers, mobile phones or tablets.

3.3   EjsS for the Online Laboratories

Technological advances have implied that the software surrounding virtual and remote laboratories has got increasingly complex. Creating a graphical user interface to operate with the laboratory is not trivial, especially for those who do not have fluent programming skills. Fortunately, EjsS offers high-level graphical interfaces to create computer simulations. Users with little knowledge of computer programming can focus their effort on defining the mathematical model of the lab, and how the graphical user interface should look like. Any other aspect of the implementation is automatically generated. EjsS is part of the OSP, which provides free online resource collections through the ComPADRE digital library. Among these resources, users can find more than 500 applications created with EjsS: http://www.opensourcephysics.org.

3.4   UNILabs for Deploying the EE

UNILabs (University Network of Interactive Labs) is a web 2.0 of interactive laboratories, a Learning Management System (LMS), where any university can freely deploy their laboratories and offer them to their students [10]. It was created in 2013, and nowadays it offers more than 20 remote and virtual laboratories distributed in 31 courses from a variety of specialties (automatic control, mathematics, optics, automation, linear systems, modern and classical physics, among others). UNILabs is based on Moodle, which offers excellent features to manage the experimentation resources by the teachers, and it includes the EJSApp Moodle plugins to deploy the EjsS laboratories.

The UNILabs web page (http://unilabs.dia.uned.es) includes more information about the courses and universities that collaborate.

4   The Furuta Pendulum Online Laboratory

The virtual laboratory consists of a simulation of the pendulum Eqs. (1) and (2) (where β is the angle described by the rotary arm, and α is the angle described by the pendulum.), developed in EjsS. The simulation incorporates a 3D representation and a few inputs to interact (see Fig. 2. The user interface is as simple as possible, and allows only the most basic actions: change the position of the pendulum, the reference, and stop, pause or resume the simulation.

$$
\ddot{\beta} = \frac{h_2 h_3 \cos^2 \alpha \sin \alpha \dot{\beta}^2}{H}
$$
$$
\frac{h_3 \cos \alpha \left(-B_1 \dot{\alpha} + h_5 \sin \alpha\right)}{H}
$$
$$
\frac{-h_4 \dot{\beta} \left(B_0 + 2 h_2 \dot{\alpha} \cos \alpha \sin \alpha\right)}{H}
$$
$$
\frac{-h_4 \left(-\tau + h_3 \dot{\alpha}^2 \sin \alpha\right)}{H}
$$
$$(1)$$

$$
\ddot{\alpha} = \frac{\left(h_2 \sin^2 \alpha + h_1\right) h_2 \cos \alpha \sin \alpha \dot{\beta}^2}{H}
$$
$$
\frac{\left(h_2 \sin^2 \alpha + h_1\right) \left(-B_1 \dot{\alpha} + h_5 \sin \alpha\right)}{H}
$$
$$
\frac{-h_3 \dot{\theta}_0 \cos \alpha \left(B_0 + 2 h_2 \dot{\theta}_1 \cos \alpha \sin \alpha\right)}{H}
$$
$$
\frac{-h_3 \cos \alpha \left(-\tau + h_3 \dot{\theta}_1^2 \sin \alpha\right)}{H}
$$
$$(2)$$

The remote laboratory has two main components: (1) the server side, developed in LabVIEW, that accepts connections using a component called JIL [3] and (2) the lab application in EjsS, whose variables are linked those LabVIEW. The remote lab interface is very similar to the virtual one, but it presents a webcam video streaming of the pendulum instead of the simulated 3D view.

The virtual and remote laboratory user interface has been streamlined as much as possible without losing usability. This is the crucial difference with previous designs that are bloated, including lots of tools and visualizations focused on a specific task.

As a comparison, Fig. 3 shows an "old style" user interface for this laboratory. The old interface needed to anticipate to the users' needs. Thus, it showed lots of elements for visualizing variables and several menus with related advanced options.

One important feature of the laboratory is that it enables to change the controller of the plant. This has been done in previous laboratories as well, but using a specific language to the domain of application, which needs a special server for

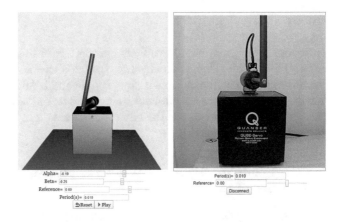

Fig. 2. Virtual (left) and remote laboratory (right)

its execution (see [4] for a previous successful implementation of this idea). In this work, this task has been faced with a novel approach being the controller defined in JavaScript. Hence, the same code can interact without modification with the simulation on the browser and with the real system. In the later case, the JavaScript code is executed in a sandbox in LabVIEW.

Furthermore, the controller needs not to be "written" at all, and the laboratory developed in EjsS does not have an editor or options to load and save files. Everything (from the controller to the visualization) can be constructed using the blocky language from the environment, as the next section illustrates.

Thus, the user interface is general, simple and can be used for many different purposes and assignments without any modification.

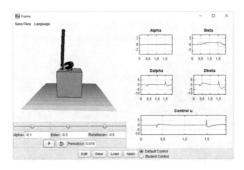

Fig. 3. Old style laboratory with a complex graphical user interface

5   Illustrative Experiments

This section presents some experiments students can perform using the Furuta pendulum remote laboratory. Thanks to the EE, students can not only carry out these technically complicated experiments in a simple and intuitive way, but as a flexible environment, they are encouraged to investigate and test their own controllers. In this way, teachers can use the environment as a tool for applying inquiry-based learning.

5.1   Keeping the Pendulum Upwards While the Rotatory Arm Follows Changes of the Reference

In many control engineering remote laboratories, a typical approach is to implement the controller on the server side. That is, instead of having the controller implemented on the client side, sending periodically the control signal to the server, the code of the controller is sent once to the server, and it is executed there.

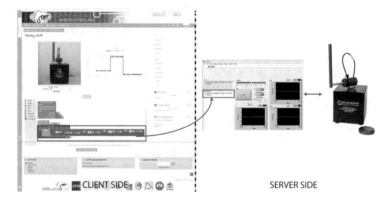

Fig. 4. Client and server side of the experimental setup when the position controller experiment is executed

The visual experimentation environment discriminates, with no need for user intervention, between the code to be executed locally and the code to be executed remotely. Figure 4 shows an experiment with local and remote code to control the pendulum and to visualize the outputs.

The plant has a default controller that puts the pendulum upwards and a student controller that can be replaced. Thus the first step is to disable the default controller and the second one to replace it with a linear feedback controller (showed in detail in Fig. 5a). The code that replaces the controller's function is the code that will be remotely executed (highlighted in red), while the rest is locally executed.

StudentCode

```
u = (((4 * (beta - betaRef) - 35 * alpha) + 1.5 * dbeta)
- 3 * dalpha);
```

(b) Server side

(a) Client side

Fig. 5. Detailed view of the experiment code and the controller received in the server

Figure 4 presents a screenshot of the server computer. It shows the LabVIEW program that controls the input and outputs of the Furuta pendulum system. The remote code that is developed in the environment and sent automatically to the server is highlighted in red. These two elements are also illustrated more clearly in Fig. 5: at the left, the controller designed with Blockly, at the right, the code that arrives at the server once the experiment has been executed. Once the controller code is received, the server runs that code in every execution step and sends the new control action value to the pendulum.

The experiment designed with Blockly can be seen in Fig. 5a. The "Create Chart" block, located at the top, is used to define the graph. It will draw the traces of the reference of the angle described by the rotary arm ("betaRef") and its real measure ("beta") in function of the time ("t"). Every 100 ms, a measurement will be made. After this definition, there is another set of blocks in the experiment: the "start data collection" block initiates the data acquisition for the visualization; the "set" block modifies a flag to indicate to the server that a user-defined controller will be used; and finally, the "replace function" block is used to replace the controller function. The content of this block will be sent to the server and used as a controller in the remote lab. To do this, the blocks will be translated into JavaScript code and sent to the server once the experiment is run.

At the top right part of the left image (client side) in Fig. 4, the graph obtained after the execution of the experiment can be seen. Changes in the angle reference, made during the execution of the automatic experiment using the lab's graphical user interface, are visible by looking at the steps in the graph. It was changed to 1.48 rad first, and then to −0,38 rad.

Thanks to this experiment, it can be seen how the parameters set for the controller are appropriate, since, in addition to making the pendulum stay in equilibrium, the system is able to follow the benchmarks marked with efficiency. The IP camera, along with the graphs defined in the experimental environment, allows users to corroborate this fact. Also, it is effortless to test other parameters and see that, either they are not as effective as those used in the experiment, or they make the system unstable, and the pendulum falls.

5.2   Creating the Base Position Controller with the Swing up Feature

Following the same process as in the previous experiment, it is possible to design a more complex controller. This controller is capable of not only to control the position of the base, but also to swing it up if the pendulum drops. This way, if the controller detects that the pendulum is far away from the region of attraction, it will run the code that will push it up. Otherwise, it will execute the code seen in the previous experiment.

For the implementation of this controller based on the design of Åström [1], instead of using traditional blocks, a JavaScript code evaluation block has been used. Being a controller that requires more code for its implementation, the use of blocks can make the creation process difficult, so the experimental environment offers the additional possibility of writing code to advanced users.

As the system raises the pendulum automatically, it is sufficient to, firstly, send a controller in which the control signal is always 0 ($u = 0$). This will drop the pendulum and allow the required controller to be checked.

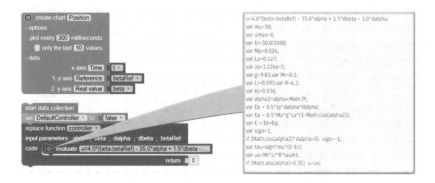

Fig. 6. Swing up and position control experiment

Figure 6 illustrates the experiment created. It has the same structure as the previous experiment, but the controller has been defined in JavaScript. For this, instead of using the blocks to set variables and equations, an "evaluate" block is used. In this block, users can write JavaScript code (see the bottom part of the defined experiment). To check its performance, it is sufficient to observe through the IP camera how the pendulum is lifted when the experiment is run, and the controller begins to operate. Once lifted, the behavior is the same as in the previous experiment.

6   Conclusions

Most of nowadays virtual and remote laboratories are closed applications which offer experimentation tasks purely based on interactivity, which limits the students' work with the labs. This paper presents an experimentation environment capable of overcoming these limitations. To this end, the environment adds new functionality to interactive laboratories, such as: (1) the possibility of designing automated experiments with a visual and easy-to-use language, solving the limitations of interactive experimentation; (2) the ability to add student-designed graphs to the visual interface of the environment, giving new possibilities for visualizing results; and (3) the ability to redefine functions and modify variables the designer might not have considered interesting at the time, opening laboratories to new experimental tasks that were not previously possible. To show the experimental environment usefulness, two different experiments are presented for a Furuta pendulum remote laboratory.

Acknowledgment. This work has been supported by the Spanish Ministry of Economy and Competitiveness under the project CICYT DPI2014-55932-C2-2-R.

References

1. Åström, K.J., Furuta, K.: Swinging up a pendulum by energy control. Automatica **36**(2), 287–295 (2000)
2. Brinson, J.R.: Learning outcome achievement in non-traditional (virtual and remote) versus traditional (hands-on) laboratories: a review of the empirical research. Comput. Educ. **87**, 218–237 (2015)
3. Chacon, J., Vargas, H., Farias, G., Sanchez, J., Dormido, S.: EJS, JIL Server, and LabVIEW: an architecture for rapid development of remote labs. IEEE Trans. Learn. Technol. **8**(4), 393–401 (2015)
4. Chaos, D., Chacón, J., Lopez-Orozco, J.A., Dormido, S.: Virtual and remote robotic laboratory using EJS, MATLAB and LabVIEW. Sensors **13**(2), 2595–2612 (2013). http://www.mdpi.com/1424-8220/13/2/2595
5. Frerich, S., Kruse, D., Petermann, M., Kilzer, A.: Virtual labs and remote labs: practical experience for everyone. In: Engineering Education 4.0, pp. 229–234. Springer (2016)
6. Guzman, J.L., Costa-Castello, R., Dormido, S., Berenguel, M.: An interactivity-based methodology to support control education: how to teach and learn using simple interactive tools [lecture notes]. IEEE Control Syst. **36**(1), 63–76 (2016)
7. Heradio, R., de la Torre, L., Dormido, S.: Virtual and remote labs in control education: a survey. Ann. Rev. Control **42**(Supplement C), 1–10 (2016)
8. Heradio, R., de la Torre, L., Galan, D., Cabrerizo, F.J., Herrera-Viedma, E., Dormido, S.: Virtual and remote labs in education: a bibliometric analysis. Comput. Educ. **98**, 14–38 (2016)
9. Potkonjak, V., Gardner, M., Callaghan, V., Mattila, P., Guetl, C., Petrović, V.M., Jovanović, K.: Virtual laboratories for education in science, technology, and engineering: a review. Comput. Educ. **95**, 309–327 (2016)
10. Sáenz, J., Chacón, J., De La Torre, L., Visioli, A., Dormido, S.: Open and low-cost virtual and remote labs on control engineering. IEEE Access **3**, 805–814 (2015)
11. de la Torre, L., Sanchez, J.P., Dormido, S.: What remote labs can do for you. Phys. Today **69**, 48–53 (2016)
12. Trower, J., Gray, J.: SIGCSE. In: Proceedings of the 46th ACM Technical Symposium on Computer Science Education, pp. 677–677. ACM (2015)

TRIANGLE Portal: An User-Friendly Web Interface for Remote Experimentation

Almudena Díaz-Zayas[✉], Alberto Salmerón Moreno, Gustavo García Pascual, and Pedro Merino Gómez

University of Málaga, Andalucía Tech, Málaga, Spain
almudiaz@lcc.uma.es
http://morse.uma.es

Abstract. As opposed to what other tools in the FIRE community, such as jFed [1], offer to the experimenter, Triangle Portal provides a simplified access to the definition of experiments and can coexist with the current FIRE tools. The way to integrate the Triangle Portal as part of existing experimentation testbeds is to offer a set of "canned" scenarios that will be specified by the Portal users during the phase of definition of their testing campaign. The testbed will then use the API REST provided by the Portal to access the information provided by the experimenter, such as the scenario, the application under test or a file containing the user actions to be replayed during the tests. The testbed will use this information to run the testing campaign and return the test results to the Portal.

Keywords: Testbed · Remote experimentation · Testing mobile apps

1 Introduction

FIRE (Future Internet Research and Experimentation) is an initiative from the European Commission integrated by a number of testbed facilities, including testbed for smart cities, drones, multimedia, wireless, as well as a unique testbed for underwater Internet of Things among others. These platforms provide an excellent Future Internet infrastructure focus and by definition are at the early stages of the lifecycle of a new Future Internet product of service. However, and despite during the last years the estimated mobile traffic growth has been used as motivation for many wireless testbeds, this increase has not reflected on the number of users of FIRE wireless testbeds. The main reason for this situation is the design of the FIRE testbeds themselves. Current testbeds are too focused on network configuration and have very complex and sophisticated configuration mechanisms, while the experimenters are not familiarised with the complex setup of the network resources and most of the time end up just using the default configuration. From our experience during the federation of

© Springer International Publishing AG, part of Springer Nature 2019
M. E. Auer and R. Langmann (Eds.): REV 2018, LNNS 47, pp. 110–117, 2019.
https://doi.org/10.1007/978-3-319-95678-7_12

PerformLTE [2] testbed, we can say that most part of the efforts were centred on providing access to all the low level parameters which have impact on the transport performance of the user traffic, however final users of the testbed do not know how to set up these parameters to generate a consistent experimentation scenario, which can be frustrating for them. Moreover, one gap within FIRE, identified by the FUSION project whose objective is to connect SMEs to FIRE testbeds, was the desire of start-ups and innovators to conduct user experience (UX) testing.

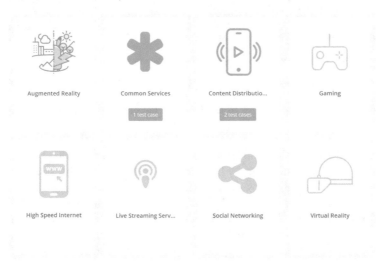

Fig. 1. Uses cases for apps

2   TRIANGLE Portal

The primary objective motivation of the TRIANGLE Testbed [3] is to promote the testing and benchmarking of mobile applications and devices in Europe as the industry moves towards 5G and to provide a pathway towards certification in order to support qualified mobile developments using FIRE testbeds as testing frameworks. As part of this project the University of Málaga has developed the TRIANGLE Portal. The main idea that underpins the development of the TRIANGLE Portal is to ensure that experimenters are not overwhelmed by the complexity of the overall testing testbed by being exposed to its full set of details. So, in order to fully understand the testbed details, the experimenter will need multi-disciplinary knowledge (protocols, radio propagation, software, etc.). To avoid this Triangle Portal provides a proper abstraction of underlying networking technologies by offering high-level canned end-to-end network scenarios. A single high-level scenario abstracts several concrete test configurations (e.g. mobility includes car at 50 Km/h and car at 100 Km/h), with the actual

Fig. 2. Test cases associated to the features belonging to different use cases

network configuration that will be applied. As summary, the Portal offers a common approach to measure the QoE of applications offering the same or similar features under the same network conditions.

The first release of the TRIANGLE Portal is focused on the testing of Android apps. In this release developers can upload their apps, a file containing an app user flow which will be replayed during the tests and a declaration of the use cases supported by the apps. An App User Flow is a sequence of user actions that can be performed on an App running on a mobile device. These actions include, but are not limited to, tapping on a UI element, entering text on a text field, swiping on a scrollable list.

Current uses cases identified are shown in Fig. 1. For each one of the uses cases it has been identified a set of features. By selecting which features are supported by the App the user will indirectly indicate test cases specifications will be applicable when they opt for standard experimentation. The standard experimentation is based on the execution of a predefined set of experiments which are focused on the testing the features declared in each one of the uses cases selected by the user. For example in Fig. 2, for the use case "Content Distribution Streaming Services" the developer have selected two features "Download content for offline playing" and "Non-Interactive Playblack". Each one of this feature has a test case associated. For the use case "Common Services" the developer has declared only the "Login and Logout" feature, which has also a test case associated. For each standard test case the user has to provide an App user flow.

After that, the Portal users can define their own experimentation campaigns to test the features of their app (see Fig. 3) in a particular configuration of the testbed. For these campaigns, users have to configure high-level options such as the scenario of the test, the device on which the test will be carried out,

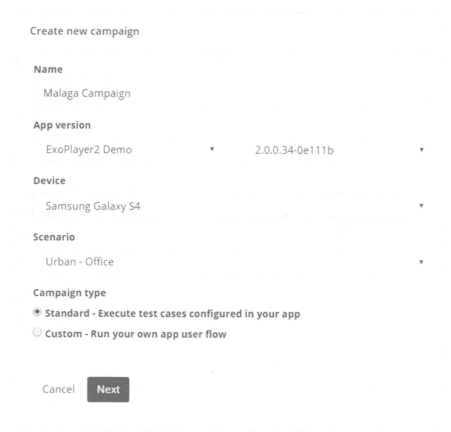

Fig. 3. Configuration of a campaign

the app and the version. The campaign can be based on the standard test case implemented to test the standard features declared previously or can be based on custom tests for which the user will have to provide an app user flow to reproduce the behaviour under test in this custom experiment.

The final target of the experiments is to compute the KPIs (Key Performance Indicators) which are applicable to the features under test. KPIs are the indicators that will tell App developers how well their Apps are doing, according to several pre-defined metrics. Examples of KPIs would be Time to login or Video streaming throughput. Not all KPIs may be applicable to every App, e.g. a social media app should not be tested against KPIs that measure video streaming performance. This is why App developers have to select which features are applicable to their Apps. A KPI is computed using the values obtained from measurement points during the execution of a test. The Portal also provides a code snippet of these measurement points that should be inserted by the developer inside his app to retrieve the measurement needed to calculate the KPIs

Fig. 4. Execution of a campaign

associated to the features declared. The concept of measurement point will be introduced later in this paper when the concept of instrumentation library is introduced.

2.1 How the Portal Information is Managed to Run the Experiments?

The app developer will select from a list of uses cases and its associated features (see Fig. 1), which ones apply to his or her app. Each feature will be tested according to a set of KPIs. It is this point at which they come into play the app user flows and the measurements points. An "app user flow" is a sequence of user actions that can be executed automatically to test a feature. When possible, the app user flows will be asked once per feature, so that users do not have to enter separately for each KPI, or in groups. For instance, if there is a post photo feature on which several KPIs can be measured (for example power consumption and successful login ratio), they will only be asked for a single app user flow.

In order to compute the KPIs, the app developer must define how some of the required measurements can be obtained in his or her app. Depending on the KPI, the user will be able to provide these measurement points providing the particular user action within the app user flow which will activate the feature and inserting the measurement point set into the source code of the app using the instrumentation library.

Internally one campaign defined by the user is composed by several test cases. A test case is defined using the information provide by the user during the creation of the campaign: the App version, the device, one concrete test configuration (scenario), and one of the features to be tested. A test case univocally defines all the possible variables that can be set during a test. Each time the same Campaign is executed, all its test cases will be executed, as shown in Fig. 4.

3   Instrumentation Library

The application instrumentation library is a library provided by the TRIANGLE project to app developers, in order to extract measurements from inside their applications. The measurements performed through the instrumentation Library will be stored along other measurements gathered during a test case execution. This Library provides the necessary measurement points for running the test specifications defined within the TRIANGLE project, and computing the corresponding KPIs and metrics. In addition, the Library allows app developers to log additional measurements outside of the ones defined within the project, and store them with the rest of the measurements. The measurements that app developers will be able to get from the Portal will include both custom and standard measurements. At the moment, this library is available only for Android applications. The same library can also be used in Unity applications for Android.

The instrumentation library will be used by app developers to provide measurement from within their own apps. These measurements are necessary to compute KPIs. The KPIs themselves belong to a particular app feature, e.g. login or post picture. The measurements also belong to these features, and can be used in one or more KPIs for that feature. Finally, the features are grouped by the use case to which they belong, e.g. live streaming services or social networking (see Fig. 1).

3.1   Standard Measurements

As part of the TRIANGLE project, a set of standard measurements have been defined. These measurements are used to compute the KPIs defined for the TRIANGLE test cases. The package/class hierarchy of the instrumentation library provides a clear path to the appropriate measurement, so that it is possible to find the appropriate method/function to call easily.

3.2   Custom Measurements

The library provides means to app developers to provide additional measurements, called custom measurements. These measurements will be parsed and stored alongside the rest of the measurements, but they will not be included as part of any standard KPI computation. To distinguish them from regular measurements, all these measurements are organized into a special use case called Custom. When logging a custom measurement, the Library user can define to which feature and measurement it belongs. This affects the classification of the measurements, when stored in the measurements database. In addition, the user can provide zero or one arguments for a custom measurement.

4   Backend

All the information is stored in the backend of the Portal. The Portal also includes a REST API that provides access to its backend. This API allows external testbeds to request information stored in the backend database, or update it. Outside of the Portal, the primary user of this REST API is the TRIANGLE testbed. TRIANGLE testbed uses the API to fetch the details of campaigns to be executed, and to upload the results once they are finished.

All API calls return a JSON object with data about the requested resource. The REST API largely adheres to the HATEOAS (Hypermedia as the Engine of Application State) principle. In practice this means that JSON responses include URLs that point to other related resources. For instance, when querying a single application, the JSON response includes a list of its versions with some information such as their Id and version code, as well as a URL to the resource of that application version, in order to request more details. The REST API provides access to the following resources:

- Devices: Provides information about devices available in the Testbed, on which the app will be tested.
- Users: Provides information about the user of the Testbed.
- Apps and their versions: Provides information about the apps and their versions. The user may have more than one app, and more than one version of that app. Therefore, he or she must select the one that will be tested.
- Features: Provides a subset of all the features/KPIs applicable to the app, if the developer is interested only in testing part of the app.
- Test cases: Provides a list of test cases that are applicable to the app.
- Scenarios: One of the high-level scenarios defined in the TRIANGLE project, which hides the complexity of the parameters that must be configured in the Testbed equipment.
- Campaigns: List all the campaigns defined by the user.

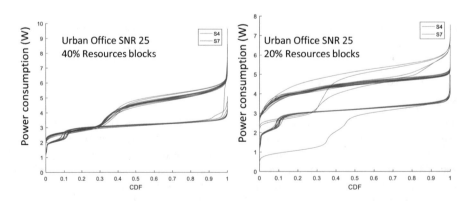

Fig. 5. Results obtained during the testing of YouTube app in different devices and network scenarios

5   Conclusions

The current implementation of the portal has been validated through the testing of YouTube. These tests have enable to obtain an initial baseline of the performance of YouTube in terms of power consumption, network performance and device resource usage in each one of the scenarios offered and to identify new features of interest for future releases of the Portal. Figure 5 shows the differences, in terms of power consumption, for two different devices (Samsung Galaxy S4 and Samsung Galaxy S7) and two different scenarios. The tests were repeated 20 times and after analyzing all the measurement collected we can conclude that the radio part of the device has a clear impact of the power consumption of the device and that scenarios with reduced radio resources available have a higher impact of the consumption for devices belonging to medium-range flagship smartphones.

References

1. jFed: a Java based framework to support testbed federation. http://jfed.iminds.be/
2. Díaz-Zayas, A., Garca-Prez, C. A., Recio-Prez, l.M., Merino, P.: PerformLTE: a testbed for LTE testing in the Future Internet. In: Proceedings of the 13th International Conference on Wired and Wireless Internet
3. Díaz-Zayas, A., et al.: Triangle: 5G applications and devices benchmarking. In: 2016 FIRE book: a Research and Experimentation based Approach, River Publishers 2016 Communications, vol. 9071, pp. 313–326. Springer, Heidelberg (2015)

"Hands-on-Remote" Laboratories

Frantisek Lustig[✉], Pavel Brom, Pavel Kuriscak, and Jiri Dvorak

Faculty of Mathematics and Physics, Charles University, Prague, Czech Republic
frantisek.lustig@mff.cuni.cz, brpav@seznam.cz,
pavel.kuriscak@gmail.com, jiridvorak@centrum.cz

Abstract. The contribution brings a new solution for remote laboratories. Benefits and effectiveness of hands-on, virtual, and remote labs have been discussed for decades. We have been developing professional and DIY real remote labs for 15 years. Despite our efforts, real remote experiments are unfortunately not so easily feasible as traditional hands-on ones. Experimenting at schools split into three isolated approaches: traditional hands-on labs (including PC-aided experiments), virtual experiments (also simulations, applets), and remote experiments. Recently, some effort to integrate these approaches has appeared (e.g. integrated e-learning strategy) although the integration with hands-on experiments is still missing. This state can be explained by high complexity of remote labs and the other limits (budget, availability of HW and SW solution, etc.). Our aim is to show that real remote labs can be easily performed as traditional hands-on labs, and they can be created with both professional equipment and cheap hardware components like Arduino. The goal is an introduction of a new experiment type "hands-on-remote" that is simultaneously hands-on and remotely controlled without need for further modifications. Students themselves may create such experiment or just observe the setup prepared by their teacher, and even operate it remotely by their mobile devices (BYOD). Students may access the experiment from the classroom, school building, and perhaps from their homes after school. Examples of remote labs based both on professional measurement system iSES and Arduino-Uno platform will be presented. Beginners need only an Arduino Uno board with sensors, our freely downloadable "Remduino Lab SDK".

Keywords: Hands-on experiment · Remote experiment · Arduino · iSES

1   Introduction

1.1   Hands-on Virtual and Remote Laboratories

Hands-on experiments are a lifelong part of natural science education. Due to proliferation of computer technologies, some experiments became computer aided (CAE) or computer-based (CBE). In the past, these were distinctly separated from the traditional hands-on counterparts. Nowadays we almost don't acknowledge the separation at all - almost all experiments are computer-assisted in some way without any specific focus on the fact. With the rise of the Internet came virtual laboratories. Relative ease of development in Java, Flash etc. enabled many developers to create new types of

© Springer International Publishing AG, part of Springer Nature 2019
M. E. Auer and R. Langmann (Eds.): REV 2018, LNNS 47, pp. 118–127, 2019.
https://doi.org/10.1007/978-3-319-95678-7_13

experiments. The experiments became interactive, supported with animations, simulations (see PhET [1]) or even modelling environments (Easy Java Simulations (see [2]). Increased internet bandwidth made remote laboratories possible. At first, they were mostly distributed as specialized client applications, but with advances in web browser technologies they mostly became browser-based.

New technologies and possibilities started experts' discussions and brought new research questions. This debate has stayed unresolved for decades and it's widely accepted conclusion is that all three approaches have no significant difference in effectiveness for the educational process outcomes (see e.g. Ma and Nickerson [3], Lang [4]). American Association of Physics Teachers (AAPT) presented five science lab objectives: 1. experiencing the art of experimentation by students, 2. development of their experimental and analytical skills, 3. support for conceptual learning (mastering of basic physics concepts), 4. understanding the basis of knowledge in physics (based on the inference of experiment and theoretical model outcomes), and 5. development of collaborative learning skills (see AAPT [5]). Accreditation Board for Engineering and Technology (ABET) compiled a list of goals and advantages of remote labs: e.g. operation 24/7/365, possibility of access to costly or dangerous experiments, an introduction of students to the real world of science and technology (with an opportunity to visit more than one laboratory), accessibility for students from developing countries and other people excluded from laboratory work for various reasons, (for others see [6]). The most frequent conclusions of existing works may result in a recommendation that we should use and combine more strategies considering the fact that various technologies address both common and different learning objectives. Therefore some authors have proposed to integrate all strategies mentioned above, e.g. the integrated e-learning strategy (INTeL, see [7]).

1.2 Hardware and Software Solution for Real Remote Labs

Remote experiments naturally started appearing with the advent of internet around 1991. First papers on this topic can be found e.g. in [8–10] and many others. First remote experiments of authors were created around 2001–2002 and they were based on Java runtime browser plugin [11]. Unfortunately, security issues started reducing their usability and ultimately rendered JRE obsolete as web technology around 2013. JavaScript took the dominant role of website interactivity and the experiment interfaces had to be completely reworked.

The number of remote labs worldwide was changing: In 2004 there were approximately 60 remote laboratories, in 2006 the number rose to 120 and in 2011 there were over 300 remote laboratories created – see [12]. In 2016 only 61 remote laboratories were mentioned in [13]. The GO-LAB project [14] offers only 63 remote experiments (25-November-2017). Our project E-laboratory iSES [15] provides 18 online remote labs. It uses our own software solution "iSES Remote Lab SDK" [16] as well as another notable project RemLabNet [17].

More projects in the overview use various software and hardware platforms (see WebLab Deusto from Universidad de Deusto in Spain [18], Remote Experimentation Laboratory RExLAB [19], UNILabs [20] and many others). Most of the remote experiments are based on LabVIEW hardware, e.g. [21, 22], however LabVIEW seems to be too complicated for beginners and non-programmers (see Sect. 3.4). Often, remote experiments are based on DIY development platforms with microcontrollers, usually Arduino and Raspberry [25–29]. Recently we have seen some remote experiments based on FPGA development platform [32].

What is the progress in the development and dissemination of remote labs? Why the number of real remote labs is so low? We can identify several possible reasons: Remote laboratories are too expensive, they require permanent service and it is hard to create new ones. After 2010 many of the existing remote experiments started to disappear. Universities started to actively block access with passwords and made their remote experiments accessible to their own students only. We feel that too many institutions concentrate on RLMS (Remote Laboratory Management Systems), clouds [23], Massive Open Online Courses (MOOCs) and Massive Open Online Labs (MOOLs) [24] etc., but no new remote experiments were created recently!

2   Goals, Motivations for a Change

We have been developing remote laboratories for more than 15 years. We have performed numerous demonstrations, workshops, conference presentations, and published many papers. We launched a free-access website (www.ises.info) that received considerable interest from both teachers and students. Despite the number of web-page hits our experiments have received, we cannot say that we were completely successful in bringing more remote experiments into Czech schools. Currently there are approx. 10 bigger remote laboratory projects in Czech Republic, that use our "iSES Remote Lab SDK", and around 10 small remote laboratories with one trial experiment. We would like to expand this network to more schools to make full use of RLMS, cloud, MOOC and other technologies. We have to admit, that number of remote experiments both in Czech Republic and worldwide is relatively small. Why are educators hesitant to create more remote experiments even though remote labs are 24/7 accessible and enable students to interact with expensive, dangerous or otherwise inaccessible equipment?

The reason might be explained by a relative difficulty of creation of such experiments. The creator has to be sufficiently qualified not only in physics education, but he also has to have sufficient expertise in measurement equipment, computers, internet, networking etc. This probably exceeds an average physics teacher's abilities.

We consider ourselves to be pioneers in the remote experiment field (our first remote experiment went online in 2002 and is still available - see "Water level control", (http:// kdt-34.karlov.mff.cuni.cz/en/mereni.html). Just like some other authors, we observed that it is very important to be physically present near the experiment at first, even though it's remote. If a student sees the experiment physically and subsequently tries to control it remotely from his smartphone or PC, it completely changes his perception. She/he can no longer "suspect" that the experiment might be a "fake simulation" or a video-on-demand. This new approach to remote experimenting is a main point of our contribution.

Another factor that probably hinders the remote experiment development is a compli-cated measurement technology, which often requires specialised hardware and software (usually not freely-accessible). In the beginning we based our experiments on our own ISES hardware and "iSES Remote Lab SDK" software. Partly because no other solution was available, partly because we wanted to differentiate our approach from other main branches like LabVIEW. Our aim was to use a universal client - web browser, which can run on almost any device today. The ISES hardware required a PCI card. Later we developed ISES-USB solution, which simplified the connection to a PC. We could have been satisfied with our professional experiments with many web page views and positive feedback, but we really wanted a greater proliferation of remote experiments to other educational institutions.

At first we developed the support for standard measuring equipment like COM/USB enabled multimeters and other measuring devices. Immediately after that we tried to build remote experiments based on a very popular and easily accessible Arduino plat-form. Experiments based on Arduino platform have already been published. The simplest solution has used only Arduino board with a LAN shield. Further experiments have used Arduino and Raspberry boards [25–29]. Our approach uses Arduino Uno board only as an A/D or D/A converter, connected to a PC via USB port. The control and data pre-processing is carried out in our new software kit "Remduino Lab SDK".

This software kit is the result of a gradual development. At first, our software kits were relatively complex and specialized. Only very qualified users were able to use them, so for all applications in schools and other institutions, we had to develop the appropriate software ourselves. Later we managed to gradually simplify the software so that it could be used as a development kit. Users can simply "glue" together an experi-ment interface from prepared components like Input, Output, Record, Export or Video Transfer. This software kit still supports the whole hardware ranging from ISES systems, COM/USB devices to Arduino boards.

3 "Hands-on-Remote" Experiment

3.1 "Hands-on-Remote" Experiment with ISES Systems

Hardware
Modular system iSES (internet School Experimental Studio) communicates with different measurement platforms. The newest system ISES-LAN and ISES-WIN includes a PC with Windows 10 operating system (not Arduino, Raspberry Pi etc.). Both feature 2× analog input, 1× analog output channel, 5× digital outputs/inputs with sampling frequency of 100 kHz. Software "iSES Remote Lab SDK" can handle up to 4 ISES-LAN or ISES-WIN units (see Fig. 1).

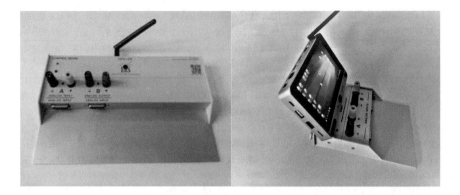

Fig. 1. New measure systems ISES-LAN and ISES-WIN

ISES has 20 modules and sensors for Physics, Chemistry and Biology for remote or local experiments: input modules/sensors (temperature, voltmeter, ammeter ...), output modules/sensors (relay, repro, booster, etc.). For more details and technical parameters about ISES modules see www.ises.info.

Software
Server-side part of a remote experiment consists of experimental hardware connected to a dedicated computer that runs the MeasureServer and optionally the ImageServer application. The MeasureServer provides two-way communication with the hardware, while ImageServer distributes the video stream captured by the webcam. Since the real-time data transfer is realized using the WebSocket technology, a web server has to be also installed and running. In the most of our experimental setups, the web server runs on the same dedicated computer and provides both user interface in a form of a webpage and relays the WebSocket connection to the MeasureServer. Clients connect to the experiment using regular web browsers.

To create a user interface of an experiment we have prepared the library of approx. 20 JavaScript components (widgets) - "iSES Remote Lab SDK". This library is freely distributable under Creative Commons license. Individual widgets are highly configurable and provide many thoroughly documented options, which allow even non-programmers to build a complex measurement and control interface with data and video transfer. Among built-in features, users have access to a real-time spline interpolation, simple processing, export of data in various formats, graphical output and other sophisticated functionality. Widgets use standard web elements, so any web developer can readily modify the default design and fit them to the appropriate page as needed.

"Hands-on-Remote" Easy Experiments with iSES
For non-experienced experiment designers we provide a collection of prepared simple experiments like remote analog record of one quantity (e.g. temperature), remote analog control of one channel (e.g. current booster), remote digital inputs and outputs, time dependence of two or more quantities, XY dependence of input and output quantities, data record, data export, webcam stream etc. The examples have the simplest possible

code and mostly use default settings for all components. These simple examples can be arbitrarily merged and combined, so even beginners are able to rapidly develop complex interfaces. They can immediately control their own remote experiments via mobile phones or tablets. This set of examples can be accessed online on http://www.ises.info/index.php/en/systemises/collection. Detailed description can be found in [30, 31].

3.2 "Hands-on-Remote" Experiment with Arduino Platform

Hardware

For simple "hands-on remote" experiment we propose to use Arduino-Uno board only as an A/D and D/A converter that is connected via USB port to a PC with the Windows OS (see Fig. 2).

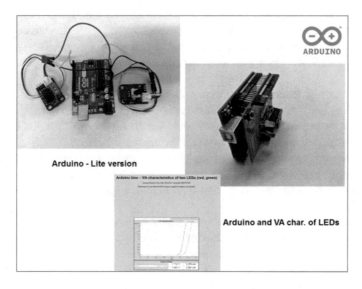

Fig. 2. Arduino and "Remduino Lab SDK".

Software

Lite version of "Remduino Lab SDK" supports the measurement on the input pin A0 (e.g. for temperature sensor) and the control of one digital output pin D3 (e.g. for relay). Full version of the kit supports all different arduino boards with all pins read/write access. Both versions provide data record and export function (e.g. to MS Excel), and the full set of functions of the "iSES Remote Lab SDK".

"Hands-on-Remote" Easy Experiments with Arduino

Examples from the freeware "Remduino Lab SDK - lite version" will be presented on example projects like the volt-ampere characteristics of LEDs, capacitor charging and discharging, mass-on-spring oscillations or a simple automatization project "heating and cooling".

3.3 "Hands-on-Remote" Experiments with Students in Classroom

"Hands-on-remote" experiments with iSES were presented at several Czech conferences in 2017. Participants were surprised how quickly one can build a remote experiment that is simultaneously traditional hands-on and remote. At the REV2018 conference we want to present this approach as well. Since September 2017 the platform has been tested at two secondary schools in the Czech Republic.

One of them is a school from the EDUCAnet network. Schools within this network have a common specialization and heavily utilize mutual network connection. EDUCAnet received a generous European grant to build a pilot remote laboratory with hands-on-remote experiments. These experiments would be shared with other schools in EDUCAnet network. The project will be realised in 2018.

Before the grant application, the school performed a pilot testing. We would like to present short video excerpts from these lessons. Since the school has its specialization in information technologies, most of the students had no problems with connection to the experiments. Students were fascinated that such simple hardware like Arduino (with "Remduino Lab SDK") may be controlled remotely with their mobile phones, tablets, etc.

Through this experience we understood how important the physical presence is for students. They need to observe a remotely controlled experiment personally at first. As creators of such experiments, we always could see the experiments locally, therefore let students to experience this as well for their further motivation.

As the teacher presents an experiment in the classroom, students can immediately operate it as "near-remote" during the lesson from the school building. For this purpose it is possible to limit the access to the experiment to a specific group of students or to the school intranet.

After school, the students can easily access the experiment as "far-remote", try it out again as a review of the lesson, or download and process their own experimental data to complete a "remote-experiment homework".

We are aware of the fact that students were most probably interested in the actual remote-control technology rather than in physics. But we believe that once they become engaged in remote experiments enough, the physics will get their attention as well.

Many students were not familiar with the Arduino platform. After finding out how cheap and easily accessible it is, almost everyone went to buy his own board. Subsequently, they were able to run their own simple examples of remote experiments that they downloaded as a part of the "Remduino Lab SDK" from www.ises.info. Since the software is open-source, the most advanced and active students started to modify it, finalize their own first hands-on-remote experiments and share them with their peers. This is what we regard as the greatest accomplishment - students became remote-experiment creators themselves!

During 2018 we can expect more interesting results after the completion of the European grant.

3.4   Comparison to Alternatives

Our "Hands-on-remote" laboratories might resemble e.g. a LabVIEW Remote Front Panel that enables to simply turn a local experiment into a remote one. LabVIEW Remote Front Panel, however, requires full installation of the LabVIEW Run-time Engine (large installation files, limited rights for installations in students' labs, lack of support by the majority of web browsers or mobile devices like smartphones) [33]. Moreover, LabVIEW is a commercial paid product, therefore it's less suitable to beginners and non-programmers.

Our new "hands-on-remote" experiments do not require any additional installations, remote experiment can be published and shared in approx. 10 min. Hands-on-remote experiments are accessible via smartphones and other mobile devices.

4   Conclusions

The proposed approach addresses both crucial advantages and disadvantages of real hands-on and remote laboratories like the sense of presence, belief in a "remote black box", development of design, experimental and scientific skills, etc. We believe that performing hands-on experiments that are simultaneously remotely controlled is an important aspect of modern experimenting at school.

Arduino platform and the "Remduino Lab SDK" cannot be used for precise or advanced measurements because of limited sampling frequency, ADC-resolution, and safety. But it was designed in order to be used by many beginners.

Since 2002 (before Arduino) we have been developing a professional platform for remote laboratories with the system iSES. On our webpage at http://www.ises.info/index.php/en/laboratory you can find a showcase of professional experiments based on "iSES Remote Lab SDK".

We are convinced that students' attitude to remote experiments has been changed positively after students had the opportunity to see (or perhaps to create) the real experiments at school during the lesson. Therefore we expect that our new approach proposed above might give rise to easy remote experimenting with all its advantages.

We believe that remote experiments with Arduino and the "Remduino Lab SDK" might become the dreamed Massive Open Online Labs (MOOLs), which are used by many participants and featuring a greater number of real remote experiments.

Acknowledgements. The work was supported by The Small and Medium Enterprise ISES, RNDr. Frantisek Lustig, Prague, Czech Republic.

References

1. PhET simulations. https://phet.colorado.edu/en/simulations/category/new. Accessed 03 Dec 2017
2. Easy Java Simulations. http://fem.um.es/Ejs/. Accessed 03 Dec 2017
3. Ma, J., Nickerson, J.V.: Hands-on, simulated, and remote laboratories: a comparative literature review. ACM Comput. Surv. **38**(3), 1–24 (2006)

4. Lang, J.: Comparative study of hands-on and remote physics labs for first year university level physics students. Teach. Learn. J. **6**(1), 1–25 (2012)
5. AAPT. http://www.aapt.org/Resources/policy/goaloflabs.cfm. Accessed 03 Dec 2017
6. Rosa, A.: The challenge of instructional laboratories in distance education. In: ABET Annual Meeting. ABET, Baltimore (2003)
7. Schauer, F., Ozvoldova, M., Lustig, F., Cernansky, P.: Integrated e-learning—new strategy of the cognition of real world in teaching physics. In: World Innovations in Engineering Education and Research iNEER, Innovations 2009, USA (2009). Special Volume
8. Aktan, B., Bohus, C., Crowl, L., Shor, M.H.: Distance learning applied to control engineering laboratories. IEEE Trans. Educ. **39**(3), 320–326 (1996)
9. Schumacher, D.: Student undergraduate laboratory and project work. Eur. J. Phys. **28**(5) (2007). (Editorial to the special issue)
10. Schauer, F., Lustig, F., Dvorak, J., Ozvoldova, M.: An easy-to build remote laboratory with data transfer using the internet school experimental system. Eur. J. Phys. **29**, 753–765 (2008)
11. ISES WEB Control—software pro vzdalene laboratore se soupravou ISES, 2002–2012, in Czech only, unpublished. http://www.ises.info/old-site/index.php?f=relizace_vzdexp. Accessed 03 Dec 2017
12. Gröber, S., Vetter, M., Eckert, B., Jodl, H.-J.: Experimenting from a Distance-Remotely Controlled Laboratory (RCL). Eur. J. Phys. **28**, 127–141 (2007)
13. Matarrita, C.A., Concari, S.B.: Remote laboratories used in physics teaching: a state of the art. In: Proceedings of the 13th International Conference on Remote Engineering and Virtual Instrumentation, REV 2016, Madrid, Spain, pp. 376–381 (2016)
14. GO-LAB. http://www.golabz.eu/. Accessed 03 Dec 2017
15. e-laboratory project iSES. http://www.ises.info/index.php/en. Accessed 03 Dec 2017
16. Dvorak, J., Kuriscak, P., Lustig, F.: iSES Remote Lab SDK—internet school experimental studio for remote laboratory software development kit, business and license agreement: SME RNDr. Frantisek Lustig. U Druhe Baterie 29, 162 00 Praha 6, 2013. http://www.ises.info/index.php/en/systemises/sdkisesstudio. Accessed 03 Dec 2017
17. REMLABNET. http://www.remlabnet.eu. Accessed 03 Dec 2017
18. WebLab Deusto. https://weblab.deusto.es/weblab/client/?locale=es#page=experiment&exp.category=Visirexperiments&exp.name = visir. Accessed 03 Dec 2017
19. Remote Experimentation Laboratory RExLAB. http://relle.ufsc.br/. Accessed 03 Dec 2017
20. UNILabs. http://unilabs.dia.uned.es/. Accessed 03 Dec 2017
21. Ko, C.C., et al.: Development of a web-based laboratory for control experiments on a coupled tank apparatus. IEEE Trans. Educ. **44**(1), 76–86 (2001)
22. Bauer, P., et al.: Survey of distance laboratories in power electronics. In: Power Electronics Specialists Conference 2008 (PESC 2008), pp. 430–436. IEEE (2008)
23. Schauer, F., Krbecek, M., Beno, P., Gerza, M., et al.: REMLABNET—open remote laboratory management system for e-experiments. In: Proceedings of the REV 2014, Porto, Portugal (2014). ISBN: 978-1-4799-2025-9
24. Salzmann, Ch., Piguet, Y., Gillet, D.: MOOLs for MOOCs, a first edX scalable implementation. In: Proceedings of the REV 2016. 13th International Conference on Remote Engineering and Virtual Instrumentation, pp. 240–245. Madrid, Spain (2016). ISBN: 978-1-4673-8245-8
25. Kaluz, M., Cirka, L., Valo, R., Fikar, M.: ArPi lab: a low-cost remote laboratory for control education. In: IFAC Proc. Vol. 47(3), 9057–9062 (2014). https://doi.org/10.3182/20140824-6-ZA-1003.00963, (http://www.sciencedirect.com/science/article/pii/S1474667016430430). ISSN: 1474-6670, ISBN: 9783902823625

26. Sobota, J., Pišl, R., Balda, P., Schlegel, M.: Raspberry Pi and Arduino boards in control education. In: 10th IFAC Symposium Advances in Control Education, pp. 7–12. University of Sheffield, Sheffield (2013). https://doi.org/10.3182/20130828-3-uk-2039.00024

27. García-Zubia, J., Angulo, I., Hernández, U., Orduña, P.: Plug & play remote lab for microcontrollers: WebLab-Deusto-pic. In: 7th European Workshop on Microelectronics Education, Budapest, Hungary, pp. 28–30 (2008)

28. Suwondo, N., Sulisworo, D.: Hands-on learning activity using an apparatus for transient phenomena in RC circuit based on Arduino UNO R3-LINX-Labview. iJOE **13**(01), 116–124 (2017). https://doi.org/10.3991/ijoe.v13i01.6317

29. Cvjetkovic, V.M., Stankovic, U.: Arduino based physics and engineering remote laboratory. iJOE **13**(01), 87–105 (2017). https://doi.org/10.3991/ijoe.v13i01.6375

30. Krbecek, M., Schauer, F., Lustig, F.: EASY REMOTE ISES—environment for remote experiments programming. In: Aung, W., et al. (eds.) Innovations 2013: World Innovations in Engineering Education and Research, pp. 80–101. iNEER, Potomac (2013)

31. Ozvoldova, M., Schauer, F.: Remote Laboratories in Research-Based Education of Real World Phenomena, 157 p. Peter Lang. Int. Acad. Publ. Frankfurt. F. Schauer, Villach (2015). ISBN: 978-3-631-66394-3

32. Toyoda, Y., Koike, N., Li, Y.: An FPGA-based remote laboratory: implementing semi-automatic experiments in the hybrid cloud. In: 13th International Conference on Remote Engineering and Virtual Instrumentation (REV), Madrid, pp. 24–29 (2016). https://doi.org/10.1109/rev2016.7444435

33. NI. Last Update 31 Mar 2017. Available: http://digital.ni.com/public.nsf/allkb/151BE12C055F57CE86257043006CB4B3. Accessed 01 Dec 2017

Virtual Power-Line Communications Laboratory for Technology Development and Research

Asier Llano Palacios[1(✉)], Xabier Osorio Barañano[1], David de la Vega Moreno[2],
Itziar Angulo Pita[2], and Txetxu Arzuaga Canals[1]

[1] ZIV Automation, Zamudio, Spain
{asier.llano,xabier.osorio,
txetxu.arzuaga}@zivautomation.com
[2] Communications Engineering, UPV/EHU, Bilbao, Spain
{david.delavega,itziar.angulo}@ehu.eus

Abstract. Power-Line Communication (PLC) consists in transmitting communications signals through the power line. The electricity network is a complex
communication medium with properties that depend on the topology of the grid
and on the usage pattern of the connected devices. Not only the variable distur-
bances of the communication channel, but also the market requirements for
continuous evolution, enforce extensive laboratory testing for technology
improvement. A typical laboratory setup for PLC technology testing involves
several analogue elements, and requires long and expensive testing tasks. This
article demonstrates how these procedures can greatly benefit from the virtualization concept to increase the testing speed and repeatability, and to reduce the
operation and maintenance costs.

Keywords: Laboratory · PLC · Virtualization

1 Introduction and Objective

Power Line Communication (PLC) technologies are used to communicate signals
through the power line. Low Voltage PLC is a solution extensively adopted by utilities
to create a communication network for the electricity meters, as it does not require a
dedicated communication medium deployment. The drawback of this solution is that
the electricity network is very variable when considered a communication medium, as
the transference function and noise sources depend on the consumption patterns of the
electricity users. The variability of the medium and the market requirement for constant
evolution motivate continuous technology improvement and validation.

Efficient PLC technology testing is a very complex issue that presents a big challenge
to PLC technology development and research. This complexity is usually based on the
following factors:

- Measuring Frame Error Rate (FER) values for different noise and channel conditions
  with the appropriate precision requires sending and receiving thousands of frames,
  which is a very time-consuming task.

© Springer International Publishing AG, part of Springer Nature 2019
M. E. Auer and R. Langmann (Eds.): REV 2018, LNNS 47, pp. 128–135, 2019.
https://doi.org/10.1007/978-3-319-95678-7_14

- Repeatability of results is a major concern, as laboratory setups are exposed to great degrees of variance, including aging, human error and environmental factors, such as temperature, humidity or background noise.
- During PLC testing sessions, minor setup changes usually require an action from a technician. Avoiding this manual operation may be desirable in order to reduce operation cost and human error probability, but the automation of this process requires additional hardware elements and maintenance, increasing the complexity of the solution.
- Digital Signal Processing (DSP) algorithm implementation in real products usually requires optimization effort in order to fit in the limited resources of the real production hardware (HW). In order to test proofs of concept, DSP implementation may require optimization efforts before checking if they have any benefit.

These problems may have a negative impact in development costs, testing quality, and design decisions, due to the limited knowledge of the PLC implementation performance. The solution described in this paper, which is called Virtual PLC Lab, is a system that replicates all the analogue elements of a PLC laboratory in digital technology. The validity of the results provided by this virtualized laboratory has been checked for certain PLC configurations by comparing the results with the ones provided by independent certification laboratories.

2   Physical PLC Laboratory

There are many testing setups, depending on the particular concept to test for PLC, but they share similar equipment in different configurations. Most of the test setups are intended to test single physical link communications in which all the variables of the link are under control: impedance, channel response and noise. The architecture of these setups is depicted in Fig. 1. Apart from the Device Under Test (DUT), this setup involves many analog devices: PLC filters to avoid external interferences; Line Impedance Stabilization Networks (LISNs) in order to separate communication signals from power supply while controlling the impedance; coupling devices or attenuators that transfer the signals between LISNs; noise generator to introduce controlled noise into the communication; and a reference transmitter of proven interoperability. Other single link setups are variations of this one, slightly modifying the source of the signal, introducing different coupling devices or proposing tools to analyze the signal quality. Examples of this architecture can be found in documents [1–4]. Other setups to test higher level protocols require many devices and sometimes repeat this kind of LISN structure with more coupling links.

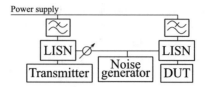

Fig. 1. Example of single physical link test setup architecture

Figure 2 depicts in more detail the typical laboratory setup for PLC technology testing involving a transmission and a reception device, interconnected through two LISNs and introducing additive noise by means of a waveform generator.

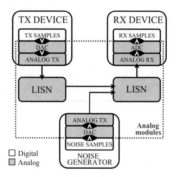

Fig. 2. Typical laboratory PLC laboratory test setup

The transmission device, the reception device and the noise generator usually implement DSP technologies, and are therefore digital devices which use mathematic algorithms applied to sampled digitized signals. The only elements of these devices that have an analog nature are the Digital to Analog Converters (DACs), Analog to Digital Converters (ADCs) and analog frontends.

Replacing the analog elements by a digital module appears to be a very natural solution considering that all the end modules are digital; that analog elements can be simulated with high precision using DSP techniques; and that for laboratory testing environments, testing elements as ideal and as deterministic as possible is highly desirable.

3   Virtual PLC Laboratory

The Virtual PLC Lab contains a digital medium that replaces all the analog elements of a testing environment.

3.1   Virtual PLC Modem

Figure 3 depicts the designed solution, which consists of a platform independent PLC Modem module running on a multiplatform Signal Processing library that deals with the platform-specific issues so that the modem modules can be oblivious of them. This library may be connected, in the case of a real device, to a physical medium via an analog frontend base on ACD/DAC; or in the case of a simulated device, to a digital medium provided by the Virtual PLC Lab by means of a communications socket inside a computer. This architecture guarantees that the physical and virtual devices behavior will be equivalent. Modems for different PLC standards such as PRIME 1.3.6, PRIME 1.4, G1 FSK and G3-PLC have been integrated with this architecture in order to be tested with the Virtual PLC Lab.

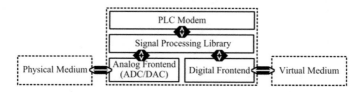

Fig. 3. Virtual PLC device

3.2 Virtual PLC Medium

This tool is able to run multiple Virtual PLC modems in the same computer and connect them through a virtual digital medium with configurable characteristics, such as attenuation, noise patterns and transfer function models. Each Virtual PLC Modem is a separated process inside the simulation environment that communicates with the Virtual PLC Medium through internal communication sockets. The Virtual PLC Medium is made of a flexible graph composed of simulation elements and signal nodes. The simulation elements can have any number of inputs and any number of outputs. They are intended to process the inputs and generate outputs in a flexible way. The nodes are the points to interconnect elements or external modules (like modems). If the nodes have more than one signal input they are additive. Many elements have been developed, like Finite Impulse Response (FIR) and Infinite Impulse Response (IIR) channel responses, white gaussian noise generator, tonal noise generator, recorded signal injector, signal recorder, simple attenuation, resampling,... The main idea after this design is to be able to have the equivalent tools of the analog laboratory but with the flexibility of the digital technology.

Figure 4 depicts the same architecture of Fig. 1 executed with the virtual laboratory with 2 nodes in charge of adding the echo, the noise and the response of the other device and 4 elements in charge of providing the channel response and noise.

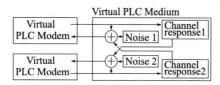

Fig. 4. Example of simple virtual PLC laboratory architecture

Different topologies may be configured according to the test to be executed. Physical link tests usually involve only 2 Virtual PLC Modems with different configurations with synthetic channels and noises, or recorded ones, sometimes introducing some resampling to simulate clock deviation or introducing non linear effects like quantization or saturation. These tests are simple in structure but require high precision as physical magnitudes are under test. Testing higher level protocols require more modems in order to create networks and the architecture of the virtual medium gets more complicated. In this situation simulation performance is critical to maintain the simulation faster than reality when the number of nodes increases.

4   Results

The Virtual PLC laboratory has been used for many tests and improvements of the PLC technologies in PRIME, G3-PLC and G1 FSK. This article will focus on the comparison of the results of the measurements of the Performance Certification tests of G3-PLC technology. During the certification process of some devices of G3-PLC technology it was possible to compare the results of the Virtual PLC lab and the Certification laboratory.

4.1   G3-PLC Performance Certification Test Results

The devices that are covered by the G3-PLC certification, all the mandatory prerequisites for submission, and the procedures to be applied by the Authorized Tests Laboratories for the G3-PLC certification are defined by the Certification Program Executive Committee (CPEC) of the G3-PLC Alliance [1].

Prior to submission for certification, some prerequisites about functional requirements must be met by the DUT. The G3-PLC certification covers interoperability, conformance and performance. For performance testing, the G3-PLC specific requirements that are met by the device must be reported in the Protocol Implementation Conformance Statement (PICS) document which has to be provided by the vendor to the Authorized Test Laboratory prior to certification. In this document, PICS values have to be provided to assert the level of performance the device under test should be able to provide. The value provided by the implementers for each PICS must respect the defined validity range.

Performance testing allows checking that the implementation fulfills the performance requirements with respect to physical robustness (against noise, attenuation, etc.). Each test returns a verdict that can be PASS or FAIL.

Table 1 represents a summary of the key performance PICS values and the real measurements performed in the certification laboratory. These values have been extracted from the certifications notes of real G3-PLC equipment certification process.

Table 1. Summary of G3-PLC Performance Certification PICS and measurements

Description	Encoding	Value	Measurement (FER @ Value)	Result
SNR with Additive White Gaussian Noise (AWGN) for Frame Error Rate (FER) <5%	ROBO	−2.0 dB	3.7% @ −2.31 dB	PASS
	DBPSK	1.8 dB	3.5% @ 1.5 dB	PASS
	DQPSK	5.5 dB	4.3% @ 5.21 dB	PASS
	D8PSK	10.8 dB	3.7% @ 10.51 dB	PASS
	ROBO_C	−3.8 dB	4.4% @ −4.02 dB	PASS
	BPSK	0 dB	3.3% @ −0.15 dB	PASS
	QPSK	3.6 dB	3.5% @ 3.3 dB	PASS
	8PSK	7.8 dB	4.8% @ 7.59 dB	PASS
Duration of impulsive noise that can support with Frame Error Rate (FER) <5%	ROBO	5600 µs	4.7% @ 5600 µs	PASS
	DBPSK	1900 µs	3.7% @ 1900 µs	PASS
	DQPSK	1050 µs	4.3% @ 1000 µs	PASS
	D8PSK	150 µs	3.7% @ 150 µs	PASS
	ROBO_C	7000 µs	3.1% @ 7000 µs	PASS
	BPSK	3000 µs	4.4% @ 3000 µs	PASS
	QPSK	1600 µs	4.8% @ 1600 µs	PASS
	8PSK	400 µs	4.3% @ 400 µs	PASS

4.2   Comparison of the G3-PLC Technology Results with Virtual PLC Lab

Figure 5 includes the graphs with the results of FER for different signal to noise ratios SNR and modulation scheme for Additive White Gaussian Noise (AWGN). The 8 lines represent the results by the Virtual PLC Lab, while the 8 points (represented with the same symbol) represent the measurements performed by an external independent laboratory during the G3-PLC certification. Both modems, the physical one and the virtual one, are executing the same version of decoding software.

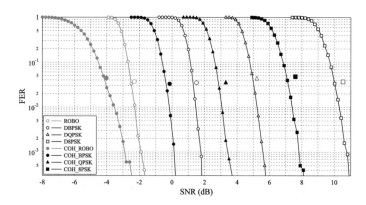

Fig. 5. Results of the virtual PLC lab simulation with AWGN

Figure 6 provides similar graphs comparing the results of the Virtual PLC Lab (lines) and the G3-PLC certification laboratory (points), for the impulsive noise.

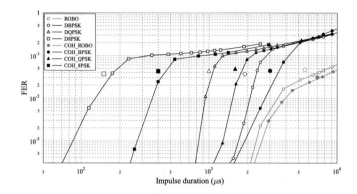

Fig. 6. Results of the virtual PLC lab simulation with Impulsive noise

Both performance results comply with the G3-PLC certification process documented in [1], but for the case of the Virtual PLC lab everything is measured with digital technology inside a computer. Comparing the graphs with results from physical and virtual laboratories, in both situations the real laboratory gives a bit worse results for a small margin: lower than 0.6 dB for AWGN and around 20% for impulsive noise (0.8 dB). These are particularly good results considering the dynamic range of both magnitudes and the high variability of the FER around the 5% threshold.

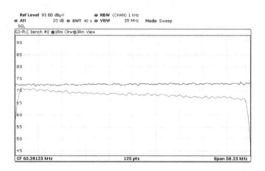

Fig. 7. Physical laboratory channel response

The small deviation between the real and virtual environment may be a consequence of non-ideal behavior of the analog frontend of the physical devices and non-ideal behavior of the physical laboratory equipment. As an example, Fig. 7 is part of the G3-PLC certification report: the AWGN test for ROBO_D. It shows a blue line that represents the noise level and a green line that represents the signal level. According to the G3-PLC certification process, this should be a test with a flat channel and white Gaussian noise, but the limitation of the analog tools makes it have a difference of 4 dB between the lower and higher frequencies of the PLC signal transfer function. This is a natural consequence of the tools being used and it is a good quality measurement procedure according to the certification industry standards.

5 Conclusions

Several PLC Modem technologies have been implemented and tested, such as G1, G3-PLC and PRIME, and the use of The Virtual PLC Lab digital approach has provided multiple advantages compared to the conventional laboratory analog approach, such as:

- Increased testing speed: Since current high end computers have higher performance than most PLC devices, testing speed can be hundreds of times faster than a test with real physical devices.
- Easy test replication: As digital algorithms are deterministic and repeatable, the test conditions can be replicated with precision. This is particularly useful for regression testing during the development process of the technology or for comparison of performance of different design options, which allows incremental performance improvement of PLC solution designs.
- Reduced development costs: Since the PLC modems have platform independent code and can be emulated in a computer; there are no platform-specific development costs. There is also no need for transitional hardware or prototypes that may become obsolete shortly afterwards.
- Fully automated tests: No human interaction is required with testing devices, in such a way that multiple PLC topologies under different network conditions can be tested in a fully automated way.

Although this approach has many advantages, it cannot completely replace a physical setup, which is needed in order to discard potential mismatches in the results due to integration issues or incorrect adaptation between digital and analog signals. Therefore, in a final stage, the results of the virtual laboratory must be verified with final device performance tests.

Nevertheless, the benefits of testing speed, easy replication, reduced development costs, automation and the possibility of reducing the number of expensive certification rounds makes the Virtual PLC Lab an attractive opportunity to improve the research and development of PLC technologies.

References

1. G3-PLC Alliance: Certification test procedures for G3-PLC certification v0.0.8 (2014)
2. IEC: CISPR 16-1-2:2014+AMD1: 2017 CSV Consolidated version. Specification for radio disturbance and immunity measuring apparatus and methods—Part 1–2: Radio disturbance and immunity measuring apparatus—coupling devices for conducted disturbance measurements (2014)
3. PRIME Alliance Technical Working Group: PRIME 1.3.6—Specification for PoweRline Intelligent Metering Evolution (2013)
4. ETSI: ETSI TS 103 909 V1.1.1. Power Line Telecommunications (PLT) Narrow band transceivers in the range 9 kHz to 500 kHz Power Line Performance Test Method Guide (2012)

Development of an Automatic Assessment in Remote Experimentation Over Remote Laboratory

Abderrahmane Adda Benattia[1(✉)], Abdelhalim Benachenhou[2(✉)], and Mohammed Moussa[2(✉)]

[1] University Ibn Khaldoun of Tiaret, Tiaret, Algeria
abderrahmane.addabenattia@univ-mosta.dz
[2] Université de Mostaganem, Mostaganem, Algeria
{abdelhalim.benachenhou,mohamed.moussa}@univ-mosta.dz

Abstract. In the last few years, performing experimentation over remote laboratory became a reality. In fact, for managing a large class, this new situation needs some appropriate pedagogical scheme. Else, the entire system must be reviewed. We have developed a method to set up an automatic assessment in remote experimentation over a flexible remote laboratory. Automatic assessment in remote experimentation can provide important benefits including: improved situation awareness, more accurate management for large class, and improved overall performance. We present a new approach to set up online experimentation under a flexible remote laboratory. In fact, we develop a particular pedagogical scheme for remote experimentation. On the basis of this analysis, we developed an automatic assessment system to evaluate remote experimentation for a large number of students. Hardware architecture consists of servers; some measurement instruments and electronic circuits, where the software architecture is based on web development tools. Furthermore, to embed pedagogical material for our system, we have used Moodle LMS platform. We illustrate an example of remote experimentation prototype which allows students to manipulate a dipole circuit workbench remotely. A set of students can connect to the remote lab at any time and perform manipulations and obtain personal and appropriate results from the shared equipment. The system includes auto test with customized interfaces for accessing to the same platform. Some results of manipulation are presented.

Keywords: Online experimentation · Automatic assessment · Remote laboratory
Moodle LMS

1 Introduction

Remote experimentation is a new progress for the improvement of education process where educational devices can be accessed remotely via web. The concept of accessing instrumentations remotely has been started since the invention of Internet technology. Actually, many higher education institutions offer a variety of web-based manipulation environments called remote laboratories (RLs), that allow remote manipulation of a physical experiments [1], most of them operate with commercial solutions such as LabView application [2, 3]. Moreover, their architectures follow specific and distinct technical

© Springer International Publishing AG, part of Springer Nature 2019
M. E. Auer and R. Langmann (Eds.): REV 2018, LNNS 47, pp. 136–143, 2019.
https://doi.org/10.1007/978-3-319-95678-7_15

implementations, i.e. there is no standard solution, allowing upgradeability and reconfiguration with different instruments and modules [4].

Our approach for developing the system, compared to others, focuses on the flexibility and the reusability of experiment devices; besides, to enlarge the number of users, we have implemented a customized user interface using web programming tools. Furthermore, we have implemented an automatic assessment to give more autonomy and active experimentation for developing competencies and skills.

2   Remote Lab Structure

This section contains a brief overview of our system; user can access to online experimentation laboratory as hardware structure which is managed by software architecture. Figure 1 illustrates the architecture of our proposed remote experiments laboratory setups. The LMS (Moodle) server allows the remote client to access according to an educational scenario in order to handle electronic instrument and measurement methods, perform various manipulations such as controlling a signal generator or a power supply, adjusting an oscilloscope display or any other operation made available by the interface.

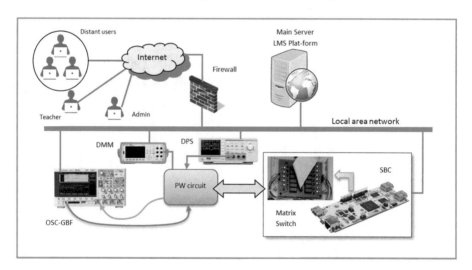

Fig. 1. The system architecture

2.1   Hardware Architecture

In this section, we outline the architecture of our system and describe the key mechanisms responsible for an efficient and reliable execution of online experimentation. Our system is designed to deliver online access to experimentation on Electronics in multiuser access environment. The hardware setup includes a set of measurement instruments, a practical work circuit, a manipulation server as a single board computer (Pcduino) and a switches matrix, these components are connected to a local area network; the whole system is accessible remotely via the main server (Fig. 2).

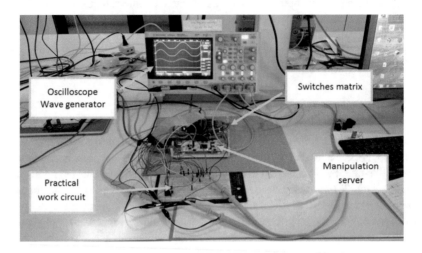

Fig. 2. Real hardware setup

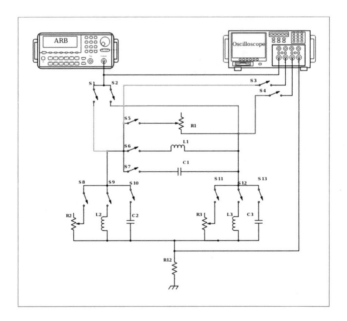

Fig. 3. Multi-variant practical work circuit

The practical work circuit is designed with multiple combinations of connection, which are established thanks to the flexible switches matrix, this one is managed by the manipulation server. A variant of experimentation consists of establishment of connection between components of the experimental circuit and appropriate measurement instrument as shown in Fig. 3.

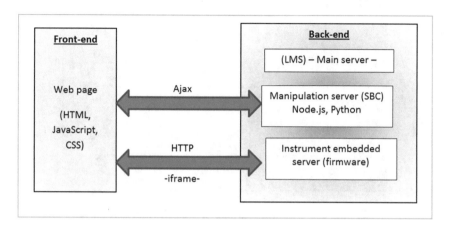

Fig. 4. Software architecture

2.2 Software Platform

The software platform is divided into two parts, back-end and front-end. Both parts communicate with other over Internet. Users access the laboratory from a web page which is the front end, it is developed using HTML5, CSS3, and JavaScript; on the other hand, back-end is implemented with Node.js [13], and Python [15]; it's composed of the main server, manipulation server and instrument embedded servers (Fig. 4).

The main server is the unique access point after authentication. The manipulation server is embedded under Linux OS on a single board computer pcduino, it manage the matrix switch to create a configuration of the experimental circuit by accessing pingpio's files [14], it communicate with the some components such as potentiometer over SPI bus, using a parametric program implemented with Python, this program submit user arguments for the circuit components such as resistor values, these arguments are sent from client to server using Ajax technique. Other measurement instruments such as oscilloscope are accessible using appropriate firmware through "iframe" on the same web page using URL.

3   A Case Study, Dipole Characterization

User access to the online experimentation through unique access point via Moodle LMS, it is provided as a test including embedded answers using cloze questions. The experimentation is about the characterization of a dipole circuit as shown in Fig. 5.

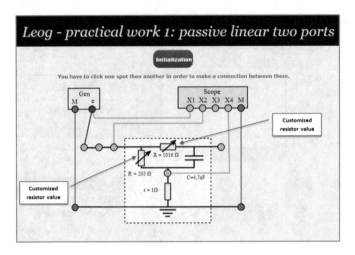

Fig. 5. User interface for a particular dipole characterization

To accomplish the experimentation according to a pedagogical scenario, the distant client wires a variant of dipole circuit to an oscilloscope and a wave generator, performs measurements and calculates (Z) impedance matrix elements, and submits results from a web page as shown on Fig. 7.

$$\begin{bmatrix} v_1(\omega) \\ v_2(\omega) \end{bmatrix} = \begin{bmatrix} z_{11}(\omega) & z_{12}(\omega) \\ z_{21}(\omega) & z_{22}(\omega) \end{bmatrix} * \begin{bmatrix} i_1(\omega) \\ i_2(\omega) \end{bmatrix} \Leftrightarrow \begin{cases} v_1 = z_{11}i_1 + z_{12}i_2 \\ v_2 = z_{21}i_1 + z_{22}i_2 \end{cases} \tag{1}$$

Theoretically, Eq. (1) is the Frequency-Domain impedance matrix representation; student has to determine all elements "z_{ij}" by modifying wires of dipole circuit (Fig. 5), observe and note measurements displayed on scope interface (Figs. 6 and 7). In this case, capacitor $C = 4.7\ \mu F$, resistor $R = 1\ k\Omega$, $R' = 200\ \Omega$, and frequency $f = 1\ kH$.

$$z_{11} = \left.\frac{v_1}{i_1}\right|_{i_2=0} \tag{2}$$

$$z_{22} = \left.\frac{v_2}{i_2}\right|_{i_1=0} \tag{3}$$

Student then calculates z_{ij} elements using measurements, for example, Eq. (2) determines the impedance z_{11} which is the ratio of the voltage at port1 to the current in the same port when port 2 is open, where according to Eq. (3) z_{22} impedance is the ratio of the voltage at port 2 to the current in the same port when port 1 is open ($i_1 = 0$). Student performs the same calculations for other z_{ij} values.

Given a circuit configuration, technically in background on loading the web page, the front-end software sends automatically the circuit configuration code to the flexible switches matrix via the manipulation server using "Ajax" technique, which is implemented in JavaScript [5]. Besides, submission of data input for the appropriate practical

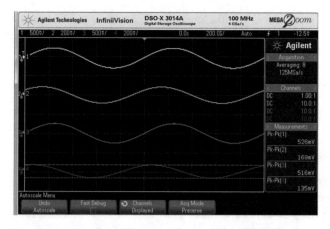

Fig. 6. Scope display to calculate impedances z_{12} and z_{22}.

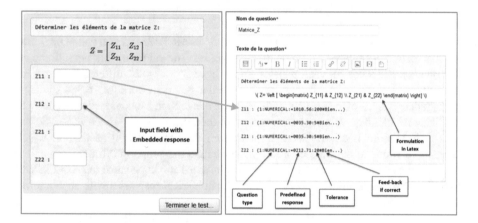

Fig. 7. Input field implementation for user response in cloze question

work circuit (resistor values) allows measurement instrument embedded server to make a customized measurements, send back and display the results. We use "iframes" to access scope web page using appropriate URL.

3.1 User Interface

The principle pedagogical challenge in remote experimentation is how to manage a large class of users when performing manipulation. Facing this issue, we envisage customizing the content of the user interface by implementing specific parameters for each user; the process involves duplicating questions then implementing automatic assessment according to each specific configuration. This solution contributes to make feasible the managing of a large class.

Figure 5 illustrate the web page content; it consists of a set of electronic components such as resistors and capacitor which compose the dipole, a wave generator and an oscilloscope. Each student gets the same interface with appropriate resistor values, i.e. on the scope, it will be displayed a specific plotted graph different from others.

Figure 6 illustrates the scope display according to a variant of resistor value, for the same values of capacitor and frequency.

Figure 6 shows a result of measurement of z_{12} and z_{22} impedances elements, according to resistance values $R = 1$ kΩ and $R' = 200$ Ω, capacitor $C = 4.7$ µF, and frequency value $f = 1$ kH.

3.2 Automatic Assessment

The automatic assessment technique is used to resolve the issue of large class evaluation in remote experimentation. To implement automatic assessment, both question and response are implemented in a customized web page. In Moodle LMS, we can use "cloze" question to embed answers in the script of the question. The user manipulates the experimentation according to a specific pedagogical scheme. Thus, each user performs his particular experimentation and gets his own results and will be assessed automatically by the system.

In the hardware setup, we use dual potentiometer MCP4231, 5 kΩ, coded on 7 bits. Each student manipulates a combination of two resistor values of this potentiometer; this one has 128 steps for each resistor, so we have $2^7 * 2^7 = 2^{14}$ different variants for the same manipulation, in this way we can give this practical work for 16,384 students, each student has his own assessment using cloze question in Moodle test; so we have to duplicate the cloze question for each variant of the test.

Using Moodle LMS, we have implemented the manipulation as test, in which students manipulate the dipole circuit and visualize the scope in "iframe", complete mathematical calculations and thus inputs results as a 4×4 impedance matrix (Z). With Moodle "cloze" question, it's possible to make numerical responses with such a tolerance; a score is given to each response.

In Fig. 7, we illustrate an example of "cloze" question in Moodle test, regarding "z_{11}" value, student has to input $z_{11} = 1010$, with a tolerance about $\pm 20\%$ to obtain the full score, otherwise, score will be null.

4 Conclusions

This paper describes a new approach to conduct remote experimentation in case of large class of users. A customized user interface is implemented to manipulate multiple variants of a single practical work board; the experimentation is about the characterization of a dipole circuit. An automatic assessment system is implemented so that each user will be evaluated according to a specific variant of the dipole circuit. The results obtained show a high degree of flexibility and interactivity in term of automatic assessment for multiple variants of the experimental circuit. In the future and according to the obtained results, we envisage to implement more circuits with different component in order to expand the variants number for a given configuration setup.

References

1. Gustavsson, I., et al.: On objectives of instructional laboratories, individual assessment, and use of collaborative remote laboratories. IEEE Trans. Learn. Technol. **2**(4), October–December 2009. https://doi.org/10.1109/TLT.2009.42

2. Lima, N., Viegas, C., Alves, G., García-Peñalvo, F.J.: VISIR's usage as an educational resource: a review of the empirical research. In: Proceedings of the Fourth International Conference on Technological Ecosystems for Enhancing Multiculturality (TEEM 2016) Salamanca, Spain, 2–4 November 2016, pp. 893–901. ACM, New York (2016). https://doi.org/10.1145/3012430.3012623

3. Limpraptono, F.Y., Faradisa, I.S.: Development of the remote instrumentation systems based on embedded web to support remote laboratory. In: Proceedings of Second International Conference on Electrical Systems, Technology and Information 2015 (ICESTI 2015). Lecture Notes in Electrical Engineering, vol. 365 (2015). https://doi.org/10.1007/978-981-287-988-2_60

4. Da Costa, R.J.G.: An IEEE1451.0-compliant FPGA-based reconfigurable weblab. Ph.D. Thesis Information Sciences and Technologies, University of Coimbra, Portugal, January 2014

5. https://www.w3schools.com/jquery/ajax_ajax.asp

6. García-Zubía, J., Hernández-Jayo, U.: LXI technologies for remote labs: an extension of the VISIR project. In: REV 2010 Proceedings, Stockholm, 29 June–2 July 2010

7. Tawfik, M., Sancristobal, E., Martin, S.: Virtual Instrument Systems in Reality (VISIR) for remote wiring and measurement of electronic circuits on breadboard. IEEE Trans. Learn. Technol. (2012). https://doi.org/10.1109/TLT.2012.20

8. Farah, S., Benachenhou, A., Neveux, G., Barataud, D.: Design of a flexible hardware interface for multiple remote electronic practical experiments of virtual laboratory. Int. J. Online Eng. (iJOE) **8** (2012). http://dx.doi.org/10.3991/ijoe.v8iS2.2004

9. Romero, S., Guenaga, M., García-Zubía, J., Orduña, P.: Automatic assessment of progress using remote laboratories. Int. J. Online Eng. (iJOE) **11**(2) (2015). http://dx.doi.org/10.3991/ijoe.v11i2.4379

10. Farag, W.: An innovative remote-lab framework for educational experimentation. Int. J. Online Eng. (iJOE) **13**(2) (2017). https://doi.org/10.3991/ijoe.v13i02.6609

11. Duran, L.B., Duran, E.: The 5E instructional model: a learning cycle approach for inquiry-based science teaching. Sci. Educ. Rev. **3**(2) (2004)

12. Achour, M., Betz, F.: Manuel PHP. In: 1997–2017 PHP Documentation Group. https://secure.php.net/manual/fr/index.php

13. https://nodejs.org/

14. Programming the pcDuio. https://learn.sparkfun.com/tutorials/programming-the-pcduino/

15. https://www.python.org/

Demonstration: Using Remotely Controlled One-Way Flow Control Valve for Speed Regulation of Pneumatic Cylinder

Brajan Bajči[✉], Slobodan Dudić, Jovan Šulc, Vule Reljić, Dragan Šešlija, and Ivana Milenković

Faculty of Technical Sciences, University of Novi Sad, Novi Sad, Serbia
brajanbajci@uns.ac.rs

Abstract. This paper presents a way of speed regulation of double acting pneumatic cylinder, using remotely controlled one-way flow control valve. For this purpose, at the Faculty of Technical Sciences in Novi Sad, an experimental setup is developed and implemented. The control system is realized using the following hardware components: Arduino Yun for system control, step motor coupled with one-way flow control valve for speed regulation, directional control valve and double acting cylinder. CEyeClon Viewer is used as a software component for remote access to the system. In addition, the user application is developed. The application allows setting the input parameters and starting the experiment. After experiment execution, the measured results, which represent the times needed for the cylinder extraction as well as average speeds, are shown in the form of diagram. Using Web camera, user can monitor the whole system during experiment execution.

Keywords: Remote control · Speed regulation · One-way flow control valve

1 Introduction

Remote control is a necessary part of modern production and will be integrated into the forthcoming changes in the implementation of Industry 4.0. In smart factories, people, products, machines, and data, have permanent communication. The real world of production is enriched with virtual information and networking. With these technologies, companies will be able to make better informed decisions and achieve better operations, and quickly meet customer demands.

Changes in production technology and automation lead to the development of cyber technologies, which ultimately involve computers, mobile phones and other devices, integrated into the smart network using the Internet. One of the possibilities to support the user interaction with industrial operations is to use distance/remote control of the equipment [1]. On the other side, distance learning provides lifelong learning for the staff, which will be mandatory for smart factories [2].

Remote experiments are widely used in the fields of mechatronics and robotics [3–6]. In the domain of pneumatics in industry, there are many situations where

© Springer International Publishing AG, part of Springer Nature 2019
M. E. Auer and R. Langmann (Eds.): REV 2018, LNNS 47, pp. 144–152, 2019.
https://doi.org/10.1007/978-3-319-95678-7_16

actions can be performed remotely [7, 8], such as shutting down the compressed air, regulating the speed of the drives, closing or opening the valves, regulating the pressure in the system, etc. In some cases, the remote control is mandatory when actuators are hard accessible or unsafe to reach.

This paper describes the experimental setup that is used for remote speed regulation of double acting pneumatic cylinder using one-way flow control valve.

2   Basics of Speed Regulation of Pneumatic Cylinder

A one-way flow control valve, shown in Fig. 1, regulates the speed of the pneumatic cylinder. With this valve, it is possible to regulate the air flow during the extraction and/ or retraction of the cylinder. In this way, the piston speed is actually regulated.

(a)   (b)   (c)

Fig. 1. One-way flow control valve: (a) symbol, (b) component (c) installation on the cylinder

The one-way flow control valve is mounted on the cylinder exhaust port (Fig. 1c) reducing compressed air flow through this port. "Meter out" principle is applied for the regulation of the cylinder speed [9]. Compressed air freely enters the cylinder through supply port. The piston is held between two air cushions, which are generated as a result

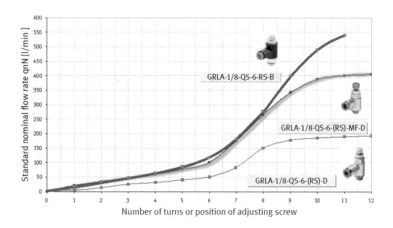

Fig. 2. Influence of the number of rotations of adjusting screw of one-way flow control valve on the air flow [10]

of the pressure and air resistance. This leads to the higher pressure in the piston rod chamber of the pneumatic cylinder and lower extracting speed.

Dependence of the air flow through the one-way flow control valve from the number of rotations of adjusting screw is shown in Fig. 2. As examples, charts are given for three different groups of one-way flow control valves. As can be noticed, this function is not linear.

3   Development of the Experimental Setup

In order to test the remote speed regulation of the pneumatic cylinder, the experimental setup, shown in Fig. 3, was developed.

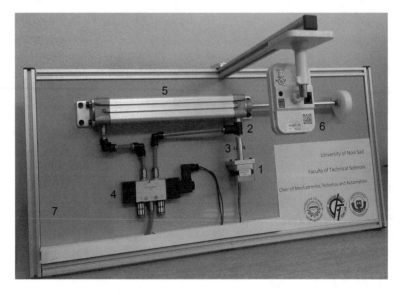

Fig. 3.   Experimental setup: 1. step motor, 2. one-way flow control valve, 3. mechanical coupling, 4. directional control valve, 5. double acting cylinder, 6. web camera and 7. frame

Electro-pneumatic control scheme of the system is given in Fig. 4. Directional control valve, type VUVS-L20-M52-MD-G18-F7-1C1 (Fig. 4, position 1V1) is used to control compressed air flow from the compressor (Fig. 4, position P) and the service unit (Fig. 4, position 0Z) to the cylinder chambers. As is said, the one-way flow control valve, type GRLA-1/8-QS-6-RS-B (Fig. 4, position 1V2) is mounted on the pneumatic cylinder, type DNC-32-200 (Fig. 4, position 1A) exhaust port. Also, the reed sensors, type SME-8-K-LED-24 (Fig. 4, positions 1S1 and 1S2) are installed on the cylinder and are used to detect the end positions of the cylinder. One-way flow control valve is mechanically coupled to the step motor (Fig. 4, position M). The motor is used to open and close the flow control valve, i.e. to control the number of rotations of adjusting screw of the valve. With opening and closing the one-way flow control valve, the speed of the cylinder is regulated.

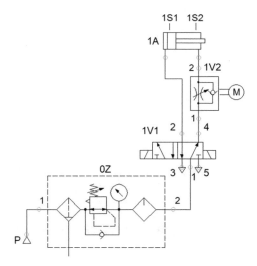

Fig. 4. Electro-pneumatic control scheme

With the described electro-pneumatic control scheme, it is possible to realize the speed regulation of double acting pneumatic cylinder, in one direction, using remotely controlled one-way-flow control valve. In this experimental setup, the regulation of the cylinder speed is performed only for the extraction of the cylinder. In the opposite direction, there is no speed regulation.

3.1   Calculating the Number of Rotations of the Motor

In order to regulate the cylinder speed, firstly, it is necessary to calculate dependence of the percentage of openness of the one-way flow control valve and number of rotations of the motor. The type of the one-way flow control valve used in this research is GRLA-1/8-QS-6-RS-B (Fig. 2). The dependence of the air flow and number of rotations of the adjusting screw for the one-way flow control valve is also shown in Fig. 2 (red line). Based on this, a new graph is created (Fig. 5), and the number of screw rotations is given depending on the required openness of the one-way flow control valve. This number is actually the same as the number of step motor rotations.

A new obtained function represents the fourth degree polynomial function (1). In this way, it is determined how many rotations the step motor should execute to provide the correct percentage of openness of the one-way flow control valve.

$$y = -0.00000025x^4 + 0.00008001x^3 - 0.00856789x^2$$
$$- 0.42072844x - 0.07762238.$$

(1)

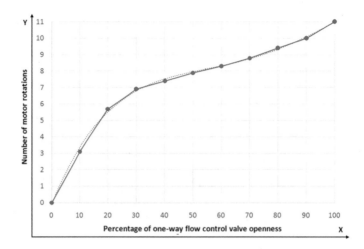

Fig. 5. Number of step motor rotations as a function of percentage of openness of one-way flow control valve

4   Remote Control of the System

The ways of communication during remote control of the system are shown in Fig. 6.

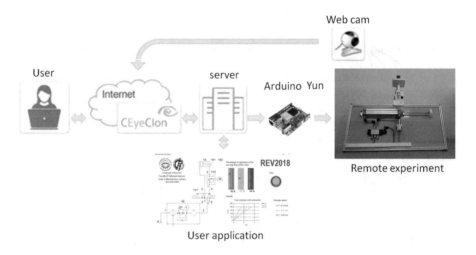

Fig. 6. Communication within the experimental setup

A remote access to the experiment is performed using CEyeClon Viewer. Users can access to the device remotely over Internet, using CEyeClon platform [11]. This platform shows some advantages compared to others that are used for remote control. Firstly, the security check is at a high level, unlike, for example, Team Viewer. In addition, there is no need to disable Windows Firewall or Antivirus, as in the case with standard

LabView applications. Finally, adding a new experiment on CEyeClon is very simply and easy. However, the platform shows some disadvantages compared to others. For example, creating a learning lectures and tests is easier with Moodle.

To connect with remote experiment, firstly, the user has to install CEyeClon Viewer on his computer. After receiving the username and the password from the administrator, logging in to the system is possible. At the home page, a selection of the experiment can be done. In the next step, the user connects to the remote desktop located in the laboratory and access to the experiment, which is controlled via microcontroller Arduino Yun. Using Web camera, user follows the experiment execution in the real time.

The flowchart of the experimental execution is shown in Fig. 7. The first two steps, START and LOGIN, are related to the CEyeClon Viewer software. Within the VALID DATA step, the username and password are verifying. If the entered data are correct, the next step SETTING EXPERIMENT PARAMETERS includes setting values in percentage of the one-way flow control valve openness.

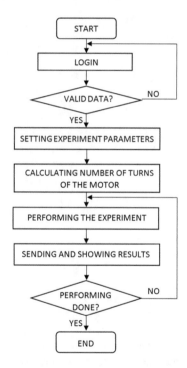

Fig. 7. Flowchart of the experimental execution

The software automatically calculates the number of rotation of the step motor. Then the experiment can be performed. During the experiment execution, the pneumatic cylinder extracts and the time needed to reach the end position is measured. The measured value is sent back to the user computer. The process is repeated three times, for the three different percentages of openness of the one-way flow control valve, set by the user. At the end of the experiment, the results (time for piston rod extraction) are displayed in the user application in the form of a diagram.

4.1   User Interface

In order to start the experiment and perform measurements, it is necessary to set the desired parameters using a developed user application. The mentioned application is presented in Fig. 8. The application is divided into four parts:

– electro-pneumatic control scheme,
– field for setting the percentage of openness of one-way flow control valve,
– start button and
– the diagram of the results - time of piston rod's extraction.

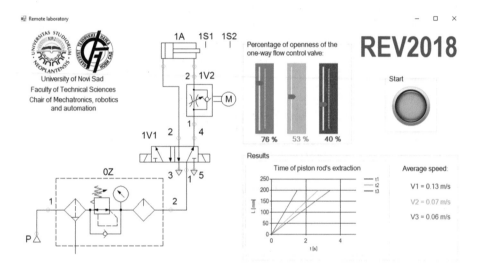

Fig. 8. User application

Three vertical sliders: blue, green and red, in the upper right corner, represents the percentage of openness of the one-way flow control valve. This parameter can be easily set by moving the slider up or down. By pressing the large green button, the values from the computer are sent to the controller, and the controller starts the experiment execution. The rotations of the step motor are transferred to a remotely controlled one-way flow control valve. When the step motor/one-way flow control valve is set in the appropriated position, the pneumatic cylinder extracts. The cylinder extraction time is measured and the data is sent to the computer. In the application, it is possible to regulate the speed for three extractions of cylinder.

By pressing the green button, user can start experiment and to clearly see impact of the speed regulation.

At the end of the experiment, the measured results are shown in the form of diagram at the bottom right corner of the user application as well as average speed. The average speed is calculated automatically for each cycle separately. The horizontal axis represents the measured time needed to reach the end positions of the cylinder. The variable on the vertical axis is fixed and represents the stroke length of the cylinder, which is 200 mm. The colours of the diagrams correspond to the colours of the sliders as well as colours of average speeds.

When all three cycles are completed, the step motor returns to the starting position and one-way flow control valve is fully closed.

5   Conclusion

For the realization of high-quality educated engineers, it is essential to continuously make efforts in the teaching modernization. For example, it can be done by developing remote laboratories, which raises the educational processes on a higher level.

Here is shown one way to obtain and merge knowledge from the field of remote engineering by developing a setup for speed regulation of double acting pneumatic cylinder using remotely controlled one-way flow control valve. It can be implemented in every system where it is necessary to remotely regulate the speed of double acting pneumatic cylinder.

References

1. Zurcher, T., Rojko, A., Hercog, D.: Education in industrial automation control by using remote workplaces. In: Proceedings of IEEE 3rd Experiment@ International Conference Online Experimentation, Exp. at 2015, Ponta Delgada, Portugal, 02–04 June 2015
2. Restivo, M.T., Cardoso, A.: Exploring online experimentation. Int. J. Online Eng. (iJOE) **9**, 4–6 (2013). ISSN: 1861–2121
3. Zurcher, T.: Distance education in energy efficient drive technologies by using remote workplace. In: Proceedings of IEEE 11th International Conference on Remote Engineering and Virtual Instrumentation, REV 2014, Porto, Portugal, 26–28 February, pp. 139–143 (2014)
4. Gadzhanov, S.D., Nafalski, A., Nedić, Z.: Remote laboratory for advanced motion control experiments. Int. J. Online Eng. (iJOE) **10**, 43–51 (2014). ISSN: 1861–2121
5. Bjekić, M., Božić, M., Rosić, M., Antić, S.: Remote experiment: serial and parallel RLC circuit. Proceedings of 3rd International Conference on Electrical, Electronic and Computing Engineering, IcETRAN 2016, Zlatibor, Serbia, vol. AUI1.2.1-6, 13–16 June 2016
6. Cauhé, E., et al.: RRLab: remote reality laboratory to teach mechanics in schools. In: IT Innovative Practices in Secondary Schools: Remote Experiments, Chap. 6, Sect. 2, pp. 119–140. Universidad de Deusto, Bilbao (2013)
7. Reljić, V., et al.: Remote control of pneumatic circular manipulator using CEyeClon platform. In Proceedings of IEEE 4th Experiment@ International Conference Online Experimentation, Exp. at 2017, Faro, Portugal, 26–28 February, pp. 139–143 (2014)
8. Šešlija, D., Šulc, J., Dudić, S., Milenković, I.: Remote control of pneumatic actuator via wireless communication. In: Proceedings of 16th International scientific conference on industrial systems, IS 2014, Novi Sad, Serbia, 15–17 October, pp. 107–111 (2014)

 9. Šulc, J., Šešlija, D., Srndaljčević, V.: Influence of pneumatic cylinder damping on consumption of compressed air. In: Proceedings of 3rd Regional Conference Mechatronics in Practice and Education, Subotica, Serbia, MECHEDU 2015, 14–16 May, pp. 175–178 (2015)
10. Flow-rate-optimized one-way flow control valves GRLA – with integrated QS push-in fitting, Festo (2017). https://www.festo.com/net/SupportPortal/Files/8180/PSI_224_3_en.pdf
11. CEyeClon Education (2017). http://ceyeclon.com/en/presentation/education

Demonstration: Virtual Lab for Analog Electronic Circuits

K. C. Narasimhamurthy[(✉)], Ankit Sharma, Shorya Shubham,
and H. R. Chandan

Siddaganga Institute of Technology, Tumakuru, India
kcnmurthy@gmail.com, ankit34567@gmail.com,
shorya007shubham@gmail.com, chandanhr71997@gmail.com

Abstract. The Objective of this Virtual Lab is to give an access to student community to learn the concepts of Analog Electronics circuits by *Experiential Learning*. Virtual lab, allows user to analyze basic analog electronic circuits by controlling the experiments remotely and measuring necessary nodal voltages virtually. In phase-I of virtual lab, experiments on rectifiers and regulators has been considered. User while performing experiments through virtual lab will have the similar experience of conducting it live. The Virtual Lab will provide complete information about the circuit under test, design specifications, waveforms at various test points and tabular column with expected values. User will be able to make the necessary measurements at the test points using National Instruments (NI) Analog Discovery-2 NI Edition and other interfacing circuitry. Video tutorials will guide the user to perform experiments without any difficulties. Virtual lab is designed by keeping students community in mind and made options to vary many circuit parameters. This virtual lab gives a platform for the students to solidify their concepts learned in the classroom by repeatedly doing experiments and with variations in the circuit parameters. *Anyone* from *Anywhere at Anytime* can access the circuits to perform experiments on Half wave, Full wave and Bridge rectifiers along with option to perform voltage regulation.

Keywords: Virtual lab · Analog electronic circuits · Analog Discovery-2

1 Introduction

Engineering education is adopting innovative techniques to make teaching learning more effective. Techniques include use of ICT tools in classroom teaching, Flipped classroom, portable lab concepts etc. Students of institutes offering such innovative techniques will get the benefit. However, other students are deprived of such innovative learning initiatives due to various constraints. One of the possible ways to provide such students an opportunity to access state-of the–art facility is through Virtual Lab.

To enable an experiential learning of Analog Electronic Circuit courses, it is necessary that students must visualize the working of the circuits by conducting the experiments in the classroom itself. This can be made possible using virtual lab, by demonstrating the circuit behavior for various inputs and measuring circuit response at

© Springer International Publishing AG, part of Springer Nature 2019
M. E. Auer and R. Langmann (Eds.): REV 2018, LNNS 47, pp. 153–158, 2019.
https://doi.org/10.1007/978-3-319-95678-7_17

desired nodes. Laboratory courses of Analog Electronic Circuits is considered to be difficult compared digital circuits counterparts by student community, as earlier involves many components, devices and one needs to know their behavior at different voltages, frequencies of operation. In phase-I of virtual lab, experiments involving rectifiers and regulators are presented. There are few online remote and virtual labs available [1–3]. In vlabs [1] user can simulate analog experiments; however the user will not feel the real hardware challenges and real time signals. In VISIR [2], user can build basic analog circuits and conduct the experiments, however there no guidelines for effective usage of various instruments in the online remote lab. In *v*Labs [3] allows students, faculty and staff to access the university's software anytime from anywhere. There is no provision for outsiders to access software and there are no hardware circuits for remote access. Even in corporate world [4] the concept of Virtual lab is being used to provide access to the latest equipment remotely.

2   Implementation of Virtual Lab

Innovative and the challenging virtual lab are implemented to give access to conduct experiments of analog electronic circuits. In this section, implementation process of the virtual lab for basic analog experiments like rectifiers and regulators is discussed.

2.1   Virtual Lab Access

Virtual lab Phase-I, consists of analog circuits: half wave, full wave and bridge rectifier circuits followed by C-filter and regulator. While conducting experiments on Virtual lab, user will have similar experience as though experiments are being conducted in live. Analog Discovery-2 NI Edition and Arduino Mega microcontroller board along with few open source software tools are used in developing Virtual Lab. Analog Discovery-2 NI Edition a Digilent product is exclusively developed for National Instruments. Analog Discovery 2 is a USB oscilloscope and multi-function instrument that allows users to measure, visualize, generate, record, and control mixed-signal circuits of all kinds. Arduino Mega a microcontroller board based on the ATmega1280 is used for selecting the desired circuit for experimentation. User can access the Virtual lab through Laptop or Smart phone. User need to install Team Viewer which allows remote connection to our Virtual lab system without having to install any software. Figure 1 shows working model of Virtual lab.

As shown in Fig. 1, any one having smart phone/personal computer can easily access the virtual lab. Once the access permission is granted, user gets the complete control of the circuit and this gives freedom to conduct the experiment at the desired pace and enables to take relevant measurements to analyze the circuit.

2.2   Conduction of Experiments Using Virtual Lab

User gets the access to conduct the experiments after formal registration process. Upon getting the access to virtual lab, the complete control of all 30+ possible variations in rectifier and regulator based experiments ranging from simple rectification to complex

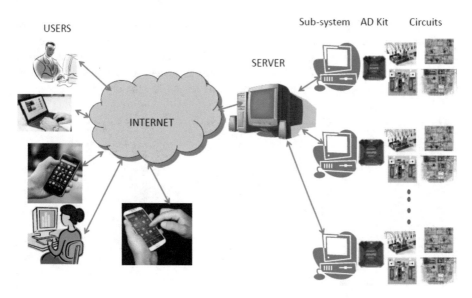

Fig. 1. Virtual Lab Phase-I conceptual block diagram.

load regulation analysis can be performed. Analog Discovery-2 along with WaveForms 2015 has made this possible. WaveForms 2015 is the virtual instrument suite for Analog Discovery 2 - NI Edition devices. WaveForms 2015 has following features: Oscilloscope, Waveform Generator, Static I/O, Network Analyzer, Spectrum Analyzer etc. Waveforms 2015 is versatile enough to make the relevant measurements to analyze the electronic circuits. Table 1 indicated the hierarchy of circuits available for half wave rectifier and regulators.

Table 1. Hierarchy of Half wave rectifier and regulator based experiments accessed in Virtual lab phase-I.

Rectifier	Rectification	C-Filter	Regulator	Line regulation	Load regulation
Half wave rectifier	R load	C = 100 uF	7805	With rectified input	No load
					Load 1
					Load 2
		C = 470 uF	7812	With rectified input	No load
					Load 1
					Load 2

Similar hierarchy is available for Full wave rectifier and Bridge rectifier circuits. Hardware design is optimized to use same components during conduction of similar type of experiments, i.e. same capacitor will be used when C-filter experiment with HWR/FWR or Bridge is being performed at different time.

Screenshots of Virtual lab phase-I during the conduction of Half wave rectifier and regulator experiments are shown in Fig. 2. Using easily navigable options, user can select any experiment for conduction. In each selected circuit, provisions are made to vary key parameters that affect the circuit performance and to visualize its effect by measuring the response at appropriate test points to analyze the circuit behavior.

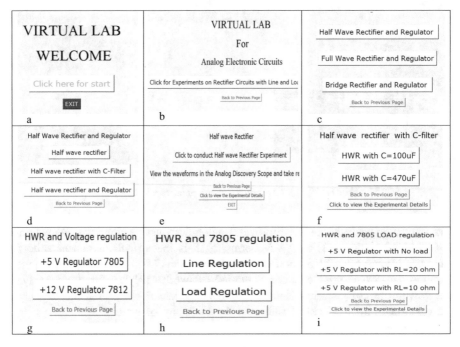

Fig. 2. Screenshots of Virtual lab at various stages: (a) welcome window, (b) analog electronics virtual lab, (c) options for circuit selection, (d) various HWR circuit options, (e) conduction of HWR circuit, (f) HWR with C filter options, (g) option for regulator selection, (h) Selection of line or load regulation, (i) 7805 regulator with different resistive loads.

Once conduction button of the selected circuit is pressed, 230 V, 50 Hz AC signal will be applied to step-down transformer for 10 s only. This is to prevent unnecessary power dissipation and to prevent circuits from damage. Upon conducting a particular circuit, waveforms of the selected nodes of the circuit will display on the oscilloscope of WaveForms 2015 for analysis. User need to capture these waveforms by selecting "STOP" option of the oscilloscope. Depending on the circuit analysis, it is possible to measure parameter like maximum voltage, average voltage, peak to peak voltage, frequency etc. At any stage of the conduction, it is possible to navigate to "previous page" or "HOME page". Web link of the videos will assist user to comfortably use the virtual lab.

In every stage, a detailed description about the circuit diagram, specifications, design equations, waveforms at test points, tabular column with expected values are provided for analysis. User can give a pause for the experiment at any stage of conduction to measure the voltages; take readings before proceeding to next stage. Figure 3 shows the one such detail for half wave rectifier with C- filter. Highlight of Fig. 3 is tabular column with expected values and measured values. As noticed the measured values slightly differ from the expected values, now user can interpret the possible reasons for deviations in the readings. Hence user gets the real time analysis of the circuits. User can take the screenshots during the conduction of experiments for preparing reports.

Half wave rectifier with C Filter

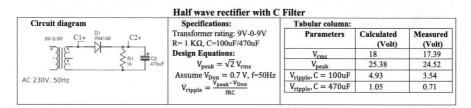

Circuit diagram	Specifications:	Tabular column:		
	Transformer rating: 9V-0-9V R= 1 KΩ, C=100uF/470uF **Design Equations:** $V_{peak} = \sqrt{2}\, V_{rms}$ Assume $V_{Don} = 0.7$ V, f=50Hz $V_{ripple} = \frac{V_{peak} - V_{Don}}{fRC}$	**Parameters**	**Calculated (Volt)**	**Measured (Volt)**
		V_{rms}	18	17.39
		V_{peak}	25.38	24.52
		$V_{ripple}, C = 100uF$	4.93	3.54
		$V_{ripple}, C = 470uF$	1.05	0.71

Fig. 3. Details of HWR with C filter having circuit, specifications, design equations and tabular column.

2.3 Analysis of Analog Electronic Circuit Experiments Using Virtual Lab

Learning of Electronic circuits is all about understanding the circuit behavior and analyzing its working by conducting experiments to measure appropriate voltages and comparing them with the expected values. In virtual lab, even though circuits are already available, appropriate parameters need to be measured to understand the circuit behavior. As there are many parameters to be measured for circuit analysis, Fig. 4 shows screenshots of input and output waveforms at different stages of half wave rectifier and regulator based experiments starting from simple rectification, C-filter, 7805 line and load regulation.

In each of the experiment different parameters are measured. In Fig. 4(a) maximum input voltage and maximum output voltage, average output voltage and input frequencies are measured. Figure 4(d) 7805 line regulation shows different input and output voltages of the regulator. Till certain input voltage regulator output has not regulated. But for input voltages 7.53 V and 15.39 V output is regulated to 5.58 V. This demonstrates the line regulation 7805 regulator. Figure 4(e) and (f) shows output voltage of 7805 for two different loads. Output voltage decreases from 5.58 V at no load to 4.56 V at 20 Ω and then to 3.56 V at 10 Ω load. This illustrates the load regulation of 7805 regulator. Notice that in each screenshot the values of input and output voltage values will enable user to know the circuit behavior. Similar analysis can be done for full wave and Bridge circuits. And comparison across three types of rectifiers will give solid good understanding of analog electronic circuits.

From the above readings of half wave rectifier and regulation circuit following concepts can be analyzed.

I. I–V characteristics of diode,
II. Conversion of AC to DC voltage by the rectifier,

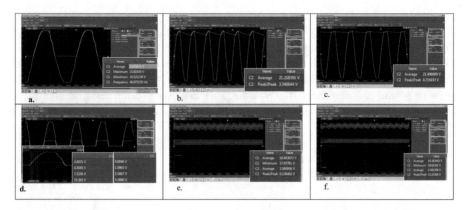

Fig. 4. Waveform of half wave rectifier and regulator circuits: (a) Rectifier output, (b) Output ripple voltage with C = 100 uF, (c) Output ripple voltage C = 470 uF, (d) Line regulation of 7805, (e) Load regulation of 7805 with load of 10 Ω, (f) Load regulation of 7805 with 20 Ω load.

   III. Effect of capacitor's value on magnitude of ripple voltage,

   IV. Noise in the regulated DC voltage,

    V. Line and load regulation plots of regulator.

3   Conclusions

Virtual lab Phase-I is successfully implemented for analog electronic circuits like half wave, full wave and bridge rectifier and 7805 and 7812 voltage regulators. User can remotely access the experiments, conduct investigation to analyze the behavior of the circuit virtually measuring the voltages. User can conduct the experiments any number of times to understand the circuit behavior. Virtual lab can also be used in classroom as **Experiential learning platform** while teaching the theoretical concepts. In future, Virtual lab will be expanded to include, experiments on BJT amplifiers, wave shaping circuits, applications of Linear ICs, oscillators and filters.

Acknowledgment. Authors thank Director, Principal and Management of Siddaganga Institute of Technology, Tumakuru for financially supporting this project. We thank 2016 batch Telecommunication Engineering students and TCE faculty members for using the virtual lab and giving valuable feedback.

References

1. http://vlab.co.in/
2. http://ohm.ieec.uned.es/portal/?page_id=76
3. http://www.uml.edu/IT/Services/vLabs/
4. http://vlabs.hpe.com/

Poster: Wireless Sensor Network to Predict Black Sigatoka in Banana Cultivations

Andrés Subert-Semanat[✉]

Universidad de Oriente, Santiago de Cuba, Cuba
asubert@uo.edu.cu

Abstract. Black Sigatoka is a disease caused by the fungus *Mycosphaerella fijiensis Morelet* and is considered worldwide as one of the most destructive diseases in the musaceae, as it produces foliar necrosis in the leaves of banana and banana plants, and consequently the fruits do not possess the necessary characteristics for their consumption or for an effective commercial activity. This disease has spread to all areas where plantains or bananas are grown. The goal of this paper is the use of prediction models of Black Sigatoka outbreak in banana cultivation in order to design a wireless sensors network for remote monitoring of climatic variables related to the disease, allowing, this way, scheduling of fungicide treatments for early removal.

Keywords: Prediction model · Precision agriculture · Black Sigatoka
Wireless sensor network · Remote monitoring

1 Introduction

Consuming of bananas has great significance in feeding of people. It is an essential part of the daily diet of citizens of one hundred tropical and subtropical countries. The banana is ranked as the fourth most important crop in the world after rice, wheat and corn.

Black Sigatoka is a disease caused by Mycosphaerella fijiensis morelet and it is regarded worldwide as one of the most destructive diseases in Musaceae[1] because it produces leaf necrosis on leaves of banana plants fruits (Fig. 1) and therefore do not have the necessary characteristics for consumption or for effective commercial activity. This disease has managed to expand to all cultivation areas, that is, why it has been fought with systematic treatment with fungicides and cultural practices. Cost fungicide applications represent a large percentage of production cost of banana and such measures entail harmful effects to the environment. On the other hand, cultural practices alone are not enough to control the disease.

Alternatively model-based programs have developed for early prognosis of possible outbreaks of this disease in order to make more efficient application of fungicides. These models are based on the correlation between the disease and weather variables that such

---

[1] Musa (Musaceae scientific name) are a family of monocot plants known for their fruit (bananas and plantains).

© Springer International Publishing AG, part of Springer Nature 2019
M. E. Auer and R. Langmann (Eds.): REV 2018, LNNS 47, pp. 159–164, 2019.
https://doi.org/10.1007/978-3-319-95678-7_18

Fig. 1. Banana leaf affected with Black Sigatoka.

as temperature, humidity, solar radiation, precipitation and wind speed. The favorable conditions for the proper development of banana plants are those that follow:

- Warm weather
- Constant humidity in the atmosphere at a level of 90 to 95%
- Growth stops at temperatures below 18 °C.
- Damage occurs at temperatures below 13 °C and over 45 °C [1].

2   Bioclimatic Forecasts

The main problem of Black Sigatoka management is that this is a very persistent disease, which over time has become resistant to the main agrochemicals (fungicides) that are used worldwide to combat it.

The main problem of Black Sigatoka management is that it is a very persistent disease, which over time has become resistant to the main agrochemicals (fungicides) that are used worldwide to combat it.

The disease sampling systems seek to make the application of fungicides more efficient based on the early prognosis of possible outbreaks of this disease. In this regard, various forecasting or early warning techniques have been developed, on the one hand manual sampling methods and on the other bioclimatic forecasts. Commercially, two manual sampling systems have been applied to decide the applications of fungicides in plantations of bananas, the Stover method modified by Gauhl and the method of biological warning or prediction.

2.1   The Method of Stover Modified by Gauhl

The method of Stover modified by Gauhl [2] is based on the quantification of the state of development of the disease, according to the symptoms it causes in the affected plants (type and number of lesions, number of affected leaves, percentage of affected leaf area, younger leaf infected, weighted average of infection and number of functional leaves).

The method consists of making a visual estimate of the diseased leaf area in all the leaves of nearby plants to flourish and that are representative in the population of the garden.

2.2   The Biological Warning System

The biological warning system is based on the analysis of biological and climatic descriptors for the timely application of fungicides, in periods in which the severity of the disease begins to increase and the climatic conditions lead to a favorable development of the pathogen.

The notice is based on weekly observations of symptoms in the young leaves of actively growing plants. Arbitrary coefficients are assigned to the three youngest leaves, according to the incidence and severity of the disease and with them two variables are calculated: the Gross Sum (SB) and the State of Evolution (EE) [3]. The SB refers to the present state of the infection and is an arbitrary value that increases with the advance of the symptoms and the youth of the leaves. The EE is calculated using the SB and the leaf emission rate of the plants.

With these variables and methods it is possible to perform the applications of agrochemicals by biological signaling, that is, that it will be fumigated only when the disease warrants it and if these variables behave with critical values; so the application of the fungicide must be carried out with haste to cut the advance of the disease because the conditions are favorable for the development of the same. It also allows knowing if the fungicide achieved progression or regression of the pathogen, if it is possible to change it in time and conclude which is the ideal pesticide for the phytosanitary combat.

2.3   Bioclimatic Forecasts

The principle of operation of for these models is the lookup of correlation functions between the climatic variables and the disease severity, which is achieved by applying iterative methods. In order to accomplish it statistical and data mining analysis is carried out where independent variables are the climatic variables and dependent ones are the biological, both registered chronologically.

The climatic variables with the highest correlation with the predictive models biological data as the Black Sigatoka epidemiology as a target are the air temperature, evapotranspiration, solar radiation, wind speed and relative humidity.

Three forecast models [3–5], applied in different regions of the Americas were analyzed. From them [5] was chosen to be oriented to a wider range of banana varieties.

2.4   Model Description

The climatic variables chosen to be measured were temperature, intensity and duration of rainfall, intensity and duration of wetting of the leaf and Piché evaporation (Ep). With the data of maximum temperature (Tmax) and minimum (Tmin) daily were calculated the sum of daily rates of evolution of the disease (SV) using (1) with which weekly accumulations were obtained.

$$SV = 7.18\,\text{Tmax} + 79.16\,\text{Tmin} \tag{1}$$

Recording equipment for measurements were used, located in a standard weather house, it comprised thermometers for maximum and minimum temperatures, a Mechel MT 1500 thermohydrograph, a Mechel UM 8100 rain gauge, a Piché evaporimeter and a Woelfe-type humectograp.

The data of rainfall amount in mm and rainfall duration in minutes were accumulated for periods of 7, 10 and 14 days and submitted to correlation analysis with the EE records obtained in the same week and between 1 and 8 weeks after; with the results of these analyzes a matrix of correlations was constructed. The same procedure was followed with the records of intensity and duration of the wetting of the leaves. The statistical analyzes were carried out with the statistical program STATISTICA [6]. Figure 2 shows the curves of the State of evolution (EE4H), observed and calculated as a function of rainfall accumulated during 14 days, 5 weeks before. Weighed Piché Evaporation (EvPp) was considered 4 weeks before. For the calculation of the latter it was determined that the model equation must follow the law that appears in (2)

$$EE4H_{cal} = 6.002\left(S_{14d} * LL_{mm}\right) - 183.95EvPp + 1732.57 \tag{2}$$

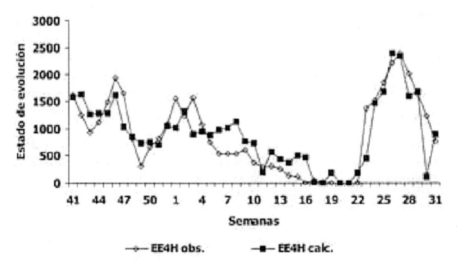

Fig. 2. Evolution State (EE) curves calculated according to accumulated rainfall for 14 days.

3 Remote Perception of Bioclimatic Variables

Wireless sensor networks can contribute to the analysis of crop health status by means of regular monitoring at specific sites of physiological conditions of plants, as well as collecting information on various climatic variables, such as temperature, precipitation, humidity, hours light, radiation levels, wind and evaporation.

Wireless sensor networks were selected among the remote monitoring technologies, its best features are scalability, low cost, auto configuration capacity and low consumption.

For application of WSN in resolving an precision agriculture issue, the methodology displayed in Fig. 3 [7] fit all design requirements but several recommendations in [1, 8] were included.

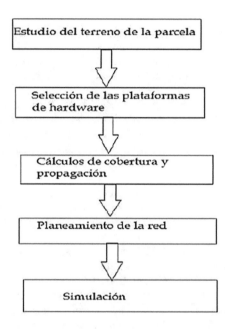

Fig. 3. Methodology for WSN application to precision agriculture issue.

4   Considerations for Wire Sensor Network Application

Sensors were needed for measuring the following primary variables: relative humidity, temperature, rainfall intensity and wind velocity. Once the variables are received in the sink processing terminal, all prediction model equations are applied in order to forecast the evolution state of Black Sigatoka disease.

For the calculations of coverage and of propagation, signal attenuation models in presence of vegetation was considered, so as the, the possible obstructions of Fresnel's zone and the network topology that better fit terrain geometry.

5   Conclusions

One hectare of banana-cultivated terrain in a farmer cooperative was selected as a test bed. Libelium was preferred as the company equipment provisioning because it allowed more freedoms for the WSN's design as well as the capability to select the frequency band and transmission technology. They network comprised 8 nodes with a maximum distance between nodes in 54,70 m and radiolink antenna's height were about 3,67 m. The nodes were distributed in a mesh topology which proved better performance and throughput.

References

1. Espinoza, E.A.A.: Diseño de una red inalámbrica de sensores para monitorear un cultivo de plátano en el distrito de Mala, Pontificia Universidad Católica del Perú (2011)
2. Gutierrez, J.: Sistema de adquisición y análisis de información para el diagnóstico de Sigatoka Negra, Sistema de adquisición y análisis de información para el diagnóstico de Sigatoka Negra.Thesis (2012)
3. Freitez, J.A.: Desarrollo de un Modelo Predictivo del Brote de Sigatoka Negra para las Plantaciones de Plátano al Sur del Lago de Maracaibo, vol. 9, pp. 191–198 (2007)
4. Hernández, L., Hidalgo, W., Linares, B., Hernández, J., Romero, N., Fernandez, S.: Estudio preliminar de vigilancia y pronóstico para sigatoka negra (Mycosphaerella fijiensis Morelet) en el cultivo de plátano (Musa AAB cv Hartón) en Macagua-Jurimiquire, estado Yaracuy. Rev. Fac. Agron. **22**(4), 325–329 (2005)
5. Pérez Vicente, L., Mauri Mollera, F., Hernández Mancilla, A., Abreu Antúnez, E., Porras González, A.: Epidemiología de la Sigatoka negra (Mycosphaerella fijiensis Morelet) en Cuba. I. Pronóstico bioclimático de los tratamientos de fungicidas en Bananos (Musa acuminata AAA), Rev. Mex. Fitopatol. **18**(1), 15–26 (2000)
6. www.statsoft.com>Products>STATISTICA Features. http://www.statsoft.com/Products/STATISTICA-Features. Accedido 30 Nov 2017
7. Vizcaíno Espinosa, I.P.: DISEÑO DE UNA RED DE SENSORES INALAMBRICOS APLICADOS A LA DETECCIÓN DE PESTICIDAS EN AGUAS DE ZONAS BANANERAS. Universidad Rey Juan Carlos, Machala, Ecuador (2011)
8. Daniel, V.V.: Diseño de una red de sensores inalámbrica para agricultura de preci-sion, Pontificia Universidad católica del Perú, Lima, Perú. Thesis (2011)

Poster: Influence of the Direction of Movement of Earth-Moving and Construction Machines on the Stability of Remote Control Data Transmission via Mobile Communication Channels

Tatyana Golubeva[1](✉), Sergey Konshin[1], Sergey Leshchev[2], Natalia Mironova[2], and Boris Tshukin[2]

[1] Almaty University of Power Engineering and Telecommunications, Almaty, Kazakhstan
ya_nepovtorimaya@mail.ru, ots2@yandex.com
[2] National Research Nuclear University MEPhI, Moscow, Russia
leshev.sergey@yandex.ru, viktort@mail.kz, tsh-k22@mail.ru

Abstract. In the article authors give research results of the influence of the movement direction of earth-moving machines on the stability of the remote control data transmission through mobile communication channels. The authors carried out an extensive analysis of the state of research in this and related fields. If movement direction of earth-moving and construction machines changes with respect to the nearest radio transmitting tower of the mobile communication base station, theoretically an instability in data transmission of data over mobile communication channels will appear. To establish the admissibility of such instability, experiments were performed in real conditions, the results of which are presented in the article.

Keywords: Remote control · Mobile communication
Earth-moving and construction machines

1 Context

The authors are conducting research on the development of an automated control system of work of earth-moving and construction machines.

Similar systems are being developed in another area - robotics. For example, article [1] examines how several cars on the robots' principle travel on paths with limited curvature in a plane that contains a priori unknown areas. Such robots are anonymous to each other and do not use communication facilities. Any of them has access to the current minimum distance to the region and can determine the relative positions and orientations of other robots within the final and specified visibility range. The analyzed article presents a distributed strategy of navigation and pointing, according to which each robot autonomously approaches the desired minimum distance to the region with constant maintenance of a given safety margin, robots do not collide with each other and do not get into clusters, and the entire team eventually moves on the corresponding equidistant curve with a speed exceeding a predetermined threshold, forming a kind of

© Springer International Publishing AG, part of Springer Nature 2019
M. E. Auer and R. Langmann (Eds.): REV 2018, LNNS 47, pp. 165–172, 2019.
https://doi.org/10.1007/978-3-319-95678-7_19

rapid barrier along the perimeter of the region. In addition, according to the authors, this strategy provides an effective sub-uniform distribution of robots along an equidistant curve. The article also offers a mathematically rigorous justification for the proposed strategy, its effectiveness is confirmed by extensive computer models and experiments with real wheeled robots.

In the world, there are very few researchers who studies a movement of earth-moving and construction machines. But there are many researchers who actively studies on various areas of research on vehicle movement control. Recent advances in the field of vehicle communications make it possible to implement transport sensor networks [2], that is, a collaborative environment where mobile vehicles equipped with sensors of a different nature (from toxic detectors to still cameras) interact to implement monitoring applications. In particular, interest to an active monitoring of cities, when vehicles constantly perceive events from city streets, autonomously process sensory data (for example, when recognizing license plates) and, possibly, send messages to vehicles in their vicinity to achieve a common goal (for example, for that policemen could track movement of concrete cars) is increasing. This complex environment requires new solutions compared to more traditional wireless sensor nodes. In fact, unlike conventional sensor nodes, vehicles demonstrate limited mobility, do not have strict limitations on processing power and storage capacity. The article describes MobEyes, which is an effective middleware specifically designed for proactive monitoring of cities and uses the mobility of nodes for opportunistically diffuse analysis of sounding data between neighboring vehicles and to create a low-cost index for querying monitoring data. The authors of the analyzed article researched consequences of simultaneous operation of several mobile subscribers from a single communication repeater, estimated the network load and overall stability of the system, and led the results of MobEyes verification in difficult urban conditions.

The article [3] is devoted to the study of vehicle movement. The application of the opportunistic networks proposed in the analyzed article on the basis of mobile devices will make it possible to realize an effective ability to search and collect information about routes and exchange information within joint nodes. The analyzed article investigates with mobile (pedestrian and automobile) and stationary terminals in urban areas, which are interested in collecting information obtained from several sources. In particular, each terminal is aimed at retrieving data items in a limited area of interest centered around the node's position. Because data items can change over time, all nodes should strive to access the latest version. In addition, for mobile terminals, the area of interest is a concept of time, changing due to the dynamic behavior of the nodes. The goal of the article being analyzed is to estimate the amount of information which each node can collect by using simple distributed data collection and exchange through opportunistic communications between neighboring nodes. In particular, the analyzed article analyzes the effect of node density, the different combination of cars and pedestrians, as well as the amount of nodes memory. In addition, the improvement in the use of memory management policies based on location is assessed, as well as the effect of adding several ideal nodes whose mobility is described by unlimited Brownian motion. To this end, the authors of article being analyzed develop a simulator based on mobility and radio programs derived from UDelModels tools. Their preliminary findings emphasize that simple memory management schemes based on location effectively use nodes with a limited amount of memory. In addition, an increasing the randomness of moving nodes has a beneficial effect on the average performance of all node types.

The main idea of research is to bring to the full autonomy of each earth-moving machine, working in a quarry or on a construction site. This is especially important in radioactively contaminated areas, where earthmoving and construction machines must operate autonomously, without the presence of drivers. As a result, it is planned that the earth-moving and construction machines will perform all the work, avoiding obstacles, not colliding with each other, becoming robots to some extent. At the same time, there must be a control room from which the operator remotely manages the tasks performed by each earth-moving machine.

Similar researches, but approximately in related or other areas are carried out around the world. For example, active research on intelligent antenna systems is carried out now [4]. In mobile communication, intelligent antenna systems using antenna arrays in combination with adaptive signal processing techniques at a basic level increase the bandwidth, coverage and efficiency of trunk lines.

The authors of analyzed article ascertain the fact that in all cases, the beta distribution can be used for empirical representation of the spatial signature of correlations. Authors also revealed that the direction of arrival is not much changed with movement in a suburban environment.

For the organization of the radio communication channel and transmission of information from earth-moving and construction machines, the authors studied the electromagnetic situation in the mining pit area [5]. Also, the authors investigated the tasks of positioning in measurements for control the operation and management of earth-moving and construction machinery [6]. We carried out experimental study of the capabilities of the 3G-324M cellular communication protocol for transmitting data from sensors and digital video cameras intended for installation on earth-moving and construction machines, as well as control signals in the system of control and management of earth-moving and construction machines [7].

The data transfer channel from the earth-moving and construction machines to the control room and the reverse remote control data transfer channel must be highly reliable and operative. It is clear that these channels must be wireless radio channels. Studies [8–12] carried out by the authors have shown that it is most optimal to realize the remote control data transfer via mobile communication channels. Besides due to a small amount of transmitted information, there are enough standard GSM channels

2   Purpose

If the direction of movement of earth-moving and construction machines changes relatively to the nearest radio transmitting tower of the mobile communication base station, theoretically there will be an instability in the transmission of data over mobile communication channels. To establish the admissibility of such instability, it was necessary to conduct experiments under real conditions.

Quite recently an article appeared in the literature [13], which with great tension can be called the prototype of the system being developed now. The development of surveillance and monitoring systems in our case can be a very difficult task, since these systems should be designed taking into account the environment to be monitored. Good surveillance systems should have dynamic characteristics, for example, surveillance cameras that are mobile and can move around the monitored territory. The article being analyzed describes one of the scenarios that require such mobile monitoring. A large building with a large number of levels was monitored, which entailed a high cost of installing many cameras in many places. The authors of the article being analyzed conclude that monitoring such a large area will also be a problem for security officers, since they will have to spend too much time for patrolling all places. Other scenarios that require dynamic surveillance systems include hazardous areas, such as areas with explosions and fire hazards, or areas contaminated with toxic gases. Another use case includes areas that can't be accessible to people. To solve these problems, the authors of the article being analyzed offer dynamic monitoring systems based on the movement of cameras using a Wi-Fi car remote control. A remotely controlled vehicle is capable of moving independently in a zone remote from the operator. Users can be in the next room, in a neighboring building or even in another country, controlling the movement of the vehicle through the Internet. Some mechanism sounding is necessary to help the user to locate the current location of the car and effectively navigate. Thus, the camera attached to the vehicle also acts as eyes, showing the front view of the vehicle, which the user sees through the browser in real time. This system uses Wi-Fi as the only means of communication to connect the vehicle to the server. A web application that provides an interface for the user to interact with the vehicle also provides streaming in close proximity for monitoring, that is, using some other cameras installed in the area.

To study the effect of the direction of movement of earth-moving and construction machines depending on the location of tower of the base station of mobile communication, following experiments were conducted. For the purity of the experiments, we chose a section of the motorway free from other machines, with a station of mobile operator, located 100 m away from the road and serving this section. At the chosen section of the Iliysky tract of Almaty region, the length of 2.4 km, a sector antenna of the cellular operator, which is equally remote from the end points of our route and is located at 100 m from the road, operates. Earth-moving and construction machines move along the road, approaching and moving away from the antenna. The data was collected by the mobile station from mobile phones in the earth-moving and construction machines using the downloaded information from the FTP server of the cellular operator. The connection of the sector antenna to the FTP server was organized by means of radio relay antennas installed at a distance of 13 km from each other in the line of sight.

3   Approach

The following equipment and software were used for the experiment: two Prestigio Wize L3 smartphones with 3G support, laptops and Garmin GPS navigation receivers for positioning in space, Terms Investigation program for taking measurements of mobile GSM signals, processing and outputting statistics.

Sector antennas serving the road with moving EMCMs, operate in the frequency range from 1710 to 2200 MHz and are designed to work with 3G networks. The directional diagram of these antennas is 65°.

Models of radio-relay antennas used in the experiment have a capacity of 34 Mbit/s, which exceeds the maximum possible capacity of the UMTS technology of 14.4 Mbps.

All the data obtained experimentally, were processed in a software way and displayed in the table. The data were selected into separate groups by the speed of movement of the mobile station, towards the antenna and from the antenna. Only those parts of the data were selected where data transfer was recorded. The moments of establishing and closing the connection were ignored. For each group, the expectation, the mean square deviation, and the variability coefficient were calculated.

4   Actual Outcomes

As a result of the experiments it was found out that the coefficient of variability at speed of movement of 21–40 km/h and 81–100 km/h is the smallest and the stability of data transmission is higher. Histograms 1, 2 have been constructed according to the received data (Figs. 1 and 2).

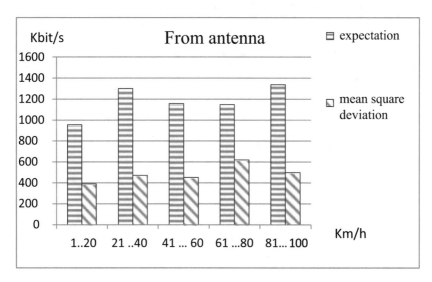

Fig. 1. The histogram of the effect of the speed of subscriber's movement on the data transmission when moving from the antenna

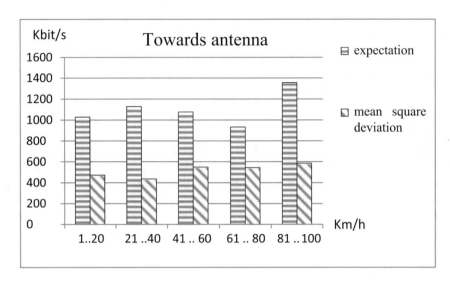

Fig. 2. The histogram of the effect of the speed of subscriber's movement on the data transmission when moving towards the antenna

The results obtained in Figs. 1 and 2 show that when the EMCM moves with a cellular phone in the direction from the antenna, the data transfer rate is more stable, and the standard deviation is much smaller than when moving to the antenna.

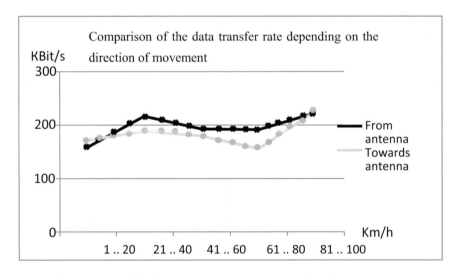

Fig. 3. Comparison of the data transfer rate depending on the direction of movement

5   Conclusions

Figure 3 shows a comparison of the data transfer rate depending on the direction of movement of earth-moving and construction machines.

It follows from Fig. 3 that at movement speed of 21–81 km/h from the antenna, data transmission is more intensive at 100–200 kbit/s, which will allow using the mobile communication channel for reliable organization of voice and service information transmission in the management system of work of earth-moving and construction machines.

References

1. Semakova, A., Ovchinnikov, K., Matveev, A.: Self-deployment of mobile robotic networks: an algorithm for decentralized sweep boundary coverage. Robotica **35**(9), 1816–1844 (2016). ISSN: 0263-5747, eISSN: 1469-8668. https://doi.org/10.1017/S0263574716000539, WOS: 000407196800002
2. Lee, U., Magistretti, E., Gerla, M., Bellavista, P., Corradi, A.: Dissemination and harvesting of urban data using vehicular sensing platforms. IEEE Trans. Veh. Technol. **58**(2), 882–901 (2009). ISSN: 0018–9545, eISSN: 1939-9359. https://doi.org/10.1109/TVT.2008.928899, WOS:000263639300032
3. Corrente, G., Gaeta, R., Grangetto, M., Sereno, M.: Local data gathering using opportunistic networking in a urban scenario. In: Proceedings of the 8th ACM Symposium on Performance Evaluation of Wireless Ad Hoc, Sensor, and Ubiquitous Networks (PE-WASUN 2011), Miami, Florida, USA, 03–04 November 2011, pp. 57–64 (2011). ISBN: 978-1-4503-0900-4. https://doi.org/10.1145/2069063.2069074, WOS:000303787200008
4. Kavak, A., Torlak, M., Vogel, W.J., Xu, G.G.: Vector channels for smart antennas - measurements statistical modeling, and directional properties in outdoor environments. IEEE Trans. Microw. Theory Tech. **48**(6), 930-937 (2000). ISSN: 0018–9480. https://doi.org/10.1109/22.846719, WOS:000087794500007
5. Golubeva, T., Zaitsev, Y., Konshin, S.: Research of electromagnetic environment for organizing the radio channel of communication of operation control system of earthmoving and construction machines. In: 19th International Symposium on Electrical Apparatus and Technologies (SIELA 2016), Bourgas, Bulgaria, 29 May–1 June 2016, pp. 1–4. IEEE (2016). https://doi.org/10.1109/SIELA.2016.7543006, WOS:000382936800035
6. Golubeva, T., Konshin, S.: Improving of positioning for measurements to control the operation and management of earth-moving and construction machinery. In: 13th International Conference on Remote Engineering and Virtual Instrumentation (REV), UNED, Madrid, Spain, 24–26 February 2016, pp. 112–115. IEEE (2016). 978-1-4673-8245-8/16/$31.00. https://doi.org/10.1109/REV.2016.7444450
7. Golubeva, T., Zaitsev, Y., Konshin, S.: Research of 3G-324M mobile communication protocol in the management and control system of work of earth-moving machines and data transfer. In: 2016 10th International Symposium on Communication Systems, Networks and Digital Signal Processing (CSNDSP), Prague, Czech Republic, 20–22 July 2016, pp. 1–3. IEEE (20016). https://doi.org/10.1109/CSNDSP.2016.7573995, WOS:000386781300099
8. Golubeva, T., Pokussov, V., Konshin, S., Tshukin, B., Zaytsev, Y.: Research of the Mobile CDMA network for the operation of an intelligent information system of earth-moving and construction machines. In: MobiWIS 2017. Lecture Notes in Computer Science, vol. 10486, pp. 193–205. Springer, Cham (2017). ISSN: 0302 -9743, ISBN: 978-3-319-65514-7. https://doi.org/10.1007/978-3-319-65515-4_17

9. Golubeva, T., Konshin, S.: The research of possibility of sharing use of wireless and mobile technologies for organizing the radio channels of operation control system of earthmoving and construction machines. In: 2016 International Conference on Intelligent Networking and Collaborative Systems (INCoS), Ostrava, Czech Republic, 7–9 Sept 2016, pp. 9–14. IEEE (2016). https://doi.org/10.1109/INCoS.2016.24, WOS:000386596100002

10. Golubeva, T., Zaitsev, Y., Konshin, S.: Research of WiMax standard to organize the data transmission channels in the integrated control system of earth-moving machines. In: 2016 17th International Conference of Young Specialists on Micro/Nanotechnologies and Electron Devices (EDM), Erlagol, Russia, 30 June–4 July 2016, pp. 91–95. IEEE (2016). Electronic ISSN: 2325–419X. https://doi.org/10.1109/EDM.2016.7538701, WOS:000390301500024

11. Golubeva, T.V., Kaimov, S.T., Kaiym, T.T.: Mathematical and computer modeling of movement of the executive mechanism of the adaptive multipurpose operating part of earthmoving and construction machine. In: Proceedings of the IRES 10th International Conference, Prague, Czech Republic, 27 September 2015, pp. 42–45 (2015). ISBN: 978-93-82702-05-4. Int. J. Electr. Electr. Data Commun. 3(11) 25–28, ISSN: 2320–2084. http://www.iraj.in/journal/journal_file/journal_pdf/1-198-144854009625-28.pdf

12. Golubeva, T., Zaitsev, Y., Konshin, S.: Improving the smart environment for control systems of earth-moving and construction machines. In: 2016 IEEE 4th International Conference on Future Internet of Things and Cloud (FiCloud), Vienna, Austria, 22–24 Aug 2016, pp. 240–243. IEEE (2016). ISBN: 978-1-5090-4053-7. https://doi.org/10.1109/FiCloud.2016.42, WOS:000391237900034

13. Zaini, N.A., Zaini, N., Latip, M.F.A., Hamzah, N.: Remote monitoring system based on a wi-fi controlled car using raspberry Pi. In: IEEE Conference on Systems, Process and Control (ICSPC), Bandar Hilir, Malaysia, 16–18 December 2016, pp. 224–229. IEEE (2016). ISBN: 978-1-5090-1181-0. https://doi.org/10.1109/SPC.2016.7920734, WOS:000405609400042

Virtual and Remote Laboratories

Development and Implementation of Remote Laboratory as an Innovative Tool for Practicing Low-Power Digital Design Concepts and Its Impact on Student Learning

Shatha AbuShanab[1,2(✉)], Marco Winzker[1(✉)], and Rainer Brück[2(✉)]

[1] Bonn-Rhine-Sieg University of Applied Sciences, Sankt Augustin, Germany
{shatha.abushanab,marco.winzker}@h-brs.de
[2] University of Siegen, Siegen, Germany
rainer.brueck@uni-siegen.de

Abstract. Since power dissipation is becoming a significant issue and requiring more consideration in the early design stage, circuit designers must now be experienced in low-power techniques to enhance designing digital systems. Therefore, when teaching low-power design techniques in electrical and computer engineering education, a tool or a method must be made available that enables students to estimate the power dissipation of their digital circuits during the design process. This contribution presents a novel approach, the low-power design remote laboratory system that has been developed at the Bonn-Rhine-Sieg University of Applied Sciences to estimate the power dissipation of a digital circuit remotely via the internet using physical instruments and providing real data. The design takes place at abstraction level and the real data is measured at the low level from the hardware devices. The low level provides more information, which is required for accurately measured values that are hidden at the high level. The technical performance results on using the remote system show that the students are enabled to implement their digital design and to meet the performance targets of reliability as well as to observe almost all influencing factors on the design's power dissipation.

Keywords: FPGA · Low-power digital design · Remote laboratory

1 Introduction

The increasing demand for low-power digital systems especially in portable devices requires circuit design engineers who can design low-power electronic systems [1–4]. Engineering studies are an applied science and, therefore, for most courses, there are laboratories that provide the students with a better understanding of theoretical concepts. The development of the internet and of communication technology adds new methods for teaching and training, enabling the e-learning approach to occupy a larger part of academic learning methodologies. Recently, the laboratory environments have changed due to the development of technologies that have opened many doors in education, and

© Springer International Publishing AG, part of Springer Nature 2019
M. E. Auer and R. Langmann (Eds.): REV 2018, LNNS 47, pp. 175–185, 2019.
https://doi.org/10.1007/978-3-319-95678-7_20

resultantly, remote laboratories have been developed in several universities to obtain realistic results from physical instruments that can be accessed via the internet. These laboratories provide flexibility for the location of both teacher and student [5–7]. This research is an example of such a laboratory where the low-power design remote system for teaching low-power techniques has been developed. This remote system at the Bonn-Rhine-Sieg University of Applied Sciences assists the students in gaining their learning outcomes by proving the concepts and the theories and by minimizing the gap between engineering education and the industry's requirements for low-power design systems.

This paper is structured as follows: Sect. 2 introduces power dissipation in digital systems and describes the categories of proposed tools for estimating the power dissipation. Section 3 presents the low-power design remote laboratory system, illustrates the instructions and the procedures involved in a client accessing and interacting with the remote experiment and presents technical performance results of the remote laboratory. Conclusions and future work can be found in the final section.

2   Power Dissipation in Digital Systems

Most components in digital systems are fabricated using CMOS (Complementary Metal Oxide Semiconductor) technology. The main sources of power dissipation in digital CMOS circuits are static power and dynamic power [1–4]. Static power dissipation is caused by leakage currents in the digital circuits and is expected to increase with the smaller geometry of transistor's size and also with higher operating temperatures [1, 4]. Dynamic power dissipates when the digital circuit is active during changing states of the logic gates in a digital circuit. In CMOS circuits, dynamic power is the main sources of power dissipation, given in Eq. 1 [1, 3].

$$P_{dynamic} = S * C_L * V_{dd}^2 * f \tag{1}$$

Where $P_{dynamic}$ represents the dynamic power dissipation of a digital circuit; S represents the average number of transitions across the entire digital circuit per clock cycle; C_L represents the capacitance of a digital circuit; V_{dd} represents the supply voltage, and f represents the clock frequency.

2.1   Categories of Power Estimation Methods

Power estimation can be defined as the method for calculating energy and power dissipation at a particular level in the design process. Studies dealing with the problem of power estimation have proposed different methodologies to estimate the power at a level of the design stage transistor-level, logic gate-level, register transfer-(RT) level and system-level.

- Transistor-Level Approaches. The power of a digital circuit can be estimated at the transistor-level by measuring the average current flow from the circuit power source. SPICE [8] and PowerMill [9] are examples of transistor-level simulators. Alipour, Hidaji, and Pour [10] propose the method, which uses models for transistors in simulated techniques to achieve an almost accurate model for each transistor. However, the method is, in practice, not feasible for large digital circuits, and the input vectors are too long. To reduce the length of the input vectors, Marculescu and Ababei put forward a technique that causes approximately 5% loss in the accuracy of the power estimation [11]. Furthermore, the sizes of transistors are also relevant in the accuracy of the power estimation in digital design so tools are available that provide transistors size characteristics [12–14].

- Logic Gate-Level Approaches. The power dissipation of a digital circuit at the logic gate-level is estimated for a gate at each transition by calculating the parasitic gate and the wire capacitances and by obtaining each gate's switching activity, based on probabilistic methods or statistical methods. The obtained information is stored in the module library for components of digital circuits. Nocua, Virazel, Bosio, and Girard present a hybrid power estimation at the gate-level [15]. It is based on a hybrid approach that obtains information from different abstraction levels. The data is measured by using SPICE to determine the transistor-level parameters under different conditions such as polarization, temperature and load values. Additionally, an activity file is generated in which all the switching activities per gate are recorded.
- Register Transfer-Level Approaches. The power dissipation at RT-level is estimated for each component in a digital circuit by developing a model that is more complex than at the gate-level. The frequently used RT-level components are arithmetic components, multiplexers, comparators, registers, multiply-accumulate units, and ALUs. The architecture for each component has a uniformly frequent primitive cell. Bruno, Macii, and Pocino present a power estimated method for a digital circuit at the RT-level that is based on using a set of LUTs (Look-Up Tables) for each component [16]. In that method, the power values of the LUTs are created using a gate-delay simulator, a circuit-level simulator or a transistor-level simulator. The power dissipation can then be estimated by summing the power dissipation for each component in a digital circuit.
- System-Level Approaches. Research from both industry and universities have presented approaches for achieving low-power dissipation. They use either a direct measurement techniques or power model-based methods. In general, the power estimation techniques from the underlying hardware provide accurate results [17]. Becker, Huebner, and Ullmann discuss measuring the power during real runtime applications for an implemented digital circuit on an FPGA [18]. The measuring equipment includes a computer, control software and an oscilloscope with a computer interface. The shunt resistor is connected to an opened bridge on the board for measuring the input current of the FPGA. The oscilloscope is used to sample the voltage over the shunt resistor and to store the data in its memory. The measured data is saved in a file needed for calculating the power using the MATLAB software.

3 Low-Power Design Remote Laboratory System

The main difficulties with most previous power estimation approaches result from trying to imitate a physical implementation of digital circuits in high-level models. The details of digital circuits are based on probability and related mathematical formulae that cannot provide all the digital circuit characteristics, such as the effect of temperature on semi-conductor devices and of time constraints of the circuit design. Therefore, the previous estimation approaches suffer from lower accuracy and more complexity, making them difficult to use in educational laboratories.

The low-power design remote system at the Bonn-Rhine-Sieg University of Applied Science has been developed and designed to overcome some of these difficulties. It is an online workbench where students can estimate remotely the power dissipation of their digital circuits with physical instrument. The remote system is based on an FPGA (Field Programmable Gate Array) technology as a design platform where the digital circuits are implemented by students to simplify performing the experiment and to allow the students obtaining accurate results. Furthermore, it uses image processing to apply a real application in teaching the laboratory experiments trying to solve the input signal pattern problem of power estimation techniques.

3.1   Development and Implementation of the Low-Power Design Remote Laboratory System

There were many challenges and limitations in the development and implementation of the remote system: what kind of features are required, how educational concepts can be included in the laboratory experiments, how remote experiments can be performed and how the interface can be created for a real-time experiment. Dealing with these challenges is closely related to understanding the functionalities and structures of the physical equipment with which the students interact through the web applications of the remote system. The general architecture of the remote system is composed of three parts:

- Remote client. An internet connection is used for the user access to and interaction with the remote experiments.
- Remote server and database. The remote server allows the client to access and share the remote instruments that connect and communicate with the server. The database assists in the management of tasks, such as user account authentication, access time, and statistics process.
- Laboratory experiment. The laboratory contains the physical components of the remote system within which the students' input and output data are organized using the remote server. The following elements are present in the remote system:
  - A platform for implementing the student's digital design (FPGA). A specially designed FPGA-board is used to meet image processing requirement [19].
  - An image generator that provides the input image required to be processed by the implemented digital circuit. In this system, the BeagleBone Black (BBB) is used.
  - An image/video grabber. This device provides the output image display required to verify the functionality of the implemented digital circuit.
  - A digital multimeter (DMM). This current meter is necessary to measure the core current so that the power dissipation of the implemented circuit can be calculated.

The remote laboratory system is based on a client-server structure using a TCP/IP connection. The remote system with its components is shown in Fig. 1.

Fig. 1. The low-power design remote laboratory system.

3.2   Instructions for Low-Power Design Remote Experiments

A major effective technique for designing a digital circuit when teaching low-power design experiments is to use an abstraction level to represent the design. The method for estimating the power dissipation is practiced on and measured from the hardware. For the FPGA technologies used, the abstraction level is a common and independent level that eases, enhances, and simplifies designing a circuit with resources increasing following to Moore's Law. Hardware description language (HDL) is adopted to accomplish a digital circuit that describes a large digital circuit using a text-based language without the need for schematics. The proposed digital circuits are compiled to fix the errors faced. At this point, the students make pin assignments and timing constraints that are essential for building a functioning digital design. This step is performed locally on a student's computer without internet connection. The students who successfully complete the design stage can transfer to the remote system (hardware level) to verify the functionality and to estimate the power dissipation correctly. The following describes the next procedures in the remote system:

1. Authentication Webpage. Once a student activates the IP address of the remote server webpage in the address bar of any standard web browser, the web browser loads the start webpage of the remote experiment: an authentication webpage. The remote server compares the entered name and password with the stored data in the database, and if matched, the student/client is permitted to access the remote system and the upload webpage is activated. While a student is carrying the remote experiment, no other students are permitted to access the remote experiment at the same time. The maximum time available for executing a remote experiment is therefore limited and can't exceed the limitation set (approx. 2 min).

2. Upload Webpage of Remote Experiment. Via the upload webpage remote experiment, the student can upload the compiled SRAM Object File (.sof) that is required to implement his/her digital circuits in the FPGA device and also upload the image to be processed by the implemented digital circuits. The uploaded files transfer via the TCP/IP model to the remote server, enabling the remote experiment to be performed. The remote server sends the .sof file to program the FPGA device. The input image is sent to the BBB to provide the image for the input video of the FPGA-board. An image/video grabber imports the output (processed) image from the FPGA-board and saves it in the remote server. The supply voltage of the FPGA core is considered to be stable; only the core current is measured using a DMM that has a USB connected to the remote server to transmit the reading to the server.

3. Webpage of Experimental Results. At the end of the remote experiment, the experimental results are given as graphical instrument interfaces, as shown in Fig. 2. The available components on the output interfaces are:

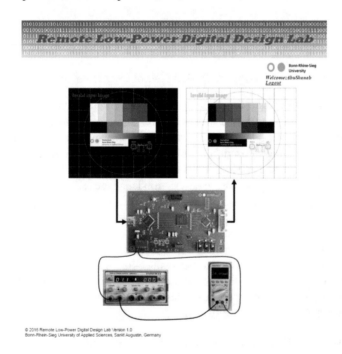

Fig. 2. The webpage of the experimental results from the remote experiment.

(a) Input image (upper left corner in Fig. 2). The input image is either uploaded by the student or the remote system uses a default image.

(b) Output image (upper right corner in Fig. 2). This image is the output from the experiment that is processed by the student's implemented digital circuit.

(c) FPGA-board (middle device in Fig. 2). In Fig. 2, Altera FPGA Cyclone V technology is used in implementing the FPGA-board.

(d) A DC power supply (lower left corner device in Fig. 2). It shows the voltage core that is applied to the FPGA device.

(e) DMM (lower right corner device in Fig. 2). It measures the core current of the FPGA device; another value of the experiment output.

Each activity in the remote system is designed to improve teaching the concepts and takes into account creating a realistic impression; the students realize that the other side is not a software package but that they are controlling the instruments and taking measurements from physical devices. Figure 3 shows the representation of the design on abstraction and hardware level.

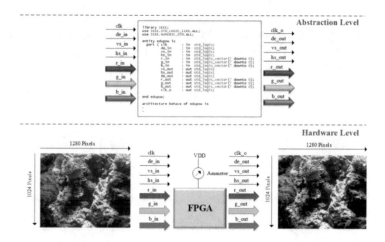

Fig. 3. A description on abstraction level is implemented on hardware level.

3.3   Brief Discussion of Low-Power Design Experiments

This section displays a brief discussion and description for a part of the low-power experiments applied to different levels in the design process: specification level, abstraction level, technology level, and environment level. To achieve the highest power reduction, the power should be taken into account at each level of the design process. However, the abstraction level has more potential for power reduction.

A finite impulse response (FIR) high pass is selected and designed to be a part of laboratory exercises for teaching low-power techniques. FIR is one of the most fundamental filters used in image and video processing applications. Figure 4 shows an n-tap FIR filter consisting of n delay elements, n multipliers, and $n - 1$ adders or accumulators. $x(n)$ is the input, C_0 to C_n are the tap coefficients, and $y(n)$ is the output.

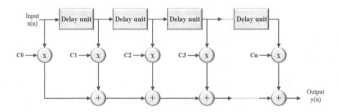

Fig. 4. n-taps FIR filter diagram.

Another laboratory exercise is similar to a delay signal. It uses a chain that is implemented in n-array so the input signals are transmitted to the next row of the array for nth times, satisfying utilization of the FPGA bases on the size of array. Finally the nth row is transmitted as output signals, as illustrated in Fig. 5. The input signals are 24 bits for RGB colors and 3 control signals; 8 bits for each color.

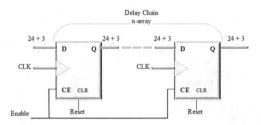

Fig. 5. The delay chain of the digital circuits contains n array for each input signals.

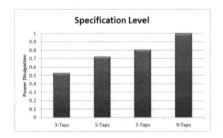

Fig. 6. Specification level.   **Fig. 7.** Abstraction level.

- **Specification Level.** Figure 6 shows the power dissipation of 3-taps, 5-taps, 7-taps FIR filters that are implemented on and measured from the Altera FPGA Cyclone V. The experimental results prove that the digital design's power dissipation is based on the number of operations in the algorithm; the numbers of taps should reflect the switched capacitances that determine the activities with respect to the circuit's size and complexity, which is established according to the specification level at the design process.

- **Abstraction Level.** The digital circuit in this laboratory exercise uses a chain for a delay signal exercise. Once a digital circuit has no delay, two circuits contain only FFs (Flip-Flops), and two other contain memory cells (RAM). The elements for the delay can be selected in the abstraction level in the design stage. Figure 7 shows the experimental results for the power dissipation of each circuit design implemented on the Altera FPGA Cyclone IV.
- **Technology Level.** The influence of CMOS technology at the technology level of the design stage provides another learning objective. Figure 8 shows the results from the previous exercises when the FIR filters are implemented on different CMOS technologies: the Altera FPGA Cyclone IV and the Altera FPGA Cyclone V.
- **Environment Level.** The influence of static power dissipation is another learning objective. Figure 9 shows the experimental data for the implemented 3-taps, 5-taps, 7-taps and 9-taps FIR filters on the FPGA Altera Cyclone IV with different environmental temperature, approx. 20 °C and approx. 1 °C.

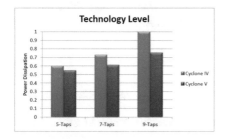   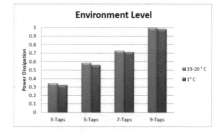

Fig. 8. Technology level.   **Fig. 9.** Environment level.

All the experimental results show that the specification of the design, the resources used in the digital circuits, the technology used and the environmental temperature model are independent and free to be used at a particular level in the design process. Thus, using of the remote laboratory system enables the students to implement their digital design and to meet the performance targets of reliability while observing almost all factors influencing power dissipation.

4   Conclusion

This contribution presents a new educational laboratory—the low-power design remote laboratory system—that can be used in electrical and computer engineering education as well as in other areas that require the teaching of low-power techniques of digital circuits design. This remote system allows the following:

- The low-power design theories can be practised with a real laboratory system that uses the high abstraction level at the design stage and obtains the results at the hardware level. The power estimation of the system occurs at the hardware level to provide the information needed for correct estimation.

- The remote system can be accessed independent of place and time and as supplementary class material, and it also supports the sharing of resources within a university as well as between universities.
- The remote client layer uses a standard browser to perform the remote experiment and thus reduces the requirement for additional software on the student's computer and decreases the required time needed by the students to learn additional operation.
- The students do not need to use the remote instruments all the time while working on the remote experiment. They only need time on the remote instruments when they have the completed design with the information required for implementation on the hardware device.
- Low-power techniques can be applied at different levels of the design process: specification, abstraction, technology, and environment level. The engineering students can observe nearly all influencing factors on the power dissipation through the levels during the remote laboratory exercises.

The first remote laboratory prototype has been developed and implemented. In the next step, the system will be evaluated by analyzing how the students use it.

References

1. Rabaey, J.M.: Low Power Design Essentials. Series on Integrated Circuits and Systems. Springer, New York (2009)
2. Arora, M.: The Art of Hardware Architecture. Springer, New York (2012)
3. Bhunia, S., Mukhopadhyay, S.: Low-Power Variation-Tolerant Design in Nanometer Silicon. Springer, New York (2011)
4. Henzler, S.: Power management of digital circuits in deep sub-micron CMOS technologies. Springer series in advanced microelectronics, vol. 25. Springer, Dordrecht (2007)
5. Feisel, L.D., Rosa, A.J.: The role of the laboratory in undergraduate engineering education. J. Eng. Educ. **94**(1), 121–130 (2005). https://doi.org/10.1002/j.2168-9830.2005.tb00833.x
6. Hofstein, A., Lunetta, V.N.: The laboratory in science education: foundations for the twenty-first century. Sci. Educ. **88**(1), 28–54 (2004). https://doi.org/10.1002/sce.10106
7. Ma, J., Nickerson, J.V.: Hands-on, simulated, and remote laboratories: a comparative literature review. ACM Comput. Surv. **38**(3), 7-es (2006). https://doi.org/10.1145/1132960.1132961
8. SPICE manual. http://bwrc.eecs.berkeley.edu/Classes/IcBook/SPICE/. Accessed 2017
9. Huang, C.X., Zhang, B., Deng, A.-C., et al. The design and implementation of PowerMill. In: Proceedings of the 1995 International Symposium on Low Power Design, ISLPED'95, pp. 105–110 (1995)
10. Alipour, S., Hidaji, B., Pour, A.S.: Circuit Level, Static Power, and Logic Level Power Analyses. IEEE, Piscataway (2010)
11. Marculescu, R., Ababei, C.: Improving simulation efficiency for circuit-level power estimation. In: Proceedings of 2000 IEEE International Symposium on Circuits and Systems. ISCAS 2000, Geneva, pp. 471–474 (2000)
12. Posser, G., Flach, G., Wilke, G., et al.: Gate sizing using geometric programming. Analog Integr. Circ. Sig. Process **73**(3), 831–840 (2012). https://doi.org/10.1007/s10470-012-9943-3
13. Shacham, O., Azizi, O., Wachs, M., et al.: Rethinking digital design: why design must change. IEEE Micro **30**(6), 9–24 (2010). https://doi.org/10.1109/MM.2010.81

14. Posser, G., Flach, G., Wilke, G., et al.: Tradeoff between delay and area in gate sizing using geometric programming, pp. 1–4
15. Nocua, A., Virazel, A., Bosio, A., et al.: A hybrid power modeling approach to enhance high-level power models. In: 2016 IEEE 19th International Symposium on Design and Diagnostics of Electronic Circuits and Systems (DDECS) (2016). https://doi.org/10.1109/ddecs.2016.7482453
16. Bruno, M., Macii, A., Poncino, M.: RTL power estimation in an HDL-based design flow. IEE Proce. Comput. Digital Tech. **152**(6), 723–730 (2005). https://doi.org/10.1049/ip-cdt:20045181
17. Bellosa, F.: The benefits of event-driven energy accounting in power-sensitive systems. In: Proceedings of the 9th Workshop on ACM SIGOPS European Workshop: Beyond the PC: New Challenges for the Operating System, pp. 37–42 (2000). https://doi.org/10.1145/566726.566736
18. Becker, J., Huebner, M., Ullmann, M.: Power estimation and power measurement of Xilinx Virtex FPGAs: trade-offs and limitations. In: Proceedings of 16th Symposium on Integrated Circuits and Systems Design, SBCCI 2003. IEEE (2003). https://doi.org/10.1046/j.1528-1157.44.s.5.1.x
19. Schwandt, A., Winzker, M.: Modular evaluation system for low-power applications. In: IEEE International Conference on Electronics, Circuits and Systems (ICECS) (2017)

Remote Labs for Electrical Power Transmission Lines Simulation Unit

Kalyan B. Ram[1(✉)], Panchaksharayya S. Hiremath[1(✉)], M. S. Prajval[1(✉)],
B. Karthick[1(✉)], Prasanth Sai Meda[1(✉)], M. B. Vijayalakshmi[1(✉)],
and Priyanka Paliwal[1,2(✉)]

[1] Electrono Solutions Pvt. Ltd., #513, Vinayaka Layout, Immadihalli Road, Whitefield,
Bangalore 560066, India
{kalyan,arun,prajvalms,karthick,prasanth,
viji}@electronosolutions.com, priyanka_manit@yahoo.com
[2] Department of Electrical Engineering, MANIT, Bhopal, India

Abstract. Electrical Power Transmission Lines are used to transferElectrical
Energy from the Generating site such as power plant to and Electric Substation.
The interconnected lines that facilitate this are known as Transmission Network.
These transmission lines are different from that of the high voltage substations to
the consumers, which is referred to as distribution network. Study of transmission
lines is an important topic in the courses offered under Electrical Engineering
program. This includes understanding of certain fundamental concepts related to
Electrical Power Transmission Lines such as Ferranti effect, ABCD parameters
and their analysis, Transmission lines efficiency and calculation of losses, Modelling of Transmission lines (T and Pi sections), Surge Impedance loading and so
on. Typically, Transmission Line Simulators are used to study the concepts
mentioned above. Students perform experiments on these Transmission Lines
Simulators that represent the equivalent model of the real Electrical Power Transmission Lines running for several kilometres of distances are the experimentation
platforms on which several aspects of Transmission lines are studied. In the
context of several Engineering Institutions, simulators such as these are a scarce
resource and typically a given institution would have at most 1 such setup. This
is because of the cost and size of these simulators.

Keywords: Remote labs · Transmission lines · Simulator · Computer controlled
Experiment setup · Electrical power transmission · PLC

1 Introduction

Transmission line is a two port system interfacing a generator circuit at the sending end
to a load at the less than desirable end. The PI section block implements a single phase
transmission line with lumped parameters in PI section. For transmission lines the
resistance, capacitance and inductance are uniformly distributed along the line. The
distributed parameter line is obtained by cascading several identical PI sections.

© Springer International Publishing AG, part of Springer Nature 2019
M. E. Auer and R. Langmann (Eds.): REV 2018, LNNS 47, pp. 186–196, 2019.
https://doi.org/10.1007/978-3-319-95678-7_21

In engineering education, laboratories represent an important academic resource as they provide practical training in addition to the fundamental theories taught in lectures. Utilization of such laboratory infrastructure in several Engineering universities and colleges in India is less than 20% due to factors such as time, security, safety, availability of skilled staff to train students to use these labs. Majority of colleges cannot afford such expensive hardware lab setup too. In order to overcome expensive hardware lab setups and utilize the existing lab setup more efficiently and to improve the technical education, Electrono solutions Pvt. Ltd., Bangalore has developed Remote Labs for Transmission Lines at MANIT, Bhopal.

2   Objectives of Transmission Lines Remote Labs

As discussed in papers published with regard to the need for Remote Labs and their benefits, Laboratory experiments setups, specifically setups such as Electrical Power Transmission Lines Simulator Unit and others that are normally scarce resources have very limited accessibility to students. A typical Electrical Power Transmission Line Simulator Unit consists of

(1) Generating Station model: Consists of a Motor – Generator set up representing the Electricity Generating Station along with Grid supply, Synchronization Relay and protection devices such as Earth Fault Relay.
(2) Transmission Lines model: Consists of PI sections that represent Transmission lines of several kilometres in length.
(3) Distribution Station model: Consists of switching devices and R, L, C loads.

Students perform several experiments on this setup such as

(1) Evaluate ABCD Parameters
(2) Ferranti Effect
(3) Surge Impedance Loading
(4) Study Of Voltage Regulation
(5) Study Of Efficiency.

In most cases, there would be a group at least 4 students performing the experiment on each experiment setup. In such cases, the entire class would be divided into batches and will have to perform the experiment on this setup. It so happens that a student would get to do (mostly see others do) the experiment only once in the entire semester (only once in the entire engineering course as they are not repeated in the following semesters) and are expected to deliver results during the exams! The reason why students (learning under such circumstances) have very little practical knowledge is because of this condition wherein they hardly get to perform hands on experimentation during their entire Engineering course!

The purpose of setting up this Remote Labs for Electrical Power Transmission Lines Simulation Unit is to address these challenges and help students get access to the experimentation setup in addition to their regular lab hours.

3 Approach

Remote Labs for Electrical Power Transmission Lines Simulation Unit consists of a Generating station model, Transmission Lines model and the Distribution station model just as the conventional Transmission lines setup. The difference is that this can be accessed from a remote location too. This setup is a fully computerized and all the setup related parameters could be set from a remote computer screen.

The first screen shown below represents Synchronous window. This screen facilitates synchronizing two power supplies, namely Grid supply and Alternator supply. Grid supply is the one that is provided from mains supply and Alternator or Generator supply is the one that is the generating power from DC motor generator setup. The DC Motor – Alternator coupled set up is used to generate electricity in a controlled environment.

There are two options provided to synchronize the power supplies. One is Manual Sync and second one is Auto Sync. Synchronization is nothing but matching of both the frequencies and Amplitudes (Fig. 1).

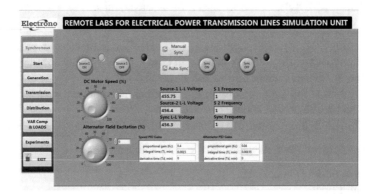

Fig. 1. Synchronous window

(1) Manual Sync

In this process we need to vary speed of the DC motor by varying speed we can control Frequency of the Generator setup and by varying Alternator Field Excitation we can control applied voltage of the generating unit.

(2) Auto Sync

For this process closed loop PID controllers are designed with appropriate sensors providing feedback, which can measure the system parameters in real time and provide control commands accordingly.

Start window provides overview and real-time values of Generation, Transmission and Distribution units (Figs. 2, 3 and 4).

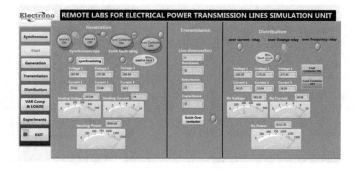

Fig. 2. Start window

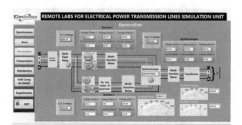

Fig. 3. Generation unit UI screen

Fig. 4. Generation unit equipments

The hardware setup of Generating station consists real time values of electrical power parameters from Grid supply and Alternator supply. The Synchronization of both these electrical sources is achieved using syncroscope. The real time values of pre and post synchronization can be captured and analyzed accordingly.

The length of each and every PI section represents 30 km and each phase consisting of six no. of PI sections. The length of each phase represents 180 km of transmission lines.

The Fig. 8 is that of Distribution station. Transmission station output will be the input to the distribution station. This screen measures the Transmission output parameters. At the receiving station, the high voltages from the transmission unit are stepped down to the levels that could be fed to the loads connected to the distribution unit. Due to the dynamics in the load variations connected to the distribution units, protection unit which can sense the over voltage, under voltage, over current, under current, over frequency and under frequency and so on is used. This device protects the distribution system infrastructure from failure by causing a trip and disconnecting the incorrect load accordingly.

Since there are several inductive and capacitive loads in a distribution networks of utilities, the phase difference between voltage and current is created (known as power factor) that increases the reactive power component in the distribution grid leading to reduced efficiencies. In order to reduce the reactive power, compensators such as VAR compensators are used (Figs. 5, 6, 7, 8 and 9).

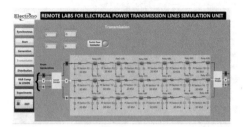

Fig. 5. Transmission unit UI screen

Fig. 6. Transmission unit equipment

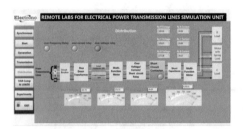

Fig. 7. Distribution unit UI screen

Fig. 8. Receiving/distribution unit equipment

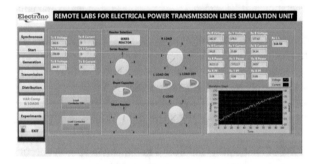

Fig. 9. RLC load and VAR compensator UI screen

The combination of Loads and VAR compensator are mentioned below.

a. Induction Load with Shunt capacitor compensation

Here we are compensating Inductive load with shunt capacitor compensation.

Current Zero crossing: 33 ms (one of the zero crossings represented)
Voltage Zero crossing: 33 ms (one of the zero crossings represented)
Time Difference: 0 ms – Current & Voltage in phase
Phase difference: 0° (Current & Voltage in phase) (Fig. 10).

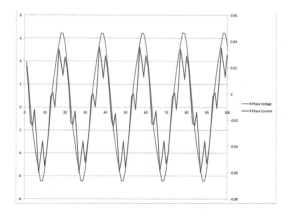

Fig. 10. Inductive load with shunt capacitor compensation

b. Capacitive Load

Current Leads Voltage – Therefore this is a capacitive load (Fig. 11)
Current peak: 26 ms
Voltage peak: 30 ms
Time Difference: 4 ms – Current leads Voltage
Phase difference: 72° (Current leads Voltage) (Note: 4 ms × 18 deg/ms = 72°).

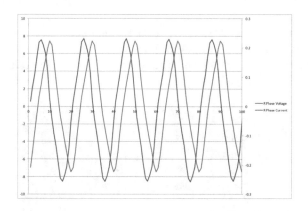

Fig. 11. Capacitive load

c. Inductive Load

Voltage Leads Current – Therefore this is inductive load
Voltage peak: 39 ms
Current peak: 38 ms
Time Difference: 1 ms – Voltage leads Current
Phase difference: 18° (Voltage leads Current) (Note: 1 ms × 18 deg/ms = 18°).

The following list of experiments could be performed using this setup.

1. Evaluate ABCD Parameters
2. Ferranti Effect
3. Surge Impedance Loading
4. Study Of Voltage Regulation
5. Study Of Efficiency (Fig. 12).

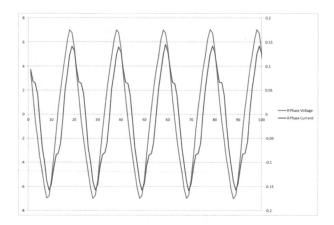

Fig. 12. Inductive load

ABCD parameters (Fig. 13) describes the relation between the sending end and receiving end voltages and currents. A more intuitive understanding would be to calculate the sending end and receiving end parameters with known impedance, admittance and gains (Figs. 14 and 15).

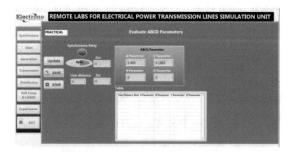

Fig. 13. ABCD parameters

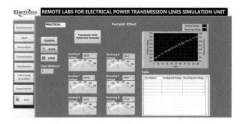

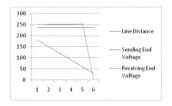

Fig. 14. Ferranti effect experiment

Fig. 15. Ferranti effect graph with data

For Long Transmission lines

1. $A = \sqrt{Zoc/Zoc - Zsc}$
2. $B = A\ Zsc$
3. $C = A/Zoc$
4. $A = D$

Ferranti effect experiment - Ferranti effect is when current drawn by distributed capacitance of the line itself is greater than the current associated with the load at receiving end of the line. This capacitor charging current leads to voltage drop across the line inductor of the transmission system, which is in phase with the sending end voltages.

Here we are plotting data between sending end and receiving end of the transmission line network at different no of cascading PI sections.

Plotted data represents Line distance, Sending End Voltage and Receiving End Voltage. Here total line distance is 180 km. Each line consisting of six no of PI sections and length of each PI section is 30 km (Fig. 16).

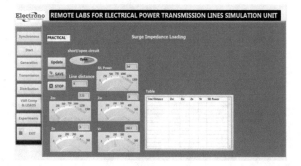

Fig. 16. Surge impedance loading experiment

This experiment is for Surge Impedance Loading. Long transmission line have distributed inductance and capacitance as it is inherent property. When line is charged, the capacitance component feeds reactive power to the line while inductance components absorbs the reactive power. Due to the dynamic variation in the inductive loads connected to the grid and disconnecting from the grid, it causes voltage/current surges accordingly (Fig. 17).

$$I2\omega L = V2\omega C$$

$$V/I = \sqrt{L/C} = Zn$$

$$Zc \text{ or } Zn = \sqrt{ZocZsc}$$

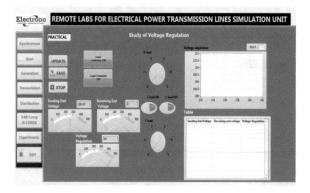

Fig. 17. Study of voltage regulation experiment window

The above screen is the experiment for Study of Voltage Regulation. The percentage of voltage difference between no load and full load voltages of a system with respect to its full load voltage (Fig. 18).

$$\% \text{ regulation} = |Vr(NoLoad)| - |Vr(Load)|/|Vr(Load)| * 100$$

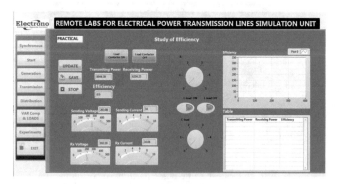

Fig. 18. Study of efficiency experiment window

The above image representing experiment window of Study of Efficiency. The Efficiency is power delivered at the receiving end and power sent from the sending end.

%efficiency (η) = *Power delivered at the receiving end/Power sent*
from the sending end ∗ 100

4   How to Access This Setup Remotely?

By fulfilling the university formalities, you will receive the login id and password, by using Electrono cloud Software you can book your slot for a particular time and then you can take access to the lab where you can conduct your experiments remotely. For more details about the slot booking refer [1].

5   Actual or Anticipated Outcomes

The setup is integrated to the Server system for remote access and the students with valid username and password can access the setup accordingly. It is anticipated that with this system being used more often, there would be better practical exposure and hands on learning among students. The knowledge of building such remotely accessible systems could also be imparted.

Currently, this system being new and technologically more advanced in comparison to the existing laboratory equipments has created a sense of awe among students, teachers and lab attendants.

6   Conclusion

1. With Remote Labs for Electrical Power Transmission Lines Simulation Unit there each of the students with access credentials have access to the Laboratory setup 24 × 7, 365 days (excluding the maintenance down time).
2. Students get to access the Physical Hardware through internet that enables hands on experimentation experience. They also get to use the setup all be themselves by booking the user slots accordingly. In addition to enhanced learning facility, this enables the students to prepare themselves for the examinations better.
3. Awareness and Knowledge of Remote Technologies helps students prepare for the jobs of the future as more and more remotely operated systems would come up in future.
4. Remotely operable systems facilitate more date generation and data processing which is inherent to the technology used to build remote labs. This drives more data analytics which is again the futuristic learning and job trends.

References

1. Kalyan Ram, B., et al.: A distinctive approach to enhance the utility of laboratories in Indian academia. In: 2015 12th International Conference on Remote engineering and Virtual Instrumentation (REV), pp. 238–241 (2015)
2. Gaikwad, V., et al.: Laboratory setup for long transmission line. Int. Res. J. Eng. Technol. (IRJET) **04**(03), March 2017
3. Halkude, S.A., Ankad, P.P.: Analysis and design of transmission line tower 220 kV: a parametric study. Int. J. Eng. Res. Technol. (IJERT), 8 August 2014
4. Deka, J., Sharma, K., Arora, K.: Study of 132 kV transmission line design and calculation of its parameters. Int. J. Electr. Electron. Eng. (IJEEE), January–June 2016
5. Yindeesap, P., Ngaopitakkul, A., Pothisarn, C., Jettanasen C.: An experimental setup investigation to study characteristics of fault on transmission system. In: International MultiConference of Engineers and Computer Scientists, Hong Kong, 18–20 March 2015

Sustainability of the Remote Laboratories Based on Systems with Limited Resources

Galyna Tabunshchyk[1(✉)], Tetiana Kapliienko[1], and Peter Arras[2]

[1] Zaporizhzhia National Technical University, Zaporizhia, Ukraine
galina.tabunshchik@gmail.com, bragina.zntu@gmail.com
[2] Faculty of Engineering Technology, KU Leuven, Leuven, Belgium
peter.arras@kuleuven.be

Abstract. The sustainability of remote laboratories is considered both from the point of view of sustainable system architecture and of sustainable teaching outcomes. The goal of the work is to provide an approach for making remote laboratory systems build on components with limited (software) resources more sustainable, meaning more long-lasting, more economic, more efficient. We consider the whole of the sustainability of the development process, sustainability of system architecture and sustainability of teaching outcomes.

Keywords: Sustainability · Remote experiments · Raspberry Pi
Reliability study · Systems with limited resources

1 Introduction

Sustainability is a widely used term and refers to the capacity of something to last for a long time, in other words the capacity to support, maintain or endure.

To reach sustainability in the global world we need good education to prepare all children for a world which is durable. The ideas of endless growth in consumption and unlimited natural resources has been countered by the reality of limits in the production of fuels, food, fresh water and other natural resources.

The paradox that teaching engineers on the one hand is going for newer, faster, better, more productive systems and on the other hand going for more sustainable, people- and earth friendly techniques brings also responsibilities towards the way we use systems in education itself.

A variety of remote experiments are developed based on small minicomputer systems, such as Raspberry Pi, Arduino etc. [1–3] for reasons of low cost and simplicity. But the key feature of the minicomputer systems is their limited system resources [4], which leads to stricter requirements for end products based on them, such as: small amount of consumed RAM and low power consumption, modularity, simplified tools for configuring communications, support for a wide range wireless and sensor technologies. During development there should be paid special attention if such system can be sustainable from the point of view of functionality (due to limited resources) and from the point of view of sustainable results (efficiency of code).

Here we consider the possibilities to make remote laboratories more sustainable as from the side of development process as from the side of teaching outcomes. We will first define sustainability in remote labs and next look a case study in Zaporizhzhia National Technical University (ZNTU).

2   Sustainability of Remote Laboratories

As common architecture of remote laboratory contains from web-interface, controller and software for the hardware [5–8], and it developed according to the iterative life cycle [9] so dealing with sustainability of the remote laboratories you should deal with different approaches (Fig. 1).

Fig. 1. Different vectors in sustainability research

2.1   Sustainability of Systems with Limited Recourses

If to look at sustainability from the point of code it means that "the software you use today will be available - and continue to be improved and supported - in the future" which can be achieved by different approaches [10]. Even from this point of view, sustainability has many dimensions that relate organization to information systems and to software engineering. Also sustainability in development (Fig. 2) should be considered separately as it is about efficiency and balancing the needs on the short and long term [11].

Software architectures play a major role in large-scale systems' sustainability (that is, economical longevity), vastly influencing maintenance and evolution costs. It's thus desirable to measure the sustainability of a software architecture to determine refactoring actions and avoid poor evolution decisions [12, 13].

There are several areas in which software sustainability needs to be applied: software systems, software products, Web applications, data centers, etc. All software for these areas nowadays use a different architecture and for sustainability assessment it is more convenient to use an integrated approach which considers as well requirements as architectural design as source code.

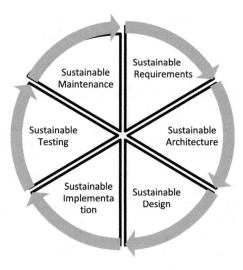

Fig. 2. Sustainability in software development

2.2 Sustainability Through Embedded Verification of Systems with Limited Resources

When using systems with limited resources it is feasible and a good practice to build an IT-system to diagnose the concept system. There are numerous of approaches for diagnostics of technical systems [14, 15], but they are not used for the web-oriented systems.

To have the possibility for a flexible exchange of parts and for constant improvement of the remote laboratories a verification module should be provided.

We are considering remote laboratories with web interface, for which there was defined a verification model:

$$MV = (interprototype, E_d, DB, rmv) \tag{1}$$

$$rmv : interprototype \rightarrow E_d \times E_h \times DB \mid \forall\, e_a, e_a \in \{rp(a), rm(a)\}, e_a \rightarrow e, e = \{e_d\}, \tag{2}$$

To determine the testing object in the Web-based systems verification model, an element e_d is highlighted. e_d is a resource or document, available over the network and identified by a unique URI. e_d is determined by the function:

$$rd : C_A \rightarrow E_d \mid \forall c \in C_A : e_d = rd(c), e_d \cup E_d \tag{3}$$

where rd – communication of atomic concepts and the information system web resources; C_A – atomic concepts set; c – concept; E_d – set of the information system web-resources; e_d – information system element.

For verification of web-oriented systems [16], the concept of a "functional unit of the system" was singled out - an elementary structural component of an information system (IS) that implements a complete functional block, for verification of which one or more automated or automatic test tests can be developed:

$$F = (E_d, P, r) \mid \forall\, e_\mathrm{d}, e_\mathrm{d} \in E_\mathrm{d}, \exists p, p \in P, \tag{4}$$

where F – set of the system functional units; P – test cases set; p – test case.

The functional unit (FU) is selected depending on the type of IS. For example, if the developed IS requires communication with a database or accessing a remote application having a different data format, the FU is the function of communicating with the database (remote application).

Based on the information systems verification model and the ontological model for the adaptive web environment that provide the basic concepts of IS and allow them to be related to the requirements for quality and testing tools, it is necessary to take into account the characteristics of systems with limited resources to create an appropriate verification model. It is also necessary to expand the notion of FU in order to automate the process of testing not only software, but also hardware and data transfer between elements of a system with limited resources.

It is necessary to modify this model for use in conjunction with a dynamic verification model to test the transmission of data between elements of a system with limited resources (Wi-Fi, Bluetooth, Ethernet, etc.).

The peculiarity of modern embedded systems is that the architecture of the system also becomes adaptive, for which we determine that the set of explicit UP(u) and implicit UM(u) user requirements includes not only a number of elements of the prototype interface, but also a set of requirements for hardware elements. We take a lot of resources for E, the set of adaptive hardware elements will be denoted by E_h, then for the adaptive built-in system we will assume.

In order to extend the concept of FU with the goal of automating the testing of hardware and data transmission between elements of a system with limited resources, it was proposed to integrate a set of tests for verifying hardware in (3) $P_h = \langle I, O, O_s \rangle$. The purpose of this complex is to test the correctness of the received data (O) from the hardware in response to a sent request containing a certain set of input data (I) by the black box method. The received set of answers O is compared with the reference value O_s and with the communication protocol, on the basis of their correspondence, a decision is made as to the correctness of the functioning of the hardware under test.

In accordance with the foregoing, many FU systems with limited resources will take the form:

$$F = (E_d, E_h, P, P_h, r) \mid \forall\, e_\mathrm{d}, e_\mathrm{d} \in E_\mathrm{d}, \exists p, p \in P, \exists p_h, p_h \in P_h, \tag{5}$$

where E_h – set of adaptive elements of information system hardware; P_h – set of test cases to verify the hardware; p_h – test case for hardware verification.

To calculate the transmission period for data from the i-th minicomputer system, we use:

$$\eta_i' = (1 + \omega_i)(\tau V log_i + \sigma Verr_i)(\frac{\tau}{Varch_i} - \frac{1}{\upsilon_{net}}),\qquad(6)$$

where η_i' – the period of archival files transfer from the i-th object of tests for storage in the database; ω – safety factor; τ – predictable device operating time; Vlog – reserved memory for verification results; Verr – the emergency file volume; Varch – the archive file volume; σ – allowable number of minicomputer systems failures; υ_{net} – average data rate over the network.

The results of verification can be represented as Y_i in the form of sets

$$Y_i = \ <Y_{Log}, \ Y_{Res}, \ Y_{resp}, \ Y_{err} > ,\qquad(7)$$

where Yi – verification results; Y_{Log} – verification progress log; Y_{Res} – verification result; Y_{resp} – response time; Y_{err} – information about errors found.

Given these advantages, the model for the verification of systems with limited resources MV will take the form:

$$MV = (interprototype, E_d, E_h, DB, rmv, Y, \eta),\qquad(8)$$

$$rmv : interprototype \rightarrow E_d \times E_h \times DB \ | \ \forall \ e_a, e_a \in \{rp(a), rm(a)\}, e_a \rightarrow e, e = \{e_d\},$$

where *interprototype* – many elements of the Web-based systems interface prototype; *DB* – data store; *rmv* – communication of the prototype elements with web-resources and the database of the information system; *ea* – elementary unit of the information system interprototype; *rm(a)* – implicit user requirements; *rp(a)* – explicit requirements (profile) of the user; *e* – subset of information system resources.

2.3   Sustainability of Learning Outcomes

Sustainability in education covers a lot of different issues as well.

One can consider the sustainability of the knowledge transfer to students, the durability of buildings or equipment, the ability to stimulate a sustainable (lifelong learning) attitude in students and staff.

In this case we look at the sustainability of computerized teaching aids. More and more knowledge transfer to students is supported by software-tools. Building expertise in a tool and creating the necessary didactics for using it requires (a lot of) time. Unfortunately however, due to the ever continuing change of computers, OS, programming languages, the specialist dedicated software tools are very often outdated and become obsolete and as such not very sustainable. Even commercial software tools suffer from this same problem. While newer versions of a software is often not more than a newer/fancier user interface, these changes divert attention from what educators really want to put in the tools as knowledge or expertise. This is a challenge when using software tools [16–18].

This shows the necessity for a sustainable construction of the software tools: tools should easy to maintain, open for change and very well documented. These 3 conditions too often clash with the reality of software coding, being often not very well documented and sometimes complex because of efficiency of programming.

Therefore for sustainable development of software tools we put effort in modular building blocks, low cost computing tools with more limited resources, as such forcing developers to keep it simple, documentation, and a multidisciplinary team (teachers, technicians, programmers) to streamline and watch over the sustainability (adaptability and future use) of the development.

2.4   Case Study. Interactive Platform for Embedded Software Development Study

The developed Interactive platform for Embedded Software Development Study (ISRT) [6] consists of a number of smaller dedicated experiments which allow students to study and experiment on different aspects of embedded systems and communication tools over the internet. The aim is to prepare students for IoT (Internet of Things). The series of experiments include experiments on the manipulation of components (LED-lights, stepper motors), on communication (mobile phone manipulation), on security (face detection through image detection) and programming (in C++, Python).

The ISRT was built to let students experiment and self-study the different aspect of programming and controlling. After the self-study, students get a project assignment in which they use the different skills they acquired using the ISRT. Evaluation of the learning outcomes was done on the project results.

If to look at sustainability of the development process from the point of code it means that "the software you use today will be available - and continue to be improved and supported - in the future" which can be achieved by different approaches. Even from this point of view, sustainability has many dimensions that relate organization to information systems and to software engineering. Furthermore sustainability in development should be considered separately as it is about efficiency and balancing the needs on the short and long term.

Software architecture plays a major role in large-scale systems' sustainability (that is, economical longevity), and is vastly influencing maintenance and evolution costs. It is thus desirable to estimate the sustainability of a software architecture to determine refactoring actions and avoid poor evolution decisions. The peculiarity of embedded systems is that the (hardware) architecture of the system also becomes adaptive and it is difficult to foresee all errors which can occur when replacing old components. So the authors suggest a verification method for systems with limited resources which allows to eliminate risks of failure caused by reusing (software) components [19] (Fig. 3).

For the experiments, the architecture of the ISRT system at ZNTU was modified. The connected hardware of the laboratory was distributed between three Raspberry Pi-devices, and the functionality was expanded with an administrative panel for specifying the functional units. Software architecture was also distributed which allows to increase the number of experiments and their efficiency.

Sustainability is achieved by using a modular approach in the conception of the software, allowing the re-use of blocks of code. Secondly, the option was made to use widespread programming languages (C++, Python) - which is not always the best solution for the tasks - but eliminates the risk of becoming obsolete soon.

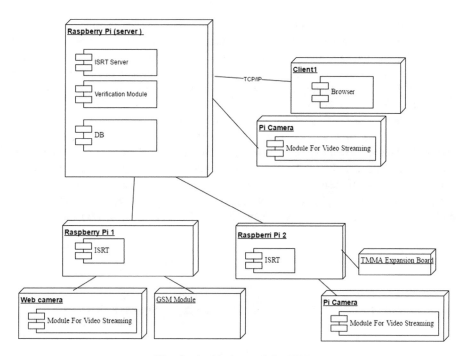

Fig. 3. Architecture of the ISRT

Flexibility of the remote experiments can provide necessary sustainability, as functionality of such systems can be increased only by increasing the number of hardware involved into the experiments. So it can decrease the risks of system failure after upgrading of the system. In Table 1 is the results of the system verification after implementation of the model of verification to the ISRT system.

Table 1. Results of the system verification after implementation of the model of verification

Information system resource number	Test results		
	Hardware, p_h	Software, p_e	Loss of data during transmission (%)
1	Passed	Passed	0
2	Passed	Passed	0
3	Passed	Passed	0
4	Passed	Passed	0
5	Passed	Passed	0
6	Passed	Failed	0
7	Passed	Passed	0
8	Passed	Passed	0
9	Failed	Failed	100

For sustainability of learning outcomes, the initial evaluation methods for students were assessed on efficiency on their contribution to the learning process. A shift between formal and informal evaluation methods was experimented to ensure that the ISRT steers the learning process of the students towards a project driven approach. To eliminate the risks of losing teaching content and outcomes the system model of evaluation was modified.

In the new approach, students are questioned after each module to show their knowledge. This formal intermediate test forces students to reflect on their knowledge and skills. This interrogation doesn't results in marks, but is solely used as a motivator. It is done intermediate during the project assignment: not at the start, as not all students maybe completed the use of the remote experiments, not at the end as at that time students cannot benefit any longer from the feedback (Fig. 4). After completion of the project a formal test with marks is organized in the form of a presentation and an exam.

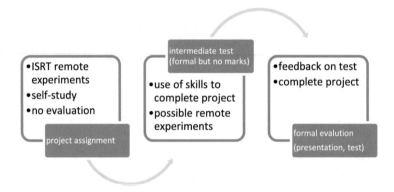

Fig. 4. Project assignment

3   Conclusions

Sustainability of the remote laboratories is a multidimensional issue. In the article the authors provides an analysis of different aspects that influence on sustainability from sustainable system architecture to sustainable learning outcomes.

Acknowledgement. This work is carried out within the framework for the scientific research work carried at the Software tools department of Zaporizhzhia National Technical University "Informational system of diagnostics of mini-computer systems in the multi-component environment" (Governmental registration number 0117U000615), party with the support of Erasmus + KA2 project Internet of Things: Emerging Curriculum for Industry and Human Applications ALIOT 573818-EPP-1-2016-1-UK-EPPKA2-CBHE-JP and Erasmus + KA1 project.

References

1. Gaglio, S., Lo Re, G. (eds.): Advances in Intelligent Systems and Computing. How Ontologies Make the Internet of Things Meaningful, p. 349. Springer, Cham (2014)
2. Bell, C.: Beginning Sensor Networks with Arduino and Raspberry Pi, p. 372. Apress, New York (2013)
3. Warren, G.: Raspberry Pi Hardware Reference, p. 248. Apress, New York (2014)
4. Parkhomenko, A., Gladkova, O., Sokolyanskii, A., Shepelenko, V., Zalubovskiy. Y.: Implementation of reusable solutions for remote laboratory development. Int. J. Online Eng. (iJOE) 12(7), 24–29 (2016)
5. Parkhomenko, A., Gladkova, O., Ivanov, E., Sokolyanskii, A., Kurson, S.: Development and application of remote laboratory for embedded systems design. Int. J. Online Eng. (iJOE) 11(3), 27–31 (2015)
6. Tabunshchyk, G., Van Merode, D., Arras, P., Henke, K., Okhmak, V.: Interactive platform for embedded software development study. In: Auer, M., Zhutin, D. (eds.) Online Engineering and Internet of Things, pp. 315–321. Springer, Berlin
7. Henke, K., Fäth, T., Hutschenreuter R., Wuttke, H.D.: GIFT-an integrated development and training system for finite state machine based approaches. In: Online Engineering & Internet of Things, pp. 743–757. Springer, Berlin, Cham, Jan 2018
8. Henke, K., Vietzke, T., Hutschenreuter, R., Wuttke, H.-D.: The remote lab cloud "GOLDi-labs.net". In: Proceedings of 13th International Conference on Remote Engineering and Virtual Instrumentation REV 2016, Madrid, pp, 37–42, February 2016
9. Arras, P., Henke, K., Tabunshchyk, G., Merode, D.V.: Iterative pattern for the embedding of remote laboratories in the educational process. In: 12th International Conference on Remote Engineering and Virtual Instrumentation (REV 2015), Bangkok, Thailand, pp. 52–55, 25–28 February 2015
10. Approaches to Software Sustainability. Software Sustainability Institute. https://www.software.ac.uk/resources/approaches-software-sustainability
11. Calero, C., Piattini, M.: Part 1. Introduction to green in software engineering. In: Green in Software Engineering, pp. 3–27. Springer, Berlin (2015)
12. Koziolek, H., Domis, D., Goldschmidt, T., Vorst, P.: Measuring architecture sustainability. IEEE Softw. 30(6), 54–62
13. Penzenstadler, B., Raturi, A., Richardson, D., Calero, C., Femmer, H.: X systematic mapping study on software engineering for sustainability (SE4S). Franch Universitat Politècnica de Catalunya
14. Subbotin, S., Oliinyk, A., Skrupsky. S.: Individual prediction of the hypertensive patient condition based on computational intelligence. In: Proceedings of the Conference on Information and Digital Technologies, IDT 2015, Zilina, pp. 336–344. Institute of Electrical and Electronics Engineers, Zilina, 7–9 July 2015
15. Oliinyk, A.A., Subbotin, S.A., Skrupsky, S.Y., Lovkin, V.M., Zaiko, T.A.: Information technology of diagnosis model synthesis based on parallel computing. Radio Electron. Comput. Sci. Control 3, 139–151 (2017)
16. Tabunshchyk, G.V., Kapliienko, T.I., Shytikova, E.V.: Verification model for the systems with limited resources. Radio Electron. Comput. Sci. Control 4 (2017)
17. Arras, P., Van Merode, D., Tabunshchyk, G.: Project oriented teaching approaches for e-learning environment. In: IEEE 9th International Conference on Intelligent Data Acquisition and Advanced Computing System, IDAACS (2017)

18. Horniakova, V., Arras, P., Kozík, T.: Challenges of using ICT in education. In: 9th International Conference on Intelligent Data Acquisition and Advanced Computing System (IDAACS), IEEE (2017)
19. Arras. P., Kolot, Y., Kuna. P., Olvecky, M., Simon, M., Tabunshchyk, G.: New teaching approaches in engineering. In: Kuna, P., Olvecky, M., Kozik, T., Nitra, U.K.F. (eds.) New Teaching Approaches in Technology, p. 264 (2017) ISBN: 978-80-558-1148-2; EAN: 9788055811482
20. Tabunschik, G.V., KaplIEnko T.I.: Patent. Ukraine. 116912, MKV G06F 19/00 Elektronna Informatsiyna sistema dlya gnuchkoi verifIkatij vbudovanih sistem; zayavnik i patentovlasnik ZaporIzkIy natsIonalniy tehnIchniy unIversitet. Zayavka u201612897. Opubl. 12.06.2017, N 11 (in Russian) (Electronic system for flexible verification of the embedded systems)

What Are Teachers' Requirements for Remote Learning Formats? Data Analysis of an E-Learning Recommendation System

Thorsten Sommer[✉], Valerie Stehling, Max Haberstroh, and Frank Hees

Cybernetics Lab, IMA/ZLW & IfU, RWTH Aachen University, Aachen, Germany
{thorsten.sommer,valerie.stehling,max.haberstroh,
frank.hees}@ima-zlw-ifu.rwth-aachen.de

Abstract. Teachers often have their own professional requirements for e-learning systems. However, these are often only subliminal known. In times of Industry 4.0 and AI, teachers e.g. in engineering are also confronted with the need to teach increasingly complex concepts. Those are only two of the reasons why an e-learning recommendation system has been developed to support teachers in choosing an e-learning format. To better understand the perspective of the teachers, the central question is: What are the teachers' requirements for the e-learning formats examined here? After a introduction to the recommendation system, the analysis of the collected data is explained. Based on recommendations given in the past, we examine which requirements have led to a clear recommendation or to the advice against individual formats. Among the formats considered here, virtual reality and simulations are the most recommended on average, as they are best suited to the teacher's requirements. Subsequently, the profound results in the areas of virtual laboratories, virtual reality, simulations and gaming-based solutions will be presented and discussed. However, the results also show how diverse the requirements are. The recommendation for e-learning developers and companies is therefore: e-learning solutions should be adaptive for teachers and students. Finally, it can be concluded that in the future teachers will have to use a mix of different e-learning solutions in order to be able to teach the increasingly complex world of tomorrow.

Keywords: Recommender system · Requirements · E-Learning
Remote solutions

1 Introduction

Teachers have personal and professional demands on e-learning systems, often without being aware of this fact cf. [1, 2]. This way, e-learning systems often are in use that do not fully meet the actual requirements and circumstances of the teaching and learning scenarios. If a teacher is not an expert on the subject, it is usually rather difficult to make a well-directed, personal selection [3]: there are approximately 223 million search results for websites [3] and tens of thousands of specialist books on the subject of e-learning. For this reason, an e-learning recommendation system has been developed at RWTH Aachen University within the project "ELLI – Excellent Teaching and Learning in

© Springer International Publishing AG, part of Springer Nature 2019
M. E. Auer and R. Langmann (Eds.): REV 2018, LNNS 47, pp. 207–216, 2019.
https://doi.org/10.1007/978-3-319-95678-7_23

Engineering Sciences" [3, 4]. The system determines the teacher's requirements for a system by means of a questionnaire and then recommends a variety of possible formats.

As Tan et al. explain, teaching and learning are changing [5]: In the past, the focus lay mainly on facts that had to be learned, but today and in the future learning how to understand and solve complex problems is becoming more important. This becomes even more apparent in the context of Industry 4.0: to meet the constantly changing requirements of this paradigm, students have to learn how to integrate lifelong learning in their daily work and life cf. [6, 7]. An example for the growing complexity is the increasing degree of automation, where an individual employee can now control an entire factory instead of being responsible for a single machine cf. [8, 9]. This however, requires further cognitive abilities and skills that proceed the knowledge about the functionalities of each machine within the factory. In addition, humans and robots are going to work together in teams in the near future [10]. This also leads to new demands on teaching and learning. Thus, it may be necessary to teach the principles of modern robotics to a wider audience in order to understand the limits of robots. Furthermore, Stehling et al. show that the cohorts in the courses of study in Germany are currently increasing: Course sizes vary from around 10 to 30 to a maximum of 1,500 to 2,000 students at RWTH Aachen University cf. [11]. Courses of this size have different requirements for the e-learning formats used.

The aforementioned recommendation system supports teachers in the described situations. This makes it easier for teachers to react to new requirements, for instance. However, the actual requirements of teachers to e-learning systems are virtually undisclosed. Therefore, the central question of this paper is: What are the teachers' requirements for the e-learning formats examined here? For this reason, the actual recommendations of the past are analysed: Which requirements led to a recommendation and which to advise against individual formats?

In the further course of this paper, the developed recommendation system will be explained first. This is followed by an explanation of the statistical analysis of given recommendations, the results obtained and the discussion of these results. At the end, a conclusion is given with a summary and the outlook.

2   Underlying Recommender System

In order to be able to support the lecturers in the best possible way, an easy-to-use recommendation system has been developed in the project "ELLI – Excellent Teaching and Learning in Engineering Sciences". After considering various technical solutions cf. [4, 13], the system has been developed in the language Go [14] and operated with the MongoDB database cf. [15]. The source code is available as open source [16] and can be operated as a Docker container.

The teachers find a link to a questionnaire on the front page. The 18 questions aim at identifying the teacher's requirements for a specific lecture of theirs. The questionnaire can be viewed online in the recommendation system (www.elearning-finder.net) and in the open source code[1]. Most of the questions are answered by a three-part scale:

---

[1] https://github.com/Cybernetics-Lab-Aachen/Re4EEE/blob/master/staticFiles/Data.xml.

"Yes, the described functionality must be included", "the described functionality must not be included" as well as "the described functionality is irrelevant". An exception to this is question nine: It asks whether the e-learning format should replace or support the on-site event. Question 6 asks for the number of students and is therefore also not answered with the three-part scale.

All responses are stored anonymously in the database. Finally, the teacher receives his recommendation. For this purpose, well-known e-learning formats have been combined into 16 groups, for instance "VR-Based Learning" and "Online Whiteboards". Each group receives a rating based on the answers given. The given recommendation is also saved along with the answers. The questions and the rating logic have been developed and evaluated in several expert workshops and subsequently adapted accordingly.

The recommendation system has been open to the public since 27. May 2015 and has been collecting anonymous data ($N = 283$) from teachers using it ever since: The system has been used 74,759 times since then. It is available in English as elearning-finder.net and in German as elearn-o-mat.net. Since then, the system has been continuously enhanced. The stated goal is that the recommender independently identifies new e-learning trends on the Internet and presents them to teachers. For this purpose, data science processes were investigated and evaluated that can analyse data from social media as well as from websites. Based on the collected data, we analysed the requirements of the teachers. The analysis methods and results of the analyses are presented and discussed in the following chapters.

3   Analysis of Teachers' Requirements

The data records were analyzed with the statistics software "R" and "RStudio". The 283 available recommendations were complete, resulting in a population size of $N = 283$. To focus on specific topics, the groups "VR Labs", "VR-Based Learning", "Game-Based Learning" and "Simulation-Based Learning" were considered for this study. These groups have in common that they can be used as a remote solution. Other e-learning formats can also be used remotely, but were not explicitly considered for this evaluation: These include MOOCs that require their own consideration, as they tend to replace attendance-based teaching. Similarly, learning management systems, wikis and blogs that can be used to support teaching but often do not work as a standalone solution.

As mentioned above, the recommendations of the past serve as a foundation for the investigations. Thus, the groups of e-learning formats from the recommendations were analyzed based on their position in the recommendations. This way, it is possible to statistically evaluate how often which of these groups took a particular place in the recommendations. Secondly, it was investigated how well these groups met the requirements of the teachers. Two extremes are of special interest: Which groups are unsuitable for teachers and which groups are particularly suitable? The lower and upper 30% matches were considered for this. These intermediate results from the second step lead to the third analysis: Here, all requirements from the questionnaires that have led to a poor or good result can be qualitatively examined. This triad of analysis steps led to the results in order to answer the question presented at the beginning. In order to support the qualitative analysis, several heat map visualizations were created to illustrate patterns in the requirements.

4   Results

This section introduces the statistical location parameters of the positioning and compliance analysis to describe the sample. Also, the qualitative results are subsequently shown. First of all, the "VR-Based Learning" group is the sixth most common format, close followed by simulations. Virtual laboratories are recommended as the ninth most common e-learning format and game based solutions reach the eleventh place. Figure 1 shows this and also illustrates the mean value and distribution.

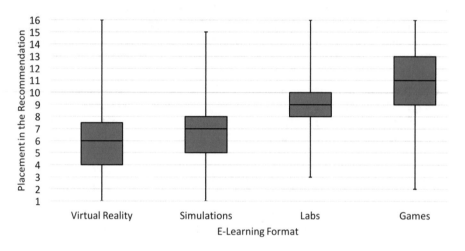

Fig. 1. Placement of e-learning formats in the recommendations

For further analysis it is important that a wide distribution is given. As shown in Fig. 1, this is the case. This is important for the following analysis: The distribution shows that different requirements lead to different recommendations. Figure 2 shows how many percent of the teachers' requirements are met by the different formats.

The next step of the analysis yields the heat maps shown in Figs. 3, 4, 5, and 6. It shows how the 18 questions were answered by the teachers and how frequently the answers were. The darker the areas are, the more frequently an answer was chosen. Answer 1 means that a teacher needs the corresponding requirement. Answer 0 stands for the fact that this requirement is not desired. The answer * indicates that this requirement has no relevance for the teacher. Question six was omitted because it cannot be evaluated in this way. In question nine a valid answer is either 1 or 0.

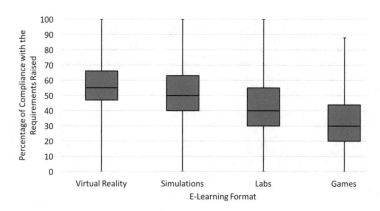

Fig. 2. Percentage of compliance with the requirements raised

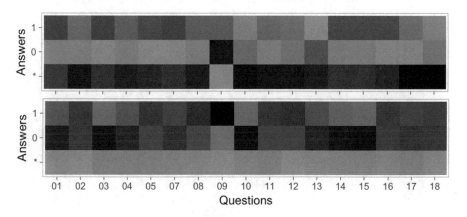

Fig. 3. Heat maps for virtual reality (first the lower 30%, then the upper 30%)

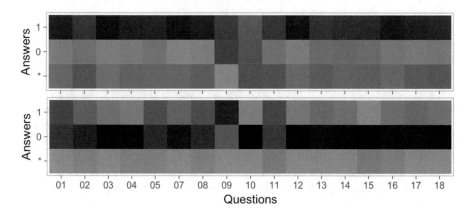

Fig. 4. Heat maps for laboratories (first the lower 30%, then the upper 30%)

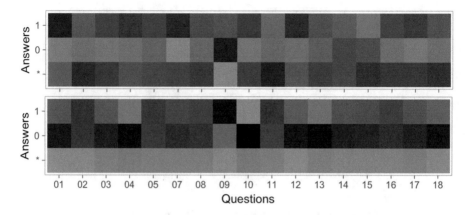

Fig. 5. Heat maps for simulations (first the lower 30%, then the upper 30%)

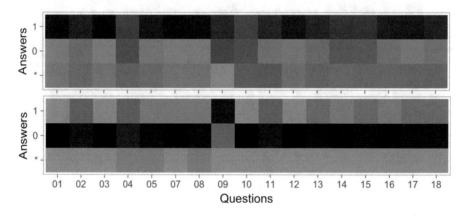

Fig. 6. Heat maps for game-based formats (first the lower 30%, then the upper 30%)

In the next step, the heat map was qualitatively examined for conspicuous patterns. In the following these patterns are shown and interpreted. Teachers who are not concerned about most of the requirements should therefore not use a virtual reality setting. Teachers who consider all requirements for their course to be met cannot use labs and games. Question 9 deals with the requirement regarding the presence of students. Most teachers want to use e-learning formats to support attendance teaching.

More in-depth insights show interesting details for the individual groups. For virtual reality (cf. Fig. 3), teachers do not want to embed videos (Q1), students should not be able to comment on the content (Q3), content should not be placed in the cloud (Q10), students should not be assigned roles (Q13) and higher mathematics should not be used (Q14, Q15). Among other factors, this combination leads to a recommendation for virtual reality.

In the case of laboratories (cf. Fig. 4), the most important requirement is that virtual and remote laboratories do not run in the cloud (Q10). Accordingly, the requirement would have to be that they run on the university's infrastructure (Q11). Interestingly, some teachers contradict here because the answers are relatively homogeneously distributed. It is also important to the lecturers that the students cannot comment on the content (Q3), the students do not have a live chat (Q4), the laboratories should not contain a function for evaluating the students (Q12), mathematical formulas should not be able to be displayed (Q14, Q15) and the laboratories should not be designed in an explorative way (Q18).

The analysis for the simulations (cf. Fig. 5), on the other hand, shows a more elaborate pattern, especially if the requirements do not fit. Teachers want to embed videos (Q1) and downloads (Q7) in simulations and these are intended to replace on-site activities (Q9). Furthermore, simulations should provide a function for the assessment of students (Q12). The requirements to recommend simulations with 70–100% are no videos (Q1), no live chats (Q4), support for on-site teaching (Q9) and no cloud (Q10).

Finally, regarding game-based formats (cf. Fig. 6), teachers are requested not to display videos (Q1), students should not be able to comment on content (Q3), downloads should not be possible (Q7), on-site teaching should be supported (Q9), solutions should work without a cloud (Q10) and no evaluation of students should take place (Q12). All results are discussed and interpreted in the following section.

5   Discussion

The different findings are discussed and interpreted in this section. For the groups considered here, it can be said that the most likely choices for the average teacher are virtual reality and simulations, and least game-based formats. At the same time, each of the formats considered here can be the ideal solution for an individual teacher. This also emphasizes the importance of such a recommendation system, since there is no one uniform e-learning solution for all teachers.

Although virtual reality is the most recommended of the four formats, the biggest obstacle at present is that this technology has not been widely spread. Because these devices are rarely found in households, it is difficult to use such solutions for remote teaching. Instead, such devices would have to be available in the university, e.g. in computer pools. However, virtual reality and especially augmented reality offer interesting possibilities in many areas of teaching. Due to the continuous development of this technology, entry barriers and costs are reduced, so that the propagation will probably progress quickly.

When considering the requirements of the teachers it is noticeable that some requests are not necessarily didactically useful. For example, the requirements that students are not allowed to comment on content and that live interactions are not permitted. Commenting on the content trains students in critical thinking and promotes scientific work. Preventing live interactions in the age of WhatsApp is at least worth discussing. These responses could come from the fact that the recommendation system might be more likely to be used by teachers who are rather conservative in their attitude towards digitalization. It should also be noted that the sample is not representative of all teachers.

For the formats considered here, it can be seen that most of the lecturers do not want to use cloud services or that the universities do not allow this. Many teachers do not want to use videos, students should not be able to comment on content and the e-learning formats should support on-site teaching.

As the detailed results show, the requirements of the teachers are diverse. It can also be assumed that experienced teachers may not use the recommendation system and that the requirements would be even more diverse in this respect. Developers of e-learning formats should therefore ensure that the tools are adaptable or that functions can be switched on and off as required.

6   Conclusion

Teachers' requirements for selected e-learning systems were investigated in order to be able to give recommendations to developers, for example. Surprisingly, most of the teachers state that they do not want or are not allowed to use cloud services. Companies and developers in the context of e-learning should consider this fact and provide adequate solutions. It was also unexpected that virtual reality solutions are the most recommended format among the groups considered here. This field will pose new challenges to universities, teachers and students, but will also offer them new opportunities in teaching and learning. A few details in the analysis of the data suggest that not all teachers are capable of handling the challenges of digitalization: One example is the desire for online solutions that are not operated in a cloud, but also not within the university. In times of WhatsApp, the desire to prohibit students from interacting live is antiquated. Universities need to train their teachers in order to keep up with the pace of innovation: Even if not every new trend automatically makes sense for teaching, it should be possible to evaluate the new possibilities.

The results and their statistical characteristics show how diverse the requirements of the teachers are. This shows how important such recommendation systems are for additional support: There is no one suitable e-learning format that fits every teachers' needs. Consequently, purely static information is not sufficient to provide teachers with comprehensive information on this topic.

7   Outlook

Since there is no one e-learning solution for all teachers, adaptive formats are promising for the future cf. [17]. Learning management systems are a suitable model: after decades of further development, they are already adaptable to the extent that they are recommended most frequently. These systems can be configured as required and adapted to teachers' and students' needs.

In addition, the e-learning recommendation system can also be enhanced. Thus, it would be conceivable that the system would point out inconsistent or untypical answers. Such a change is controversial, however: the recommendation system would manipulate the opinion of the teachers. In the future, however, teachers will have to accept and prepare for a broader e-learning mix in order to meet the complex demands of the information age. With the advent of the fourth industrial revolution, interrelationships become even more complex. Beyond even that, it is hard to imagine how a teacher would like to teach e.g. modern, complex AI systems without the targeted use of adequate e-learning technologies.

Acknowledgment. This work is part of the project ELLI, "Excellent Teaching and Learning in Engineering Sciences," and was funded by the federal ministry of education and research ("BMBF"), Germany.

References

1. Plumanns, L., Sommer, T., Schuster, K., Richert, A.S., Jeschke, S.: Investigating mixed-reality teaching and learning environments for future demands: the trainers' perspective. In: Proceedings of the 18th International Academic Conference, London, UK, pp. 596–609 (2015)
2. Anicic, K.P., Divjak, B., Arbanas, K.: Preparing ICT graduates for real-world challenges: results of a meta-analysis. IEEE Trans. Educ. **60**(3), 191–197 (2017)
3. Sommer, T., Stehling, V., Richert, A.S., Jeschke, S.: The e-learning-finder: an e-learning recommendation system for engineering education teachers. In: E-Learning Excellence Awards: An Anthology of Case Histories 2016, pp. 207–215. Academic Conferences and Publishing International Limited, London (2016)
4. Sommer, T., Bach, U., Richert, A.S., Jeschke, S.: A web-based recommendation system for engineering education e-learning solutions. In: Proceedings of the 9th International Conference on e-Learning (ICEL), Valparaiso, Chile, pp. 169–175 (2014)
5. Tan, A.Y.T., Chew, E., Mellor, D.: To infinity and beyond: e-learning in the 21st century. In: 2016 IEEE Conference on e-Learning, e-Management and e-Services, Langkawi, Malaysia, pp. 156–161 (2016)
6. Schuster, K., Groß, K., Vossen, R., Richert, A.S., Jeschke, S.: Preparing for Industry 4.0: collaborative virtual learning environments in engineering education. In: Engineering Education 4.0: Excellent Teaching and Learning in Engineering Sciences, pp. 477–487. Springer, Cham (2016)
7. Abdelrazeq, A., Janßen, D., Tummel, C., Richert, A.S., Jeschke, S.: Teacher 4.0: requirements of the teacher of the future in context of the fourth industrial revolution. In: Conference Proceedings of the 9th International Conference of Education, Research and Innovation, Seville, Spain, pp. 8221–8226 (2016)
8. Jiang, J.-R.: An improved cyber-physical systems architecture for Industry 4.0 smart factories. In: Proceedings of the International Conference on Applied System Innovation, pp. 918–920, Sapporo, Japan (2017)
9. Rong, W., Vanan, G.T., Phillips, M.: The internet of things (IoT) and transformation of the smart factory. In: 2016 International Electronics Symposium Proceedings, Denpasar, Indonesia, pp. 399–402 (2016)
10. Müller, S.L., Schröder, S., Jeschke, S., Richert, A.S.: Design of a robotic workmate. In: Digital Human Modeling. Applications in Health, Safety, Ergonomics, and Risk Management: Ergonomics and Design, Vancouver, Canada, vol. 10286, pp. 447–456 (2017)
11. Stehling, V., Bach, U., Richert, A.S., Jeschke, S.: Teaching professional knowledge to XL-classes with the help of digital technologies. In: Engineering Education 4.0: Excellent Teaching and Learning in Engineering Sciences, pp. 77–90. Springer, Cham (2016)
12. Johnson, L., Adams, S.B., Estrada, V., Freeman, A.: NMC Horizon Report: 2015 Higher Education Edition. The New Media Consortium, Austin (2015)

13. Sommer, T., Bach, U., Richert, A.S., Jeschke, S.: A web-based recommendation system for engineering education e-learning systems. In: Proceedings of the 6th International Conference on Computer Supported Education (CSEDU), Barcelona, Spain, pp. 367–373 (2014)
14. Meyerson, J.: The Go programming language. IEEE Softw. **31**(5), 101–104 (2014)
15. Kang, Y.-S., Park, I.-H., Rhee, J., Lee, Y.-H.: MongoDB-based repository design for IoT-generated RFID/sensor big data. IEEE Sens. **16**(2), 485–497 (2016)
16. Sommer, T., Kupper, R.: An E-Learning Recommendation System for Engineering Education Teachers (2017). https://github.com/Cybernetics-Lab-Aachen/Re4EEE
17. Agadzhanova, S., Tolbatov, A., Viunenko, O.: Using cloud technologies based on intelligent agent-managers to build personal academic environments in e-learning system. In: 2nd International Conference on Advanced Information and Communication Technologies (AICT), Lviv, Ukraine, pp. 92–96 (2017)

Evaluating Remote Experiment
from a Divergent Thinking Point of View

Cornel Samoila[1,2(✉)], Doru Ursutiu, and C.A. Neagu

[1] "Transylvania" University of Brasov, Brasov, Romania
{csam, udoru}@unitbv.ro, andrei.c.neagu@gmail.com
[2] Technical Science Academy of Romania, Bucharest, Romania
[3] Science Academy of Romania, Bucharest, Romania

Abstract. In recent years, education talks have shifted from the areas: "*classes*," "*schools*", "*universities*" to the "*learning environment*". This is because the Internet has introduced "*virtual environment*" into education. The map of study places has been significantly changed, adding to temporal, spatial and geographical learning, learning in an environment without geographical and temporal boundaries. The term "*learning environment*" now includes the classes, schools and universities we are talking about above, but also the virtual environment with its own classes, schools and universities, or other forms of education that have not been seen before (e.g. MOOC's). This process has determined a strong modification of the current teaching theories of learning.

The paper examines how an important learning element in the "*learning by doing*" category called "*remote experiment*", which belongs to the virtual environment, contributes to the development of creative thinking. As is know, the first who underline the importance of divergent thinking in creativity was Guilford. After him, the development of this subject has become exponential. The paper analyzes, using statistically methods, how the elements of divergent thinking such as fluency, originality, and flexibility are found in a remote experiment work, and how these independent variables can be regarded as producing "*treatment effects*", that is they can increase the potential of creativity.

Keywords: Divergent thinking · Treatment effects · F ratio · Blind variation
Null hypothesis

1 Introduction

According to Guilford (1961), many other researchers (Meyrs and Torrance - 1964, Treffinger and Bahlke - 1970, Mansfield, Busec and Krepelka - 1978, Renzulli - 2000) considered divergent thinking as a decisive element of creativity and advocated for its stimulation with special training programs. A number of researchers wanted to be able to identify the creatively-gifted people and so were launched some creativity tests (TTCT-Torrance-1964, Divergent Thinking Tests of Wallach and Kogan -1965, Runco - 1986, etc.). The specific aspects of the divergent thinking tested were: fluency, originality and flexibility (Meyrs and Torrance - 1964, Feldhusen, Treffinger and Bahlke - 1970, Convington, Davis and Olton 1974, Baer 1988, Renzulli 2000). In our opinion, Torrance is the researcher who studied in deep the problem and developed the

© Springer International Publishing AG, part of Springer Nature 2019
M. E. Auer and R. Langmann (Eds.): REV 2018, LNNS 47, pp. 217–225, 2019.
https://doi.org/10.1007/978-3-319-95678-7_24

best test of creativity (TTCT - Torrance Test of Creative Thinking) [2–4]. The definition of creativity from which he started is:

"...a process of becoming sensitive to problems, deficiencies, gaps in knowledge, missing elements, disharmonies and so on: identifying the difficulty, searching for solutions, making guesses or formulating hypotheses and possibly modifying and retesting them and finally communicating the results"

Starting from this definition, Torrance developed two types of tests: verbal and figurative. He believes that these tests not only measure creativity but can serve as a tool for improving it. Through them it is possible to individualize the training for each student in the direction of increasing the creative potential. The TTCT test - from which the idea of this paper started, uses 5 subscales (Table 1):

Table 1. The TTCT test subscales

Nr. Crt.	Subscale	Explaining the role
1	Fluency	Is appreciated the number of relevant ideas issued on a given topic
2	Originality	Is appreciated the number of non-communal ideas that come out of the ordinary ideas category
3	Flexibility	Is appreciated the number of given answers and their variety
4	Abstractness of titles	Is appreciated the ability to label a process which indicates abstraction power
5	Resistance to premature closure	Is appreciated the power of the evaluated subject to have an *"open mind"* even when the theme seems to have been exhausted

In the concrete case approached in the paper, the last two subscales cannot be considered as independent variables because:

- A *"remote experiment"* is an applicative singularity. Although it is part of a well-defined field, it cannot be labeled in abstracted way;
- At the same time a remote experiment does not record *"premature closure"*. It is not an exhaustive theme but a strictly defined theme.

As a result, in the paper, from the above scales, three independent variables, namely scales 1, 2, and 3, were considered for testing.

2 Reasoning

The virtual environment that has been added to the classical educational environment has produced, as mentioned before, a strong change in the angle of approach to instructive actions. From static, conservative places where is used *knowledge consumption* - is recorded, through the intervention of the virtual environment, a step-by-step shift towards combining *consumption* with *production* of knowledge. In order that this production process to be truly effective it is necessary to determine a creative attitude of students, which has led to the invention of direct means of stimulating "*creative thinking*".

The introduction of the remote experiment (direct or simulated) using the virtual environment has progressed gradually. From the stage of direct reproduction of experiments from the "*face to face*" laboratories up to the stage of experiments specifically designed for learning in the virtual environment (e.g. VISIR). Remote experiential networks and specific learning technologies for the virtual environment have emerged technologies that have imposed a rapid completions in the "*learning by doing*" theories (reviving Kolb and Nonaka's theories, updated for the new virtual environment).

Is remote experiment a source of stimulating creativity at an individual level? It has been demonstrated that the stimulation of creativity occurs when the following domains are intersected (Fig. 1):

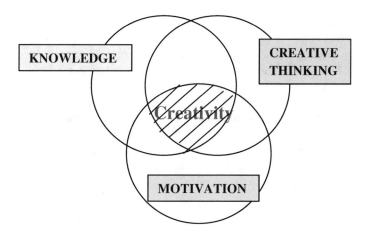

Fig. 1. Diagram of confluences that determine creativity (Amabile 2011)

Taken separately these three components in turn have to fulfill some conditioning. We will analyze these components directly in the relationship with the remote experiment (particularize), although the mentions which we will make are also valid in general terms. Each remote experiment transmits a set of knowledge and requires a set of knowledge to be driven. In order to stimulate creativity using remote experiments, two conditions are required:

- the user has the set of knowledge needed to understand "*in-deep*" the experiment and its manipulation;
- the user must have the ability to combine in new ways the disparate elements of the remote experiment that he is offered. The condition refers both to the set of knowledge necessary for understanding and maneuvering as well as to the set of knowledge transmitted by the experiment.

The above remarks are related to *"knowledge"* (Fig. 1).

As for *"creative thinking"* (Fig. 1), the remote experiment fits perfectly into Steinberg's appreciation of *"triarchic thinking"*, especially as regards:

- **Selective comparison:** remote experiments in the virtual environment can refer to distinct problems created on distinct themes. They can also refer to experiments related to the same parameter, with different software solutions and different experimental schemes. These differences are inherently imposed by the different levels of university endowment, by the software used predominantly in each environment, by the level at which knowledge is transmitted in each university, and often by the specificity of the cultural environment. As a result, the *"selective comparison"* function becomes implicit when you want to get acquainted with all the similar experiments on the network. The student will instinctively make comparisons and will be able to classify experiments according to their own criteria, depending on their level of knowledge.
- **Selective combination:** the student looks critically at the similarities and differences between experiments in the virtual environment that have the same final experimental objective (e.g. Ohm's law). He can combine these concepts in original ways by creating his own remote experiment. At this stage, critical/analytical thinking acts as the student makes the difference between the ideas of different remote experiments and can appreciate their qualities and weaknesses.
- **Selective encoding:** following the remote experiments that the virtual environment provides for the same experimental subject, the student instinctively separates the relevant and irrelevant information, establishing its own value system on the subject. This operation will help in his career to work only on the relevant issues chosen from a many other issues that could lead him towards insignificant ways, but after sustaining effort and time consuming.

The third domain of *"motivation"* (Fig. 1) is intrinsic included in remote experiments. Only motivated students can investigate on-line experiments similar to those received in the classroom. Unmotivated students will be summed up to solve the experiment received from professor without trying to see what others did on the same subject.

The question is whether the intellectual processes described above are all done consciously. The answer is not. The best explanatory basis to justify this statement is provided by the Campbell model of Creative Darwinism (1960), [6]. It can be said that in the student's mind, the processes described above generate *"blind variations"* in the same way that genes can generate random mutations. This theory adapts happily when we are trying to explain the role of remote experiment in stimulating creativity. It was found that only the *"learning by doing"* theory is not convincing enough. It does not provide a sufficient explanatory basis for supporting the way in which original ideas are generated. In addition, the Campbell model defines the environment as the one that determines these *"blind variations"*. They produce bad ideas and good ideas in a random way, just as they happen through mutations and recombination in genetic theory. So the environment can stimulate creativity and the remote experiment is such an environment that acts in a random way on the minds of students and in some determines the increase of creativity [7–9].

3 Experiment

Based on the above, we intend to evaluate the creative potential of some students by evaluating, with statistical methods, the creative thinking processes determined by the environment called the "*remote experiment*". We have focused on the assessment of divergent thinking and have chosen as independent variables, fluency, originality and flexibility. The experiment is described below, with the provision that it was conducted online and referred to a remote work installed on the iLab-MIT server existed in the CVTC (Center for Valorization and Transfer of Competence) of Transylvania University in Brasov. Students responded to questions about the mentioned three independent variables presenting their own ideas and were scored with notes from 1 to 20. It can be seen from the list of notes (Table 2) that the fluency allowed the giving of big grades that the originality had the lower marks and the flexibility is mostly located at the top of the scoring average.

Table 2. Test notes

	Treatment levels		
	a_1	a_2	a_3
	18	3	4
	15	5	10
	12	8	11
	16	2	7
	19	7	13
A_i	80	25	45
$\bar{A}_i$	16	5	9
$\sum\limits_{j}^{s}\left(AS_{ij}\right)^2$	1310	151	455

There were set up three groups (three because we have three independent variables) of 5 students each. One group was subjected to the request to describe the experiment by making the most complete appreciations on the solutions contained (fluency level - a_1). The second group was asked to present original solutions other than those used in the experiment for certain stages of the experiment and even for the whole experiment (originality level - a_2). The third group was asked to propose other methods to measure the main parameter existed in the experiment (flexibility level - a_3). Following the responses received on each level, the following notes were recorded (we use Keppel notations [1], including the form of his calculation tables) - Table 2.

It is established "*between group sum of squares*": $SS_A = \dfrac{\Sigma(A)^2}{s} - \dfrac{(T)^2}{a \cdot s} = 310$ (1)

It is established "*within group sum of squares*": $SS_{S/A} = \Sigma(AS)^2 - \dfrac{\Sigma(A)^2}{s} = 106$ (2)

It is established $SS_T = SS_A + SS_{S/A} = 416$ and the average $\bar{T} = 10.00$ (3)

The deviation scores analyze is referred to the deviation of grades of each student's from the overall average $\bar{T}$. And these deviations are divided "*between group deviations*": $(\bar{A}_i - \bar{T})$ and "*within group deviations*": $(AS_{ij} - \bar{A}_i)$. The total deviation is:

$$(AS_{ij} - \bar{T}) = (\bar{A}_i - \bar{T}) + (AS_{ij} - \bar{A}_i)$$ (4)

This relationship leads to (Table 3):

Table 3. Deviation scores

AS_{ij}	$(AS_{ij} - \bar{T})$	$(\bar{A}_i - \bar{T}) + (AS_{ij} - \bar{A}_i)$	
Level 1			
18	$18 - 10 = 8$	$16 - 10 = 6$	$18 - 16 = 2$
15	$15 - 10 = 5$	$16 - 10 = 6$	$15 - 16 = -1$
12	$12 - 10 = 2$	$16 - 10 = 6$	$12 - 16 = -4$
16	$16 - 10 = 6$	$16 - 10 = 6$	$16 - 16 = 0$
19	$19 - 10 = 9$	$16 - 10 = 6$	$19 - 16 = 3$
Level 2			
3	$3 - 10 = -7$	$5 - 10 = -5$	$3 - 5 = -2$
5	$5 - 10 = -5$	$5 - 10 = -5$	$5 - 5 = 0$
8	$8 - 10 = -2$	$5 - 10 = -5$	$8 - 5 = 3$
2	$2 - 10 = -8$	$5 - 10 = -5$	$2 - 5 = -3$
7	$7 - 10 = -3$	$5 - 10 = -5$	$7 - 5 = 2$
Level 3			
4	$4 - 10 = -6$	$9 - 10 = -1$	$4 - 9 = -5$
10	$10 - 10 = 0$	$9 - 10 = -1$	$10 - 9 = 1$
11	$11 - 10 = 1$	$9 - 10 = -1$	$11 - 9 = 2$
7	$7 - 10 = -3$	$9 - 10 = -1$	$7 - 9 = -2$
13	$13 - 10 = 3$	$9 - 10 = -1$	$13 - 9 = 4$
Sum 0	0	0	

For "*between group deviation*" each level has the same deviation scores: 6 for level a_1, -5 for level a_2 and -1 for level a_3. It is also noted that for each group the sum of the deviation scores is zero.

Under these experimental conditions, it is clear that the null hypothesis is written:

$$H_0: \quad (\mu_1 - \mu) = (\mu_2 - \mu)\ldots\ldots(\mu_n - \mu) = 0 \tag{5}$$

and it means that no effects can be identified in the population as a result of the applied treatment. Any different value leads to the alternative hypothesis: H_1: *all μ's is not equal*. Although $(\bar{A}_i - \bar{T})$ there are deviation scores higher than zero in the example, they are insufficient to justify the conclusion that treatment effects in the population can be noticed. The only suitable indicator for this conclusion is the ratio:

$$F = \frac{between\ group\ variance}{within\ group\ variance} \tag{6}$$

much more useful for verifying the validity of the null hypothesis. It is known that when its values are lower than "1" H_0 is true, and when its values are greater than "1" the alternative hypothesis H_1 is true.

For F calculation, is used the relationship:

$$F = \frac{MS_A}{MS_{S/A}} \tag{7}$$

where:

- MS_A represents the combination of the treatment effects at which is added the variance errors;
- $MS_{S/A}$ represents only variance errors.

For the calculation of these sizes, we need the value of the "*degree of freedom*". This is defined as the difference between the number of independent observations and the number of the estimated population. In our example we have a = 3 representing the treatment conditions and s = 5 representing the number of subjects analyzed. As a result:

$$df_A = a - 1 = 2$$
$$df_{S/A} = a \cdot (s - 1) = 12 \tag{8}$$

using the previously calculated values are thus determined:

$$MS_A = \frac{SS_A}{df_A} = 155$$
$$MS_{S/A} = \frac{SS_{S/A}}{df_{S/A}} = 8.83 \tag{9}$$

$$\text{Result: } F = \mathbf{17.55} \tag{10}$$

Even if the value is sensibly higher than "1" before drawing the conclusions, it is necessary to set the critical value for F (F') and compare it to the calculated value. From the tables (Fisher & Yates-1953) for the value $F(2;12)$ (i.e. F ($d_{nominator} = 2$, $d_{denominator} = 12$), at $d_{denominator} = 12$ there are several values for $d_{nominator} = 2$ (Table 4):

Table 4.

$d_{denominator}$	α	Values $F(2,12)$ when $d_{nominator} = 2$
12	0.25	1.16
	0.10	3.18
	0.05	4.75
	0.025	6.65
	0.01	**9.33**
	0.001	18.6

The rule of decision is as follows: H_0 shall be rejected if $F_{calculated}^{0.01} \geq 9.33$. Obviously, $F = 17.55$ fulfills this condition, so it is possible to reject H_0 and accept the alternative hypothesis H_1, which means that in the population has experienced "*treatment effects*" following the remote experiment.

4 Conclusions

The statistically processed experimental data shows us:

1. Students, who freely accepted to respond at test questions, are interested in creativity and in its assessment at individual level;
2. Evaluation of only one remote experimental work has been done only to explain the proposed approach and cannot yet be generalized. For generalization, it should be accepted that the method is consistently applied to all the remote work that students have in the program;
3. The small number of students who have responded to the test is still a reason to avoid generalizing conclusions about the creative potential of the tested students;
4. If we now refer strictly to the experiment, we can see that the chosen independent variables (the same as those chosen by Torrance) can produce "*treatment effects*", so they can serve as a hint of creative individual potential, and may in particular serve as a means of screening thinking to students;
5. It is intended to continue the evaluation with the method on multi-levels until a general conclusion can be drawn.

Acknowledgement. The paper was written with the sustaining of CVTC (Center of Valorization and Transfer of Competence) from Transylvania University of Brasov and Brasov branches of "Technical Science Academy" and "Science Academy" from Romania.

References

1. Keppel, G.: Design and Analysis: A Researcher's Handbook, p. 658. Prentice Hall, Upper Saddle River (1973). Editor: Jenkins, J.J., ISBN: 0-13-200030-X
2. Benedek, M., Mühlmann, C., Jauk, E., Neubauer, C.A.: Assessment of divergent thinking by means of the subjective top-scoring method: effects of the number of top-ideas and time-on-task on reliability and valid. Psychol. Aesthet. Creat. Arts. **7**(4), 341–349 (2013). https://doi.org/10.1037/a0033644
3. Guo, J.: The development of an online divergent thinking test. Doctoral Dissertations 1304. University of Connecticut Digital Commons@UConn. http://digitalcommons.uconn.edu/dissertations/1304
4. Silvia, P.J.: Subjective scoring of divergent thinking: examining the reliability of unusual uses, instances, and consequences tasks. Thinking Skills Creativity **6**(1), 24–30 (2011). http://www.sciencedirect.com/science/article/pii/S1871187110000295
5. Lemons, G.: Diverse perspectives of creativity testing: controversial issues when used for inclusion into gifted programs. J. Educ. Gifted **34**(5), 742–772 (2011). https://doi.org/10.1177/0162353211417221, http://jeg.sagepub.com
6. Simonton, D.K.: Creativity as blind variation and selective retention: is the creative process darwinian? Psychol. Inquiry **10**(4), 309–328 (1999)
7. Sawyer, R.K.: Educating for innovation. Thinking Skills Creativity **1**, 41–48 (2006)
8. De Haan, R.L., Diane Ebert-May, D.: Teaching creativity and inventive problem solving in science. CBE Life Sci. Educ. **8**(3), 72–181 (2009). https://doi.org/10.1187/cbe.08-12-0081
9. Chena, A., Dong, L., Liu, W., Li, X., Sao, T., Zhanga, J.: Study on the mechanism of improving creative thinking capability based on Extenics. Procedia Comput. Sci. **55**, 119–125 (2015)
10. Sak, U.: Selective problem solving (SPS): a model for teaching creative problem solving. Gifted Educ. Int. **27**, 349–357 (2011)
11. Sawyer, R.K.: The future of learning: grounding educational innovation in the learning sciences. In: The Cambridge Handbook of the Learning Sciences, 2nd edn. Cambridge University Press, Cambridge (2014)
12. Kim, K.H.: Can we trust creativity tests? A review of the torrance tests of creative thinking (TTCT). Creativity Res. J. **18**(1), 3–14 (2006)

"Electromagnetic Remote Laboratory" with Embedded Simulation and Diagnostics

Franz Schauer[✉], Michal Gerza, Michal Krbecek, Das Sayan,
Mbuotidem Ime Archibong, and Miroslava Ozvoldova

Faculty of Applied Informatics, Tomas Bata University in Zlin, 760 05 Zlin, Czech Republic
fschauer@fai.utb.cz

Abstract. This paper presents new form of remote experiment "Electromagnetic Remote Laboratory", which is designed for students and educational purposes and integrated in REMLABNET system. The remote experiment was built using universal and reliable Intelligent School Experimental System (ISES). The main new feature of the remote experiment "Electromagnetic Remote Laboratory" is the output of information in data sets, which gives the possibility to further analyze this data to obtain insight into theory of electromagnetism of Faraday. Another new feature of the remote experiment is the embedded synchronized simulation, which further deepens the insight into Faraday's theory of electromagnetism.

Keywords: Electromagnetic Remote Laboratory · ISES
Embedded simulations and diagnostics

1 Introduction

In this paper, we present the solution and function of our new remote laboratory experiment "Electromagnetic Remote Laboratory" as the case study for a new functionality, namely two-level diagnostics and embedded synchronized simulation of the measured Electromagnetic phenomenon. Within this context, we will highlight the validity of Faraday's law of electromagnetic induction using the new components, we have built previously and we introduce a new ISES control board USB ISES [1].

2 State of Art of ISES Physical Hardware

For the purpose of explanation of new components, let us briefly touch ISES properties and basic component parts of remote experiments, based on ISES.

2.1 Physical Hardware

The basic approach of our remote experiments is finite state machine approach with the controlling. psc file and as a hardware of our experiments we use proven Intelligent School Experimental System (ISES). All experiments of our laboratory are to be found

© Springer International Publishing AG, part of Springer Nature 2019
M. E. Auer and R. Langmann (Eds.): REV 2018, LNNS 47, pp. 226–233, 2019.
https://doi.org/10.1007/978-3-319-95678-7_25

in Remote Laboratory Management System (RLMS) REMLABNET (http://remlabnet.eu). More details about these component parts of our approach towards remote experimentation are to be found in our recent monograph [2] and on web page http://www.ises.info. For orientation it may be shortly mentioned that ISES consists of several basic components, first, it is the Analog-to-Digital or Digital-to-Analog converter configured as the PCI 1202 interface card in the computer. Second, it is universal control board and third, a rich set of sensors for natural sciences. ISES possesses two analogue and binary input channels and programmable analogue output. The physical hardware formed by transducers, board and various modules of ISES, see Fig. 1 (more can be found in [2] and http://www.ises.info).

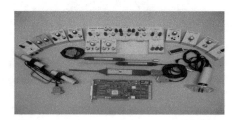

Fig. 1. ISES professional hardware set of various physical modules with the ISES board.

2.2   The Measureserver Unit

Besides, the Measureserver (MS) unit is the engine of the whole remote measuring process. MS helps establish effective connection between the physical hardware components and the users while there are online and offline. It serves as the mathematical model used in implementing control programs by the PSC script file and the measurement logic of finite state machine principle.

More so, within the physical hardware the MS establishes connection with the software driver, which is PCI 1200 interface card. At this stage, data are passed via the pins and it is translated with the help of Analog-to-Digital or Digital-to-Analog converter. These pins serve as inputs and outputs which are positioned at the universal control board connected to the system. During measurement by the user, the MS receives signal via TCP/IP (transmission Control Protocol or Internet Protocol) on the internet and respond to the input of the user. In Fig. 2 is the Measureserver of ISES remote experiment with its most important parts, described in figure caption. It is sufficient to mention that both new components, namely diagnostic module of both levels I and II including embedded simulation with a mathematical solver are integrated in the Measureserver (more information to be found in [2]).

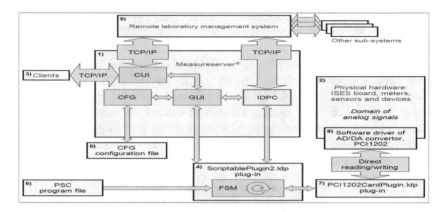

Fig. 2. Formation of software and hardware components communicating with the Measureserver unit are listed below: (1) Measureserver with configuration (CFG) and graphical user interface (GUI) components and the newly integrated intelligent data processing component (IDPC), (2) ISES physical modules set building the RE, (3) Registered clients connected via the RE web pages, (4) Plug-in building the RE logic by the finite-state machine (FSM), (5) Configuration file which delivers the initial parameter, (6) Control program file creating the RE structures and logic, (7) Plug-in communicating with the PCI 1202 AD/DA converter drivers, (8) PCI 1202 AD/AD converter drivers and (9) REMLABMET platform managing all remote laboratories.

3   Aim of "Electromagnetic Remote Laboratory"

The remote experiment "Electromagnetic Remote Laboratory", gives the possibility; first, to prove the validity of the Faraday's law of electromagnetism, giving the relation of the magnetic flux Φ and electromotive voltage (emv) ε, second, to quantify the induced emv on the angular frequency of the rotating coil remotely and third, to get acquainted with the electromagnetic phenomena.

In summary, the induced emv given by Faraday's law is shown below in the equation

$$\varepsilon = -\frac{d\Phi}{dt} \tag{1}$$

Where ε is the induced emv measured on the loop, and Φ is the magnetic flux through the loop, which is $\Phi = B.S \cos \alpha$, and $\alpha = \omega t$ is the instantaneous angle between the vector of the magnetic induction B and S vector area of the coil.

The electromotive voltage induced on the rotating coil with N windings, rotating with the constant angular velocity ω is then

$$\varepsilon = -BSN\frac{d\cos(\omega t)}{dt} = BSN\omega\sin(\omega t) \tag{2}$$

The magnetic flux via a loop might change due to some of the reasons: first, the change of the magnitude of the magnetic induction $B(t)$ over time, second, the change of the dimension of the area $S(t)$ and third, the change of the magnetic induction vector $\boldsymbol{B}$ and the area $\boldsymbol{S}$.

Besides, the coil rotates with the constant angular frequency $\omega = 2\pi/T$, where T is the period of revolutions. By integrating Eq. (2) the validity of Faraday's law is confirmed. See Fig. 3.

$$\int_0^{T/2} |\varepsilon|\,dt = \int_0^{T/2} NBS\omega \sin(\omega t)\,dt = 2NBS = const. \tag{3}$$

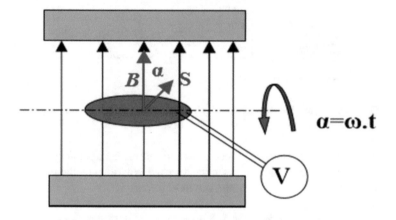

Fig. 3. Rotation of the coil in a uniform magnetic field.

4   Embedded Simulation in "Electromagnetic Remote Laboratory"

More so, as the part of the solution of embedded simulations in our ISES RE we used the mathematical solver built in RE Measureserver and its PSC file. The solver provides solving of a wide range of arithmetic operations and solves differential equations [3, 4]. The example of the arbitrary use is shown in Fig. 5 [4], here was used for simple plotting of calculated quantities according Eqs. (1–3) (Fig. 4).

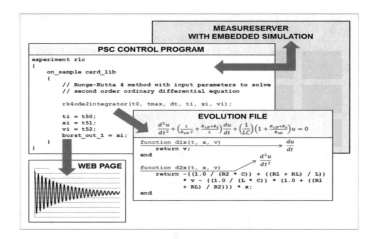

Fig. 4. The PSC file and Measureserver which enable arithmetical operations and differential equations in the ISES remote experiment Measureserver.

The embedded simulation of "Electromagnetic Remote Laboratory" is shown, see Fig. 5. It gives opportunity for the users to measure the experiment remotely and compare them with the synchronized simulation, see in Fig. 6.

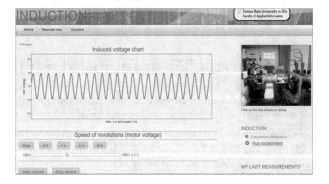

Fig. 5. The above diagram is the webpage of the remote laboratory "Electromagnetic Remote Laboratory". The page can be accessed via this link (http://www.remlabnet.eu/).

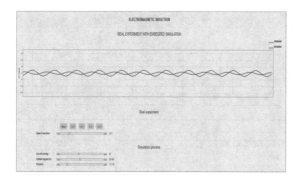

Fig. 6. Simulated and measured induced electromotive voltage of the rotating coil. The blue line is the measured voltage, the red line is the simulated one. In down panel are controls for adjusting amplitude, frequency and phase of the simulation.

5   Two Level Diagnostics I and II

Furthermore, the embedded diagnostic which consist of Ist embedded diagnostics and IInd embedded diagnostics are all configured in the Measureserver. It contributes significantly to the remote experiment especially, when experiment is running. The embedded diagnostics communicate constantly with the physical ISES modules connected to the ISES control board. The below diagram, shows the function and description of the embedded diagnostics in the Measureserver (Figs. 7 and 8).

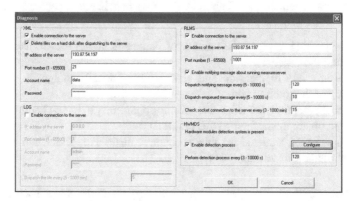

Fig. 7. The Ist Embedded diagnostics. The first embedded diagnostics helps in establishing connection between the remote experiment online and the server.

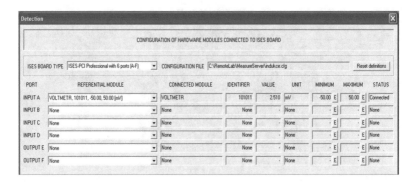

Fig. 8. The IInd embedded diagnostics incorporated in the Measureserver. At this stage, all the ISES physical hardware attached to the ISES control board are initialized in the Measureserver, indicating if the connection is correct.

6   Conclusions

This remote laboratory, "Electromagnetic Remote Laboratory" was designed and assembled. It further provides learners the opportunity to get familiar with the electromagnetic phenomena and quantities like magnetic flux, the induced electromotive voltage and to further obtain results and data which are in conformity with Faraday's law of electromagnetic induction.

Finally, the summary of our research and the results obtained from the remote laboratory "Electromagnetic Remote Laboratory" shown here in this paper has the following benefits.

1. Students are able to prove the validity of Faraday's law of electromagnetism and understand the concepts.
2. Students are able to measure and analyze the induced voltage on the angular frequency of the rotating coil.
3. The remote laboratory provides students the opportunity to interact with real experiments and obtain the desired results online.
4. Most importantly, students are able to obtain realistic data, which can be imported to Excel file or html file smoothly.

Acknowledgment. The authors acknowledge the support of the Swiss National Science Foundation (SNSF) – "SCOPES". Also, the support of the Internal Agency Grant of the Tomas Bata University in Zlin is highly appreciated.

References

1. Gerza, M., Schauer, F.: Intelligent processing of experimental data in ISES remote laboratory. Int. J. Online Eng. **12**(3), 58–63 (2016). (ISSN 1861-2121)
2. Ozvoldova, M., Schauer, F.: Remote laboratories-in research-based education of real world. In: Schauer, F. (ed.), p. 157. Peter Lang, Int. Acad. Publ. Frankfurt (2015). ISBN 978-80-224-1435-7
3. Krbecek, M., Schauer, F.: Optimization of remote laboratories for mobile devices. In: 2016 International Conference on Interactive Mobile Communication, Technologies and Learning (IMCL), San Diego, CA, pp. 1–2 (2016). https://doi.org/10.1109/imctl.2016.7753759
4. Gerza, M., Schauer, F.: Advanced modules diagnostics in ISES remote laboratories. In: IEEE 10th International Conference on Computer Science and Education (ICCSE 2015), Brunel, 22–24 July 2015, pp. 583–589. Univ London, Cambridge (2015)

Smart Grid Remote Laboratory

Kalyan B. Ram[1(✉)], S. Arun Kumar[1], Manish Ahlawat[2], Sanjoy Kumar Parida[3],
S. Prathap[1], Preeti S. Biradar[1], and Vishnu Das[1]

[1] Electrono Solutions Pvt. Ltd., #513, Vinayaka Layout, Immadihalli Road
Whitefield, Bangalore 560066, India
{kalyan,arun,prathap,preeti,vishnu}@electronosolutions.com
[2] NI India, Bangalore, India
manish.ahlawat@ni.com
[3] Department of Electrical Engineering, IIT Patna, Patna, India
skparida@iitp.ac.in

Abstract. A "smart" grid is capable of providing power from multiple and widely distributed sources such as from wind turbines, concentrating solar power systems, photovoltaic panels and so on. Further, Energy storage devices such as Batteries and Fuel cells would be an integral part of Electricity grid in addition to conventional power generation techniques. A smart grid uses digital technology to improve reliability, security, and efficiency of the electric grid system. Existing electric infrastructure in several parts of India is aging and it is being pushed to do more than it was originally designed to do, as the demand for electricity is increasing encompassing diverse sectors from Manufacturing to Agriculture and now with increased efficiencies of Electric cars built by companies such as Tesla, the demand for Electricity is expected to rise multi fold. Modernizing the grid to make it "smarter" and more resilient through the use of cutting-edge technologies, equipment, and controls that communicate and work together to deliver electricity more reliably and efficiently would naturally be the necessary infrastructural upgrade to facilitate such a change. Consumers can better manage their own energy consumption and costs because they have easier access to their own data. Utilities also benefit from a modernized grid, including improved security, reduced peak loads, increased integration of renewable, and lower operational costs. Smart grid Remote lab is developed and equipped in-order to provide a platform for the students and research scholars to undergo their project works remotely. Here the energy generated by modules from both solar and wind sources installed over the roof top will be fed to our control system and will be switched between ON Grid and OFF Grid modes. The system is interfaces with LabVIEW and its user interface design allows students/researchers to observe the real-time electrical parameters during all levels of its operation.

Keywords: Remote labs · Smart grid · Wind turbine · Solar panel
System control · LabVIEW

1 Introduction

INDIA is one of the fastest growing country in the world. Despite of economic growth of the country still facing few basic problems, scarcity of electricity is one of them. Most

© Springer International Publishing AG, part of Springer Nature 2019
M. E. Auer and R. Langmann (Eds.): REV 2018, LNNS 47, pp. 234–243, 2019.
https://doi.org/10.1007/978-3-319-95678-7_26

of the rural areas have no access to electricity. From the past decade India doubled its energy generations, but its grid systems looses more than 20% of the generated power. Smart grid is a digital technology where intelligence built into the electric grid, it allows two way communication between the utility and the consumers. A Smart Grid is capable of providing power from multiple and widely distributed sources such as from wind turbines, solar power systems, photovoltaic panels, hydro, natural gas and so on. Since all renewable energy sources vary with respect to time, a smart grid must be capable of flexibly storing electric power for later use like batteries. A smart grid uses digital technology to improve reliability, security, and efficiency of the electric system.

Laboratories play a crucial role in enhancing learning among students by providing platform for experimentation leading to development of skills, trigger ideas and help simulate physical system on to digital platforms by creating digital twins, help them to deepen their understanding through relating theory and practice, illustrating and validating the analytical concepts and so on. In traditional laboratories, the user interacts directly with the hardware by performing physical activities. (ex.: connecting complex wires, turning knobs, pressing buttons, and recording the sensory feedback. Utilization of such laboratory infrastructure in several Engineering universities and colleges in India is less than 20% due to factors such as time, security, safety, availability of skilled staff to train students to use these labs. Majority of colleges cannot afford the expensive hardware lab setup too. In order to overcome expensive hardware lab setups and utilize the existing lab setup more efficiently and to improve the technical education, Electrono solutions Pvt. Ltd., Bangalore has developed Remote Labs for Smart grid at Indian Institute of Technology, Patna.

2 Smart Grid

A Smart Grid is an interconnected system of information, communication technologies and control systems used to interact with automation and business processes across the entire power sector encompassing electricity generation, transmission, distribution and the consumer. It's been called "electricity with a brain". The evolution towards Smart Grid would address these issues and transform the existing grid into a more efficient, reliable, and safe and less constrained grid that would help provide access to electricity to all.

2.1 Need for Smart Grid

Smart grid refers to an intelligent electricity supply chain from main power plant to consumers. Ensuring reliable and efficient electricity delivery process is very necessary. Hence intelligence is required in the process of electricity delivery starting from generation, transmission to distribution. The basic concept of smart grid is to add intelligence in the existing infrastructure, by communicating between the utility and consumer, continuous monitoring and analysing the system.

3   Remote Lab

"I hear - I forget; I see - I remember; I do - I understand." Attributed to Confucius, these words may not have been as relevant to his times as they are to the world today. With technology having made inroads into every sphere of modern life, the need of the day is to have engineers who have "hands-on" experience in facing the challenges of today's fast-paced technologies. The easiest way for technical students to gain this "hands-on" experience is by the use of laboratories designed for the desired field. In order to overcome expensive hardware lab setups and utilize the existing lab setup more efficiently thereby improving on the quality of technical education, more and more technical institutions around the world are turning to Virtual and Remote Laboratories.

3.1   Need for Up-gradation of Existing Labs

An established remote lab could offer a platform which can aid the students to access the lab beyond the limitations posed by the conventional labs. It can also serve to realize the functionality of systems with different specifications, make or model for a given requirement when designed aptly. The aim of this paper is to evaluate the feasibility of remote labs in Indian Engineering Institutions from academic course-defined experiments to research and industrial consultancy requirement.

4   Approach

The Smart grid Remote lab is developed and equipped in-order to provide a platform for the students and research scholars to undergo their project works remotely. The Setup consists of 1 KW Solar Panels (2 Monocrystalline and 2 Poly crystalline 250 W each) and 1 KW Wind turbine placed on the roof top. For Weather station Wind speed and wind direction sensor, Radiation sensor and temperature Sensors are placed on roof top, camera is placed for live feed and remote operations as shown in the Figs. 2 and 3.

The main components used for the lab set up are listed below

– Programmable Automation Controller NI CRIO 9082 is been used along with the following controller modules

- Analog Voltage Input Module C Series NI 9225,
- Analog Current Input Module NI C Series 9227 with accessories & Connectors NI 9971,
- Digital Input Output Module NI C Series 9375,
- Wireless Node Gate way Module NI 9795,
- RS 232 Serial Programmable Wireless Node,
- Modular Graphical Programming Development Platform LabVIEW,
- PV System-1 KW & Battery Bank 100AH Make: Zytech Solar Model Number PV 250 P,
- Wind Turbine Electrono Integrated Model Number ELNOWD1012,
- Energy Meter/Smart Meter RIG: Schneider Model Number PM 1500,

- DC & AC Distribution Cables and Wires Havels/Finolex/Equivalent,
- Sensors & its Connecting Accessories: Electrono Solutions Model: ELNOPS01,
- Work Station HP Z238/Dell Precision T 3500,
- Remote & Cloud Access: Electrono Solutions (Fig. 1).

Fig. 1. Solar panels and wind turbine installed at the site

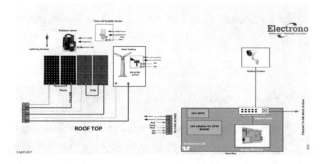

Fig. 2. Top level block diagram of roof top

The lab setup block diagram and smart grid control room components are shown in Figs. 3 and 4.

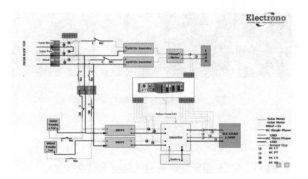

Fig. 3. Top level block diagram of Lab setup

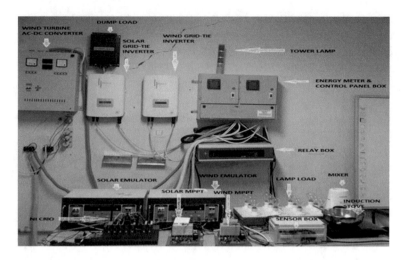

Fig. 4. Smart grid control room containing the systems

The Entire System is been designed to operate in three Different modes:

1. ON GRID Mode
2. OFF Grid Mode
3. Emulator Mode

The user interface for the mode selection is as shown in Fig. 5.

Fig. 5. Mode selection window

4.1 ON Grid Mode

By default system will be in on grid mode. ON grid set up consists of Grid-Tie Inverters (GTI) Energy Meter and control panel Box. Here the energy produced by the solar panel and wind turbines after passing through the grid-tie inverters are directly feed to the main grid.

All the connections are going through the Energy meter and control panel box where switching to ON Grid mode occurs in response to the command from the user interface software and also all the energy parameters (voltage, current, power) that both solar and wind energy sources combined is providing to the main grid is known. All the parameters will be read in the software end using rs232 to usb converter.

User interface Software screen for the on grid mode is as shown in Fig. 6.

Fig. 6. User interface for ON grid mode

4.2 OFF Grid Mode

When user want to take the access to the setup remotely or physically, user can switch to off grid mode by giving command from the user interface software. Off grid mode consists of

- Energy Meter and control panel Box
- MPPT
- Sensor box
- Tubular batteries
- Single phase Inverters
- Loads (lamp load, mixer, induction stove)

In this mode, the energy produced by the solar panel and wind turbines are stored in batteries for future usage. Outputs of solar panel and wind turbine are going through the Energy meter and control panel box When system is switched to off grid mode both Solar and wind outputs will be connected to two separate MPPTs and output of each is made parallel and feed to 48Vdc batteries. Now the battery output is given to single phase Inverter inputs and at the output of which all the single phase loads are connected. All electrical parameters corresponding to the loads applied can be monitored at the energy meter at off grid side of Energy Meter and control panel Box. User interface Software screen for the off grid mode is as shown in Fig. 7.

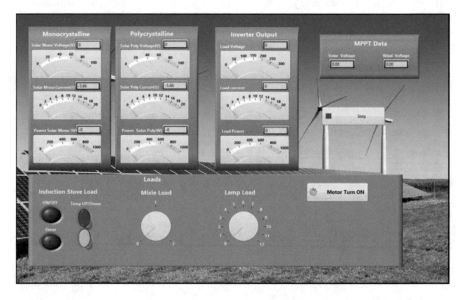

Fig. 7. User interface for off grid mode

4.3 Emulator

Emulator is a hardware or a software that make one system to behave like another system. The Emulator is a programmable power supply designed to emulate solar panels and

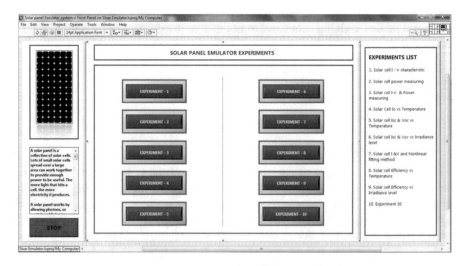

Fig. 8. User interface for solar emulator

wind turbine. With fast transient response, the emulator will responds to change in load conditions and maintains the output on IV characteristics of the selected panel for a given ambient condition. Emulator is a flexible instrument designed to emulate the output of solar panels and wind turbine from different manufacturers, variations due to time of the day, effect of season and different geographical locations of installation (Figs. 8 and 9).

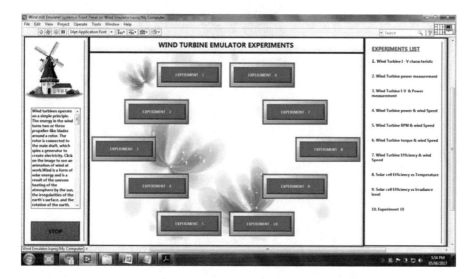

Fig. 9.   User interface for wind emulator

Emulator is open platform given freedom to the research scholar to design their model and test with the emulator which behave exactly as solar panels and wind turbine. Two programmable DC power supplies are provide as Solar and wind Emulators to emulate the working of solar PV panels and wind turbine provided. The Solar PV Emulator is a programmable power supply designed to emulate solar panels. With fast transient response, the emulator responds to change in solar parameters like solar irradiance, temperature and maintains the output on IV characteristics and PV characteristics of the selected panel for a given ambient condition. Wind turbine emulator mimics the behavior of wind turbine for hardware level simulations. This system is a programmable DC power supply which outputs DC voltage in response to the speed of wind, turbine speed (rpm) for a fixed turbine blade specs. All controlled as per the speed reference calculated by solving the mathematical model of wind turbine. Researcher can execute the mathematical models of their newly developed or modified wind turbine and can simulate the speed/power of profile of turbine for different wind speeds.

5   How to Access This Setup Remotely?

By fulfilling the university formalities, you will receive the login id and password, by using Electrono cloud Software you can book your slot for a particular time and then you can take access to the lab where you can conduct your experiments remotely. For more details about the slot booking refer [1].

IIT PATNA REMOTE LAB ARCHITECTURE FOR SMART-GRID

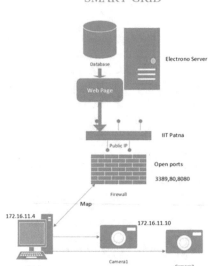

6 Actual or Anticipated Outcomes

Since the system is designed to provide Remote access for students, researchers and the faculty members to work in requirements related to Smart and Micro grids, the anticipated outcome is that with increased awareness, this system would be used to perform more experiments, capture and analyze data, design and develop systems and solutions to address the needs of Smart and Micro Grids.

This setup is open platform which can be accessed from any corner of the world with internet. To begin with we have developed few experiments where students can do in all the three modes.

- In on grid mode students can see the real time live working of the solar panel and wind turbine. Historical data is available in the system where students can analyze the power generated by the solar and wind with weather condition of the Patna.
- In off grid mode generated power from the solar and wind turbine is stored in the battery where students can do various experiments with variable loads.
- In emulator mode, system will be completely disconnected from the solar panels and wind turbine, by using emulator students can understand the complete functionality of the solar panels and wind turbine working at various climate. At present we provide around 4 experiments each. Where students can do the experiments and understand the system.

7   Conclusion

Smart grid Remote lab is developed and equipped in-order to provide a platform for the students and research scholars to utilize the laboratory infrastructure round the clock all through the year.

The system is designed keeping in mind the exposure required for the budding engineers to have a hands-on practice and experience in line with the field implementation of such systems. This platform facilitates students to understand the working of solar panel and wind turbine, inverter, smart meter, sensors, smart grid and perform experiments accordingly.

References

1. Kalyan Ram, B., et al.: A distinctive approach to enhance the utility of laboratories in Indian Academia. In: 2015 12th International Conference on Remote Engineering and Virtual Instrumentation (REV), pp. 238–241 (2015)
2. Norwegian University of Science and Technology. https://www.sintef.no/globalassets/sintef-energi/sintef_smartgrid_lab_a5_lr.pdf
3. Gungor, V.C., et al.: Smart grid technologies: communication technologies and standards. IEEE Trans. Ind. Inform. **7**(4), 529–539 (2011)
4. Yan, Y., Qian, Y., Sharif, H., et al.: A survey on smart grid communication infrastructures: motivations, requirements and challenges. IEEE Commun. Surv. Tutor. **15**(1), 5–20 (2013)
5. Cao, Y., Jiang, T., He, M., et al.: Device-to-device communications for energy management: a smart grid case. IEEE J. Sel. Areas Commun. **34**(1), 190–201 (2016)

e-LIVES – Extending e-Engineering Along the South and Eastern Mediterranean Basin

Manuel Gericota[1,2(✉)], Paulo Ferreira[1], André Fidalgo[1], Guillaume Andrieu[3], Abdallah Al-Zoubi[4], Majd Batarseh[4], and Danilo Garbi-Zutin[5]

[1] School of Engineering, Polytechnic of Porto, Porto, Portugal
{mgg,pdf,anf}@isep.ipp.pt
[2] University College of Southeast Norway, Kongsberg, Norway
[3] Université de Limoges, Limoges, France
guillaume.andrieu@unilim.fr
[4] Princess Sumaya University for Technology, Amman, Jordan
{Zoubi,m.batarseh}@psut.edu.jo
[5] International Association of Online Engineering, Vienna, Austria
dgzutin@ieee.org

Abstract. The number of students in the higher education system in South and Eastern Mediterranean Basin countries more than doubled in the last 15 years [1]. This positive step forward creates important difficulties for universities forced to handle overcrowded classes. In STEM (Science, Technology, Engineering, and Mathematics) related courses, one promising solution involves the development of accredited e-engineering courses, as a very convenient and efficient way of dealing with the constantly surging number of students.

The e-LIVES (e-Learning InnoVative Engineering Solutions) project, a recently approved Erasmus+ program project whose consortium includes European Institutions of higher education from France, Portugal, Belgium and Spain, and from the South and Eastern Mediterranean Basin, Algeria, Jordan, Morocco, and Tunisia, aims to address the problem by providing solutions based on e-engineering. Profiting from the experience gained with the EOLES (Electronics and Optics e-Learning for Embedded Systems) course, a fully online e-engineering third-year accredited Bachelor degree course, the long-lasting result of a previous TEMPUS program project, the EOLES project [2], the consortium hopes to provide the knowledge and the tools needed for partner countries to become autonomous in the development and accreditation of their own e-engineering courses.

Keywords: e-Engineering · Remote laboratories · e-Learning accreditation

1 Introduction

According to the UNESCO Institute for Statistics, the number of worldwide students enrolled in tertiary education more than doubled in 15 years, growing from 94.5 million in 1999 to almost 207 million in 2014, a gross enrolment ratio change from 19% to 34%. However, these global figures hide major differences between regions. While in Europe

M. E. Auer and R. Langmann (Eds.): REV 2018, LNNS 47, pp. 244–251, 2019.
https://doi.org/10.1007/978-3-319-95678-7_27

the higher education gross enrolment ratio is around 75%, in the South Mediterranean Basin countries the values range from 28% to 38% [1]. Nonetheless, in the last few years, higher education enrolment has been on the rise in less well-off countries of this region, as shown in the graphic of Fig. 1.

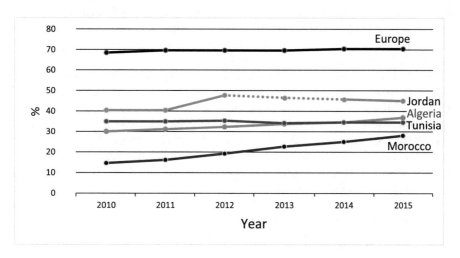

Fig. 1. Gross enrolment ratio in tertiary education between 2010 and 2015 in the countries involved in the e-LIVES project [3]

The graphic of Fig. 1 shows the number of students enrolled in higher education in Morocco doubled in the last five years while in Algeria it incremented by a third in the same period. On the contrary, in Jordan and Tunisia, the number remained stable but rather below the desirable value, mainly when compared to the European rates [3].

This positive step forward poses important challenges to universities struggling with lack of resources while forced to handle overcrowded classrooms.

One promising solution involves the extensive development of nationally accredited e-learning undergraduate and postgraduate courses. E-learning is a very efficient solution to cope with higher education access massification while meeting multiple students' profiles. For instance, students with low economic resources and/or living in distant isolated areas may pursue their higher education studies without leaving their homes and running into temporary expensive relocations. Another case is students looking for lifelong learning training. Usually, these students work and therefore may not attend classes during regular hours. In this case, e-learning provides them with the flexibility they need to study after-hours and to progress at their own pace, pursuing their studies without interrupting their working careers. And even regular students may blend their learning to avoid some overcrowded classes.

The long-lasting purpose of the e-LIVES (e-Learning InnoVative Engineering Solutions) project, a recently approved Erasmus+ program project, is to generate on Partner Countries a more committed and professional environment ready to introduce new forms of flexible learning into daily training activities and to create and manage accredited e-engineering courses.

2   The Framework

The development of an e-learning implementation strategy is one of the national priorities of the educational policy set by Algeria, Jordan, Morocco, and Tunisia, the Erasmus+ Partner Countries involved in the e-LIVES project. Depending on the country, the implementation of this commitment is at different stages. While in Tunisia, the development of e-learning in the higher education system through the reinforcement of the capabilities of existing Universities has been assigned to the Virtual University of Tunis (VUT), established in January 2002, in Algeria the Government is only now preparing legislative texts regarding the regulation of distance education. In between, the Morocco officials have passed in 2014 a law authorizing e-learning training (blended or fully online) while in Jordan, the ministry of higher education places the matter of developing standards for the assessment of quality in e-learning as a top priority. In 2009, a new higher education regulatory law actively supported the development of education. Currently, Jordan universities are deeply involved in the creation and implementation of their own e-learning strategies, and in urgent need of knowledge support.

Indeed, to build a fully online course is a difficult process requiring universities to address a range of heterogeneous aspects:

- the construction of an efficient economic model;
- the national accreditation of the e-learning courses in the absence of appropriate legislation and sometimes despite the support of Government officials;
- the training of an ample pool of teachers able to understand the inherent pedagogical differences between face-to-face and distance learning and capable of mastering the authoring tools needed to produce the course contents;
- the development and build up of the technological infrastructure needed to create and deliver the online courses;
- the training of the technical staff necessary to maintain the technical infrastructure and to support the teachers on the utilization of the authoring tools;
- in the case of STEM (Science, Technology, Engineering, and Mathematics)-related courses, the development, and implementation of real-time online controlled laboratories to allow students to perform laboratory works remotely.

In STEM-related courses, sometimes referred as "hard sciences", the acquisition of practical skills normally associated to face-to-face laboratory work raise issues that are sometimes difficult to solve, hindering the development of full STEM-related online courses.

All these challenges were successfully addressed during the previous EOLES TEMPUS project [4], also coordinated by the University of Limoges and involving, among others, nine of the new e-LIVES project consortium partners. Indeed, during the project, it was designed and implemented an entirely online English-taught 3rd year Bachelor's degree in Electronics and Optics for Embedded Systems, the L3-EOLES (Electronics and Optics e-Learning for Embedded Systems) course [5]. The course, that started being offered in the school year 2014/15, is currently in its fourth edition [6]. Designed as a specialization year, this course is oriented towards a currently expanding field in the electrical and computer engineering area, the field of electronics and optics for embedded systems. This area of knowledge requires students to be able to perform experimental work to acquire the expected technical experimental skills. The execution of laboratory assignments over the Internet required the development of remotely accessible experimental laboratories enabling students to interact in real-time with real experimental setups. This innovative training has been accredited by the educational authorities in Tunisia, Morocco, and France. If this previous project can be considered as a success, the progress made is still insufficient to determine a fast growing on the offer of e-engineering courses in the short term.

By documenting a set of good practices and following a hands-on approach, the e-LIVES project aims to help partner countries' universities to build innovative e-engineering courses by themselves in a sustainable way. This ambitious objective is grounded in two main goals:

- to help universities to move through the different course design and development stages (building of a curriculum, getting the national accreditation, training teachers, create contents,…);
- to help universities to develop by themselves (from A to Z) a remote laboratory.

It is important to note that these remote laboratories can also be used in face-to-face training. Indeed, due to the exponential growth in students' number, numerous universities had to replace the laboratory works in the first year of their Bachelor degree in STEM fields by paper-and-pencil work. This project is then expected to have a direct structural impact on the higher education system modernisation of the involved Partner Countries.

3   Strengths and Weaknesses Identification and Transfer of Knowledge

The success of the project depends in a great measure of the experience of the higher education institutions from Programme Countries and their ability to understand the problems of the Partner Countries and to address the challenges facing their higher education institutions and systems. The aim is to induce the voluntary convergence of their systems with European Union standards and development in higher education, encouraging the modernisation and internationalization of their institutions, creating conditions to enlarge the access of their youth to higher education while fostering people to people contacts, intercultural awareness, and understanding.

The Program Partners have a broad experience in the creation of e-learning courses and on the application of quality management methodologies to those courses, namely by applying detailed learning analytics and data analysis, aimed at providing accurate feedback able to contribute to improving students' experience in innovative forms of education such as e- and b-engineering.

During the EOLES project, the partners from the Programme Countries successfully designed and implemented, in cooperation with the Partner Countries, the L3-EOLES course, a course in the field of electronics and optics for embedded systems [7]. This innovative e-engineering training relies on a dedicated remote laboratory hosted by three of the beneficiary Partner Countries institutions: one in Algeria, one in Morocco, and one in Tunisia.

The e-LIVES project is a follow-up step needed to ensure e-engineering sustainability and to expand it into more South and Eastern Mediterranean Basin countries. The knowledge transfer to Partner Countries remains insufficient in areas such e-learning practices and pedagogy, resources and infrastructure management and the development of remote laboratories. e-LIVES proposes a set of innovative solutions to foster this transfer, the first step being the identification of best practices in e-engineering. Programme Countries have a key role here due to their more than 10 years' e-learning experience in Europe. A set of implementation-oriented documents - tutorials, summary data sheets, practical exercises, will be produced and published in open access in the project website to ensure a wide dissemination of the results.

National dissemination workshops, with the support and participation of staff from the Programme Countries, will be organized in all e-LIVES Partner Countries. These workshops, apart from a presentation of the project and of its main results, will include lectures, training sessions and hands-on demonstrations on how to design, implement and run an e-engineering course. The workshops will take place in the last year of the project and will be one of the main dissemination tools. 15 persons (teachers, technicians but also university deans) from each Partner Country are expected to attend the workshop, which will be opened to all interested persons from the whole country. A final open dissemination conference will be organized in Jordan in the framework of the last General Assembly of the project.

To guarantee the sustainability and exploitation of all the knowledge produced during the project and its use after project's end, a Special Interest Group (SIG) will be created within the International Association of Online Engineering (IAOE), an international non-profit organization whose aim is to encourage the wider development, distribution and application of Online Engineering (OE) technologies counting more than 2,000 members distributed all over the world. The SIG will promote the development of e-engineering in the South and Eastern Mediterranean Basin, a geographical area currently poorly represented.

4   Remote Laboratories

Successful courses in the STEM areas demand students to perform experimental work to acquire technical skills in subjects like physics, chemistry, mechanical and electrical machines, or digital and analog electronics, for example. In a fully online course, like the L3-EOLES, this requires the remote access to experimental laboratories and real-time interaction with real experimental setups that are complex to implement [8].

One of the most ambitious aims of the e-LIVES project is to help partners from Partner Countries to build (from A to Z) a remote laboratory by themselves, allowing the creation of fully online e-engineering courses. This ambitious aim will be achieved thanks to the help of the European Programme Countries which have an extensive experience in the development of remote laboratory solutions [9–11]. In particular, they are expected to work together with the Partner Countries' Universities to develop an operational remote laboratory including a High-Quality Reference Remote Practical Work (HQRRPW) and the associated online lectures.

Even if the e-LIVES project is not the first one in the field of e-engineering or dealing with the development of remote laboratories, the aim of the other projects was mainly to mutualize already existing ones. e-LIVES goes a step further by helping partners to acquire the required skills to develop their own laboratories adapted to their own e-engineering courses' requirements.

The Partner Countries' Universities will benefit from a subvention under the e-LIVES project that will allow them to purchase the necessary equipment for the development of a remote laboratory and the implementation of an HQRRPW. In the process, they will be able to acquire the required skills with the continuous help of the European partners.

In each University, the remote laboratory will be tested in real conditions with a selected group of around 30 students, leading to a total close to 300 students for all the Partner Countries' Universities. The students will participate in online lectures and perform the associated laboratory assignments requiring the use of the remote laboratory and hence testing it in real conditions. Students feedback will be analyzed to help partners to improve the quality and functionality of the laboratory and of the associated works.

After successfully testing the remote laboratory each partner will be required to mutualize it, making it available to all e-LIVES partners. This sharing step involves a set of issues related to security and access management to the remote laboratory, a step where the collaboration and know-how of the LABSLAND, the European technical partner that participates in the e-LIVES consortium, plays a fundamental rule.

LABSLAND's mission is to improve technical and scientific learning by creating, promoting and managing an international network of remote experimentation laboratories that, physically located in its own premises or in the facilities of its customers, allow them to share the laboratories over the Internet, being able to give and have access to a much larger number of real experiments (non-virtual or simulated) for users around the world. LABSLAND will be in charge of facilitating the optimal integration of the developed remote laboratories in a Remote Laboratory Management System able to manage students access to the different remote laboratories while providing security, scalability, and reliability. The experience of LABSLAND team of professionals guarantees a quality support service in the development and mutualization of the remote laboratories, collaboration to technical data sheets drafting and tutorials development.

5   Conclusions

By and after the end of the project, different kinds of target groups from Partner Countries, as well as other countries of the South and Eastern Mediterranean Basin, are expected to benefit from the outputs of the e-LIVES project:

- national decision-makers (university deans and national higher education officials) are expected to have all the information needed to have their doubts about this innovative way of teaching and learning dissipated and then facilitate the national accreditation process of new e-engineering courses in their countries;
- Partner institutions are expected to have acquired all the necessary know-how to create ambitious e-engineering courses, and in particular, to have overcome all the administrative, human and material resources obstacles they face today;
- each Partner institution is also expected to have a fully operational remote laboratory ready to be used by the students enrolled not only in e-engineering courses but also in face-to-face courses, in order to partially cope with the replacement of laboratory works in the first year of their Bachelor degrees in STEM fields by paper-and-pencil work due to lack of enough resources to accommodate all of them;
- students are expected to benefit from the e-LIVES project results as the participating institutions will be able to build innovative high-quality accredited e-engineering courses suitable for different profiles of students, namely students with weak economic resources and/or living in isolated areas or students in continuing education, and therefore their number is expected to increase after the e-LIVES project;
- teachers and technical staff involved in the project activities are expected to be ready to be part of an e-engineering course by the end of the project.

The Partner Countries' Universities are expected to benefit from the outputs of the e-LIVES project in a sustainable way as they will be able to create, develop, manage and teach innovative high-quality e-engineering courses.

Acknowledgment. This project is funded by the European Commission, under agreement number 2017 – 2891/001 -001. The European Commission support for the production of this publication does not constitute an endorsement of the contents which reflects the views only of the authors, and the Commission cannot be held responsible for any use which may be made of the information contained therein.

Authors wish to thank all the administrative, technical and pedagogical teams working on this project for their support.

References

1. Six ways to ensure higher education leaves no one behind. Policy Paper 30, Global Education Monitoring Report, UNESCO (2017)
2. The EOLES project in a few words. In: Electronic and Optic e-Learning for Embedded Systems (2017). http://www.eoles.eu. Accessed 17 Nov 2017
3. Data collected from the UNESCO Institute for statistics (2017). Gross enrolment ratio by level of education. http://uis.unesco.org. Accessed 16 Nov 2017
4. Gericota, M., et al.: EOLES course—the first accredited on-line degree course in electronics and optics for embedded systems. In: Proceedings of the IEEE Global Engineering Education Conference, Mar 2015, pp. 410–417. Tallinn University of Technology, Estonia (2015)

5. Gericota, M., et al.: Combining E-technologies & E-pedagogies to create online undergraduate courses in engineering—an example of a successful experience. In: Proceedings of the 8th International Conference on Education and New Learning Technologies, Barcelona, July 2016, pp. 4209–4218 (2016)

6. Andrieu, G., et al.: Overview of the first year of the L3-EOLES training. In: Proceedings of the 13th International Conference on Remote Engineering and Virtual Instrumentation, February 2016, pp. 396–399. Universidad Nacional de Educación a Distancia, Madrid (2016)

7. Andrieu, G., et al.: L3-EOLES—electronics and optics for embedded systems course. In: Remenyi, D. (ed.) The e-Learning Excellence Awards 2017: An Anthology of Case Histories, 1st edn. Academic Conferences and Publishing International Limited, Reading (2017)

8. E-learning tools. L3-EOLES—The university of future at home! (2017). http://l3-eoles.unilim.fr/?lang=en. Accessed 23 Nov 2017

9. Garcia-Zubia, J., et al.: Easily integrable platform for the deployment of a remote laboratory for microcontrollers. Int. J. Online Eng. **6**(3), 25–35 (2010)

10. Said, F., et al.: Design of a flexible hardware interface for multiple remote electronic practical experiments of virtual laboratory. Int. J. Online Eng. **8**(S2), 7–12 (2012)

11. Sousa, N., et al.: An integrated reusable remote laboratory to complement electronics teaching. IEEE Trans. Learn. Technol. **3**(3), 265–271 (2010)

A Reliability Assessment Model
for Online Laboratories Systems

Luis Felipe Zapata-Rivera$^{(\boxtimes)}$ and Maria M. Larrondo-Petrie$^{(\boxtimes)}$

Department of Computer and Electrical Engineering and Computer Science,
Florida Atlantic University, Boca Raton, FL 33431, USA
{lzapatariver2014,petrie}@fau.edu

Abstract. Online laboratories are a broad field that includes virtual laboratories, remote laboratories and hybrid configurations. The assessment of the reliability for these systems requires the identification of the laboratory components or human actions that can lead to a possible failure in the online laboratory normal operation. Having now bigger online laboratories implementations that provide access to hundreds or even thousands of users, the identification and evaluation of failures, causes and developing countermeasures, such as recovery mechanisms or alerts, is becoming increasingly important in order to provide more reliable systems. The paper presents a model for the assessment of failures, causes and countermeasures (actions, alerts or practices) that mitigate or eliminate failures. The model was created based on common failures reported by online laboratories users and based on the testing of a remote laboratory prototype implemented specifically for this purpose. The model for the assessment of the reliability of the online laboratory systems can be used to support reliability in implementations that are based on software components, such as virtual laboratories; and also in implementations that combine hardware and software components such as remote and hybrid laboratories. The model proposes a classification of the failures and its causes in a scale of low, medium and high frequency of occurrence. A definition of a rule based system based on the laboratories constrains is presented. Finally a definition of the integration of the model with the Remote Laboratory Management System (RLMS) is presented.

Keywords: Assessment model · Failure · Online laboratories
Recovery · Reliability · Remote laboratories · Virtual laboratories

1 Introduction

Current development in Virtual and Remote laboratories have shown a lack of re-usability, interoperability in their hardware and software implementations [1,2]. There are works in the modeling of hardware failures [3] and also authors have defined models of probability to determine failure rates [4]. Some works have described the issues with the reliability in complex software systems [5] and authors such as: [6] have defined formal measures related to a system's reliability

© Springer International Publishing AG, part of Springer Nature 2019
M. E. Auer and R. Langmann (Eds.): REV 2018, LNNS 47, pp. 252–260, 2019.
https://doi.org/10.1007/978-3-319-95678-7_28

such as: the Mean Time Between Failures (MTBF) or the Mean Time to Repair (MTTR), that are useful for the classifications of the type and seriousness of the failures.

There is not reference models for the evaluation of the hardware and software components of the online laboratories. An assessment model that collects and measures different variables during the operation of online laboratory systems can help developers and administrators to incorporate the best practices that mitigate the impact on the overall system when some problem occurs. Due to the significant differences in the design and implementation of virtual, remote and hybrid laboratories, the model proposed needs to take into account elements related to distributed web applications such as cloud servers, data bases and mobile devices interfaces; an also for remote and hybrid configurations, the model should also include analysis and evaluation of the hardware components of the laboratory experiments.

The rest of the paper is organized as follows: Sect. 2 presents the identification of failures, causes and countermeasures for online laboratories. Section 3 shows the design of a rule-based system for processing failures and causes and make decisions. Section 4 presents the integration of a reliability module into a remote lab platform architecture. Finally the discussion section is presented in Sect. 5.

2   Identification of Failures, Causes and Countermeasures for Online Laboratories

To assess the online laboratories (the virtual, remote or hybrid laboratories), we assume an architecture of distributed experiments connected to a centralized Remote Laboratory Management System (RLMS). The users in this scenario are also distributed and have access through internet.

Remote laboratories use components such as actuators, sensors, controllers, processors, storage and supportive components, to articulate the remote experiment. A sensor is a device that can measure changes in the values of temperature, pressure, humidity, light, movement, smoke, among others. Some examples are: thermometers, cameras, light sensors, proximity sensors, touch sensors. Actuators perform actions that affects the environment. They can be hydraulic, pneumatic, electrical or mechanical systems; some examples are: pistons valves, electric motors, LEDs, displays, among others. Controllers, processors and storage devices include memory modules, micro-controllers, microprocessors and all the communication modules such as: WIFI devices or Ethernet cards. Supportive components help the implementation of the experiments. They can be strips, cables, resistors, diodes, logic gates or any other object necessary for the experiment operation.

Virtual laboratories are programmed in different languages such as: Java, JSP (Java Server Pages), C#, C++, ASP.net, MatLAB, LabView or HTML 5. The elements that can be evaluated in order to determine the quality and specifically how reliable is the virtual laboratory are similar to those defined to evaluate quality of software systems: stability, compatibility, interoperability, among others [7].

According to [8], failures in distributed systems occur when there is an unexpected behavior. The failures classification includes: Omission failure (server omits to respond to an input), Timing failure (Server response is correct but was not in the right or expected interval of time, those type of failures are related with performance failures), Response failure (server responds with incorrect values; can be either a value failure or a state transition failure), and finally, Crash failure (omission to produce an output, the server fails to produce outputs to all subsequent inputs until its restart).

Other failures classifications include those that are related, not with the server, but with the distributed lab experiments; it includes user interface failures, hardware failures, communication failures and, finally, the software failures.

An online laboratory system is composed of hardware and software components that provide determined services. A failure happens when the system cannot perform one or a group of tasks. The causes can be a software problem or a physical fault, leading sometimes the system to a state of error or inconsistency. Failure recovery is a process with the goal of correcting errors and move the system back to a normal or non-error state.

Recovery from a failure is done with one of two approaches [9]. The first approach, forward recovery, tries to solve the problem in the process or system through a specific action without loss of steps in the system execution, normally is a signal, a code or a patch that recovers the system.

The second approach, backward recovery, goes back to a previous state when the system was working correctly. This approach of recovery can be done in two ways: operation-based recovery, that implies the recording of all the details to restore the system to the immediately previous state; or through a state-based recovery that relies in a periodical storage of checkpoints (this procedure saves memory but has the risk of losing the activity performed between the last check point and the failure occurrence).

3   Design of a Rule-Based System to Support the Failures and Recovery Management

A rule-based system is a type system composed by: a knowledge base, composed by a set rules (IF some condition THEN some action), a database, composed by facts, an inference engine that make the rule processing (chaining process) and make decisions and, an explanation system that translate the results in a more human understandable structure [10].

An application of rule-based systems are the expert systems which use rules to make deductions or make decisions in specific problems. The general architecture of a rule-based system is presented in the Fig. 1. Implementation of these type of systems have been developed in different domains. The most common applications of these type of systems have been developed in: economics, weather

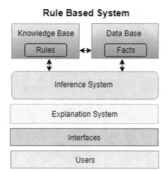

Fig. 1. General architecture of a rule based system

analysis, health applications (disease diagnosis, treatment suggestion) and general classification systems [11].

The system proposed is a rule-based system, which in this case uses as facts the most common failures and most common failure causes. Also a periodical log or report of the frequency of occurrence of an specific failure with its specific cause or causes is used to generate a more dynamic and accurate system. The second component of the rule-based system is the set of rules which validates each failure with a specific set of causes; the rule-based system makes inferences based on the validation of the rules. The Table 1 shows the most common failures for these type of systems. This information has been processed using as input some of the software and hardware failures reported by [12–14]. Also based on information retrieved from a demo of a traffic light remote experiment [16] which has been tested during a period of one year. The laboratory administrator reported several failures mostly related with connectivity, memory issues, camera problems, among others. The Table 2 shows the most common causes of failure, and countermeasures for these types of systems.

The occurrence frequency of a cause of a specific failure is a parameter obtained based on historic data about failures and its possible causes. The system takes historical data from a specific period of time defined by the user. For example, if a "sensor wrong measurement" failure has occurred recently, and a defect in the sensor is the cause, there is a counter that increases that cause frequency.

To simplify the rule-based system, 3 levels of frequency have been defined: *low*, *medium* or *high frequency* of occurrence. These results can change during the time according to the current information available in the log of reported failures and causes. The levels are returned in order from high to low frequency of occurrence.

Table 3 presents an example of the periodical relative frequency of occurrence for failures and causes. This is used as part of the facts of the system.

Table 1. Common failures online laboratory systems

ID	A. Graphic user interface	B. Online experiment components	C. RLMS (remote lab management system) (admin software, hardware component)	D. Academic systems LMS (or any other external tool)	E. Communications
1	User can not see all or parts of the on-line lab experiment	Experiment is missing all or part of the user commands	Server is not listing correctly the available online labs	Lab activities were not available in the LMS	Communication problems between the online laboratory and the server
2	User can not see all or parts of on-line lab activities	Sensor is not reading or giving wrong values (error measurements)	Server is not scheduling session	User results were not upload to the LMS	Communication problems between the user and the server
3	Online lab is total or partially unresponsive from the user interface (blocked or slow)	Actuator is not executing actions or is receiving wrong values	Server is not managing the user session	User profile were not retrieved from the LMS	Communication problems between the server and the LMS or external tool
4	User does not receive all or part of the results from the on-line lab learning Object	On-line lab Experiment controller is not reading or sending signals through the ports	Server is not login users (authentication)	LMS does not allow the visualization of the on-line lab learning Object Interface	Communication problems between the User and the LMS or external tool
5	User can not send all or part of the commands to the on-line lab	On-line lab controller is not send data to the server	Server is not storing log of the lab sessions	The external tool does send information to the server	
6	User can not see or is having problems with the quality of the video streaming	On-line lab supportive components are not working			

The rules are defined as the validation of the specific failure based on the current state of the causes (one rule is defined for each possible failure), for example: in the case of failure F_1: *Wrong value in the sensor readings*, some of the causes can be C_1: *defect sensor*, C_2: *power problems in the sensor* or C_3: *interference generated for another component*. An example of a rule in this case is:

Rf1: if current_failure $= F_1$ then get the Frequency (Causes of (F_1));

If another failure in the database has the same set of causes (occurrence frequency levels can be different), the system will report that failure as a concurrent failure that can be happening in the system because of the detection of the other failure.

Rc1: if Causes of $(F_1) ==$ Causes of (F_2) then F_1 and F_2;

A prototype of the system has been implemented in the Tool JESS [15]. The Figs. 2 and 3 show some of the facts and rules defined for the test.

4   Integration of the Reliability Module with the RLMS

The reliability module has as the main component the rule-based system. The log reports about the labs performance are stored in the RLMS and are available

Table 2. Causes of failures and countermeasures

Cause	Countermeasure	Failure ID
Client terminal browser does not support lab interface language (java, flash etc)	Verify the plugins in the browser, check permission in the browser	A1, A2, A3, A6
Security configuration of the browser blocked the components of the online lab	Check security restrictions	A1, A2, A3, A6
Terminal does not meet the minimum hardware requirements to run the lab	Improve the resources in the user terminal	A1, A2, A3, A6
Communication channel lost in the server	Verify the network configuration in the server, verify with service provider ISP if the internet service is working properly, activate the service with a backup internet connection (if is available), use the backup connection to activate alert system	A1, A2, A3, A4, A5, A6, C1, C4, C5, D4, E1, E2, E3
Communication channel lost in the user terminal	Save the current session status and wait while the session period is active	A1, A2, A3, A4, A5, A6, E2, E4
Communication channel lost in the online lab (virtual, remote or hybrid)	Notify the user, send a signal to reboot the experiment, if it does not start again then set alert for the administrator of the lab	A1, A2, A3, A4, A5, A6, B1, B2, B3, E1
Communication channel lost in the LMS or external tool	Save results in the server until is possible to make updates in the LMS, if the activity could not be load in the online lab make it unavailable until the communication is re-established and set an alert to the LMS administrator	A2, C4, D1, D2, D3, D5, E3, E4
Power issues in the server	Have a backup power source for the server, send a signal to the user informing about the problem	A1, A2, A3, A4, A5, A6, B1, B5, C1, C2, C3, C4, C5, D4, E1, E2, E3
Power issues in the client	Close the user session, send report to the server, save the last state of the user session	A1, A2, A3, A4, A5, A6, E2
Power issues in the online lab (virtual, remote or hybrid)	Isolate the lab experiment until the power is back, disable all the Online lab learning object that requires access to that experiment, notify the user about the problem and set an alert for the experiment administrator	A1, A3, A4, A5, A6, B1, B2, B3, B4, B5, E1
Sensor damaged during the use of the remote laboratory	Isolate the sensor, notify the user about the problem, set an alert for the experiment administrator	B2
Actuator damaged during the use of the remote laboratory	Isolate the actuator, notify the user about the problem, set an alert for the experiment administrator	B3
Supportive components damaged during the use of the online laboratory	Isolate the supportive components and notify the user about the problem, set an alert for the experiment administrator	A6, B6
Actuators is receiving values out of the ranges	Run a test with known values to determine if there is a problem with the actuator or if the values have been changed for an external component	B3
Controllers, memory and storage, damaged before or during the use of the remote laboratory	Run test for each of the components and if necessary isolate the experiments connected to that specific controller, notify the user about the problem and set an alert for the experiment administrator	A1, A3, A4, A5, A6, B1, B2, B3, B4, B5
Defect Sensor in the remote laboratory	Isolate the experiment and notify the user about the problem, set an alert for the experiment administrator	B2
Defect Actuator in the remote laboratory	Isolate the experiment, notify the user about the problem, set an alert for the experiment administrator	B3
The user has a slow internet connection speed	Notify the user and create a log record in the server	A1, A3, A4, A5, A6, B1, E2, E4
Interference caused by an external component	Identify the component, notify the administrator responsible of the operation of the external component	B2, B3
User does not get authorization rights	Validate user credential against the LMS, set and alert for the server administrator	A1, A2, D1, D4
Problems in the interface software of the online lab	Set an alert for the online lab administrator, after the severity of the problem is determined, disable the online lab according with the admin report	A1, A2, A3, A6
Problems in the software for the online lab controller	Set an alert for the online lab administrator, after the severity of the problem is determined, disable the online lab according with the admin report	A1, A3, A4, A5, A6, B1, B2, B3, B4, B5, E1
Problems in the software for the data storage of the online lab (traces of results xAPI and admin data)	Set an alert for the online lab administrator, after the severity of the problem is determined, disable the online lab according with the admin report	C5, D2, D5
Problems in the software of the administrative modules of the server (user, lab repository, scheduler, communications etc)	Set an alert for the server administrator, after the severity of the problem is determined, restore the software to a previous version only if needed	C1, C2, C3, C4, C5, D1, D2, D3, D4, E1, E2, E3
Streaming service down in the online lab	Restart streaming service of the online lab, notify the online lab administrator	A1, A6

Table 3. Example of periodical relative frequency of occurrence

Cause	Failure		
	F_1	F_2	F_3
C_1	10 (High)	5 (High)	0
C_2	5 (Medium)	1 (Low)	0
C_3	2 (Low)	0	4 (Medium)
C_4	0	0	2 (Low)
C_5	0	0	9 (High)
C_6	0	0	0
C_7	0	0	0

```
(deffacts initial-phase
    (phase get-failure)
)

(deffacts failure-causes-knowledge-base
        (failure-causes
        (failure 1)
            (cause1 "Defect Sensor")
            (cause2 "Power Problems")
            (cause3 "Interferrence by another device"))
)

(deffacts causes-frequency
        (failure-causes-frequency
        (failure 1)
            (cause1f 10)
            (cause2f 5)
            (cause3f 2))
)
```

Fig. 2. JESS facts definition example

```
(defrule failure-causes
    ?selection <- (failure ?choice)
    (failure-causes (failure ?choice)(cause1 ?c1)(cause2 ?c2)(cause3 ?c3))
    =>
    (printout t "Based on the Failure, the causes are " ?c1 " or " ?c2 " or " ?c3 crlf)
    )

(defrule failure-causes-frequencies
    ?selection <- (failure ?choice)
    (failure-causes-frequency (failure ?choice)(cause1f ?c1f)(cause2f ?c2f)(cause3f ?c3f))
    =>
    (printout t "For that Failure, the causes frequencies are " ?c1f " , " ?c2f ", " ?c3f  crlf)
    )
```

Fig. 3. JESS rule definition example

in the academic system, supported by the implementation of xAPI services in Online Laboratories [16]. These reports can be accessed by the reliability module in order to feed the model. If some failure is detected and associated with one or more causes and countermeasures that the system has predefined, the system will trigger those countermeasures automatically. If the solution of the problem needs a human intervention, the module will activate an alert for the lab administrator and will take some actions trying to prevent the spreading of the risk, trying to maximize the availability of the rest of the online labs connected to the RLMS.

The Fig. 4 shows the reliability module as part of the On-line laboratories architecture (Based on [17]).

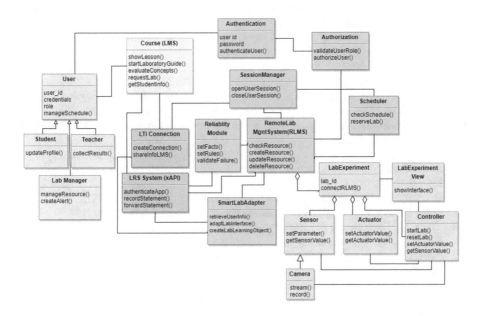

Fig. 4. Reliability module integrated in a remote laboratory design

5   Discussion

A standard model of assessment gives to developers of online laboratories tools to protect and improve the systems, making the systems more reliable, efficient and secure. Generating in users a higher level of confidence when they want to invest in this type of solutions.

The reliability model proposed is intended to be applied to different types of online laboratories, more specifically remote laboratories, virtual laboratories and hybrid configurations that use of software and hardware components. However the model do not use specific data about batch experiments or other asynchronous configurations failures information as part of its knowledge base.

Acknowledgments. This work has been supported by The Latin America and Caribbean Consortium of Engineering Institutions LACCEI, which is providing resources to make advances in this field. With the goal in the near future to have a great impact with online laboratories in the communities of the Latin America and the Caribbean region that has been affected by natural disasters such as Haiti, Puerto Rico, among others.

References

1. Gravier, C., Fayolle, J., Bayard, B., Ates, M., Lardon, J.: State of the art about remote laboratories paradigms foundations of ongoing mutations. Int. J. Online Eng. **4**, 1–9 (2008)
2. Nickerson, J.V., Corter, J.E., Esche, S.K., Chassapis, C.: A model for evaluating the effectiveness of remote engineering laboratories and simulations in education. Comput. Educ. **49**, 708–725 (2007)
3. Ramakrishnan, T., Pecht, M.: Electronic hardware reliability. In: Avionics Handbook, pp. 22.1–22.21. CRC, Boca Raton (2000)
4. Smith, D.J.: Reliability, Maintainability and Risk, 9th edn. Butterworth-Heinemann, Amsterdam (2017). ISBN 9780081020104
5. Randell, B., Lee, P., Treleaven, P.C.: Reliability issues in computing system design. ACM Comput. Surv. **10**(2) (1978). https://doi.org/10.1145/356725.356729
6. Shooman, M.L.: Probabilistic Reliability: An Engineering Approach. Mc-Graw-Hill Inc., New York (1968)
7. ISO/IEC 25010 - Systems and software engineering - Systems and software Quality Requirements and Evaluation (SQuaRE) - System and software quality models. ISO/IEC (2010)
8. Cristian, F.: Understanding Fault-Tolerant Distributed Systems. Thesis in Computer Science and Engineering, University of California, San Diego (1993)
9. Yu, T.L.: Lecture notes: operating systems concepts and theory. California State University (2010). http://cse.csusb.edu/tongyu/courses/cs660/notes/recovery.php
10. Newell, A., Simon, H.: A Human Problem Solving. Prentice Hall, Englewood Cliffs (1972)
11. Qin, B., Xia, Y., Prabhakar, S.: A rule-based classification algorithm for uncertain data. In: IEEE 25th International Conference on Data Engineering, ICDE 2009 (2009)
12. Marco, C., Prattichizzo, D., Vicino, A.: Operating remote laboratories through a bootable device. IEEE Trans. Ind. Electron. **54**(6), 3134–3140 (2007)
13. Wu, C., Hwang, J., Chladek, J.: Fault-tolerant joint development for the space shuttle remote manipulator system: analysis and experiment. IEEE Trans. Robot. Autom. **9**(5), 675–684 (1993)
14. Guimaraes, G., et al.: Design and implementation issues for modern remote laboratories. IEEE Trans. Learn. Technol. **4**(2), 149–161 (2011)
15. Jess, The Java Expert System Shell. http://www.jessrules.com/docs/45/
16. Zapata-Rivera, L.F., Larrondo Petrie, M.M.: Implementation of cloud-based smart adaptive remote laboratories for education. In: Frontiers in Education (2017)
17. Zapata-Rivera, L.F., Larrondo Petrie, M.M.: Models of collaborative remote laboratories and integration with learning environments. Int. J. Online Eng. **12**(9), 14–21 (2016)

Digital Remote Labs Built by the Students and for the Students

J. Nikhil[✉], J. Pavan, H. O. Darshan, G. Anand Kumar, J. Gaurav,
and C. R. Yamuna Devi

Dr. Ambedkar Institute of Technology, Bangalore 560056, Karnataka, India
nikhil.janardhana@gmail.com, pavanj278@gmail.com,
darshanho16@gmail.com, kumaranand15@gmail.com,
gauravprasad96@gmail.com, yamuna.devicr@gmail.com

Abstract. One of the most challenging problems faced by India in the field of Engineering Education is that most of the graduating engineers are not employable due to lack of practical skills. Practical education is one of the most important aspects in any student's learning process. Application of theoretical concepts should be taught to students for which laboratories play a prominent role. In this context student have to be taught to utilize laboratories at the first place and build things on their own in order to understand any concept. Our purpose was to design remote labs for digital system design laboratory and make students build them on their own at a very low cost. Digital system design laboratory consists of various basic level experiments like verification of logic gates, full adder, half adder etc. Our goal was to completely structure all the laboratory experiments through remote engineering and build a completely working digital remote lab.

Keywords: Practical education · Utilization · Digital system design laboratory

1 Introduction

The extensive use of internet have led to the inventions of many technologies. One such technology that has come to the limelight in the recent years is REMOTE ENGINEERING. This technology is solely built around the concept of internet and clearly shows how important internet is to human being and also the benefits one can with this concept. REMOTE laboratories that are developed with an intention to make access to laboratories easier also gives enough space for students to think and apply the learnt theoretical concepts in an extraordinary manner. It can also be a platform to a lot of students around the world who cannot afford to buy education but have the interest in understanding engineering subjects more deeply. This is definitely a step to transform education system around the globe as students need not worry about the difference in the quality of knowledge they acquire from laboratories, instead equality is preserved to each and every student. One such approach is made here in Dr. Ambedkar Institute Of Technology (Bangalore, India) where students have transformed a complete laboratory (Digital System Design), where all related experiments can be remote accessed anytime of the day and also for much lower cost. These

© Springer International Publishing AG, part of Springer Nature 2019
M. E. Auer and R. Langmann (Eds.): REV 2018, LNNS 47, pp. 261–268, 2019.
https://doi.org/10.1007/978-3-319-95678-7_29

remote labs are not a replacement to the already existing labs but an additional feature provided to it, which involves a very simple architecture. The architecture of the laboratory is as shown in Fig. 1.

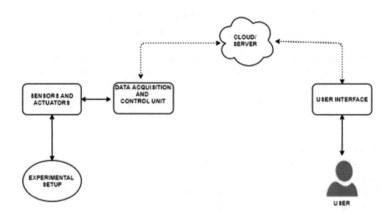

Fig. 1. Architecture of the digital system design laboratory.

2 Approach

In this section, the build of the Digital System Design lab whose description has been told towards the end of introduction is discussed in depth.

In this section we also elaborate about (1) the steps taken in building the Digital System Design lab infrastructure. (2) The hardware details and (3) the software us Comparison between conventional lab and remote lab (Fig. 2).

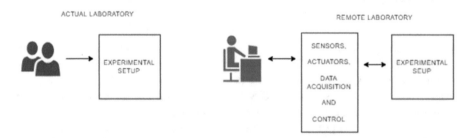

Fig. 2. Shows the basic comparison between actual and remote laboratory setups.

The actual/conventional laboratories do not have an infrastructure built between the user and the experimental setups. The users will have direct but limited access to these types of laboratories. Where as in remote lab setup the infrastructure allows the user to access the laboratories 24/7 from any region.

2.1   Steps Taken in Building the Infrastructure

2.1.1   Choosing the Laboratory

The primary step in developing the infrastructure in choosing the laboratory. Among the option that was available between Analogy, Digital and Microwave labs. Digital laboratory was selected keeping in mind that this would serve as a basic model in developing more advanced laboratory setups. The digital system design labs had less complexities compared to other labs. Also building of digital labs were economic compared to the remaining options available.

2.1.2   Selecting of Experiments

The digital system design laboratory consists of various simple digital experiments. The design laboratory consists of digital trainer kits on which experiments are conducted. The trainer kits have separate IC slots and are also made up of few general purpose connections. Therefore in order to perform experiments manual alterations in the connections have to be made. But complexities arise when the same experiments have to be performed remotely, as manual alterations in the connections on the digital trainer kit is not possible through the user interface on the computer. Hence in order to conduct the experiments remotely separate circuitry on the PCB boards have to be designed for each experiment where connections are permanently made for each experiment. Which means dedicated circuitry boards have to be prepared for each experiment as mentioned in Fig. 3. Our digital system design laboratory consists of the following experiments:

Fig. 3. Circuitry built to replace the digital trainer kit for remote conduction of experiments.

(1) verification of logic gates (2) half adder and full adder (3) full subtraction (4) Multiplexers and de-multiplexer (4) encoder and decoder (5) flip-flops (6) seven segment display (8) motor control.

The complexities were first taken into consideration and studied, later out of the already available experiments a few of them were picked based on their order of complexities and keeping in mind the time constraints and cost factors.

Experiments that were selected are:

(1) Design and verification of digital logic gates:

- AND logic gate, OR logic gate, NAND logic gate, NOR logic gate, XOR logic gate, NOR logic gate

(2) Combinational logic circuits:

- Full adder
- Half adder

(3) Real time operations

- Seven segment display control
- Motor control

2.2   Hardware Details

The hardware components used should be such that it should be operated on low voltages and should be cost effective for implementation. Apart from all the LEDs, IC's motors, PCB boards used, and arduino microcontroller is used because it is an open source prototyping platform and is one of the cheapest controllers.

The hardware used in the infrastructure is:

(1) arduino mega (2) arduino nano (3) dc power supply (4) leds (5) general purpose PCB board (6) connecting wires (7) resistors (8) dc motors (9) L293d (10) 7 segment display

Component	Type	Cost	Purpose
Arduino mega	Controller (master)	38$	(1) to give power supply to the slave controller (2) to operate with labVIEW LINX code
Arduino nano	Controller (slave)	22$	(1) to execute the arduino codes (2) bridge between the master controller and led
LEDs	Red, green, blue, white, yellow	0.1$ each	Displaying of the inputs and outputs
PCB boards	General purpose PCB	3$ each	To embed the circuit connections so, the hardware will act accordingly
Resistors	220, 1 k, 10 k ohms etc.	0.1$ each	Limit the current to the LEDs
Motors	DC or geared DC motor	10$	To control the dc motors remotely using the infrastructure
L293D	Motor driver IC	7$	To drive the dc motors
7 segment display	Common anode/cathode	14$	To observe the 7 segment output

2.3 Software Requirements

Software used:

(1) Arduino IDE is an easy to use software platform that has already built in code for various operations which can be easily modified based on the user's requirements.
(2) LabVIEW is an integrated development environment designed specifically for engineers and scientists building measurement and control systems.

ARDUINO IS CHOSEN FOR:

Simplicity.
Strong hardware – software interaction.
Code at an embedded C level.
Open source and a huge community for support.
Large database of libraries and binaries.

LABVIEW IS CHOSEN FOR:

Excellent design in form of front panel and block diagram.
Built in libraries and tools.
Precision measurement reading.
Highly customizable.

LINX: is a software upgrade developed by MakerHub to facilitate arduino control through LabVIEW.
LINX provides an interface between LabVIEW and an Arduino.
LINX requires a data connection between LabVIEW and the Arduino always.
LabVIEW sends a packet to the Arduino.
The Arduino processes the packet and performs the specified operation (usually some I/O).
The Arduino sends a response packet back to LabVIEW (Fig. 4).

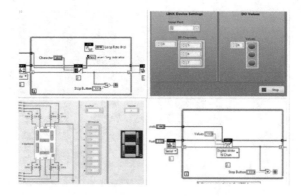

Fig. 4. Labview front panel and block diagram

3 Working

Students were divided into several groups with each group consisting of four students. Each student group was assigned with one experiment to work on. Initially each logic gate circuitry were built separately and verified individually. Later using Arduino nano a common circuit was built to conduct and verify all the logic gates on a single PCB board. The circuitry also included full adder and half adder experiments. Similarly controlling the servo motor and 7 segment display experiments were designed. The experimental kits that were designed for each experiment which consisted of LEDs, microcontrollers, ICs', resistors etc. which were interconnected together giving the required output. Since the laboratory experiments were simple, arduino boards served the purpose of controlling the designed kits. The Arduino boards are controlled by local servers (computers). All the computers of the laboratory were connected to cloud. Hence through cloud computing technology the user (student) could use the laboratory at any time of the day and from any geographical location. These labs are completely student driven where the responsibility of planning, building, maintenance and running the remote laboratory is of the students.

Instead of building separate circuitry for each experiment, few experiments were embedded into a common circuitry which is controlled by two microcontrollers (arduino) as depicted in Fig. 5. The first microcontroller which acts as a master controller (arduino mega) is programmed by LINX makerHUB in LabVIEW and is directly connected to the computer. The second microcontroller is the slave controller (arduino nano) which is programmed by arduino to execute the written code and display the outputs through the LED based on the inputs given the user in the interface. The Nano board is powered by the arduino mega. The entire system is the serial flow of commands from the user to the LEDs on the circuitry which are mediated by the two microcontrollers.

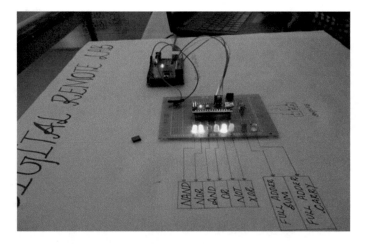

Fig. 5. Actual working model indicating the output

Figure 6 is the graphical representation indicating the quality time spent by the users in conventional labs and remote labs. The red indication mentions about the usage of conventional laboratories by users which can be maximum used till 8 h a day, moreover hands on experiments are not guaranteed in these labs as each experiment is conducted by a group of two or more students. The blue indication is about the usage of remote laboratories by users which can be maximum used up to 24 h a day, where each user/student gets to perform any experiment individually which allows each user to spend quality time on each experiment.

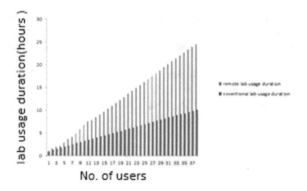

Fig. 6. Graphical Representation indicating the users vs. usage.

4   Outcomes

Completely developed laboratories that can be remotely accessed. Labs are built by the Students for the future use of students. Improved efficiency of the utilization of the lab with access up to 6500 h a year comparatively less than 2000 h previously. Improved practical aptitude of students as they get more time hands-on. Has given enough space for students for creative thinking and implementation Serves as a new subject of research and provides infrastructure for faculty and students to work on.

5   Conclusion

Remote lab has been presented an innovative solution to developing and enhancing lab practice and access for distance learners, and as a means of providing training and experience for teaching lab. Remote lab reduces work and time. It has provided user friendly screens to enter the data. It is web enabled portable and flexible for further enhancement. This project can be highly beneficial for students who want to continue their study from e-learning programs. Remote Labs provides the basic knowledge to the student who doesn't have any idea about the experiment.

References

Research Paper

1. Auer, M.E.: Virtual lab versus remote lab. In: 20th World Conference on Open Learning and Distance Education (2001)
2. Ram, B.K., Kumar, S.A., Sarma, B.M., Mahesh, B., Kulkarni, C.S.: Remote software laboratories: facilitating access to engineering softwares online. In: 2016 13th International Conference on Remote Engineering and Virtual Instrumentation (REV), pp. 409–413. IEEE (2016)
3. Pruthvi, P., Jackson, D., Hegde, S.R., Hiremath, P.S., Kumar, S.A.: A distinctive approach to enhance the utility of laboratories in Indian academia. In: 2015 12th International Conference on Remote Engineering and Virtual Instrumentation (REV), pp. 238–241. IEEE (2015)
4. Esche, S.K., Chassapis, C., Nazalewicz, J.W., Hromin, D.J.: Architecture for multi-user remote laboratories. Dynamics (with a typical class size of 20 students) **5**, 6 (2003)
5. Nickerson, J.V., Corter, J.E., Esche, S.K., Chassapis, C.: A model for evaluating the effectiveness of remote engineering laboratories and simulations in education. Comput. Educ. **49**(3), 708–725 (2007)
6. Wuttke, H.-D., Henke, K., Ludwig, N.: Remote labs versus virtual labs for teaching digital system design. In: International Conference on Computer Systems and Technologies-CompSysTech, p. 2-1 (2005)
7. Gontean, A., Szabó, R., Lie, I.: LabVIEW powered remote lab. In: 2009 15th International Symposium for Design and Technology of Electronics Packages (SIITME). IEEE (2009)
8. Auer, M., Pester, A., Ursutiu, D., Samoila, C.: Distributed virtual and remote labs in engineering. In: 2003 IEEE International Conference on Industrial Technology, vol. 2, pp. 1208–1213. IEEE (2003)
9. Gowripeddi, V.V., Ram, B.K., Pavan, J., Devi, C.Y., Sivakumar, B.: Role of Wi-Fi data loggers in remote labs ecosystem. In: Online Engineering and Internet of Things, pp. 235–249. Springer, Cham (2018)
10. Kalúz, M., Čirka, Ľ., Valo, R., Fikar, M.: ArPi Lab: a low-cost remote laboratory for control education. IFAC Proc. Vol. **47**(3), 9057–9062 (2014)

Virtual Learning Environment for Digital Signal Processing

Yadisbel Martinez-Cañete[1]([⊠]), Sergio Daniel Cano-Ortiz[1], Frank Sanabria-Macias[1], Reinhard Langmann[2], Harald Jacques[2], and Pedro Efrain Diaz-Labiste[1]

[1] Universidad de Oriente, Santiago de Cuba, Cuba
{ymartinez,scano,fsanm77,pedro.diaz}@uo.edu.cu
[2] University of Applied Sciences Düsseldorf, Düsseldorf, Germany
R.Langmann@t-online.de, harald.jacques@hs-duesseldorf.de

Abstract. At the Electrical Engineering Faculty of the Universidad de Oriente (UO) in Santiago de Cuba, for the increase and improvement of students' knowledge in the subject Digital Signal Processing difficulties are detected in the learning of this subject in Biomedical, Telecommunications and Electrical careers in the Faculty, due to the high degree of mathematical theoretical concepts that they require, which has caused lack of motivation in the students and low academic results.

The virtual learning environment is a web application developed in the Django framework, written in Python, which respects the design pattern known as Model-View-Controller, as a relational database management system, Sqlite is proposed, due to its speed, reliability and ease of use, multiplatform and multi-user. Using the observer pattern for the management of student activities, through which students can check the progress of the assimilation of the contents in each of the topics of the subject. Two user roles are considered for the system, teachers' role that is responsible for the management of educational resources, such as topics, videos, questions, or other reference materials, and students' role, who interact with each of the resources, registering the time and answers to the questions, for a post-top statistical analysis that can be shown to the student or the teacher.

Keywords: Digital Signal Processing · Web technology
Virtual learning environment

1   Introduction

Internet is becoming an important educational resource thanks to the capacity it offers us to complement the traditional format classes in the classroom environment with the possibility of allowing access to information online. The limitations of time and place have disappeared.

In addition to this possibility, it is important to take into account the effect of the interactivity that a computer tool can have on the learning process. In this sense, it is known that interactivity is an important and probably essential attribute of any learning technique that pretends success, that learning is more effective when the student can control the exchange of information.

© Springer International Publishing AG, part of Springer Nature 2019
M. E. Auer and R. Langmann (Eds.): REV 2018, LNNS 47, pp. 269–276, 2019.
https://doi.org/10.1007/978-3-319-95678-7_30

The information and communication networks, through the Internet, break barriers of time and space to develop teaching-learning activities. Information and Communication Technologies (ICT) is the main tool that organizations and educational institutions have begun to use to offer courses and virtual study programs through the web [1].

In short, the increase in training needs is calling for the creation of new networks and forms of access to education. In this sense, new technologies to support digital education are becoming one of the strategies used to do so. One of the most representative examples of the teaching media used on the web are the tutorials, which support the student's learning process, as they guide him through a series of activities that he must carry out on his own [2–4].

Studies conducted in the early nineties at the University of Deakin (Australia) [5] showed that students who had used interactive and multimedia learning techniques had a 55% gain over those students who received the classes in the traditional teaching environment. Other significant figures showed that the students learned the material 60% faster and the retention of knowledge after 30 days was between 25 and 50% higher.

In another study carried out at the Acadia University (Canada) [6] in students of an introductory physics course in the first year of university teaching, it can be seen that both the results and satisfaction at the end of the students' course. That they used interactive tools were clearly superior to those of those who received the course with classical techniques.

It is also worth noting the usefulness that this tool can have in the classroom itself, by allowing the teacher to use other materials than usual and in this way to speed up the classes and attract the attention of the students. In this sense the author in [7] has been able to verify that by using the tutorial in the classroom the attention increased and the comprehension improved, and therefore the learning.

Signal processing is an area of Electronic Engineering that focuses on the representation, transformation and manipulation of signals, and the information they contain.

Students in experimental and technical areas must be able to process physical signals, regardless of the area in which they were generated, all these signals can provide valuable information, in some crucial cases, that we need to know how to deal with.

Students in scientific and technical degrees must receive an eminently practical and applied, experimental and scientific education, which helps to understand - in the deepest sense of the word - the abstract concepts learned in the theoretical classes, which also have a high mathematical load [8–10].

This experimentation does not always occur in the classroom due to overcrowding, lack of resources, lack of time, cost of instrumentation, etc. [11].

Problem: Difficulties in the learning of the subject Digital Signal Processing in the Biomedical, Telecommunications and Electrical careers of the Faculty, due to the high degree of mathematical theoretical concepts that they require, which has caused lack of motivation in the students and low academic results.

Research Objective: Development of an interactive tutorial for the development of signal processing practices that incorporate elements of reality in subjects of various careers and at different levels, according to the students' preparation and the training objectives of each course.

A multimedia system can be Khan Academy, a non-profit organization founded by Salman Khan with the mission of providing a world-class free education for all. The site offers more than 5,000 educational videos online in a series of subject areas (including mathematics, science, economics, finance, history and art), an extensive repository of mathematical exercises, data and information in real time regarding the advances and difficulties that users present when using resources. The Khan Academy platform provides resources for students and teachers. The four main components of the site that support learning are: videos, exercises, data, and a community of users. They all work together to create what Khan calls "a personalized, refined and interactive environment for learning." However, this site is not accessible from Cuba.

In the Electrical Engineering Faculty of the UO in Santiago de Cuba there is no an educative software as didactic mediator that helps to improve the teaching learning process mainly in signal digital processing subject that allows students access faster to the content of this subject, even to research, interact with demonstrative and interactive questions to fulfill a higher development of professional abilities being useful for both activities face to face and semi face to face for the improvement of knowledge of the students.

Taking into account the previous statement, the following was determined as a general objective: Implement a Virtual Learning Environment for the subject Digital Signal Processing (VLEDSP).

2   Materials and Methods

2.1   Design of the VLEDSP

The Virtual Learning Environment for the subject Digital Signal Processing has the following functionalities according to the different user roles.

Student: They have access to all the information without privileges to modify it. This user covers the students of the Electrical Engineering Faculty, being those who interact more with the application, it is for whom the software was conceived.

- Request management
- Access to Questions
- Access to Videos
- Access to topics
- Access to the bibliography
- Access to the glossary of terms
- Access to statistics of learning

Teacher: This user is in charge of managing the resources that the software has, with the objective of keeping the content updated within it. It is the person trained in the subject matter, the teacher with permissions within the software to manage these resources.

- Management of topics
- Questions Management

- Video Management
- Management of bibliographies
- Terms glossary management
- Management of consultation materials
- Access to statistics of student learning

Administrator: You can insert, modify or delete the users of the system, as well as manage the resources of the subject

- Student management
- Teacher management
- Show list of students per career
- Show list of teachers

Next in Fig. 1 the navigation map of the VLEDSP is shown, relating the pages to which you will have access depending on the role.

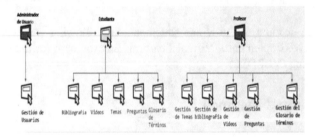

Fig. 1. Navigation map

2.2   Tools and Languages

Django is an open source web development framework, written in Python, that respects the design pattern known as Model-View-Controller. The fundamental goal of Django is to facilitate the creation of complex websites.

SQLite It is a relational database management system, it is fast, reliable and easy to use, and it is multiplatform, multi-user. The Python language is compatible with SQLite, because of the extensive set of instructions defined for the treatment of it.

Python is an interpreted programming language whose philosophy emphasizes a syntax that favors a readable code. It is a multi-paradigm programming language, since it supports object orientation, imperative programming and, to a lesser extent, functional programming. It is an interpreted language, it uses dynamic typing and it is multiplatform.

2.3   Architecture of the VLEDSP

Django is known as an MTV Framework, see Fig. 2.

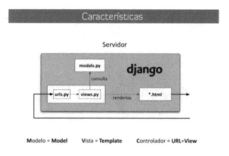

Fig. 2. MTV architecture

M means "Model" (Model), the access layer to the database. This layer contains all the information about the data: how to access these data, how to validate them, what the behavior is, and the relationships between the data.

T means "Template", the presentation layer. This layer contains the decisions related to the presentation: how some things are shown on a web page or another type of document.

V stands for "View", the layer of business logic. This layer contains the logic that accesses the model and delegates it to the appropriate template: you can think of this as a bridge between the models and the templates.

2.4 Domain Model

The design of the database of the VLEDSP consists of 5 fundamental tables Student, Teacher, Videos, Exercises and Documents, which are related between them as a relation of much to much, with the objective of registering each activity that the students perform in the learning environment and on the other hand that the teacher has proof of them. As essential attributes in these new tables are the date and time, for the statistics and reports that must be obtained from the activity of each student in the VLEDSP, see Fig. 3.

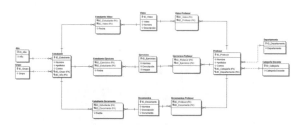

Fig. 3. Entity relationship diagram

2.5 Observer Pattern

It is a design pattern that defines a dependence of the one-to-many type between objects, so that when one of the objects changes its state, it notifies this change to all the dependents. It is a behavioral pattern, that is, it is related to functioning algorithms and assignment of responsibilities to classes and objects see Fig. 4.

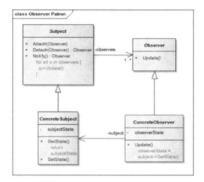

Fig. 4. Pattern diagram of observer pattern

Subject: Know your observers, which can be from 0 to N, and offers the possibility of adding and eliminating observers.

Observer: Defines the interface used to notify observers of the changes made in the Subject.

Concrete Subject: Stores the state that is of interest to observers and sends a message to its observers when their status changes.

Concrete Observer: Maintains a reference to a Concrete Subject. It stores the state of the Subject that is of interest to it and implements the Observer update interface to maintain consistency between the two states.

3   Results and Discussion

The VLEDSP is a web platform to learn through videos, which the teacher proposes to his students, who have to see in their time of independent study to learn a new topic, see Fig. 5.

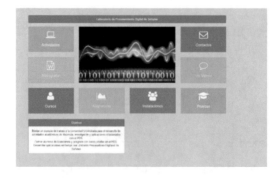

Fig. 5.   Home page of the VLEDSP

In this way each student can go at their own rhythm, pausing the video or repeating the explanation as many times as necessary without interrupting others, it will also allow solving the exercises or problems that arise for each topic by selecting one of the possible answers, see Fig. 6.

Fig. 6. Videos page by topics

As an interesting addition, the VLEDSP will allow each student to have statistics of their progress through their institutional email account, see the videos they have seen, the exercises they have completed, the areas they have not yet mastered and the points achieved.

As a complement to learning and linked to their work practice, the VLEDSP has a link to the web application for biomedical signal management WebSA 2.0 [12], which includes four types of biomedical signals: cry signal, electroencephalogram signal (EEG), electrocardiogram signal (ECG) and electroculogram signal (EOG).

4   Conclusions

With the implementation of a Virtual Learning Environment with the Django tool for the subject of Digital Signal Processing, the management by the teacher of the contents of the same is achieved, as well as, a blended teaching medium for the students of each one of the careers of the Electrical Engineering Faculty that will allow to increase the level of independent study and the self-preparation in the students, not only through videos and documents but by means of questions that are presented as a means of checking the knowledge, supporting in great way to the teaching and learning process.

Acknowledgements. Part of this research was made thanks to the financial support derived from the Webbasierte FuE-Plattform Zur Signalanalyse Project (WebSA) in collaboration with the University of Applied Sciences from Dusseldorf.

References

1. The offer of higher education through the Internet. Analysis of the virtual campuses of Spanish universities. Final report. http://www.edulab.ull.es/campusvirtuales/. 1 de junio de 2015
2. Jaimez-González, C., Sánchez-Sánchez, C., Zepeda-Hernández, S.: A web platform for creating and administering interactive online tutorials. In: Proceedings of the Canada International Conference on Education, Toronto, Canada, pp. 88–92, 4–7 Apr 2011
3. Jaimez-González, C., Sánchez-Sánchez, C., Zepeda-Hernández, S.: Creating and administering interactive online tutorials and performance evaluation tests through a novel web platform. Int. J. Cross-Discip. Subj. Educ. (IJCDSE) 2(3), 447–455 (2011)
4. Distance education: From the theory to the practice. http://www.edulab.ull.es/campusvirtuales. 1 de junio de 2015
5. Street, S., Goodman, A.: Some experimental evidence on the educational value of interactive Java Applets in web-based tutorials. In: Third Australasian Conference on Computer Science Education, pp. 94–100. Association for Computing Machinery, Brisbane (1998)
6. Williams, P., MacLatchy, C., Backman, P., Retson, D.: Studio physics report on the Acadia. Department of Physics, Acadia University (1996). http://www.acadiauca/advantage/physics.htm
7. Torrubia, G.S., Lozano Terrazas, V.M.: Algoritmo de Dijkstra. Un Tutorial Interactivo
8. Ifeachor, E.C., Jervis, B.W.: Digital Signal Processing. Addison-Wesley, Wokingham (1993)
9. Oppenheim, A., Schafer, R.: Discrete-Time Signal Processing. Prentice Hall, Upper Saddle River (1999)
10. Smith, S.: Digital Signal Processing: A Practical Guide for Engineers and Scientists. Newnes, Boston (2002)
11. Santos Peñas, M., Farias Castro, G.: Laboratorios virtuales de procesamiento de señales. Revista Iberoamericana de Automática e Informática Industrial, vol. 7, Núm. 1, pp. 91–100. Enero (2010). ISSN: 1697-7912
12. Cano-Ortiz, S.D., Langmann, R., Martinez-Cañete, Y., Lombardía-Legrá, L., Herrero-Betancourt, F., Jacques, H.: A web-based tool for biomedical signal management. In: Auer, M., Zutin, D. (eds.) Online Engineering and Internet of Things. Lecture Notes in Networks and Systems, vol. 22. Springer, Berlin (2017). ISSN 2367-3370, ISSN 2367-3389 (electronic), http://www.springer.com/gb/book/9783319643519

Online Experimenting with 3D LED Cube

Katarína Žáková[(✉)]

Faculty of Electrical Engineering and Information Technology, Slovak University of Technology,
Bratislava, Slovakia
katarina.zakova@stuba.sk

Abstract. The paper illustrates how 3D LED cube can be used for remote experimenting and for gaining basic programming skills. For this purpose two different realizations of the cube were used. Both of them have 512 ($8 \times 8 \times 8$) LEDs and use Arduino motherboard for influencing the behavior of LEDs (their switching on and off).

The aim was to control the LED cube remotely. Therefore it is connected to the web server where the supporting web application is placed. The interface for the user is very simple – one text area serves the user for entering the control code.

The developed application enables to test own codes and to learn how control loops and conditional commands work. The feedback for user is offered by video channel.

Keywords: Online experiment · 3D LED cube · Computer aided education

1 Introduction

Learning of programming brings novice students several difficulties. At the beginning they are connected mainly with basic programming elements such as loops, conditionals, arrays or recursion. It is not always easy to find a way how to facilitate these first steps for student. The most commonly used method is to let students to algorithmize as many examples as possible. This approach may not always be very interesting.

Young people do not have problems to use new technologies. As it is written in [6] "they play and have fun with video games, search for information they need on the internet, develop and maintain social relationships on social networks, and communicate with their friends and relatives via cell phones. Traditional media such as television, radio and newspapers, often become confined to a dark corner of their free time."

They also welcome more attractive ways for learning programing. It is possible to start with visual programming where user drag and drop blocks to generate executable code (using e.g. Scratch[1] or Blockly[2]).

Using multimedia objects can also make programing more attractive. Rudder et al. [11] describe an example of a web-based course for teaching programming using visualization and a gaming theme. In the course they use various multimedia components

---

[1] https://scratch.mit.edu/.
[2] https://developers.google.com/blockly/.

© Springer International Publishing AG, part of Springer Nature 2019
M. E. Auer and R. Langmann (Eds.): REV 2018, LNNS 47, pp. 277–282, 2019.
https://doi.org/10.1007/978-3-319-95678-7_31

such as animation, sound and video in order to immerse the student in an environment where learning is fun.

Next approach is to program things that work in the real world. A variety of robotics kits like e.g. LEGO Mindstorms can be used for this purpose.

Another experimenting can be connected with 3D LED cube use (see e.g. [2, 3, 12]). We try to contribute to this topic by bringing 3D LED cube as a tool for visualization of student's source code.

2   3D LED Cube Hardware

We considered 2 realizations of real LED cube laboratory model that can be seen in Fig. 1. Both of them consists of 512 LEDs in a $8 \times 8 \times 8$ array.

Fig. 1.   3D LED cube laboratory models

The first one was realized in frame of Diploma work [5] at our faculty (Fig. 1 on the left side). It is based on Arduino Uno motherboard that enables to interact with the control unit of the cube and in this way to influence the behavior of LEDs (their switching on and off).

The decision for Arduino was quite easy. Its advantage consists in its price and some other characteristics – it is multiplatform (Arduino software runs on Windows, Macintosh OS X a Linux), it enables simple programming, the software is open-source and therefore everybody can modify it and to contribute to its improvement.

For our solution we used just one color LEDs – they have blue backlight.

Later we found also the commercial product from the company Seeed Development Limited[3] that can be seen in Fig. 1 on the right side. The advantage is that the cube offers not only some predefined demonstrations but it also enables to create own codes. The

---

[3] https://www.seeedstudio.com/L3D-Cube-%288x8x8-Full-Color-Kit%29-p-2322.html.

cube is programmable over Wi-Fi network and users can use company's cloud solution for code uploading. However, it is also possible to connect the cube to Arduino motherboard that can be used for running own algorithms as well. In difference to our solution this cube can offer the whole RGB backlight. In this way the impression from experiment can be more imposing.

Our aim was to control the LED cube remotely because in this way it can serve for more students. The online experiment is available for registered users after booking their time slot via complex remote laboratory management system.

The equipment has to be connected to the web server where the supporting web application is placed. The web server is running on Raspberry PI 2 B using the standard port 80. It means that the created application can be accessed via web browser using IP address of the device.

The whole realization was developed by using just open technologies. It is part of the whole online laboratory ecosystem that was described in more details in [9, 10].

3   Web Application

Remote approach to the cube is realized via web application that is shown in Fig. 2. As it was already told the application is part of the whole portal that was developed not only for the LED cube but also other experiments. After setting the basic parameters (such as static IP address of the experiment), the user can start to use it.

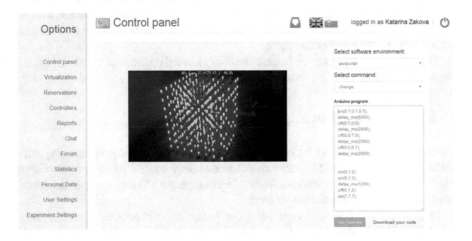

Fig. 2. Web application for remote 3D LED cube programing

As it is possible to see the application except of menu consists of two main parts. On the right side there is a simple text area where the user can enter own code that influences the behavior of the LED cube. The natural programming language for Arduino is C language. Therefore it can be immediately used also for programming the cube LEDs.

We have prepared two functions named on and off [13]. The function on(x,y,z,c) enables to switch on one LED whose position is determined by the first three input arguments of the function. The last input parameter c defines color of LED light for RGB cube. In the case of one color LED cube the last parameter is omitted. The function off(x,y,x) enables to switch off the specified LED. Since we consider $8 \times 8 \times 8$ LED cube, parameters x, y and z are integers from the interval <1, 8>.

The value of the first three input parameters can also be set to 0. Such setting influence several LEDs in one moment. For example, setting $x = 0$ substitutes 8 various commands where x varies from 1 to 8 ($x = 1, 2, 3, ..., 8$). In this way the commands on(0,0,0,c) or on(0,0,0) enable to switch on the whole cube at once (with or without specified color depending on the cube type).

In addition, we have prepared one more function – delay_ms(d). It defines how long the program should wait before executing the next command. The input parameter is represented by integer d that defines the waiting duration in miliseconds.

After submitting the inserted code by clicking on Run Command, the code is compiled and uploaded to the Arduino memory. Then, the code is executed and the user can see the result using video component in the central part of the screen.

Administrator can follow also additional output messages in HTML or JSON format informing about the process status.

The first video channel was provided by PI camera produced directly for Raspberry PI. However, the connecting cable to this camera is pretty fragile. Always when we used this experiment outside of laboratory for demonstration purposes the connection between camera and Raspberry was broken. Therefore in present time we are working on solution where this camera will be substituted by web camera that can be connected to the board by traditional USB cable.

The application also enables to save the code inserted by the user to the text file that can be downloaded to user's local computer.

4   Implementation in Education

The aim is to use both cubes for educational purposes to gain basic programming skills. Students or other interested users can practice mainly control loops and conditional commands that can be written preferably in C language. However, the basic control structures in JavaScript can also be used.

Programming various tasks students need to use different sequences of successive switching of LEDs (their turning on and turning off). In this way they can create interesting effects that can be cyclically repeated. It is natural that in such programming they need to use different conditional commands and loops.

There exist really big number of tasks that can be solved such as

- lighting every LED on the cube one-by-one to check they all work,
- turning on and off each cube layer whereby one will go from the cube left side to the right side and opposite, turning can be repeated several times - each time in other color,
- lighting LEDs step by step along all spatial cube diagonals,
- or e.g. creating effect of falling lights from the top luminous layer.

Description of various other effects can be found e.g. in [12].

The advantage of acquiring basic programming skills in this way consists in the fact that students can immediately see the visualization of their output and they do not work only with numerical values of a three dimensional array. Playing with colors can increase their impression.

5 Conclusions

The work on the experiment is not finished yet. In future we are planning to enable students to use the cube also for other programming languages – Blockly for use by kids and Python for older students.

We have also realized WebGL representation of the cube [13]. This animated model has the advantage that it is always available and it does not need the booking. However, as it is easy to understand, the real device offers more advanced experience.

Since the cube is available remotely, it can be used not only by students at the faculty but also by other interested users.

Acknowledgements. Author thanks to Patrik Novotný and Daniel Melo for their help with the cube realization, application implementation and fruitful discussions. Their help is gratefully acknowledged.

This work was supported by the Slovak Grant Agency, Grant KEGA No. 025STU-4/2017.

References

1. Arduino USB Board (Uno R3). https://solarbotics.com/product/50450/. Accessed 10 May 2016
2. Daigle, M., Hunter, R., Kreider, K.: Audio visual LED cube. Final Report, University of New Hampshire (2011). http://xenia.unh.edu/ece792_2011/projects/LEDCube/A-V%20LED%20Cube%20Final%20Report.pdf
3. Jackowski, D., Stepniewicz, J.: Playing with lights: music generation using the LED cube. In: International Computer Music Conference—"Non-Cochlear Sound" (ICMC 2012), Ljubljana, Slovenia (2012)
4. Nickson, Ch.: How a Young Generation Accepts Technology (2016). http://www.atechnologysociety.co.uk/how-young-generation-accepts-technology.html. Accessed 2 Dec 2017
5. Novotný, P.: 3D LED cube and its use in online experiments. Diploma thesis, FEI STU, Bratislava, Slovakia (2016). (in Slovak)
6. Piciarelli, F.: Young people and new technologies: what are the positive aspects? http://www.familyandmedia.eu/en/news/young-people-and-new-technologies-what-are-the-positive-aspects/. Accessed 2 Dec 2017
7. Rob, R., Panoiu, C., Panoiu, M.: Command algorithms for an $8 \times 8 \times 8$ LED cube using data acquisition board, AWER Procedia Inf. Technol. Comput. Sci. **3**, 565-570. Proceedings of 3rd World Conference on Information Technology (WCIT-2012), Barcelona, Spain. http://www.world-education-center.org/index.php/P-ITCS
8. Robins, A., Rountree, J., Rountree, N.: Learning and teaching programming: a review and discussion. Comput. Sci. Educ. **13**(2) (2003)

9. Rábek, M., Žáková, K.: Simple experiment integration into modular online laboratory enviroment. In: exp. at'17: 4th Experiment@ International Conference, Faro, Portugal, 6–8 June 2017, pp. 264–268. IEEE, Danvers (2017). ISBN: 978-989-99894-0-5

10. Rábek, M., Žáková, K.: Online laboratory manager for remote experiments in control. In: 20th World Congress on the International Federation of Automatic Control, Toulouse, France, 9–14 July 2017, vol. 50 (2017). ISSN: 2405-8963 (IFAC-PapersOnLine)

11. Rudder, A., Bernard, M., Mohammed, S.: Teaching programming using visualization. In: Proceedings of the Sixth Conference on IASTED International Conference Web-Based Education (WBED 2007), vol. 2, pp. 487–492. ACTA Press, Anaheim (2007)

12. Sauer, Ch.: $7 \times 7 \times 7$ LED Cube: A Truly 3D Display (2012). Online: http://roundtable.menloschool.org/issue17/6_Sauer_MS_Roundtable17_Winter_2014.pdf

13. Žáková, K.: The use of 3D LED cube for basic programming teaching. IFAC-PapersOnLine **49**(6), 203–206 (2016)

Management of Control Algorithms for Remote Experiments

Matej Rábek[✉] and Katarína Žáková

Faculty of Electrical Engineering and Information Technology, Slovak University of Technology,
Bratislava, Slovakia
{matej.rabek,katarina.zakova}@stuba.sk

Abstract. Remote experiments have been frequently used as an effective educational tool by various academic institutions over the last few years. This paper aims to define and illustrate some key features of a particular implementation of an online laboratory management system. More than anything, it focuses on the process of creating and testing a controller and explaining existing approaches provided by this specific online laboratory. Since the system offers a variety of real devices and simulation environments, during the development arose a need to build a generalized and user-friendly way of creating control algorithms. Teachers and system administrators are able to define robust structures, which students can later use to test their solution without any risk to the devices or the system itself.

Keywords: Remote experiment · Online laboratory
Computer aided education

1 Introduction

With the rise of remote experiments (see e.g. [2, 3]), several educational facilities came up with their own web applications to generalize the access to these experiments. This approach demands a creation of a robust interface, which connects users to the real devices via the internet. There are several possible solutions to this problem and each comes with its own advantages and drawbacks. The real challenge is to find the one that is the best suited to fulfill the exact needs of a specific implementation [4].

Although there already exist similar solutions (e.g. WebLab Deusto[1] or iLab Shared Architecture [2]) they lack a possibility to manage not only experiments but also control algorithms, that can be used to control plants. This inspired us to come up with our own solution.

The most focus was put on the system's modularity. This implies that any addition to the system must be easily realizable and no malfunctioning device could affect the overall functionality [10].

---

[1] http://weblab.deusto.es/website/.

© Springer International Publishing AG, part of Springer Nature 2019
M. E. Auer and R. Langmann (Eds.): REV 2018, LNNS 47, pp. 283–289, 2019.
https://doi.org/10.1007/978-3-319-95678-7_32

2   System Design and Architecture

2.1   Design Pattern

To simplify the process of adding or removing an experiment device, the system was designed around the star pattern with a web server representing the central node [1]. The central web server serves as an interface between users and the experiment server controlling each device (Fig. 1).

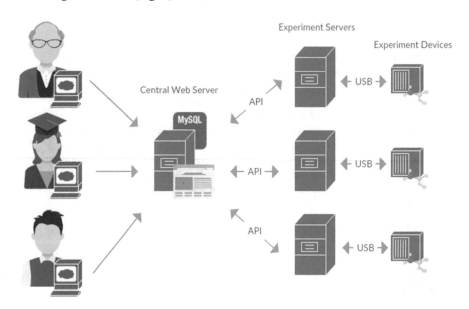

Fig. 1. Schematic representation of online laboratory interfaces and communication channels

Since the experiments are connected to the network via the internet, they are not physically bound to a certain location. This makes a possible collaboration of several scientific and educational institutions easier. The one disadvantage of this approach is that a static public IP address must be assigned to each experiment server.

It is also possible to connect two devices of the same type to the central web server, so the application can respond to an increase in the number of user requests.

2.2   Connected Devices

An "experiment" in the context of this online laboratory management system is a device connected to a designated experiment web server. The device is usually connected via the Universal Serial Bus (USB). The experiment server controlling the device must satisfy various software requirements. Luckily, the installation process was largely simplified thanks to a script which takes care of most software dependencies and configuration.

There are four different devices that were successfully integrated to the online laboratory system.

- Thermo-optical plant designed to aid with the education of basic control algorithms. It is mainly used to regulate the intensity of emitted light and heat by a light-emitting diode, light bulb and a fan [13].
- Segway or an inverted pendulum on a two-wheel undercart driven by two DC motors. This device has faster and more unstable dynamics, which makes the control more challenging. The plant can be used in courses dedicated to control of nonlinear systems.
- Towercopter is propeller attached to an electric motor on a platform restricted to a one-dimensional movement by two vertical rods. The plant can also be used in nonlinear control courses. The aim is to control vertical position of the propeller [9, 12].
- 3D LED Cube is the only connected device not designed specifically for the purposes of control studies. It was developed to illustrate and visualize simple algorithmic principles [15].

2.3   Experiment Control Requirements

In terms of experiment access and control, the proposed design supports a different set of environments to be used for each device. It is important to distinguish early on, what functionality should apply to every component and what must be specially tuned to meet the requirements unique to each device [14]. Every device is different and was designed with a distinct control environment in mind. A good online laboratory application must provide this degree of freedom to use a variety of environments or other software solutions without the loss of modularity and unified access. The developer should not be restricted but at the same time the user must be greeted by a friendly and easy to understand interface [11].

3   Implementation

There is an Apache server running PHP on both the central web server and every experiment server. In addition to that a separate MySQL database is located on each server. The process of running an experiment is somewhat complex (Fig. 2). The issued command coming from the user must be sent to the right experiment server and the measured data presented in real time to the user. Moreover, the data must also be stored on the central web server in order to be visible on later date, when the user might want to reexamine the results.

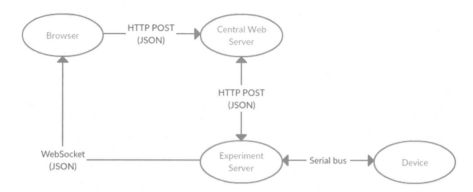

Fig. 2. Data propagation within the system – experiment lifecycle.

3.1 Simulation Environments

The runtime process of the earlier described remote experiments is supported by packages such as Matlab, OpenModelica, Scilab, etc. These software environments provide a stable and accessible way to control the devices as opposed to the alternative, which is a simple client-server approach without any intermediate level.

The system offers several simulation environments, so the students are able to choose the most appropriate tool for the assigned task. Another reason for implementing more than one simulation environment is the difference in skill and knowledge among students, so they can work with the software that they are the most familiar with.

Our attention was focused on environments that enable modelling and simulation of dynamical systems, regardless of whether the solution is commercial or not.

3.2 User Interface

The user has two choices when it comes to running an experiment. A reservation can be set up at a prior date and time or an experiment can be issued by a so-called batch mode. The main difference between these two approaches is that by using the batch mode, the user can not choose a specific device if there is more than one of the same type connected to the central web server. There is also no real-time feedback since this mode functions as a queue and runs the experiment as soon as a suitable device becomes available. The user is then notified and can examine the experiment result in the report tab, where all the previous experiment measurements are stored (Fig. 3).

Experiments can run in either open or closed loop. The open loop mode provides a way to determine the system's dynamics, which is a crucial to successfully identify the correct mathematical model.

However, since open-loop mode can only be used to collect output data based on predefined input parameters, it does not provide the enhanced level of feedback control. To satisfy this requirement, the control in closed loop is also considered. The system enables the user to interact with the experiment and influence the control process. It can

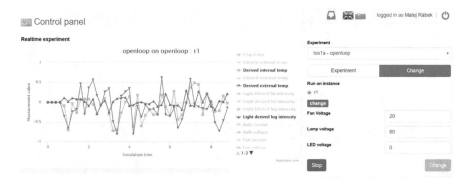

Fig. 3. Control panel showing a visual representation of data measured during the runtime of an experiment in open loop.

be done by changing the control or simulation parameters or even by modifying the whole control algorithm.

The simulation schema can be separated into two distinct instances (Fig. 4) whereby each one can have different privileges regarding its later modification. The control algorithm can be developed by all registered users and it represents just a single block in the overall schema, which can only be created or edited by someone with administrative privileges. This ensures the stability and security of the system. It protects the devices from attacks or other harmful code.

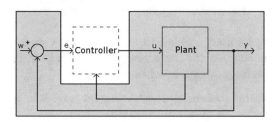

Fig. 4. Schematic block scheme of the experiment for simulation.

In the present time the system supports the definition of controller algorithm in ascii format that is defined via form text area.

In Fig. 5 one can see the administrator's controller management interface. It is divided into 4 parts. The first two sections are also available for regular users.

The users can create their own controllers in context of already defined schemas and test them on any available devices. They can either keep their created controllers private (meaning that they are the only one with access to it) or share them with others. However, before a controller is classified as public, it has to be approved by an administrator.

Additionally, there is a set of already developed public controllers to choose from. Public controllers can also be further modified to suit each user's individual needs and requirements.

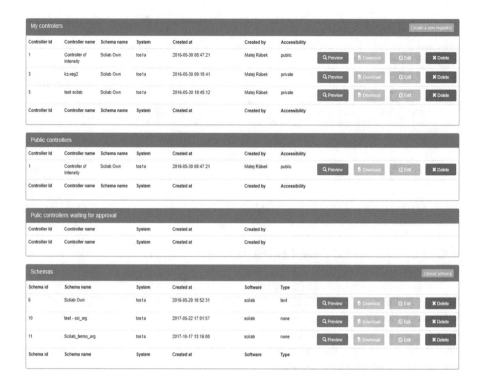

Fig. 5. User interface for controller management (admin view).

Compared to regular users, administrators can access a couple more functions. On top of the ability to approve or discard new public controllers, they can create new schemas. This is necessary every time when

- a new simulation environment or programming language is being introduced to the system,
- the user needs a modified set of controller inputs (e.g. Fig 4 shows a schema with an error input, although sometimes it might be advantageous that this input was a plant output and desired value),
- a new firmly predefined controller structure with variable parameters is offered.

4 Conclusion

Both users and experiment developers need a system they can navigate with ease to maximize their learning or working potential. At the same time, this system has to be secure and robust. No user input can threaten the overall stability and functionality. Even the devices themselves must be protected so a user does not succeed in running a code that could potentially harm them either by mistake or with malintent. A well-designed interface for control algorithms management can reduce this danger.

Acknowledgment. The work presented in this paper has been supported by the Slovak Grant Agency, Grant KEGA No. 025STU-4/2017.

References

1. Barranco, M., Rodriguez-Navas, G., Proenza, J., Almeida, L.: CANcentrate: an active star topology for CAN networks, pp. 219–228 (2004). https://doi.org/10.1109/wfcs. 2004.1377714
2. Harward, V.J., Del Alamo, J.A., Lerman, S.R., Bailey, P.H., Carpenter, J., DeLong, K., et al.: The ilab shared architecture: a web services infrastructure to build communities of internet accessible laboratories. Proc. IEEE **96**(6), 931–950 (2008)
3. Heradio, R., de la Torre, L., Galan, D., Cabrerizo, F.J., Herrera-Viedma, E., Dormido, S.: Virtual and remote labs in education: a bibliometric analysis. Comput. Educ. **98**, 14–38 (2016)
4. Lojka, T., Miskuf, M., Zolotova, I.: Service oriented architecture for remote machine control in ICS. In: 2014 IEEE 12th International Symposium on Applied Machine Intelligence and Informatics (SAMI), pp. 327–330. IEEE (2014)
5. Sivakumar, S., Robertson, W., Artimy, M., Aslam, N.: A web-based remote interactive laboratory for internetworking education. IEEE Trans. Educ. **48**, 586–598 (2005). https://doi.org/10.1109/TE.2005.858393
6. Rábek, M., Žáková, K.: Online laboratory manager for remote experiments in control. IFAC-PapersOnLine **50**(1), 13492–13497 (2017)
7. Rábek, M., Žáková, K.: Simple experiment integration into modular online laboratory environment. In: 2017 4th Experiment@ International Conference (exp. at'17), pp. 264–268. IEEE (2017)
8. Restivo, M.T., Mendes, J., Lopes, A.M., Silva, C.M., Chouzal, F.: A remote laboratory in engineering measurement. IEEE Trans. Industr. Electron. **56**(12), 4836–4843 (2009)
9. Sánchez-Benítez, D., Portas, E.B., Jesús, M., Pajares, G.: Vertical rotor for the implementation of control laws. IFAC Proc. Vol. **45**(11), 224–229 (2012)
10. Scapolla, A.M., Bagnasco, A., Ponta, D., Parodi, G.: A modular and extensible remote electronic laboratory. Int. J. Online Eng. (iJOE) **1**(1) (2005)
11. Schneiderman, B., Plaisant, C.: Designing the user interface (1998)
12. Ťapák, P.: One degree of freedom copter. In: 2016 International Conference on Emerging eLearning Technologies and Applications (ICETA), pp. 351–354. IEEE (2016)
13. Ťapák, P., Huba, M.: Laboratory model of thermal plant identification and control. IFAC-PapersOnLine **49**(6), 28–33 (2016)
14. Yu, E.S., Mylopoulos, J.: Understanding "why" in software process modelling, analysis, and design. In: Proceedings of 16th International Conference on Software Engineering, 1994, ICSE-16, pp. 159–168. IEEE (1994)
15. Žáková, K.: The use of 3D LED cube for basic programming teaching. IFAC-PapersOnLine **49**(6), 203–206 (2016)

Demonstration: Web Tool for Designing and Testing of Digital Circuits Within a Remote Laboratory

Javier Garcia-Zubia[1(✉)], Eneko Cruz[1], Luis Rodriguez-Gil[2], Ignacio Angulo[1], Pablo Orduña[2], and Unai Hernandez[1]

[1] Faculty of Engineering, University of Deusto, Bilbao, Spain
{zubia,eneko.cruz,Ignacio.angulo,unai.hernandez}@deusto.es
[2] DeustoTech - Deusto Institute of Technology, Bilbao, Spain
{luis.rodriguezgil,pablo.orduna}@deusto.es

Abstract. The tool named WebLab-Boole-Deusto allows users to design and implement a bit-level combinational digital circuit. This tool helps students during the design process step by step: truth table, K-maps, Boolean minimization, Boolean expressions, and digital circuit. Also, students can access a remote lab for implementing and testing the system designed by themselves. The remote lab is FPGA-based and it is included in the WebLab-Deusto RMLS (Remote Lab Management System). WebLab-Boole-Deusto is a web tool. Users access a web page instead of installing software on their computer (desktop application). This feature promotes its dissemination to universities and training centers.

Keywords: Digital electronics · Remote laboratory

1 Introduction

Digital electronics are part of the curricula of different engineering degrees and other technical studies. Digital electronics circuits are divided into sequential and combinational circuits, and each of those types of circuits can be divided into bit-level and word-level; bit-level ones are related to logic gates and flip-flops, and word-level ones are related to MSI Integrated Circuits. Figure 1 shows two different digital circuits: bit-level (left) and word-level (right).

The presented WebLab-Boole-Deusto tool is oriented towards bit-level combinational digital circuits. The design process at this level consists on applying a clear step-by-step methodology. Figure 2 shows the design process [1]: create the truth table, write the canonical forms, draw the K maps, obtain the minimized Boolean expressions and draw the digital circuit with AND-OR-NOT logic gates.

© Springer International Publishing AG, part of Springer Nature 2019
M. E. Auer and R. Langmann (Eds.): REV 2018, LNNS 47, pp. 290–297, 2019.
https://doi.org/10.1007/978-3-319-95678-7_33

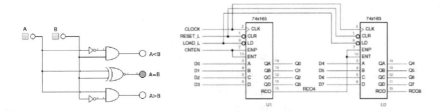

Fig. 1. Digital circuits

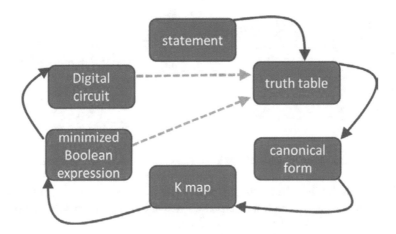

Fig. 2. Design process for bit-level combinational digital circuit

In Fig. 2 the non-dotted lines describe the design process, and the two dotted lines describe the analysis process. Every step of the design process implies that students have to apply a clear method: to obtain the truth table students need to understand the statement to express it using only a 1 (true) and 0 (false); the K map is a graphical representation of the truth table that allows minimization; to minimize a K map the student needs to apply a non-trivial algorithm; and finally, the digital circuit describes the minimized Boolean expressions with graphical symbols.

The WebLab-Boole-Deusto helps students learn how to apply the different steps of the design process because it allows the students to see all the process step-by-step. Additionally, WebLab-Boole-Deusto is able to implement the digital circuit in a FPGA-based remote laboratory, using this feature students can test the system in a real way without needing a lab in their hands and without needing to connect all the ICs.

Before describing the WebLab-Boole-Deusto tool is important to remark that WebLab-Boole-Deusto is oriented towards education, it is not industry-oriented. In industry, engineers do not use the same approach than in the classroom; they use VHDL and a FPGA or C and a microcontroller. But in education students use another methodological approach to ensure that they understand the gate level.

2   Description of WBD Tool

The WebLab-Boole-Deusto tool is a web application available for all the devices and web browsers, and it can be accessed from https://goo.gl/cjxA52.

The first step of the design process is to write the statement of the exercise and to declare the number and names of the input and output variables. The exercise in Fig. 3 is a BCD-seven-segment display decoder, and it has four inputs (e3-e2-e1-e0) and eight outputs (a-b-c-d-e-f-g segments and enable).

Fig. 3. Description of the exercise and the input and output variables

Figure 4 shows the corresponding truth table of the BCD-seven-segment display decoder. Inputs are active-high and outputs are active-low, and the enable is active-high. The truth table is filled using only mouse clicks.

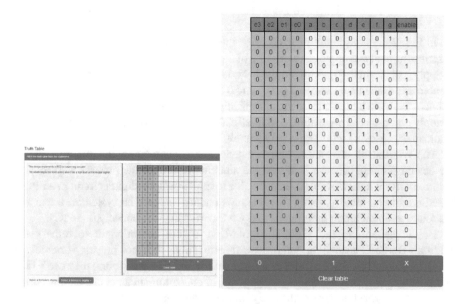

Fig. 4. Truth table of the BCD-seven-segment display decoder

In this step students can see the canonical forms under different formats: using minterms (Fig. 5) or maxterms, and using the compact form or a Boolean expression.

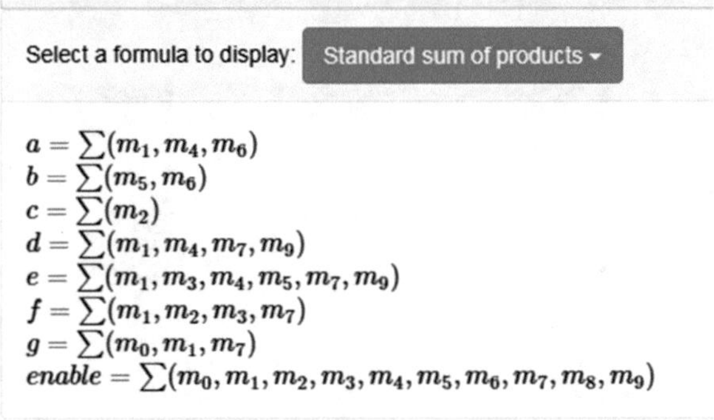

Select a formula to display:
Standard sum of products
$a = \sum(m_1, m_4, m_6)$
$b = \sum(m_5, m_6)$
$c = \sum(m_2)$
$d = \sum(m_1, m_4, m_7, m_9)$
$e = \sum(m_1, m_3, m_4, m_5, m_7, m_9)$
$f = \sum(m_1, m_2, m_3, m_7)$
$g = \sum(m_0, m_1, m_7)$
$enable = \sum(m_0, m_1, m_2, m_3, m_4, m_5, m_6, m_7, m_8, m_9)$

Fig. 5. Compact canonical form with minterms

Each column of the truth table can be converted into a K-map that can be used for minimizing the canonical form. Figure 6 shows the K-map of the column of the "f" output variable. The minimized Boolean expression is written below the K-map, and in this expression each color corresponds with a colored group in the K-map.

f
e3
e1
e2
e0
$f = (e1 \cdot \overline{e2}) + (e0 \cdot \overline{e2} \cdot \overline{e3}) + (e0 \cdot e1)$
Ocultar

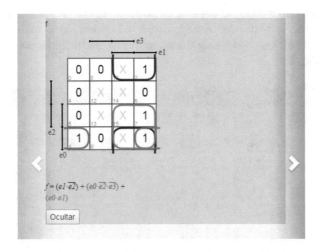

Fig. 6. Minimized Boolean expression and its K-map

Figure 7 shows the digital circuit with AND-OR-NOT gates that implement the eight outputs of the BCD-seven-segment display decoder.

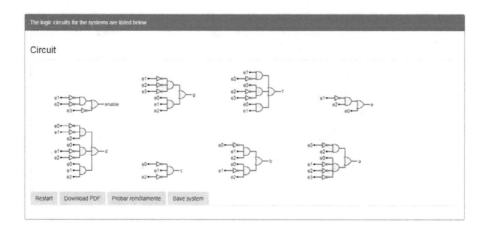

Fig. 7. Digital circuit

At this moment students should go to the lab to implement the digital circuit using IC like 7404 (NOT), 7408 (AND) and 7432 (OR). Before, after or instead of going to the classical lab, students can access the WebLab-Deusto remote lab to test the designed circuit. The remote lab has 10 switches, 4 buttons, six LEDs and 4 multiplexed seven-segment displays. The first step of the implementation of the designed digital circuit in the WebLab-Deusto remote lab consists of assigning to each input and output variable one of the available real devices in the remote lab. The four inputs of the BCD code are assigned to four switches, and the eight outputs are assigned to the second seven-segment display (see Fig. 8).

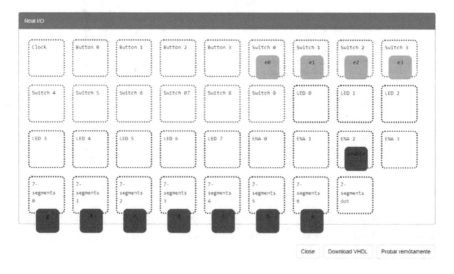

Fig. 8. Assignation between variables and devices

After this assignation students have to download a VHDL file. This file contains the description of the digital circuit using VHDL. It is not relevant at this moment if the students know what VHDL is: they simply have to download this file. The content of this file is shown in Fig. 9 and students can understand it easily.

```
        led0 : inout std_logic;
        led1 : inout std_logic;
        led2 : inout std_logic;
        led3 : inout std_logic;
        led4 : inout std_logic;
        led5 : inout std_logic;
        led6 : inout std_logic;
        led7 : inout std_logic;
        ena0 : inout std_logic;
        ena1 : inout std_logic;
        ena3 : inout std_logic;
        dot : inout std_logic
        );
end base;


] architecture behavioral of base is
]     begin
        ena2<=( not ( swi1 ) and not ( swi2 ) ) or ( not ( swi3 ) );
        seg0<=( not ( swi1 ) and not ( swi2 ) and not ( swi3 ) ) or ( swi0 and swi1 and swi2 );
        seg1<=( swi1 and not ( swi2 ) ) or ( swi0 and not ( swi2 ) and not ( swi3 ) ) or ( swi0 a
        seg2<=( not ( swi1 ) and swi2 ) or ( swi0 );
        seg3<=( not ( swi0 ) and not ( swi1 ) and swi2 ) or ( swi0 and not ( swi1 ) and not ( sw:
        seg4<=( not ( swi0 ) and swi1 and not ( swi2 ) );
        seg5<=( not ( swi0 ) and swi1 and swi2 ) or ( swi0 and not ( swi1 ) and swi2 );
        seg6<=( not ( swi0 ) and swi2 ) or ( swi0 and not ( swi1 ) and not ( swi2 ) and not ( sw:
    end behavioral;
```

Fig. 9. VHDL code of the digital circuit

In Fig. 10 students have to introduce their user and password combination (demo/demo) and then upload the previously downloaded VHDL file.

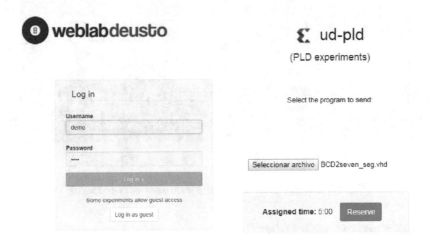

Fig. 10. Uploading the VHDL file

At this moment the remote lab is processing the VHDL file (it takes time) (Fig. 11).

Fig. 11. Processing the VHDL file

Finally, students can activate the four switches to see the number visualized in the seven-segment display. Inputs 0000 and 0101 are shown in Fig. 12.

Fig. 12. Seven-segment display for 0000 and 0101 BCD inputs

3   Comparison with Other Tools

There are hundreds of Boolean minimizer tools, but they usually have different problems:

- Many of them are not web-based applications, they are desktop applications.
- Many of them are web-based applications, but they are implemented using Java or Flash.
- Many of them do not use graphical support to draw and show the minimized Boolean expression.
- Many of them do not allow the user to introduce his own Boolean solution to test if it is minimal or not.
- The use of a remote lab to test the Boolean expression is not included in the tools.

The WebLab-Boole-Deusto tool is web-based, uses a graphical approach and includes a remote lab. As the WebLab-Boole-Deusto is included in the WebLab-Deusto RMLS, it integrates learning analytics tools. As far as we know, this is the only tool with these characteristics.

Boole-Deusto [1] is similar to WebLab-Boole-Deusto but the former is a desktop application. Boole-Deusto includes a remote laboratory as WebLab-Boole-Deusto and but it also has other features that WebLab-Boole-Deusto does not include yet.

4   Conclusion and Future Work

WebLab-Boole-Deusto is a web-based application that is oriented towards teaching digital electronics design to students of engineering, it is oriented only towards combinational bit-level digital circuits. Using this tool the students can replicate in a web page the same methodology that they are using in the classroom. This approach helps students to improve the learning process. WebLab-Boole-Deusto offers students a remote laboratory in which they can test the functionality of the designed system.

Future work will be centered in different topics: test the usability of WebLab-Boole-Deusto in the classroom using the UMUX test, allow the user to include his own minimal function to see if it is correct or not, and include finite state machines.

Reference

1. Garcia-Zubia, J.: Educational software for digital electronics: BOOLE-DEUSTO. In: Proceedings of IEEE International Conference on Microelectronic Systems Education (IEEE MSE), pp. 20–22 (2003)

Poster: An Experience API Framework to Describe Learning Interactions from On-Line Laboratories

Pedro Paredes Barragán[1(✉)], Miguel Rodriguez-Artacho[2(✉)],
Elio San Cristobal[1(✉)], Manuel Castro[1(✉)],
and Hamadou Saliah-Hassane[3(✉)]

[1] IEEC Department, UNED, Madrid, Spain
{pparedes,elio,mcastro}@ieec.uned.es
[2] LSI Department, UNED, Madrid, Spain
miguel@lsi.uned.es
[3] Science and Technology Department, TELUQ University, Montreal, Canada
hamadou.saliah-hassane@teluq.ca

Abstract. Learning Analytics is an emerging field focused on analyzing learners' interactions with educational content. One of the key open issues in learning analytics is the standardization of the data collected. This is a particularly challenging issue in online laboratories. This paper presents an implementation with one of the most promising specifications: Experience API (xAPI). The Experience API relies on Communities of Practice developing profiles that cover different use cases in specific domains. This paper presents the Online Laboratories xAPI Profile: a profile developed to align with the most common use cases in the online laboratories domain. The profile is applied to a case study (a windmill lab), which explores the technical practicalities of standardizing data acquisition. In summary, the paper presents a framework to track online laboratories and their implementation with the xAPI specification.

Keywords: xAPI · Online laboratories · Learning analytics

1 Motivation

ICT-based (Information and communication technologies) education has triggered new trends in learning, especially in the use of external tools from educational environments.

In this context, online laboratories are technological tools composed of software and/or hardware that allows students to remotely conduct their practice as if in a physical Lab. Online Laboratories are considered a useful approach to increase the skills and the interest of students in STEM (science, technology, engineering, and mathematics) [1–3].

One of the most important aspects is the collection and subsequent data analysis on the interaction of students with the on-line laboratories, generating educational meaning from learning analytics and Educational Data Mining from the interaction

M. E. Auer and R. Langmann (Eds.): REV 2018, LNNS 47, pp. 298–303, 2019.
https://doi.org/10.1007/978-3-319-95678-7_34

between the student and the laboratory in the context of a Virtual Learning Environment interaction.

The objective of this work is to define a semantic framework and establish what type of information is necessary to track students in an online laboratory and how we can store that information in a standardized and accessible way for other systems.

2 Previous Works

Some previous efforts in standardization are focused in capturing results inside the educational resource, usually in the form of a log [4–7]. Activity Streams is a specification that allows representing the sequence of actions as sentences in specific contexts. Advanced Distributed Learning Initiative (ADL) Experience API (xAPI) is greatly inspired in Activity Streams.

Experience API works by allowing statements of experience (usually learning experiences, but could be any experience) to be delivered and stored safely in an LRS (Learning Record Store). In this format, the actor is the agent of the statement; it can be a student, a teacher, a group or the Laboratory. The verb describes the action of the state, such as increased, decreased, or rotated. And the object is what the actor interacts with, like a fan, a battery, or a clock. This syntax that is used to record the learning of an actor is a JSON (JavaScript Object Notation) with a structure defined in the specification.

xAPI does not set any constraints on the vocabulary that can be used in the statements. Practitioners can create their own verbs and activity types to define domain specific vocabularies, such as serious games [8] or mobile open laboratories [9]. Additionally, xAPI allows extensions to expand the specification and fulfill new or unique requirements. The definition of new vocabularies along with these extensions is denominated a "xAPI profile". These profiles are usually developed through a xAPI Community of Practice.

3 Our Proposal

In the context of online laboratories, the user can control real or virtual actuators and see the effects of his actions. When tracking the learning process the focus could be in events or system status. Experience API allows the tracking of these events and the context (system status) in which the event occurs.

The objective is to capture the interactions between two actors (student, instructor…) or an actor and the laboratory (in both directions) and the system status in those moments.

Some examples of xAPI statements with the laboratory as an actor could be informing the student or the instructor whether the lab is ready or some component does not work properly.

3.1 Verbs

There are more general verbs applied to the experiment such as launched, initialized or abandoned (Table 1).

Table 1. General verbs

Action	Verb	Definition
Launched	http://adlnet.gov/expapi/verbs/launched	Starts the process of launching the next piece of learning content
Initialized	http://adlnet.gov/expapi/verbs/initialized	Indicates the activity provider has determined that the actor successfully started an activity
Abandoned	https://w3id.org/xapi/adl/verbs/abandoned	Indicates the activity provider has determined that the session was abnormally terminated either by an actor or due to a system failure

These verbs are defined in the ADL vocabulary (https://registry.tincanapi.com/) while other specific verbs can be defined to expand the vocabulary with the consensus of a Community or Practice. These verbs are used in a windmill lab: increased, decreased, rotated, set and queried (Table 2).

Table 2. Specific verbs

Action	Verb	Definition
Increased	http://example.com/ieee1876/onlineLabs/verbs/increased	Indicates the actor has raised the value of an actuator
Decreased	http://example.com/ieee1876/onlineLabs/verbs/decreased	Indicates the actor has diminished the value of an actuator
Rotated	http://example.com/ieee1876/onlineLabs/verbs/rotated	Indicates the actor has turned the object on an axis certain degrees
Set	http://example.com/ieee1876/onlineLabs/verbs/set	Indicates the actor has established the value of an actuator
Queried	http://example.com/ieee1876/onlineLabs/verbs/queried	Indicates the actor asked information about the value of a sensor

3.2 Objects

As with verbs, we can find a list of types of activities at https://registry.tincanapi.com/. However, in the context of online labs we need to include specific types of activities such as experiment, sensor or actuator (Table 3).

Table 3. Activity types

Activity type	Definition
http://example.com/ieee1876/ onlineLabs/activities/experiment	The online laboratory is a real or virtual apparatus and its instrumentation that can be accessed, monitored and controlled at distance
http://example.com/ieee1876/ onlineLabs/activities/sensor	The sensor is the building blocks of an online laboratory, and represents the components, which can be monitored
http://example.com/ieee1876/ onlineLabs/activities/actuator	The actuator is the building blocks of an online laboratory, and represents the components, which can be controlled

3.3 Context

An optional property that provides a place to add contextual information to a Statement. All "context" properties are optional. The "context" property provides a place to add some contextual information to a Statement. It can store information such as the instructor for an experience, if this experience happened as part of a team-based Activity, or how an experience fits into some broader activity. In our domain, we can

Table 4. Context extensions for Online Labs

Name	IRI	Definition
Lab information	http://example.com/ ieee1876/onlineLabs/ extensions/lab-info	Value is an object containing key/value pairs describing the status of sensors and actuators
Observer	http://id.tincanapi.com/ extension/observer	Context extension containing an Agent or Group object representing an agent or group who observed the experience
Mode	http://example.com/ ieee1876/onlineLabs/ extensions/mode	Indicates the mode of the system (charging or discharging)

extend context with the status of sensors and actuators while tracking any event, track the people who is observing the experiment, or registry whether the experiment is in mode "charging" or "discharging" (in the case of a windmill lab).

Observer is already defined in https://registry.tincanapi.com/ (Table 4).

4   Level of Granularity

A laboratory should provide tracking of the interactions at different levels of granularity. The granularity depends on the object of the statement (target of the interaction):

- Level 1: the object of the statement is the experiment and these verbs are defined in the experiment operational metadata.
  Example: "Pedro initialized Windmill Lab"
- Level 2: track Level 1 actions plus specific set of actions over sensors/actuators defined in the sensor/actuator operational metadata.
  Example: "Pedro queried battery"
- Level 3: track Level 2 plus all interactions in both directions (from user to experiment and from experiment to the user)
  Example: "Windmill Lab sent ok to Pedro".

5   Conclusions and Future Work

Our approach focuses on the creation of taxonomy of verbs and granularity levels to describe student's activities in the laboratory.

Our objective is to define a meaningful set of elements that describe an abstract model adapted to the online laboratory context. At the same time, this model has to be simple enough so that it fits the needs of a broad community of developers in order to be compliant and interoperable. A free version of SCORM cloud will serve as an initial repository for the prototypes developed.

From the integration of the xAPI standard by means of a case study it was possible to identify the possible registers that are generated in the LRS, in this way the teacher has the capacity to perform an analysis of the reports in order to track the traces of student learning. Additionally, the advantages of this implementation in learning environments are multiple, including:

- Operation inside and outside a Learning Management System (LMS).
- Registration of activities such as games, videos, blogs, social networks, laboratories.
- Use of mobile devices on and off line.

The flexibility of xAPI is its main strength. It can be adapted to any context and you can create verbs and new activities depending on what you need. However, the consensus of the community of practice is needed for us to use those verbs and activities with the same meaning, xAPI thus relies on the educational community to publicize and deliver standards for these recipes. This way you can share the data of different LRS and analyze the results obtained in several systems.

The information obtained from the interaction of the students with the laboratories fulfils a double task:

- teachers: to detect critical points of the laboratory design and general trends of the groups of students
- students: to follow up on their teaching-learning process and its results

We consider that online Laboratories xAPI Profile can establish some basic principles and open new research paths for analyzing interactions in Online Laboratories including M2M (machine to machine) or M2H (machine to human) activities.

Acknowledgment. This research was partly funded by the Autonomous Community of Madrid, e-Madrid project, number S2009/TIC-1650.

References

1. Bright, C., Lindsay, E., Lowe, D., Murray, S., Liu, D.: Factors that impact learning outcomes in both simulation and remote laboratories. In: World Conference on Educational Multimedia, Hypermedia and Telecommunications, vol. 2008, no. 1, pp. 6251–6258 (2008)
2. Nickerson, J.V., Corter, J.E., Esche, S.K., Chassapis, C.: A model for evaluating the effectiveness of remote engineering laboratories and simulations in education. Comput. Educ. **49**(3), 708–725 (2007)
3. Jona, K., Roque, R., Skolnik, J., Uttal, D., Rapp, D.: Are remote labs worth the cost? Insights from a study of student perceptions of remote labs. Int. J. Online Eng. (iJOE) **7**(2), 48 (2011)
4. Orduña, P., Irurzun, J., Rodriguez-Gil, L., Zubía, J.G., Gazzola, F., López-de Ipiña, D.: Adding new features to new and existing remote experiments through their integration in weblab-deusto. iJOE **7**(S2), 33–39 (2011)
5. Ferguson, R.: The state of learning analytics in 2012: a review and future challenges. Knowledge Media Institute, Technical Report KMI-2012-01 (2012)
6. Siemens, G., Long, P.: Penetrating the fog: analytics in learning and education. Educ. Rev. **46**(5), 30–32 (2011)
7. Sie, R.L., Ullmann, T.D., Rajagopal, K., Cela, K., Bitter-Rijpkema, M., Sloep, P.B.: Social network analysis for technology–enhanced learning: review and future directions. Int. J. Technol. Enhanc. Learn. **4**(3), 172–190 (2012)
8. Serrano-Laguna, Á., Martínez-Ortiz, I., Haag, J., Regan, D., Johnson, A., Fernández-Manjón, B.: Applying standards to systematize learning analytics in serious games. Comput. Stand. Interfaces **50**, 116–123 (2017)
9. Saliah-Hassane, H., Reuzeau, A.: Mobile open online laboratories: a way towards connectionist massive online laboratories with x-API (c-MOOLs). In: Frontiers in Education Conference (FIE), pp. 1–7 (2014)

Poster: Remote Engineering Education Set-Up of Hydraulic Pump and System

Milos Srecko Nedeljkovic[1], Novica Jankovic[1], Djordje Cantrak[1(✉)],
Dejan Ilic[1], and Milan Matijevic[2]

[1] Faculty of Mechanical Engineering, Hydraulic Machinery and Energy Systems Department,
University of Belgrade, Belgrade, Serbia
`{mnedeljkovic,njankovic,djcantrak,dilic}@mas.bg.ac.rs`
[2] Faculty of Engineering, Department for Applied Mechanics and Automatic Control,
University of Kragujevac, Kragujevac, Serbia
`matijevic@kg.ac.rs`

Abstract. Educational setup for demonstration of a pump operation and determination of full performance curves is presented. The upgrade of the setup has been done in order to implement electronic measurements, data acquisition, computer control, and for preparation of all necessary hardware elements for remote operation via Internet. Used equipment, rationale behind, characteristics and possibilities of the current status of the setup is discussed in detail.

Keywords: Engineering education · Laboratory setup
Pump system performance · Cavitation · Data acquisition

1 Introduction

Educational setup (Fig. 1) for demonstration of several pump operation modes and determination of its full performance curves, with additional possibility of flow meter calibration, has been in used for students' lab measurements since 2000 at the University of Belgrade - Faculty of Mechanical Engineering - Hydraulic Machinery and Energy Systems Department. It has proved itself very indicative and useful for students' practical work in the laboratory, hands-on getting to know of cavitational phenomenon, calibration, etc.

Components shown in Fig. 1 are: 1 - pump with the transparent casing, 2 - elbow, 3 - transparent pipe, 4 - transparent Venturi flow meter, 5 and 6 - valve for emptying the upper tank, 7 - scale on the calibration reservoir, V.10 - suction valve, 11 - suction pipe, 12 - lower reservoir, 13 - T-joint, V.20 - valve, 21 - pipe for filling the upper reservoir, 22 - nozzle, 23 - upper reservoir, V.30 - valve, 31 - pipe to lower reservoir.

The following laboratory exercises may be demonstrated and appropriate performance curves measured:

- Venturi flow meter calibration following volumetric method (valve V.30 is closed).
- Pump testing procedure for determination of the Q-H curves, as well as pump unit power and efficiency determination after standard ISO 9906.
- Pump cavitation test.

© Springer International Publishing AG, part of Springer Nature 2019
M. E. Auer and R. Langmann (Eds.): REV 2018, LNNS 47, pp. 304–311, 2019.
https://doi.org/10.1007/978-3-319-95678-7_35

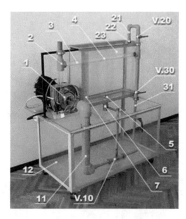

Fig. 1. Original educational setup [1].

- Pump transports water on the specific geodesic height H_{geo} = const (Fig. 2). In this case valve V.30 is closed.
- Pump circulates water in the main, lower reservoir H_{geo} = 0 (Fig. 3).
- Pump regulation with by-pass. In this case pipes on the pump pressure side are in parallel and regulation is performed by valve V.30.
- Energy efficiency issues: Comparison of pump regulation with throttling valve, pump speed and by-pass.

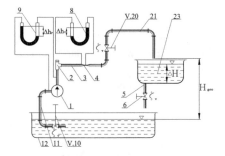 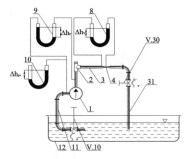

Fig. 2. Pumping water on the geodesic height H_{geo} = const.

Fig. 3. Water circulation (H_{geo} = 0).

All pressure transmitters are presented in figures with U-tubes for better physical interpretation and understanding of the pressure values and differences [2, 3].

Students first learn how to start and stop the centrifugal pump by following the next procedure:

- Check if the pump is filled with water.
- If not, open the cap on the T-joint 2 (Fig. 1).
- Close valves V.10, V.20 and V.30.
- Fill up with approximately 3.5 l of water.

- Close the cap on the T-joint 2 (Fig. 1).
- Start the pump.
- Open the valve on the pump suction side V.10.
- Open the valve(s) V.20 and/or V.30.

  The closing procedure is:

- Close the valve(s) V.20 and/or V.30. Remark: All valves on the pump pressure side should be closed.
- Close the valve V.10.
- Turn off the pump.

Pump moves water from the lower reservoir (12) through suction pipe (11) with valve V.10, pipe (3) and Venturi flow meter (4) to the T-joint (2) to both reservoirs, if valves V.20 and V.30 are open (Fig. 1).

The upgrade of this setup has been done towards more complex hydraulic system, data acquisition and computer control of the processes, as well as to the internet connection and availability for remote operation.

The intention is to create a complex experiment which could be conducted via internet, so students can measure and calculate necessary values, generate electronic report on distance and discuss obtained results with the lecturer.

2   Upgraded Test Rig

2.1   What Is Going to Be Demonstrated?

The upgraded demonstrational-educational pump setup (Fig. 4) can demonstrate operation of a pump for raising up the fluid to a geodesic height and the work of a pump for circulating the fluid, to show the influence of speed of rotation on pump work and its energy efficiency, to visualize cavitation bubbles in the pump and even on its pressure side, and to demonstrate the calibration of a non-standard flow meter of Venturi type [5, 6].

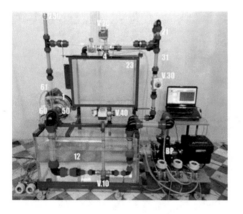

Fig. 4. Upgraded setup – front view.

Fig. 5. Photo of cavitation bubbles at pump inlet.

Students are educated how to measure pump performance characteristics for various values of pump rotational speed, to calculate and present pump performance characteristics in dimensional (speed dependent) and non-dimensional (speed independent) parameters after ISO standard [4], how to measure and calculate pump cavitation characteristics, etc.

2.2 Test Rig

The dimensions of the mobile facility are $2 \times 0.6 \times 1.5$ m. The powers of the built-in pumps are 5.5 kW and 1.3 kW with the possibility of flow rate up to 5 l/s.

Main parts, important for test rig operation, are designated in Fig. 4: 1 - main/test pump, V.10 - valve in the test pump suction pipe (i.e. between two pumps), 4 - Venturi flow meter with differential pressure measurement positions, 12 - main reservoir (tank), V.20 - valve in upper pipe, 21 - upper pipe, 23 - upper calibration reservoir, V.30 - valve in the lower pipe, 31 - lower pipe, V.40 - valve for emptying the upper reservoir, 50 - pump impeller, V.50 - valve for the first filling of the test pump with water in order to allow the operational start, 60 and 61 - pressure taps for measuring pressure difference on test pump, BP - booster pump.

2.3 What Has Been Upgraded

In general, the upgraded parts are: a booster pump (BP: GRUNDFOS NBE32-160.1/169 AF-A-BAQE-96538987) frequency regulated (the test pump already is), enlargement of lower reservoir of transparent plastics (box of acrylate - plexiglass reservoir) to accommodate a new booster pump implementation (12), several pressure transducers at appropriate places (in parallel with the classical measurements), replacement of manual valves with motorized ones (V.10, V.20, V.30, V.40 and V.50), PTZ cameras for cavitation bubbles (Fig. 5) and whole test rig operation surveillance, data acquisition and control system with connection to the internet. The instance of the first bubble in cavitation process is not predicted to be measured, but the eye-notification from the camera picture is considered to be enough.

It is, also intention to measure pumps' power and their rotational speed. The metal construction for this pump system is also changed.

Differential pressure transmitter 0–1 bar (M400D001121) is used for measuring Δp_1 for test pump (1) head measurements. For measuring Δp_2 on the Venturi flow meter (4), differential pressure transmitter 0–3 bar (M400CD01121) is used. For measuring Δp_3 for booster pump (BP) head measurements, differential pressure transmitter 0–5 bar (M400CF01121) is used. For measuring water level in the upper tank (23), differential pressure transmitter 0–0.1 bar (M400CE01121) is used. For measuring the pressure at the test pump (1) inlet, absolute pressure transmitter 0–1 bar (M401CE01128) is used. The difference between the old and new pressure transmitters is that it is adjusted to the measuring range.

Motorized valves (VXG44.25-10, DN25, kvs 10 m³/h, NP16) with valve actuator (SAS61.03) are used for: valve in the lower pipe (V.30), valve for emptying the upper reservoir (V.40), valve in the test pump suction pipe (i.e. between two pumps - V.10), valve in upper pipe (V.20), valve for the first filling of the test pump with water in order to allow the operational start (V.50).

3 Description of Work

The essential flow motion is to take the water into an inflow pipe of a booster pump (BP), boost it up to the test pump (1), and afterwards pump it through different pipes and elements to the upper reservoir or back to the original one (circulation).

The booster pump (BP) has a multiple use:

- Filling the test pump with water to allow for its start.
- Increasing the test rig potential, i.e. testing of pump 1 at higher flow rates. At previous installation only a part of the curve with smaller flow rates was possible to be tested. It is expected that the whole pump head curve will be possible to be measured.
- Making possible a series connection of the two pumps, and consequently measuring of the characteristics of the pumps in series.

After the booster pump, water flows through valve V.10, which is submerged. The water then passes one pipe elbow, goes up through the larger pipe, and then passing the second elbow flows through the flow straightener, which is not seen in the picture since it is embedded into downstream part of the elbow and the photo view angle is from its back.

Then, through a transparent pipe, water enters the tested pump (1), while inlet pressure signal is taken at position 60. Through a transparent spiral casing of the pump, the impeller may be seen (50). Passing through the impeller and outlet spiral casing, the water goes up through the transparent outlet pipe, where outlet pressure signal is taken at position 61 (the differential pressure signal Δp_1 is also taken by the pump transducer in parallel). The water then makes a turn (a T-branch with one dead-end – valve V.50), and then flows through a transparent pipe and non-standard flow-meter (4).

Finally, it comes to the main T-branch where, in the case when V.20 is completely closed and valve V.30 is completely or partially open, water passes vertically down-stream to the main tank (12).

In the case when valve V.30 is closed and V.20 is partially or fully open, water, after the T-branch, is directed through the pipe 21 to the upper tank (23). The outlet hole of the tank 23 is blocked by the valve V.40.

For several positions of valve V.30 (fully open, several mid-positions and closed), measured, sampled and calculated values of flow rate (Q) and head (H) provide pump characteristic presented (diagram not shown here due to space economy). This could be repeated for another pump speed of rotation.

4 LabVIEW Environment

LabVIEW application for remote measurement and control is under development, as analogue to already developed local one [1]. The front panels are shown in

Figs. 6, 7 and 8 (here, and in full width at the lower part of the Poster), where: 1 - end, 2 - start, 3 - record, 4 - atmospheric pressure, 5 - ambient temperature, 6 - acceleration due to gravity, 7 - efficiency, 8 - power, 9 - head, 10 - flow rate, 11 - valve position, 12 - torque, 13 - water temperature, 14 - pump maximum pressure in kPa, 15 - pump speed in percents, 16 - pump speed in rpm, 17 - simulated pump characteristic curve. Pressure transmitters' calibration curves are imported into the generated software and connected to the eight channel input module National Instruments NI-9203. USB CompactDAQ chassis cDAQ-9174 is connected to a laptop. Students can now follow the procedure described in the previous chapter and record signals from various differential pressure transmitters.

   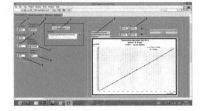

Fig. 6. LabVIEW application front panel.   **Fig. 7.** LabVIEW application first screen.

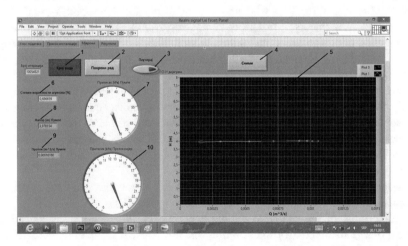

Fig. 8. LabVIEW application screen for measurements.

This LabVIEW application is simplified for the case of two differential pressure transmitters and it is presented in Figs. 6 and 8. In this simplified, application four screens exist. In Fig. 7 the first "welcome" or input data screen is presented.

In input (first) screen (Fig. 7) the following necessary data are presented: 1 - water density, 2 - acceleration due to gravity, 3 -possibility to export data, 4 - calibration curve of the differential pressure transmitter, 5 - calibration curve coefficients, 6 - acquisition card measuring frequency, 7 - calibration coefficient of the Venturi flow meter, 8 - number of samples, 9 - pump unit power. Most of these data are necessary for further measurements and data processing. Students generate differential pressure transmitters calibration curves and import them into the presented LabVIEW application.

The second important screen is the one for data acquisition (Fig. 8) where: 1 - stop, 2 - start, 3 - pause, 4 - record, 5 - chart for experimentally obtained pump characteristic curve, 6 - pump efficiency, 7 - differential pressure transmitter for measuring pump head, 8 - pump head, 9 - flow rate and 10 - differential pressure transmitter on Venturi flow meter.

Red square on screen in Fig. 8 denotes real time pump measurements. This is useful due to the fact that each new valve V.30 position needs first flow stabilizing and afterwards measurements. Pump characteristic curves are obtained for various positions of the valve V.30. Various parameters could be followed on the front panel, such as pump head and flow rate, water density, pump speed and etc.

In addition to pump characteristic curve determination, students can determine system characteristic curves. This can be done, for example, for the installation without geodesic height for various positions of the valve V.30.

5   Conclusions

This paper highlights the importance of laboratory measurements in education of students of technical faculties who are more interested about pumping systems. It may be even used by engineers from various companies. Facilities of this kind are demanding both in terms of design and construction, as well as in financial terms. For this reason, it is very interesting to make accessible basic principles of pumping plants to a wider number of attendees.

It is proved in practice that, although student groups for laboratory exercises are small, attendees don't learn all the elements planned in subject curriculum. Now, using internet and remote control of the setup, the students will still keep the "do-it-yourself" concept, but have more time to experiment and learn. They will be able to generate their reports via internet and teachers would have a possibility to follow their work and enable various access limitations.

Despite many technical possibilities, authors still encourage oral final exam where students would have a possibility to explain obtained results.

Unlike most of the available remote experiments, this setup has a considerable number of signals for control and sampling and belongs to a group of the most complex ones in the remote labs repositories. Work on LabVIEW software application development is in progress. Many safety measures are discussed and implemented in this set-up. The main goal is to have this set-up fully operational remotely. Installation will be, afterwards, integrated in Golabz.eu repository.

In fact, the possibilities of this setup are fully equal to the possibilities of "big-real" ones, both of accredited laboratories and industry. The transparency of "critical" points of flow serves for obvious learning, with still keeping the possibility for precise measurement of flow parameters. The problems of complexity of processes to be demonstrated, measured, monitored and controlled are effectively solved, connection with data acquisition system fully accomplished, and the software implementation is undergoing.

This part is now under development with colleagues from the GoLab project (http://www.go-lab-project.eu/) and is expected to be finished until mid-2018.

Acknowledgment. This work has been partly funded by the SCOPES project "Enabling Web-based Remote Laboratory Community and Infrastructure" of Swiss National Science Foundation (the project leader is Dr. Denis Gillet on behalf of the EPFL - Ecole Polytechnique Federale de Lausanne, http://www.go-lab-project.eu/partner/%C3%A9cole-polytechnique-f%C3%A9d%C3%A9rale-de-lausanne) and partly by Project TR 35046 Ministry of Education, Science and Technological Development Republic of Serbia what is gratefully acknowledged.

References

1. Nedeljkovic, M.S., Cantrak, D.S., Jankovic, N.Z., Ilic, D.B., Matijevic, M.S.: Virtual instruments and experiments in engineering education lab setup with hydraulic pump. In: Proceedings of 2018 IEEE Global Engineering Education Conference (EDUCON), Paper No. 1368, 17–20 April, 2018, pp. 1145–1152. Santa Cruz de Tenerife, Canary Islands, Spain
2. Protić, Z., Nedeljković, M.: Pumps and Fans - Problems, Solutions, Theory. Faculty of Mechanical Engineering, University of Belgrade, Belgrade (2006)
3. Ilić, D., Čantrak, Đ.: Laboratory Practicum for Fluid Flow Measurements. Faculty of Mechanical Engineering, University of Belgrade, Belgrade (2017)
4. ISO 9906: Rotodynamic Pumps - Hydraulic Performance Acceptance Tests - Grades 1, 2 and 3 (2012)
5. Nedeljkovic, M.S., Cantrak, D.S., Jankovic, N.Z., Ilic, D.B., Matijevic, M.S.: Virtual instrumentation used in engineering education set-up of hydraulic pump and system. In: Proceedings of the 5th International Conference on Remote Engineering and Virtual Instrumentation (REV2018), Paper No. 1166, 21–23 March 2018, pp. 341–348. University of Applied Sciences, Duesseldorf, Germany
6. Nedeljkovic, M.S., Jankovic, N.Z., Cantrak, D.S., Ilic, D.B., Matijevic, M.S.: Engineering education lab setup ready for remote operation - pump system hydraulic performance. In: Proceedings of 2018 IEEE Global Engineering Education Conference (EDUCON), Paper No. 1376, 17–20 April 2018, pp. 1175–1182. Santa Cruz de Tenerife, Canary Islands, Spain

Poster: "Radiation Remote Laboratory" with Two Level Diagnostics

Michal Krbecek[1], Sayan Das[1], Franz Schauer[1($\boxtimes$)],
Miroslava Ozvoldova[1], and Frantisek Lustig[2]

[1] Faculty of Applied Informatics, Tomas Bata University in Zlin,
760 05 Zlin, Czech Republic
fschauer@fai.utb.cz

[2] Faculty of Mathematics and Physics, Charles University,
Ke Karlovu 3, 121 16 Praha 2, Czech Republic

Abstract. The paper describes the remote experiment "Radiation Remote Laboratory" with two levels diagnostic system, built on ISES - Internet School Experimental System, accessible across the Internet and provided via the system REMLABNET (http://www.remlabnet.eu/). The remote experiment strives to provide the basic knowledge on γ radioactivity and/or γ radiation and its basic application laws, and parameters like its Poisson distribution, intensity dependence on distance from the point source and provides basic ideas about its absorption in various materials. Absorption in Cu on thickness of the Cu material is possible to examine in detail. Besides, this experiment serves to develop the basic knowledge for handling the radioactive materials in education and practice.

Keywords: ISES · Remote experiment · Diagnostic system · Radioactivity
Intensity · Shielding effect · Absorption

1 Introduction

Research shows that many students misinterpret the physical principles behind radiation and radioactivity and are unable to apply scientific knowledge connected with radiation and radioactivity [1]. By addressing these issues "Radiation Remote Laboratory" provided via the Internet was developed. Students and interested may understand the basic concepts as source, radiation and detector model, similar as in [2]. They can then easily understand the concepts of absorption of radiation, radioactive decay and half-life of radioisotope.

The proposed "Radiation remote laboratory", controlled as finite state machine and its core – Measureserver, provided by two level diagnosis for the purpose of the increased reliability is described in detail in [3]. The remote experiment is aimed at real measurement of the statistical distribution γ radiation, its point source properties, absorption power of Al, Fe, Pb materials and detailed dependence of absorption of Cu on thickness. As a point source of γ radiation Americium (Am^{241}) was used as the most prevalent isotope of americium in nuclear waste. Americium 241 has a half-life of 432.2 years, γ emission with energy of 59.5 keV [4]. The concept of γ radiation and its

M. E. Auer and R. Langmann (Eds.): REV 2018, LNNS 47, pp. 312–320, 2019.
https://doi.org/10.1007/978-3-319-95678-7_36

properties are important subject matter in contemporary industry and science in connection with the energetics, use in medicine and military applications [5]. As a teaching tool this experiment is contributing to teaching of radioactivity, because it is one of the first remote experiment dealing with the real measurements in the field of radioactivity [6].

2   Purpose and Goals

The whole concept of the Remote Experiment "Radioactive Laboratory" experiment comprehended to enable demonstrating the basic concept of radioactive phenomena:

– The concept of basic properties of γ radiation source and its Poisson distribution of emission,
– To study the dependence of the intensity of radiation on the distance from the source,
– To study the absorption power of different materials in general,
– To study the absorption in Cu on thickness of the Cu shielding and the stopping power of Cu.

3   Short Theory, the Experiment Setup and Students Result

3.1   Theory

Firstly, let us assume that the relative intensity of γ radiation of the point source is denoted by I [min^{-1}] (and expressed by number of registered events by Geiger-Müller tube), the intensity at $x = 0$ m is denoted by I_0, the absorption coefficient μ [m^{-1}] and the distance x [m].

Radiation Intensity on the Distance From the Source
The radiation intensity of the γ radiation of the point source depends on the distance x from the source

$$I \sim \frac{1}{x^2}.$$ (1)

Absorption Power of Materials
The absorption of γ radiation in material depends on the material thickness x as

$$I = I_0 e^{-\mu x},$$ (2)

and the corresponding absorption coefficient μ [mm^{-1}] then is

$$\mu = \frac{1}{x} \cdot (\ln \frac{I_0}{I}).$$ (3)

Poisson Statistical Distribution of γ Radiation

The Poisson statistical distribution gives the prediction of the outcome of random and independent events [7]. The probability density function of the Poisson distribution is given by

$$P(X, \mu) = \frac{\lambda^k . e^{-\lambda}}{k!}, \tag{4}$$

where λ is the average number of events per interval and k takes values 0, 1, 2.... Poisson distribution Radioactive decay will be calculated with the help of radioactive decay can be calculated easily for small No of counts according [8].

3.2 "Radiation Remote Laboratory" - Setup

The arrangement of the RE "Remote Radioactive Laboratory" is shown in the Fig. 1. As the source of γ radiation is used the certified school demonstration source DZZ GAMA Americium (Am-241). As detectors are used couple of ISES Geiger-Müller (GM) tubes. The first GM movable tube (GM1) is used for the background and source intensity–distance measurements, whereas the second, also movable (GM2) is used for absorption power of Cu, Al, Fe and Pb materials. The motion of GM tubes is by step motor drives. The whole setup uses the optical bench. Both the driving and measured signals are transmitted and collected by ISES standard hardware. The remote experiment "Radiation Remote Laboratory" is controlled by notebook FUJITSU with ISES USB module. The whole system is built on the Internet School Experimental

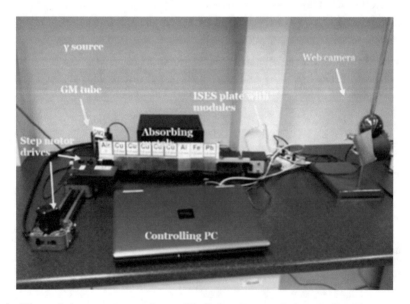

Fig. 1. View of the remote experiment "Radiation Remote Laboratory" with components described

System (ISES) components [9]. The whole experiment is organized as finite state machine and its core – Measureserver with PSC file controlling program and the web page were built using Easy Remote ISES (ER ISES) for compiling the control RE program [9]. The diagram shown in Fig. 2 present the controlling web page of RE "Remote Radioactive Laboratory".

In upper part there is the panel with the measured data output, below are the controls for adjusting the type of measurements and corresponding times, and in the lower panel is the transferred data for processing. The experiment is covered by the stream from web camera (right upper corner).

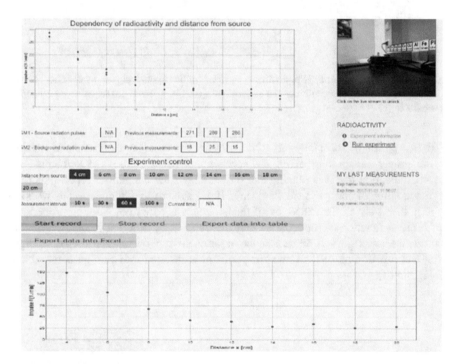

Fig. 2. Controlling webpage of the remote experiment "Radioactivity Remote Laboratory"

3.3 Two Level Diagnostic System

The first diagnostic system gives the feedback of the functionality of the experiment with the help of "traffic lights" [3]. (a) The green light indicates that the experiment is available to use, (b) the orange light indicates that the experiment is currently occupied and (c) red light indicates that the experiment is out of order (see Fig. 3).

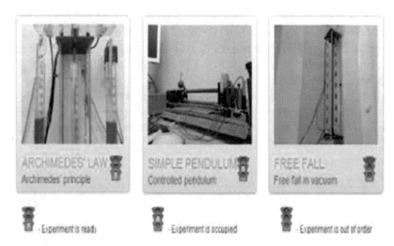

Fig. 3. Remote experiments **diagnostics I** - the "traffic lights" signal availability of the experiment (green) (a) occupation of the experiment, (orange) (b) and out of order experiment – not available (c)

Furthermore, the second system of diagnostics II is more sophisticated and watches for the proper functioning of all the components of the remote experiments. See Fig. 4b, indication that it is running and further shows the continuous report regarding the functioning of the physical hardware. If all system and functions are working properly then they are indicated, but if the apparatus is either not working properly or their limiting function disrupted, then they are denoted by different colour. On issuing an error, the message will be sent to the experiment supervisor. The proper configuration and functioning of the remote experiment is inserted into configuration table – see Fig. 4a [3].

Fig. 4. Reference list for configuration of Diagnostics II module (a), running report of diagnostics II module with proper functioning (green/light) and artificially introduced two faults (brown/dark)

3.4   Student's Experimental Results

The results of the radioactivity experiment measured by students are depicted in Fig. 1. Figure 1(a) gives the dependence radiation intensity of the γ radiation of the point source on the distance x (for collection time 30 s) with fitting according to Eq. (1) (where the upper (red) curve is with background pulses and the lower (black) one is without background pulses. In Fig. 1(b) is the absorption in Cu, plotted as the transmitted intensity of γ radiation on Cu thickness $I(d)$ with fitting by Eq. (2) (here also the upper (black) curve is with background pulses and the lower (red) one is the without background). In Fig. 1c are the examples of the Poisson statistical distribution for different average number of events λ.

In Tables 1, 2, 3 and 4 are given evaluations of the measured data.

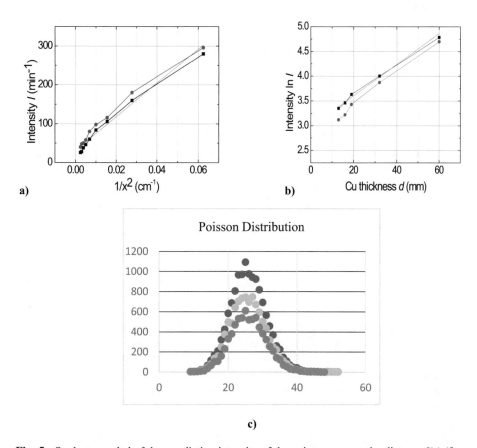

a)

b)

c)

Fig. 5. Students results' of the γ radiation intensity of the point source on the distance $I(x)$ (for the collection time 30 s) with fitting by Eq. (1) (continuous curve) (a), the transmitted intensity of γ radiation on Cu thickness I (d) with fitting by Eq. (2) (continuous curve) (b), the Poisson statistical distribution for different average number of events λ.

Table 1. Intensity of the source

Equation (1)	Intensity I	Value
$y = a\ 1/x^2$	Intensity at zero pos. I_0 [min^{-1}]	0.0625
	Coefficient a [cm^{2}s^{-1}]	4.31

Table 2. Absorption coefficient of copper (Cu)

Model	Equation (2)	Intensity I		Standard error
Equation	$y = a\ exp^{(-bx)}$	Fitting constant **a** (intensity at zero position I_0 [min^{-1}]	2.97	0.11
Reduced Chi-Square	0.0066			
Adj. R-Square	0.98			
		Fitting constant **b** (absorption coefficient μ [mm^{-1}])	0.82	

Table 3. Absorption coefficients for several metals

Materials	GM1 value N	GM2 background radiation B	Interval t (s)	Corrected pulse	Intensity I	Absorption coefficient μ [mm^{-1}]
Al (1 mm)	111	9	30	102	3.40	0.26
Fe (1 mm)	63	8	30	55	1.83	0.88
Pb (1 mm)	8	12	30	4	0.13	3.5

Also, with these tables, students can easily comprehend the radioactivity theory like the intensity-distance relationship, absorption phenomena of the material and also the randomness of radioactivity. The measured intensity value plot in Fig. 5a with respect to distance after transferring it into logarithmic value. The linear fitting of the graph gives the zeroth position intensity value and constant coefficient value, see Table 1. The figures recorded in the tables, help to improvise the idea of intensity-distance relationship theory of radioactivity among the students. Also, the absorption coefficient of the material like Cu, Al, Fe and Pb (Tables 2 and 3) can easily be calculated by using this experiment. This absorption coefficient value develops the idea of shielding property of the material among the users. They easily understand theoretical concept of it. This experiment helves the user to understand the probability of the randomness through the Poisson distribution calculation too. This paper built the practical idea of the theoretical knowledges and also provide the better acquaintance of the radioactivity properties.

Table 4. Sample of probability distribution without background pulses

Distance (X)	Mean	Cumulative	Probability distribution
38	9	4	1
21	10	5	0.9993
21	11	8	0.997748
28	12	21	0.999977
29	13	24	0.999962

4 Conclusions

From the results of measurements, it is obvious that:

- The measured intensity of radiation from point source decreases with the square of the distance as predicted by theory,
- Statistical approach helps us to limit the impact of errors on results of measurement.
- Results of absorption measurements give direct information on absorption power of various materials with respect to protection against harmful radiation,
- The results of measurements give evidence of Poisson statistical distribution of the events of emission in accordance with theory.

Along with the advantages in experimental analysis the advantages of the real remote experiment over virtual laboratories also increase the interest among the students for performing the radioactive experiments and accumulate the knowledge.

Acknowledgement. The support of the project of the Swiss National Science Foundation "SCOPES", No. IZ74Z0_160454 is highly appreciated. The support of the Internal Agency Grant of the Tomas Bata University in Zlin for Ph.D. students is acknowledged.

References

1. Eijkelhof, H.M.C.: Radiation and Risk in Physics Education. CD-ß Press, Utrecht (1990)
2. Schauer, F., Gerza, M., Krbecek, M., Ozvoldova, M.: 'Remote Wave Laboratory' with embedded simulation–real environment for waves mastering. In: Online Engineering and Internet of Things, pp. 182–189. Springer, Heidelberg (2018)
3. Schauer, F., et al.: REMLABNET III—federated remote laboratory management system for university and secondary schools. In: 2016 13th International Conference on Remote Engineering and Virtual Instrumentation (REV), pp. 238–241 (2016)
4. Holm, E., Persson, R.B.R.: Biophysical aspects of Am-241 and Pu-241 in the environment. Radiat. Environ. Biophys. **15**(3), 261–276 (1978)
5. Nwosu, O.B.: Comparison of gamma ray shielding strength of lead, aluminium and copper from their experimental and MCNP simulation result (2015)
6. Schauer, F., Krbecek, M., Beno, P., Gerza, M., Palka, L., Spilakova, P.: REMLABNET-open remote laboratory management system for e-experiments. In: 2014 11th International Conference on Remote Engineering and Virtual Instrumentation (REV), pp. 268–273 (2014)

7. Arbia, G., Griffith, D., Haining, R.: Error propagation modelling in raster GIS: overlay operations. Int. J. Geogr. Inf. Sci. **12**(2), 145–167 (1998)
8. Shamim, S., Hassan, H., Anwar, M.S.: Natural radioactivity and statistics. Lab Monogr. Introd. Exp. Phys. 206 (2010)
9. Ozvoldova, M., Schauer, F.: Remote laboratories in research-based education of real world phenomena. Peter Lang, Frankfurt am Main (2015). 183 pp. [9,44 AH]. ISBN 978-80-224-1435-7. ISSN 2195-1845

Cyber Physical Systems
and Cyber Security

Enabling Remote PLC Training
Using Hardware Models

Alexander A. Kist[1(✉)], Ananda Maiti[1], Catherine Hills[1],
Andrew D. Maxwell[1], Karsten Henke[2], Heinz-Dietrich Wuttke[2],
and Tobias Fäth[2]

[1] University of Southern Queensland, Toowoomba, Australia
`kist@ieee.org, anandamaiti@live.com,`
`{catherine.hills, andrew.maxwell}@usq.edu.au`
[2] TU Ilmenau, Ilmenau, Germany
`{karsten.henke, dieter.wuttke,`
`tobias.faeth}@tu-ilmenau.de`

Abstract. Programmable Logic Controllers (PLCs) are widely used for industrial control applications. Developing programming skills with these devices is essential for students of electrical engineering, and also for professionals wanting to learn new skills. Whilst programming is performed using software development tools, it is essential that learners have access to real hardware to test these programs in a realistic context. This paper introduces a PLC laboratory architecture and discusses how the experiment is integrated into two existing remote laboratory environments, namely GOLDi (Grid of Online Lab Devices Ilmenau) and RALfie (Remote Access Laboratories for Fun, Innovation and Education). The different approaches are compared in detail.

Keywords: PLC · Remote laboratory · Industry training

1 Introduction

Programmable Logic Controllers (PLCs) play an instrumental role in automation and control. Learning how to program these devices is not only important to students of control engineering but also to a wide range of professionals in electrical engineering. With the advent of the Industrial Internet of Things, their role will continue to increase. In this context continuing professional development and upskilling is very important. Providing access to practical and authentic development activities for professionals in industry and students can be a challenge. While some PLC programming environments are freely available, practical and meaningful learning activities require access to learning materials and physical hardware models that are being controlled.

Remote laboratories are being widely used at universities to provide remote access hardware, sometimes as alternatives to face-to-face laboratory activities. Two examples include the GOLDi Remote Lab Cloud [1] and the RALfie system [2]. The aim of this project is to develop and test an approach that allows external participants to control

© Springer International Publishing AG, part of Springer Nature 2019
M. E. Auer and R. Langmann (Eds.): REV 2018, LNNS 47, pp. 323–332, 2019.
https://doi.org/10.1007/978-3-319-95678-7_37

real hardware models via a PLC in a safe environment. The proposed architecture integrates into both the GOLDi and the RALfie system. This work discusses the overall system approach.

For GOLDi the PLC interfaces with a protection unit and gains access to the hardware model. A newly developed Remote Desktop software module allows access to the computer that hosts the PLC programming software. Users are authenticated for access via a virtual control panel to the lab server. For RALfie, the PLC connects to the electromechanical model via a microcontroller unit. The RALfie system has in-built remote desktop and camera access methods. Users are authenticated via the system and can then access the virtual control panel.

The remainder of the paper is organised as follows. Section 2 briefly discusses related work and Sect. 3 discusses the learning activity, its physical components and how users interact with the environment. Section 4 discusses the GOLDi architecture and Sect. 5 discusses the RALfie System architecture. The two approaches are compared in Sect. 6 and conclusions are presented in Sect. 7.

2 Related Work

Providing students with opportunities to interact with practical experiments is an integral part of many technical disciplines such as engineering [3]. Online and remote laboratories provide opportunities for learners to use equipment remotely. These have been available for some time and are used in different contexts. Various systems architectures are use [4]. These systems support various degrees of experiment autonomy [5]. A number of papers have reported on remote labs for teaching PLCs. Examples include a process plant that has implemented augmented reality [6] and PLC programming courses that use manufacturing cells [7]. Sheng-Jen and Hsieh discuss in [8] a web-based PLC lab for manufacturing engineering and in [9] provide a PLC simulation and tutorial system. PLC have also been used to control remote laboratories [10]. The project that is discussed in this paper focuses on a flexible system solution that provides opportunities for a PLC to control real models that can be integrated with existing RAL systems.

3 Learning Activities

The aim of this practical activity is to allow users to learn PLC programming. As for most modern control systems, programs are written in an integrated development environment and then downloaded to the controller. For the learning activity, the authors have selected the Click PLC programming software [11], as this software is freely available. Students can develop the PLC control program offline on their own computer without requiring hardware models. Later the program can be tested in the (remote) laboratory, where the program is uploaded to programming software that is running on the lab server.

The general system architecture is shown in Fig. 1. An electromechanical model, e.g. a model of a four-level lift, has a number of sensors and actuators. These are connected to a system protection unit via I/O lines. In turn, the protection unit is connected to a PLC either via individual I/O lines or Modbus. The lab server running the development environment is connected to the PLC via USB. The protection unit ensures that the model is not damaged by incorrect commands, either accidental or deliberate. Such a configuration can also be readily used for proximal teaching as shown in Fig. 2. The lift model can be seen in the background, the PLC is shown on the left and the development environment is running on the laptop shown on the right.

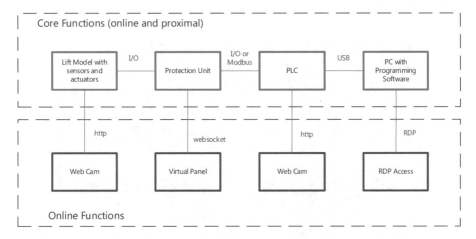

Fig. 1. PLC laboratory functions

The lift model was tested by four separate groups of two or three students at a residential school. Students who self-identified as having significant prior PLC programming experience were given access to the model, an input/output list and the programming software with basic configuration complete. The students also had access to a touch panel HMI to simulate call buttons and indicator lights for each floor.

The facilitator introduced the hardware and software and discussed suitable approaches to programming. The facilitator was on hand throughout the 8 h activity to observe or provide guidance as necessary. The four groups employed significantly different strategies, seeming dependent upon their previous experience and confidence levels. The groups also made differing levels of progress within the limited time frame. From these observations, suitable material, scaffolding activities and progress check points are being developed to support the learning process. The most difficult aspect of programming the lift is the conceptual management of call priorities and choosing the next action based on the previous and current operation states as well as the state of the call buttons. Students generally managed the design of a two-level control algorithm easily, finding the difficulty increasing as additional floors were introduced. This is an area that will require particular attention in the support material and activities.

In this context, the lift model provided an extensible activity to engage and challenge students who in some cases work with industrial automated equipment on a regular basis. It also provides an opportunity to discuss relevant standards as well as the differences between safety and control in industrial automated systems.

To make the unit accessible remotely, a number of additional functions are required. The operation of the model can be remotely observed through web cameras. The programming environment can be controlled via remote desktop access to the laboratory PC. In addition, virtual panels can show sensor values as well as virtual control inputs, e.g. call buttons for the lift. The remote lab systems also needs to address authentication and access control management. The details of how this remote access is provided differs between the two RAL architectures.

Fig. 2. Lift model and PLC

4 GOLDi System Architecture

The interactive hybrid online lab GOLDi (Grid of Online Lab Devices Ilmenau) implements a grid concept to realize a universal remote lab infrastructure. The GOLDi Remote Lab offers students a working environment that is as close as possible to a real world laboratory. Furthermore, the lab infrastructure offers students stimulus to the design of safety critical control systems.

The GOLDi Remote Lab architecture allows a large number of experiments to be performed for implementing digital control algorithms. For the functional description, GOLDi offers different specification techniques supported by noncommercial development tools for various web-based control units in the remote lab:

- a Finite State Machine (FSM) based design to describe digital automata - executed within a client-side FSM interpreter;
- a software-oriented design in C or assembler executed on microcontrollers;
- a hardware-oriented design in hardware description languages (e.g. VHDL) or schematic block design by using FPGAs; and
- a PLC-based design in various description languages commonly used in industrial applications using manufacturer specific design software.

Figure 3 provides an overview of the GOLDi Remote Lab installed at each partner installation. The infrastructure is based on a universal grid concept, which guarantees reliable, flexible and robust usage of this online lab. A more detailed description of this grid concept, as well as the main components, is presented in [1, 2].

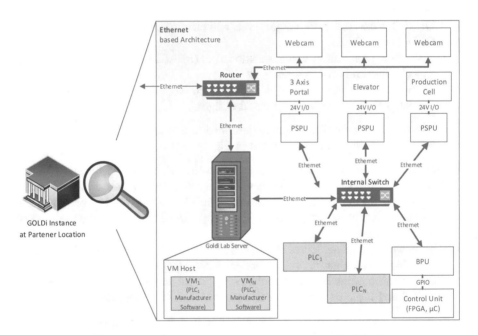

Fig. 3. GOLDi hardware setup (PLC components highlighted)

The server side infrastructure (remote lab) consists of three parts:

- an internal Ethernet-based *remote lab bus* to interconnect all parts of the remote lab;
- a *bus protection unit* (BPU) to interface the control units to the remote lab bus and to protect the bus from blockage, misuse and damage; as well as
- a *physical system protection unit* (PSPU), which protects the physical systems (the electro-mechanical models in the remote lab) against deliberate damage or accidentally wrong control commands and which offers different access and control mechanisms.

The interconnection between the web-control units and the selected physical systems during a remote lab work session (experiment), as well as the webcam management is performed by the GOLDi Lab Server as part of the remote lab infrastructure.

Currently, GOLDi infrastructure is implemented at ten universities in Armenia, Georgia, Germany and the Ukraine - based on two TEMPUS projects. This means, that each project partner received a complete remote lab. As a consequence, the network of all GOLDi remote labs was implemented as a cloud system [2]. Each GOLDi user communicates with the GOLDi cloud to access the following Web services (see Fig. 4):

- Experiment Control Panel (ECP): User interface to perform remote experiments and provide the communication with the GOLDi Partners;
- Experiment Management: Experiment booking and pre-planning;
- User Management;
- GOLDi Documentation for control units and electromechanical hardware models;
- GOLDi Design Tools:
  - Block diagram Editing and Simulation Tool (BEAST): User interface to develop block diagram based designs; and
  - Graphical Interactive Finite State Machine Tool (GIFT): User interface to develop Finite State Machine based designs.

To enable the remote PLC training described in this article, the GOLDi hardware architecture was extended as highlighted in Fig. 3:

- In addition to the GOLDi Lab Server software, the Lab Server machine now Virtual Machine (VM) host software. Each virtual machine corresponds to a PLC device connected to the internal Ethernet network. The VM host software is configured in a way that each time an experiment is started the virtual machine will be reset to an initial configuration.
- The firewall of the Lab Server machine is configured so that each virtual machine is only able to communicate with its assigned PLC device.
- Each virtual machine has the manufacturer specific development environment pre-installed and is pre-configured for the use within the GOLDi Lab.

An experiment using the PLC as a control unit is started in the same manner as any other experiment within the GOLDi Lab. When configuring the experiment the student selects the PLC as a control unit. When the experiment is started, in addition to the ECP, another browser tab is opened in which the remote desktop of the virtual machine is displayed using Apache Guacamole. As displayed in Fig. 4, the user's machine is communicating with the GOLDi cloud via HTTP. The Guacamole instance within the GOLDi cloud then translates this communication to the RDP (Remote Desktop Protocol) used by the virtual machines for remote desktop access.

Currently, work is underway to implement another GOLDi instance at the University of Southern Queensland in Australia to be able to establish additional PLC device based experiments.

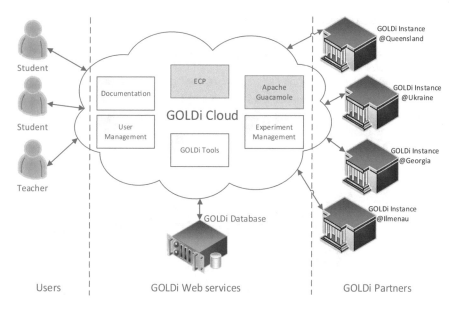

Fig. 4. GOLDi cloud architecture for PLC

5 RALfie System Architecture

RALfie was initially developed as an environment to easily integrate remotely accessible experiments for school students, and is presently being used to provide access to remote learning activities for university students. This environment provides mediated and authenticated access to remote laboratory equipment. The system separates hardware experiments from their corresponding activities: i.e. a hardware rig can have multiple different activities associated with it. Activities can use different control interfaces to the lab. It supports custom web-based, Snap![1]-based or RDP-based user client interfaces. Common to all is access to live web camera streaming feeds from the experiment. The focus here is on the RDP-based interface.

The RALfie system supports authentication and access control. Once users are authenticated for a particular activity, the experiments can be accessed directly via specific URLs. For example, `cam1-pn.ralfie.net` enables direct access the video feed of the experiment and `panel-pn.ralfie.net` allows access to the virtual panel. The user sees the activity page that shows the experiment client user interface (here using RDP) and provides access to additional information, such as video stream feeds and virtual experiment controls via popup panels.

Figure 5 shows an overview of the system. On the right hand side, the physical lift model is shown. The sensors and actuators of the model are connected to the MCU which implements protection and interface functionality. The physical system protection unit ensures that commands that are sent to the physical model are within given

---

[1] Snap! is a browser-based educational graphical programming language (snap.berkeley.edu).

parameters and tolerances and will not damage the physical model. The Virtual Interface Unit (VIU) makes the physical model inputs and outputs available via virtual panel. This includes the call buttons for the lift, for example. These are rendered as a web page.

The control unit, i.e. the PLC, is connected to the protection unit via I/O lines. Alternatively, this connection can also be achieved via Modbus. In turn, the PLC is connected via USB to the virtual PC that is hosting the development software. The virtual PC encapsulates the experiment user interface. This is remotely accessed via RDP through a web browser.

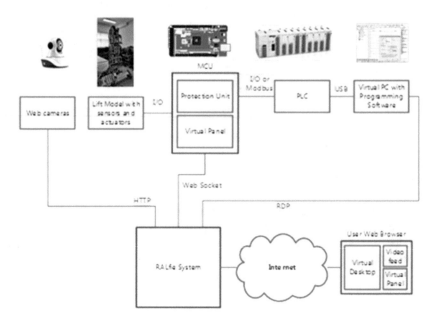

Fig. 5. RALfie PLC architecture

6 Comparison Between the Two Approaches

This section compares the two approaches in more detail. Table 1 lists the main operational tasks and briefly explains how those are managed in the two systems. As future work the authors plan investigate differences from a user perspective.

Table 1. Operational tasks

Task	GOLDi	RALfie
Experiment selection	Selection of PLC as control unit and any physical system as control object	Users access learning activities that define the experiment, UI and system parameters
Experiment access	Via a Web-based remote desktop access integrated into the GOLDi Cloud Web services	RDP access is native to the environment
Experiment management (startup, reset, etc.)	Via the ECP	Control scripts are used to reset the VM and configure the experiment MCU which in turn resets the hardware model and alarms
Video feeds	MJPG stream fetched from the Webcams located at each hardware model provided through the GOLDi Lab Server displayed in the ECP for experiment observation	Are supported natively and are authenticated when the experiment is being accessed
Experiment sensor state and virtual inputs	Animation of sensor/actuator values through the user interface. Virtual inputs are possible by a framework integrated into the ECP	Custom panels are supported natively and web socket communication is authenticated with experiment access. The virtual panel is generated the experiment MCU
Other features	Animated visual models run simultaneously with the real physical systems. Simulation for all hardware models exist and are selectable during experiment configuration	Work is being completed that also supports a virtual model of the experiment

7 Conclusions

An initial prototype has been built and tested with students during a recent residential school. The model demonstrates that the PLC training unit can be integrated with existing remote laboratory systems. Students using the system are able to develop the control programs offline on their own computers and then subsequently test the completed programs online using the electromechanical models. The authors plan to investigate differences in user experience between the two systems as future work.

References

1. Henke, K., Vietzke, T., Wuttke, H.-D., Ostendorff, S.: GOLDi-grid of online lab devices Ilmenau. Int. J. Online Eng. **12**(4), 11–13 (2016)
2. Maiti, A., Kist, A.A., Maxwell, A.D.: Real-time remote access laboratory with distributed and modular design. IEEE Trans. Industr. Electron. **62**(6), 3607–3618 (2015)
3. Feisel, L.D., Rosa, A.J.: The role of the laboratory in undergraduate engineering education. J. Eng. Educ. **94**(1), 121–130 (2005)
4. Maiti, A., Maxwell, A.D., Kist, A.A.: Features, trends and characteristics of remote access laboratory management systems. iJOE **10**(2), 30–37 (2014)
5. Maiti, A., Zutin, D.G., Wuttke, H.-D., Henke, K., Maxwell, A.D., Kist, A.A.: A framework for analyzing and evaluating architectures and control strategies in distributed remote laboratories. IEEE Trans. Learn. Technol. (2017). https://ieeexplore.ieee.org/document/8241394/
6. Márquez, M., Mejías, A., Herrera, R., Andújar, J.M.: Programming and testing a PLC to control a scalable industrial plant in remote way. In: 2017 4th Experiment@ International Conference (exp. at'17), pp. 105–106. IEEE (2017)
7. Bellmunt, O.G., Miracle, D.M., Arellano, S.G., Sumper, A., Andreu, A.S.: A distance PLC programming course employing a remote laboratory based on a flexible manufacturing cell. IEEE Trans. Educ. **49**(2), 278–284 (2006)
8. Hsieh, S.-J., Hsieh, P.Y.: Web-based programmable logic controller learning system. In: 32nd Annual Frontiers in Education, vol. 3 (2002)
9. Hsieh, S.-J., Hsieh, P.Y., Zhang, D.: Web-based simulations and intelligent tutoring system for programmable logic controller. In: 33rd Annual Frontiers in Education, FIE 2003, vol. 1, pp. T3E-23–T3E-28 (2003)
10. Bowtell, L.A., Moloney, C., Kist, A.A., Parker, V., Maxwell, A., Reedy, N.: Enhancing nursing education with remote access laboratories. Int. J. Online Eng. (iJOE) **8**(S4), 52–59 (2012)
11. AutomationDirect: CLICK PLC Programming Software. https://www.automationdirect.com/adc/Overview/Catalog/Software_Products/Programmable_Controller_Software/CLICK_PLC_Programming_Software

Towards Data-Driven Cyber Attack Damage and Vulnerability Estimation for Manufacturing Enterprises

Vinayak Prabhu[1]([⊠]), John Oyekan[2], Simon Eng[1],
Lim Eng Woei[1], and Ashutosh Tiwari[3]

[1] Nanyang Polytechnic,
180 Ang Mo Kio Avenue 8, Singapore 569830, Singapore
vinayak_prabhu@nyp.edu.sg
[2] Cranfield University, Bedford, UK
[3] University of Sheffield, Sheffield, UK

Abstract. Defending networks against cyber attacks is often reactive rather than proactive. Attacks against enterprises are often monetary driven and are targeted to compromise data. While the best practices in enterprise-level cyber security of IT infrastructures are well established, the same cannot be said for critical infrastructures that exist in the manufacturing industry. Often guided by these best practices, manufacturing enterprises apply blanket cyber security in order to protect their networks, resulting in either under or over protection. In addition, these networks comprise heterogeneous entities such as machinery, control systems, digital twins and interfaces to the external supply chain making them susceptible to cyber attacks that cripple the manufacturing enterprise. Therefore, it is necessary to analyse, comprehend and quantify the essential metrics of providing targeted and optimised cyber security for manufacturing enterprises. This paper presents a novel data-driven approach to develop the essential metrics, namely, Damage Index (DI) and Vulnerability Index (VI) that quantify the extent of damage a manufacturing enterprise could suffer due to a cyber attack and the vulnerabilities of the heterogeneous entities within the enterprise respectively. A use case for computing the metrics is also demonstrated. This work builds a strong foundation for development of an adaptive cyber security architecture with optimal use of IT resources for manufacturing enterprises.

Keywords: Data driven · Cyber security · Manufacturing enterprises
Cyber attack damage · Cyber attack vulnerability · Metrics

1 Introduction

Manufacturing has been immensely important to the prosperity of nations, with over 70% of income variations of 128 nations explained by differences in manufactured product export data alone. It is also a vital industry sector for the stability and growth of the global economy. It has 70% share in global trade and 16% share in global GDP, amounting to $12 Trillion [1, 2].

© Springer International Publishing AG, part of Springer Nature 2019
M. E. Auer and R. Langmann (Eds.): REV 2018, LNNS 47, pp. 333–343, 2019.
https://doi.org/10.1007/978-3-319-95678-7_38

As a result, cyber attacks on the manufacturing industry could be crippling to the global economy. The threat is real and its full damage potential may not have been fully comprehended yet. A recent Symantec report revealed that 20% of all cyber attacks on industry in 2014 were directed at the manufacturing sector [3]. McAfee identified that industrial networks were the most vulnerable to cyber security issues [4], a fact corroborated by the Dragonfly Espionage Malware Program incident in 2014 that affected 1000 industrial control systems [5]. In 2013, 91% of all cyber attacks took a matter of hours out of which 60% were left undetected for weeks and 53% took months to contain by which time the damage had been done [5]. As identified in a recent review paper [6], the purposes for such attacks were to alter critical data as in the Shamon attack on Aramco, impair or deny process control as in the Stuxnet attack on Iran's nuclear facility, and/or steal data as the 'Shadow Network' espionage operation.

Another type of cyber attack aims to steal intellectual property of industries via Manufacturing data. Manufacturing data' takes the form of product data (e.g. CAD models), manufacturing process data (e.g. machine parameters) and critical infrastructure data (e.g. automation controls). The loss or damage of manufacturing data can cripple not only that enterprise but also the supply chain to which it is connected. This calls for security mechanisms to ensure the privacy of the enterprise. In a manufacturing system, different entities (such as machines, workstations, etc.) have different data access and transfer protocols, operating systems and data storage systems, all with varying levels of data security. This complex heterogeneity presents a strong case for the investigation of data-driven cyber security mechanisms to ensure the trust between entities as well as confidentiality, integrity and availability of manufacturing data.

2 Literature Review

Current research in implementing cyber security for manufacturing has focused on high-level security issues, such as risk assessments and vulnerability analysis [7]. In terms of cyber security solutions, the focus has been on development of detection and mitigation mechanisms for Supervisory Control and Data Acquisition (SCADA) networks [8]. In some cases, common security issues for critical infrastructures, such as smart energy grids, water management and transportation systems, are investigated [9]. While manufacturing shares similarities with critical infrastructures, it has different requirements for cyber security, because it comprises a complex mix of design, process and control entities within its operations. Thus far, the focus has been on securing each manufacturing entity as a silo without considering entity-to-entity connectivity and critical data flows within the system [10]. Furthermore, as a result of recent advances in IoT, there is scope for autonomous cyber defense of manufacturing systems enabled by the study of data flows [11–13].

This paper focuses on developing a novel data-driven approach to quantify the extent of damage suffered by a manufacturing enterprise due to and its vulnerability to cyber attacks. Currently, the value of data that the enterprise depends on for its manufacturing operations is often underestimated based on the explicit components of cost. For example, a recent research [14] has revealed that discarded process data logs with perceived zero value can be used to automate aircraft wing inspection process, significantly reducing lead-time and enhancing productivity. In quantifying the 'real' value of data [15], has suggested a financial model that takes into consideration the cost value of replacing lost data or generating data, the economic value of the data asset to the revenue of an organization and the market value of data generated when sold in the market place. Other factors that contribute to determining the value of data in a system include its accuracy, its completeness towards an information goal, as well as its uniqueness in the information marketplace. However, in order to ensure end-to-end data protection, entity-to-entity vulnerability must be assessed using attack route-based vulnerability quantification [16] weighted according to the value of the data in entities.

3   Method

In order to optimally utilise the cyber security protection resources, this research has developed a set of empirical metrics, namely Damage Index (DI) and Vulnerability Index (VI) to estimate the extent of damage and to estimate the cyber attack vulnerability of entities in a manufacturing system respectively. The metrics are graphically represented in a System Vulnerability Map (SVM) that shows both high value data flows in relation to their vulnerabilities to cyber attacks.

Manufacturing Data Types
Manufacturing data can be broadly categorised into product, process, and control data. Product data refers to the engineering drawings or CAD models of products. Process data refers to the manufacturing programs used to produce the physical parts. Control data refers to the control parameters used by machines to execute the programs.

Damage Index (DI)
DI is to quantify the maximum extent of monetary damage that can be effected to a manufacturing enterprise after a cyber attack. The calculation of DI is based on the 'real value' of data, which is stored in each of the system entities. This research postulates that this 'real value' constitutes three components: (i) perceived or explicit component (ii) implicit component, and (iii) deciphered component.

Perceived/Explicit component is the price the enterprise has paid to obtain or create the data. This data can be directly translated into monetary figure using the below formula where n is the number of datasets.

$$Perceived\ or\ Explicit\ Component = \sum_{i=1}^{n} (Price\ to\ Purchase\ Data)_n \\ + (No.of\ hrs\ to\ create\ data_n \times Hourly\ Rate)$$

Implicit component refers to cost of non-availability of data due to a cyber attack. It is essentially the monetary loss suffered by the enterprise due to resulting downtime, loss of reputation and reduced market share. The formula to compute the Implicit component is given below where n is the number of jobs.

Implicit component (for product and process data) =

$$\sum_{i=1}^{n} \left[\underbrace{(Down\ Time_n \times Hourly\ Rate)}_{①} + \underbrace{\left(1 + \frac{Job\ Revenue_n}{Yearly\ Revenue}\right) \times Job\ Price_n}_{②} + \underbrace{\left(1.2 - \frac{Market\ Share\%}{100}\right)_n \times Job\ Price_n}_{③} \right]$$

Component ① in the equation quantifies the down time monetary loss when there is non-availability of product/process data. ② translates to the loss of reputation. The conversion involves the scaling up of the individual job price carried by the data based on the particular job revenue to the enterprise's yearly revenue ratio. ③ quantifies the reduced market share in terms percentage converted to monetary loss by proportioning the individual job price carried by the data that is compromised.

For control data, any compromise will result in downtime. The formula to quantify the implicit component carry by the control data can be simplified to solely consider the monetary loss due to downtime suffered by the machines only.

Implicit Component (for control data) $= \sum_{i=1}^{n} (Down\ Time_n \times Avg.\ Hourly\ Revenue)$

Deciphered component involves cost of compromising confidentiality of data. It includes the monetary value of critical product or process information that can be obtained by mining the stolen data. The deciphered component only involves the product and process data because control data does not carry any information related to the product or process by itself. The formula to quantify the deciphered component is as below where n is the number of jobs.

Deciphered Component $= \sum_{i=1}^{n} \left(\frac{Complexity\ Index_n}{10} + \frac{Critically\ Index_n}{5} \right) \times Job\ Price_n$

Where the Complexity Index is a combination of the number of setups needed to complete the job and the number of key dimensions in the drawing of the parts involved in the job.

No. of machine setups	1 setup	2 setups	3 setups	4 setups	5 setups and above
Score A	1	2	3	4	5

No. of key dimensions	Zero	1 to 2	3 to 6	7 to 10	11 to 15	16 and above
Score B	0	1	2	3	4	5

Criticality Index is:

Description of the product drawing	Index
Non-critical part such as spacer block	1
Standard Component such as pin, fasteners, connector, linkages, and gears	2
General component such as housing or casing	3
Customised component with unique and/or specialised design	4
Major critical component or parts with patented technology	5

After consolidating the explicit, implicit, and deciphered components into the 'real value', the figure is then normalised to a product or process Damage Index (DI) based on the average single job value. The normalised DI provides a practical gauge of the magnitude of monetary loss with respect to the business scale of the enterprise if there is a cyber attack to the system entity. The formula for consolidated DI is:

$$DI\,(for\,product\,and\,process\,data) = \frac{Explicit\,Component + Implicit\,Component + Deciphered\,Component}{Avrg\,Single\,Job\,Price}$$

For DI for control data, there is no deciphered component. Hence, it equates to the sum of explicit and implicit components normalized to yearly depreciation of the entity.

$$DI\,(for\,control\,data) = \frac{Explicit\,Component + Implicit\,Component}{Yearly\,Decpreciation\,Cost}$$

Vulnerability Index (VI)

The vulnerability of a manufacturing enterprise to cyber attacks can be quantified based on the correlation between data types, cyber security principles and Common Vulnerability Score (CVS) of each of the entities. Depending on the specificity and accuracy of each correlation, a deterministic mathematical model is derived to compute VI.

Each data type is assigned a weighted score based on the cyber security principle that focusses on three aspects: (i) confidentiality; data is accessible only to authorised entities, (ii) integrity; data is not unduly corrupted, and (iii) availability; data is available when needed. The weighted score based on these principles is assigned to each data type and consolidated into a Vulnerability Quantification Table (VQT). VQT is data type driven and should be assessed and adjusted accordingly to improve the accuracy of the computation. A sample VQT for a manufacturing system is shown in Table 1.

Table 1. Vulnerability Quantification Table (VQT)

Data	Confidentiality	Integrity	Availability	Total
Product	0.5	0.2	0.3	1
Process	0.4	0.4	0.2	1
Control	0.2	0.6	0.2	1

Common Vulnerability Score System (CVSS) is a free and open industry standard for accessing the severity of computer system security vulnerabilities [17]. CVSS assesses entities in three areas: (i) base metrics to quantify intrinsic vulnerability, (ii) temporal metrics, to quantify the evolution of vulnerability, and (iii) environmental metrics, to quantify vulnerabilities that depend on a particular implementation. For this research, the proposed VI calculation will involves the 'Impact Matrix', which is a subset of the environmental metrics and rates the impact of cyber attacks on the confidentiality, integrity, and availability of data rated as 'none', 'low' or 'high' [CVSS version 3].

After developing the VQT, each entity in the system is evaluated to obtain CVS. VI can be computed based on the deterministic mathematical model shown below:

$$V = \begin{bmatrix} V_{VI} \text{ for Product Data} \\ V_{VI} \text{ for Process Data} \\ V_{VI} \text{ for Control Data} \end{bmatrix} = Q \times C \quad \begin{array}{l} V \text{ is Vulnerability Index Matrix,} \\ \text{where } Q \text{ is Vulnerability Quantification Matrix and} \\ C \text{ is CVSS Weighted Score Matrix} \end{array}$$

Vulnerability Quantification Matrix, Q is then derived from the VQT, as shown below:

Data	Confidentiality	Integrity	Availability
Product	$Q_{1,1} = 0.5$	$Q_{1,2} = 0.2$	$Q_{1,3} = 0.3$
Process	$Q_{2,1} = 0.4$	$Q_{2,2} = 0.4$	$Q_{2,3} = 0.2$
Control	$Q_{3,1} = 0.2$	$Q_{3,2} = 0.6$	$Q_{3,3} = 0.2$

$$Q = \begin{bmatrix} Q_{1,1} & Q_{1,2} & Q_{1,3} \\ Q_{2,1} & Q_{2,2} & Q_{2,3} \\ Q_{3,1} & Q_{3,2} & Q_{3,3} \end{bmatrix}$$

CVSS weighted score matrix, C, is determined for all hardware and software components within each system entity.

$$C = \begin{bmatrix} C_1 \\ C_2 \\ C_3 \end{bmatrix}, \quad \begin{array}{l} C_1 = (Count \ of \ high \ score \ for \ Confidentiality \times 1) + (Count \ of \ low \ score \ for \ Confidentiality \times 0.5) \\ C_2 = (Count \ of \ high \ score \ for \ Integrity \times 1) + (Count \ of \ low \ score \ for \ Integrity \times 0.5) \\ C_2 = (Count \ of \ high \ score \ for \ Availability \times 1) + (Count \ of \ low \ score \ for \ Availability \times 0.5) \end{array}$$

System Vulnerability Map (SVM)

The SVM provides a graphical map of the cyber security needs of manufacturing enterprises by identifying high value data flows through its entities and the vulnerabilities of those entities thereby enabling targeted and optimized cyber security. The SVM is a graph comprising nodes representing manufacturing entities and edges representing data flow between the entities.

In the SVM, the DI and VI are denoted for each node. The higher the DI of an entity, bigger is the node diameter. Each node is further circumbanded by sectors, denoting the proportion of product, process and control data within the entity. The bands are colour coded to indicate the VI of the entity ranging from green (low vulnerability) to red (high vulnerability). Figure 1 illustrates a sample SVM that represents a manufacturing system comprising three entities where entity 2 has the highest DI whereas entity 3 has the highest vulnerability to cyber attack.

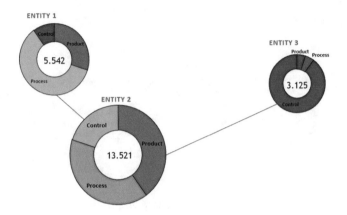

Fig. 1. Sample SVM

4   Results and Discussion

The steps to compute the DI and VI and to generate the resulting SVM for a manufacturing cell belonging to an academic manufacturing lab are given below. The cell comprises a Computer Aided Design and Manufacturing (CAD/CAM) workstation, a Distributed Numerical Control (DNC) station, a Manufacturing Execution System (MES) server, an Electro-Discharge-Machine (EDM) and a 5-axis CNC machine.

Step 1

The product, process and control data in each of the above entities is studied and the DI for each entity is computed. Data from commercial projects that the lab has undertaken for one year is used. Due to the confidential nature of the data, such as pricing, revenue generated, value of IP, etc., only the final DI components, each constituting explicit, implicit and deciphered components, are shown in Table 2 below.

Based on the above overall DIs, the nodes are drawn with varying diameters to illustrate the size proportionality to the DI (Fig. 2).

Table 2. DIs of product, process, control data and overall DI of the entities

Entity	DI (product data)	DI (process data)	DI (control data)	Overall DI
CAD/CAM workstation	3.949	2.672	2.175	8.796
DNC workstation	8.119	2.444	2.800	13.363
MES server	11.688	3.392	2.175	17.255
EDM machine	0	10.104	3.840	13.944
5-Axis CNC machine	0	8.277	3.840	12.117

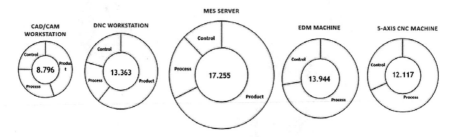

Fig. 2. Nodes sized to denote their overall DIs

Step 2

In this step, the VI is computed for each entity based on the data type bands per entity. The constituents of the VI and the values for these constituents are computed for all the entities but they are shown only for the 5-axis CNC milling machine in Table 3. The nodes are re-drawn in Fig. 3 to denote the VIs of product, process and control data as a range of colours from Green (low vulnerability) to Red (high vulnerability) using the smart charting function of Microsoft PowerPoint.

Table 3. Constituents of the VI and overall VIs for the 5-axis CNC milling machine

Entity		Major Components	CVSS Reference	CVSS Confidentiality	CVSS Integrity	CVSS Availability
5-axis CNC Milling Machine	Hardware	Simplified PC (with Ethernet Card)	http://www.cvedetails.com/cve/CVE-2004-2048	High	High	High
Model: Mikron 500U		Siemens PLC	https://www.cvedetails.com/cve/CVE-2016-2201	None	Low	None
Make: Agie Charmilles	Software	Windows XP	http://www.cvedetails.com/cve/CVE-2017-0176	High	High	High
			No. of "Complete"/High (Score=1)	2	2	2
			No. of "Partial"/Low (Score =0.5)	0	1	0
			No. of "None"/Zero (Score=0)	1	0	1
				Total CVSS Confidentiality	Total CVSS Integrity	Total CVSS Availability
			Total CVSS Score	2	2.5	2
			VI for Product Data	2.1		
			VI for Process Data	2.2		
			VI for Control Data	2.3		

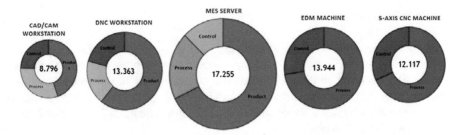

Fig. 3. Nodes redrawn to denote VIs of the data type bands using colour code

Step 3

In this step, the nodes are connected by edges to denote the presence of data flow between them. The complete graph, which is now the final SVM is shown in Fig. 4 below.

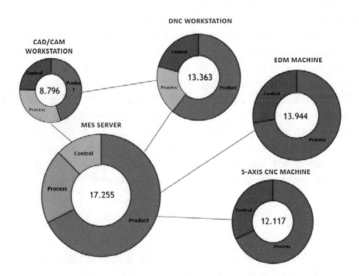

Fig. 4. Final SVM of the manufacturing cell

From the above SVM, it is clear that while MES server contains the highest value of data, it is also the least vulnerable to cyber attacks due to the enterprise cyber security solutions in place. The SVM also reveals that the control data residing in the other four entities is highly vulnerable to cyber attacks and therefore need the highest level of protection. Therefore, the SVM can be used by the IT administrators to select the type of cyber solutions and the extent of protection required to reduce the vulnerabilities of these entities and thereby the entire manufacturing cell.

5 Conclusion

Standard enterprise cyber security solutions, such as firewalls, are not adequate for manufacturing enterprises due to the heterogeneity of entities and the myriad of interfaces between the entities both internally and externally to the enterprises. Furthermore, there are no standard metrics available for manufacturing enterprises to gauge their vulnerabilities to cyber attacks and to deploy targeted and customised cyber security for optimum protection.

This work makes an attempt to develop metrics such as DI and VI to serve as tools for manufacturing enterprises to quantify the maximum damage caused by cyber attacks and their vulnerability to cyber attacks respectively. These metrics inherently take the heterogeneity of the manufacturing entities into account as shown in the paper above. Graphically representing the DI and VI into an SVM that allows IT administrators to prioritize resources to provide targeted and optimized cyber security for the enterprise. These tools also provide lifetime management of cyber security enabled by regular monitoring of the manufacturing system to continually refine and validate its DIs and VIs, while the cyber attack landscape and the enterprise itself evolve over time.

The work presented in this paper is in no manner complete and requires deepening of the study to ensure that all parameters and factors are taken into account to compute the damage and vulnerabilities of manufacturing entities to cyber attacks. Also, while expanding the use case to include all the manufacturing entities in the lab, a validation study that uses the manufacturing shopfloor of a commercial manufacturing enterprise is required to establish credibility of this work. This work suffers from the deficiency of requiring all the data to be deterministic in nature for computing DI and VI of the entities but this may not be possible in all industrial cases. A machine learning technique to stochastically determine these metrics when data is not fully available or is available as a range will be required to be developed. This work provides a solid foundation for all this work to be carried out in the future.

References

1. McKinsey & Company: Manufacturing the Future: The Next Era of Global Growth and Innovation. McKinsey Global, New York (2012)
2. World Economic Forum (WEF): The Future of Manufacturing: Opportunities to Drive Economic Growth. WEF, Switzerland (2012)
3. Symantec: Internet Security Threat Report, vol. 20. Symantec Corporation, Mountain View (2015)
4. McAfee Labs: 2012 Threat Predictions. McAfee, Santa Clara (2011)
5. Cisco: Cisco Connected Factory – Security. Infographic Report. Cisco, San Francisco (2014)
6. Wangen, G.: Role of malware in reported cyber espionage: a review of impact & mechanism. Information 6(2), 183–211 (2015)
7. Wells, L.J., Camelio, J.A., Williams, C.B., White, J.: Cyber-physical security challenges in manufacturing systems. Manuf. Lett. 2(2), 74–77 (2014)
8. Yang, W., Qianchuan Z.: Cyber security issues of critical components for industrial control system. In: IEEE International Conference on Guidance, Navigation and Control (CGNCC), Yantai, China, 8–10 August 2014

9. Dacer, M.C., Kargl, F., König, H., Valdes, A.: Network attack detection and defense: securing industrial control systems for critical infrastructures (Dagstuhl Seminar 14292). Dagstuhl Rep. **4**(7), 62–79 (2014)

10. Knowles, W., Prince, D., Hutchison, D., Disso, J.F.P., Jones, K.: A survey of cyber security management in industrial control systems. Int. J. Crit. Infrastruct. Prot. **9**, 52–80 (2015)

11. He, H., Maple, C., Watson, T., Tiwari, A., Mehnen, J., Jin, Y., Gabrys, B.: The security challenges in the IoT enabled cyber-physical systems and opportunities for evolutionary computing & other computational intelligence. In: 2016 IEEE Congress on Evolutionary Computation (CEC), pp. 1015–1021. IEEE (2016)

12. Meshram, A., Haas, C.: Anomaly detection in industrial networks using machine learning: a roadmap. In: Machine Learning for Cyber Physical Systems, pp. 65–72. Springer, Berlin (2017)

13. Thames, L., Schaefer, D.: Cybersecurity for Industry 4.0 and advanced manufacturing environments with ensemble intelligence. In: Cybersecurity for Industry 4.0, pp. 243–265. Springer International Publishing (2017)

14. Tiwari, A., Vergidis, K., Lloyd, R., Cushen, J.: Automated inspection using database technology within the aerospace industry. Proc. Inst. Mech. Eng. Part B: J. Eng. Manuf. **222** (2), 175–183 (2008)

15. Ko, J., Lee, S., Shon, T.: Towards a novel quantification approach based on smart grid network vulnerability score. Int. J. Energy Res. **40**, 298–312 (2015)

16. Ko, J., Lim, H., Lee, S., Shon, T.: AVQS: attack route-based vulnerability quantification scheme for smart grid. Sci. World J. **2014**, 1–6 (2014)

17. Common Vulnerability Scoring System (CVSS) https://www.first.org/cvss/. Assessed 1 Oct 2017

Practical Security Education on Combination of OT and ICT Using Gamification Method at KOSEN

Keiichi Yonemura[1(✉)], Ryotaro Komura[2], Jun Sato[3], and Masato Matsuoka[4]

[1] Department of Information and Computer Engineering, National Institute of Technology, Kisarazu College, Kisarazu, Chiba, Japan
yoramune@gmail.com
[2] Department of Electronics and Information Engineering, National Institute of Technology, Ishikawa College, Tsubata, Ishikawa, Japan
komura@ishikawa-nct.ac.jp
[3] Department of Electrical and Electronic Engineering, National Institute of Technology, Tsuruoka College, Tsuruoka, Yamagata, Japan
jun@tsuruoka-nct.ac.jp
[4] Kaspersky Labs Japan, Tokyo, Japan
masato.matsuoka@kaspersky.com

Abstract. Industry needs the talents who have the knowledge of OT (Operational Technology) security and ICT security and the skill of OT security and ICT security to change the current situation. KOSEN (National Institute of Technology, Japan) has the potential which can change this situation. We attempted to educate OT security for KOSEN students using KIPS (Kaspersky Industrial Protection Simulation) for confirming effect of using gamification theory. And we also examined next issue from that result. We confirmed the security educational effectiveness using two versions of KIPS to get our student's awareness to OT security. We also confirmed that validity of tactics using KIPS. Furthermore, we find out that potential of our original contents which can learn ICT security knowledge to utilize in OT security. We have future works which we need to reconsider our contents so that it will influence the KIPS score in a proper manner. At the same time, we have to develop the new educational contents which have the property that KOSEN students can learn the basic of OT security as a preliminary learning contents and the property that has global learning with basic OT security and basic ICT security. And it is ideal that we can use it for a number of engineering educational things. We keep examining the method to apply our educational tactics and we will try to foster a talent who can be active in various fields of industry.

Keywords: OT security · ICT security · KIPS · Gamification · KOSEN

© Springer International Publishing AG, part of Springer Nature 2019
M. E. Auer and R. Langmann (Eds.): REV 2018, LNNS 47, pp. 344–353, 2019.
https://doi.org/10.1007/978-3-319-95678-7_39

1   Introduction

1.1   The Current State of Industry

Industry has been faced with cyber-attacks, so the security countermeasures are urgent issue. For example, "Stuxnet", which is one of the cyber weapons on industrial equipment, used to attack nuclear facilities in Iran in 2010, and its subspecies have gone on a rampage through the industrial equipment. In Dec. 2015, major power outage was caused by black hacker's penetration of the computer network which is used by the power companies in Ukraine.

However, industry has also factors which make these countermeasures difficult. First, industrial equipment structured with view to use for a long time, so their equipment was no regard is given to connect the internet. Therefore, that is vulnerable to cyber-attacks when it connects to the internet. Second, there is the availability as the most important thing on the industrial equipment, so their administrators cannot stop that operation although they knew what they have to apply appropriate patch to their equipment so as not to affect by the computer virus. Moreover, industry has many mission-critical infrastructures such as oil, gas and power plants, factories, and so on. The system might shut down by applying that patch. Administrators fear shut down the system from updating it. Third, some industrial equipment administrators think that our equipment has not connected to the internet, so they are not affected by the computer virus and also they have not been under threat of cyber-attacks. However in fact, their equipment often connected to the internet for remote maintenance, and there is a possibility that someone put USB flash memory into their computer which has a control of their equipment.

Industry needs the talents who have the knowledge of OT (Operational Technology) security and ICT security and the skill of OT security and ICT security to change the current situation which industry is facing [1].

1.2   The Current State of KOSEN

KOSEN, which is our national institute of technology, is a national academic institution which can serve the industry needs. We have over fifty thousand students across the entire organization throughout the country, so KOSEN competes with the large scale of general university. We have a special curriculum encompassing all domains (Mechanical & Material Engineering, Electrical & Electronic Engineering, Information Technology, Biological & Chemical Engineering, Civil Engineering, Architectural Engineering, Maritime Technology and Others) of engineering expertise on specialized education with early stage from fifteen to twenty years old, because this large scale is also contributing to this completeness. It is possible to foster students which have both basic knowledge of cyber security in addition to high expertness on this early stage education. We also operate a practical lecture that OT security human resource on non-ICT engineer requires, for example, using KIPS (described below). We provide the basis of an education for special students who have high class ICT skill. Therefore, industry in our country, companies and universities are appreciative of the importance of KOSEN, because it is possible that we can foster advanced specialists which have a variety of knowledge of ICT security and OT security and the skill of OT security and ICT security that industry requires. Companies and universities have promoted a close relationship between KOSEN and them so that our students enter them directly after graduation.

1.3 Issue of OT (Operational Technology) Security Education

People, who engage the job related to the ICT security, have to protect the data, and also prevent information leakage and tampering, to do so some situation makes them select temporarily shutting down the system. Therefore, the priority on ICT security will be in the order of "Confidentiality", "Integrity" and "Availability".

However, when people engage the job related to the OT security, there is not much value in the data itself. Most importantly, it does not stop the equipment running. Therefore, priority on OT security will be in the order of "Availability", "Integrity" and "Confidentiality".

So, we have to explain students that the Availability is very important on OT security in our educational class because the main field for our students is OT, after graduating.

1.4 Purpose of This Investigation

In the next stage, industry utilizes the IoT (Internet of Things) for the purpose of improvement of effective management and competitiveness effect. However, it is not easy for us to progress the next stage, so KOSEN must contribute to the achievement of economic and social development in industry. OT security skill has been practiced by ICT security education and expansion of the knowledge of security. In addition, on OT security education which availability is most important, we need actual equipment and practical education. However, it is not easy to prepare the actual equipment so it is important to practice using simulation game considering the gamification theory, that is, KIPS.

KOSEN has two types of students, one is the non-ICT engineer with basic ICT and OT security skill, the others is the ICT engineer with advanced ICT security skill and OT security skill. In this investigation, we attempt to educate OT security for KOSEN students using KIPS, confirming effect of using gamification theory and examining next issue if we cannot confirm any effect.

We make the students of non-ICT engineer course play the KIPS because they enter industry society after graduating, so we can think that they have high level motivation. And we attempt to make them do multiple plays the KIPS (Corporation version of KIPS which seems to comparatively related to ICT security and Water Plant version of KIPS which seems to comparatively related to OT security) because we will examine teaching method more effectively for them by considering the difference between Corporation version and Water plant version, that is, ICT security version and OT security version.

2   Gamification and KIPS

2.1   Gamification Theory

The definition of gamification has been used as "the use of game-play mechanics for non-game applications" [2] and it is used in many fields such as education, business and medicine. One of the important features of learning by using gamification or game-based learning is that students actively learn problems and take solutions by facing problems. Problem solving with gamification is also noted to be an important benefit of using games in education [3, 4]. A problem-solving mechanism built with a game-based strategy enables both knowledge acquisition and its application throughout the learning process.

2.2   KIPS the Cyber Security Practice [5, 6]

KIPS (Kaspersky Interactive Protection Simulation) is the cyber security practice by team game which designed to enhance analytical skill about problems on the cyber security and the risk about latest computer system in operation (see Fig. 1). Educational targets are executive managers including business managers, departmental managers and information security administrators.

Fig. 1. KIPS (Kaspersky Industrial Protection Simulation)

The purpose of KIPS is that to prevent the profit maximally and to preserve the trust during exposure to a series of unexpected cyber threats. The aiming is that to develop and run the cyber defense tactics by selecting best suited plan out of cyber security countermeasure prepared preliminarily.

Subsequent deployment and final profit and expense of company change, in response to countermeasures against security events pouring in waves. Each team analyze a situation considering priorities of engineering against damage by cyber-attack, business and security, making a tactical decision based on uncertain information, a limited budget and feasible measure. Each scenario is decided based on security events changed by the situation each team is in, so it can not only simulate a situation occurred cyber incidents actually, but also verify the decision making based on appropriate tactics and effectivity of selected measures in real time.

KIPS has the following characteristics based on gamification theory.

1. Completing in a short time (two hours) while concentrating and having fun in the form of a game
2. Building up cooperativeness crossover the organization by teamwork
3. Train upping the autonomy and analytic skill by re-experiencing realistic security events.

3   Approach and Procedure

First, the students of non-ICT engineer course took a class of the basic ICT security using the educational contents developed by ourselves. In this class, students learned about general basic ICT security but did not learn about specialized security techniques and OT security. And they took ICT basic knowledge test after taking the class.

Second, we plan to make the same students mentioned above play two scenarios (the corporation version, see Fig. 2, and the water plant version, see Fig. 3). Originally, KIPS is suitable for non-ICT engineers, because KIPS needs no deep expertise of security. Therefore, our engineering course students can measure security skills without great effort [1]. In order to examine whether ICT basic knowledge they had already learned influence KIPS score, in other words, KIPS can measure security skills, we will compare the score of ICT basic knowledge test with the score of KIPS. Also, in order to reduce the gap between the content students learned and the content of the game, we will use the corporation version of KIPS game which seems to be comparatively close to ICT security domain.

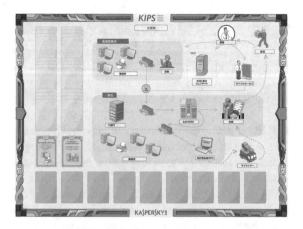

Fig. 2. Corporation version of KIPS seems to be comparatively close to ICT security domain

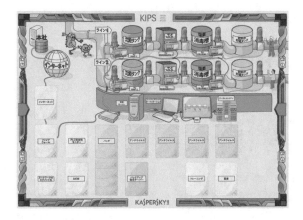

Fig. 3. Water plant version of KIPS seems to be comparatively close to OT security domain

Third, we will use the water plant version of KIPS in order to examine simply the correlation of ICT knowledge and OT knowledge that need water plant version of KIPS which seems to be comparatively close to OT security domain. The purpose of this measure is also that we examine whether ICT basic knowledge they had already learned influence KIPS score, which seems to be comparatively close to OT security domain. In other words, we compare the score of ICT basic knowledge with the score of KIPS to confirm whether KIPS, which seems to be comparatively close to OT security domain, can measure security skills.

Finally, we attempt to confirm whether we need to use multi scenarios for KIPS or not, to do so we compared the score of corporation version of KIPS with the score of water plant version of KIPS after our students play these two scenarios. If the scores of two scenarios have the strong correlation, we don't need to use two scenarios because the scenarios just measure the same skill and knowledge, however if the scores of two scenarios don't have the correlation or have the weak correlation, we can use the two scenarios and measure the different skill and knowledge, namely, for example, we can measure the ICT security skill and knowledge by using the corporation version of KIPS and we can measure the OT security skill and knowledge by using the water plant version of KIPS. To know this result and fact is that it is important for us so that we can reach to the goal we want more effectively.

4 Actual Outcomes

4.1 Outcomes of Corporation Version of KIPS

Figure 4 are the result of evaluation of game result of corporation version on KIPS. Left line chart is the results which are evaluation of KIPS score (blue line) and the score of ICT basic knowledge test (red line). Both scores had standardized. And right figure is scatter diagram of them. These results show ICT basic knowledge has not influenced the KIPS scores. Because correlation coefficient didn't show high level and had no significant.

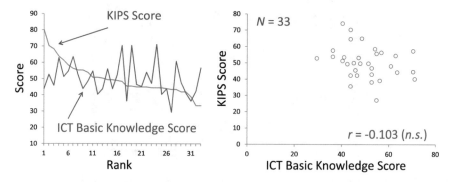

Fig. 4. Evaluation of game result (corporation version)

These results show a certain level of knowledge requires to influence KIPS score because this corporation version seems to be comparatively hard to play. Originally, KIPS is made as not to need a deep knowledge for player, so we can think players require a lot of knowledge if they want a high score. So, we have to consider to use another factor to compare the KIPS score with ICT basic knowledge score.

The other side of the coin is that this result shows us the possibility which we can consider on how our student's ICT basic knowledge can influence this KIPS score. Therefore, we have to examine making new educational contents to do that as a new measure of the educational effect.

4.2 Outcomes of Water Plant Version of KIPS

Figure 5 are the results of water plant version on KIPS, and that is made the same way as previous results. These results also show ICT basic knowledge has not influenced the KIPS score. But the coefficient showed the weak effect. And it's negative effect.

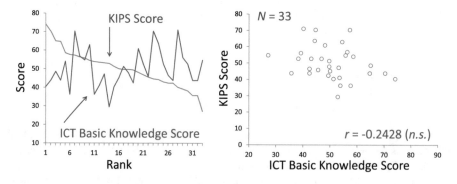

Fig. 5. Evaluation of game result (water plant version)

These results show this water plant version seem to be comparatively easy to play. So, ICT knowledge they have might be easy to influence the KIPS score comparing with the score of corporation version. However, as mentioned above, main focus of ICT security is different from the OT security. So, ICT knowledge might influence as negative effect when the players have ICT knowledge, and they don't have OT security.

These results show that playing this version make the players notice which the OT security is important and is different from ICT security if they don't know them enough. This consideration let us reconsider that OT security is important and what we have to analyze this result using a variety of factor.

4.3   Outcomes of Analysis of Two Scenario's Score

Figure 6 shows the comparison between the results of corporation version with water plant version. We cannot confirm the different between two versions of difficulty, but this result shows a certain level of week correlation. It is not only no-correlation but also not too high. So, we can use these two versions to measure ICT security skill and OT security skill. Therefore, we can use these two versions to measure the validation of our educational contents.

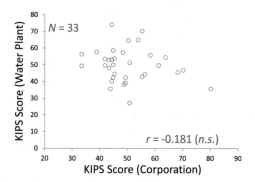

Fig. 6.   Scatter diagram of two scenario's score

5   Recommendations

Industry is facing cyber-attacks. So, KOSEN students are focused as the talents who can break through these situations because KOSEN can provide practical education of skill and knowledge of OT security and ICT security for KOSEN students [1].

KOSEN provides an early professional education as our main feature, it is developed that base to learn basic OT security skill, basic ICT security and advanced ICT security skill, and industry and university place confidence for this early professional education.

Acquisition of practical OT security skill is built upon which they have already obtained basic OT security skill and basic ICT security skill. Most of KOSEN students are non-ICT engineers and they will graduate by learning the basic OT security skill and the basic ICT security skill. In this investigation, we focused on the non-ICT engineers who involve in an industry after graduating and examine an educational effect of practical OT security skill. This is the validation of an educational effect utilized the gamification theory with KIPS, and we also examine the issues to next step. To examine the difference between the effects with two types of the KIPS, the corporation version is close to ICT security, the water plant version is close to OT security, lead to the examination of more effective the way to teach.

We examined weather ICT basic knowledge influence the KIPS score, conversely, weather KIPS can measure ICT security skill using the corporation version of KIPS which seems to be comparatively close to ICT security. As a result, ICT basic knowledge has not influenced the KIPS score. It takes reasonable ICT knowledge to influence the score because it seems that the corporation version of KIPS is comparatively hard to play. This result shows really useful issue that developing of educational content which can learn reasonable ICT knowledge will be able to measure interacted with a KIPS score, and it makes more effective educational effectiveness.

Using the water plant version of KIPS which seems to be comparatively close to OT security, we also examined weather ICT basic knowledge influence the KIPS score. Here, results also showed ICT basic knowledge has not influenced the KIPS score, but the coefficient showed the week negative effect. From the difficulty, ICT basic knowledge might influence to score. However, it seems that ICT basic knowledge worked to OT handling negatively, and as a result, it makes the players notice which the OT security is important and is different from ICT security. This notice is really useful for them to be a talent which can play a key role in industry in the future.

As a result of comparing the score of corporation version with the score of water plant version, we obtained the week correlation. This result is really interesting that it is not no-correlation and is not strong correlation. That is, ideally, there is a possibility that it derived from which it has correlation on the security front, and it has no correlation on the OT skill and knowledge and ICT skill and knowledge. Therefore, it suggests that using of two types of KIPS is useful as the indicator to assess the validity of educational contents which we will develop later.

These result shows that we have to make the new educational contents which we can measure our student's skill using KIPS. So, we can examine an appropriate method that is suited for the KOSEN students by analyzing the process while they play many times. And we get the educational knowledge which educational stage influence the KIPS score effectively, and what action and selection influence the next action and selection on playing KIPS. These knowledge allow us examine the best method to apply the KIPS that based on gamification theory to a variety of students other than KOSEN students.

We confirmed the security educational effectiveness using two versions of KIPS to get our student's awareness to OT security. And we also confirmed that validity of tactics using KIPS. Furthermore, we confirmed that potential of our original contents that can learn ICT security knowledge to utilize in OT security. And we have future works which we need to reconsider our contents so that it will influence the KIPS score in a proper manner. At the same time, we have to develop the new educational contents which have the property that KOSEN students can learn the basic of OT security as a preliminary learning contents and the property that has global learning with basic OT security and basic ICT security. And it is ideal that we can use it for a number of engineering educational things.

On the next stage, establishing an IoT secure infrastructure has been important issue in which not only industry but also a variety of the field. Next mission for KOSEN has been that make a contribution to that establishing surrounding industry which needs practical OT security on a variety of side such as competitiveness, operation and development. While ICT security has been important as an infrastructure of practical OT security, the education of ICT security skill has been comparatively easy to progress because establishing a body of knowledge supports. However, it takes real industrial equipment and practical education to practice a practical OT security, so developing the education utilized the gamification theory with KIPS has been great value. In KOSEN, we have to keep a practical OT security education using KIPS. And we also examine the education with KICS (Kaspersky Industry CyberSecurity) which is more practical, and we will try to foster a talent who can be active in various fields of industry.

References

1. Sato, J., Tansho, N., Kiyota, K., Kishimoto, S.: Gamification for education of cybersecurity in operational technology. In: The 11th International Symposium on Advances in Technology Education (ISATE 2017), 19–22 September 2017, pp. 699–703. Ngee Ann Polytechnic, Singapore (2017)
2. Deterding, S., Dixon, D., Khaled, L., Nacke, L.E.: From game design elements to gamefulness: defining "gamification". In: Proceedings of the 15th International Academic MindTrek Conference: Envisioning Future Media Environments (MindTred 2011), Tampere, Finland, 28–30 September 2011, pp. 9–15 (2011)
3. Sun, C.T., Dai, Y., Chan, H.L.: How digital scaffolds in games direct problem-solving behaviors. Comput. Educ. **57**(3), 2118–2125 (2011)
4. van der Spek, E.D., van Oostendor, H., Meyer, J.J.C.: Introducing surprising events can stimulate deep learning in a serious game. Br. J. Edu. Technol. **44**(1), 156–169 (2013)
5. Kaspersky Lab: Kaspersky interactive protection simulation—an effective way of building cybersecurity awareness among top managers and decision makers, pp. 1–3. Kaspersky Lab (2017)
6. Kaspersky Lab: Kaspersky security awareness—gamified training programs for all organizational levels, p. 2. Kaspersky Lab (2017)

SEPT Learning Factory Framework

Dan Centea$^{(\boxtimes)}$, Mo Elbestawi, Ishwar Singh, and Tom Wanyama

McMaster University, Hamilton, ON, Canada
{centeadn,elbestaw,isingh,wanyama}@mcmaster.ca

Abstract. The term learning factory covers a variety of learning environments. Each implementation of a learning factory looks differently and is used for a different purpose. This paper presents a framework for the development of a learning factory that uses Industry 4.0 technologies and has a strong experiential learning approach. This learning factory is used for training students and employees, and for conducting applied research. The facility provides modern design, prototyping, and manufacturing processes that incorporate Internet of Things, Industrial Internet of Things, and Industry 4.0 technologies. The learning factory also provides opportunities for partnerships where students design and implement projects that foster industry-student-faculty collaboration. This collaboration is expected to culminate in an environment that creates a world-class engineer impacting the adoption of new technologies at a faster pace.

Keywords: Learning factory · Industry 4.0 · IoT · Cyber-physical systems

1 Introduction

Advanced concepts in manufacturing such as Internet of Things (IoT), cloud-based manufacturing and smart manufacturing address a digitally-enhanced manufacturing environment called Industry 4.0. Although many technologies included in the Industry 4.0 concept are available for use in industry, the employees are generally not prepared for its successful implementation.

The concept of the learning factory has been developed recently to improve learning and training in manufacturing. The aim is to modernize the learning process and bring it closer to industrial practice. Several objectives have been reported in the literature, including the development of a modern, realistic manufacturing environment for training [1], the learning of interdisciplinary skills, abilities of synthesis, and adaptation to various situations [2].

Typically, these learning factories include integrated physical components such as machining and assembly, as well as digital environments [2]. In particular, they allow learning and training of the key aspects of Industry 4.0 such as cyber-physical systems, collaborative robotics, and IoT. Various learning factories have been reported in the literature [1, 3–5]. Abele et al. [1] have classified the types of learning factories to include: (1) learning factories for production process improvement; (2) teaching and learning factory concepts; (3) learning factories for re-configurability, production and layout planning; and (4) learning factories for Industry 4.0.

© Springer International Publishing AG, part of Springer Nature 2019
M. E. Auer and R. Langmann (Eds.): REV 2018, LNNS 47, pp. 354–362, 2019.
https://doi.org/10.1007/978-3-319-95678-7_40

A learning factory is an entity in continuous development. The authors in [6] noted that the complexity and effort for developing, implementing and managing production systems that execute new technological trends will increase continuously. They observed that many companies in the mechanical engineering and plant engineering field "view Industry 4.0 with caution and skepticism. Therefore, it is crucial that the benefits of these developments are demonstrated and evaluated. This situation causes an urgent demand for research and learning facilities to offer new workshops, trainings and other events to target the specific needs and production environments."

In 1994, the U.S. National Science Foundation awarded a grant to develop a "learning factory". This term, used for the first time, was referred to as an interdisciplinary hands-on senior engineering design projects with strong links and interactions with industry. A partnership of Universities in the USA collaborated in the development of practice-based curriculum that is able to provide an improved educational experience [7].

Fulfilling individual customer demands with affordable products requires flexible and adaptable production processes [3]. These forms of production control and flexible manufacturing increase the complexity of production systems. Current automation solutions cannot address these challenges. To meet these challenges and prepare future engineers for related issues, several universities have developed learning factories that deal with Cyber-Physical Production Systems. The authors of [3] list some of the learning factories developed in Germany and briefly describe the most advanced ones.

The published literature shows that the term learning factory covers a variety of learning environments. Each implementation of a learning factory looks differently and is used for a different purpose. Several of the learning factories developed in the last 20 years are used for education, for training industrial employees, and for research. Some of the first learning factories, mostly developed in the USA, have a very practical approach but do not include the digital manufacturing components that are absolutely required for implementing Industry 4.0 concepts. Many of the newer learning factories, mostly developed in Europe, have a strong focus on Industry 4.0 and demonstrate different implementation aspects. Many of these learning factories lack hands-on learning where the person being educated is more of an observer rather than a participant in the manufacturing process. Only a few learning factories implement several Industry 4.0 components and have a strong hands-on aspect. The purpose of this paper is to present a framework for the development of a learning factory that uses several technologies included in the Industry 4.0 concept and has a strong experiential learning approach.

This paper presents a framework for the development of a hands-on learning factory for training students and employees in addition to or as well as for conducting applied research. The paper illustrates the concepts and practices for obtaining the intended outcomes that define the frameworks for designing, implementing and planning the use of the learning factory. This facility is currently in development at McMaster University, School of Engineering Practice and Technology (SEPT). The design framework presented in Sect. 2. Section 3 describes the implementation framework. The planning framework is presented in Sect. 4. Section 5 describe the outcomes of the framework and lists the potential benefits and the impact of the SEPT Learning Factory in teaching, research and training.

2 Design Framework

The current tendency of engineering education towards experiential learning can be accomplished through various means that include hands-on labs, computer simulations, co-op education, industry or community projects, engineering clubs and competitions, and so on. The concept of a learning factory is a disruption of the traditional teaching approaches and incorporates many of these experiential learning applications. The SEPT Learning Factory presented in this paper is a hands-on world-class facility that provides modern design, prototyping, and manufacturing processes that incorporate Industry 4.0 technologies, IoT, and Industrial Internet of Things (IIoT) – the building blocks of Smart Systems.

With Industry 4.0 concepts in mind, the design framework of the SEPT Learning Factory addresses eight key areas: (1) standardization and reference architecture; (2) managing complex systems; (3) a comprehensive broadband infrastructure for industry; (4) safety and security; (5) work organization and design; (6) training and continuing professional development; (7) regulatory framework; and (8) resource efficiency. The achievement of Industry 4.0 goals through these eight key areas presents challenges and opportunities for educational institutions. These goals can be further distilled down to key enabling technologies considered as necessary for the Cyber-Physical Systems Learning Centre [8] associated with the SEPT Learning Factory. These key enabling technologies include (a) extensive use on electronic components such as microprocessors, microcontrollers, PLCs, sensors, and actuators; (b) development of a networking infrastructure; (c) use of software components, cloud computing, and web-based services; (d) use of artificial intelligence and data analytics; (e) use of machines that adopt to the needs of human beings using multimodal user interfaces (e.g. collaborative robots); and (f) disruptive technologies such as 3D printing. The SEPT Learning Factory has been designed to demonstrate applications of these requirements and implementations of modern technologies for the purpose of Industry 4.0 education and applied research.

3 Implementation Framework

The fundamental approach implemented in the SEPT Learning Factory is the digitization of a production line. While producing a physical object, a series of sensors collect information from the production line modules, transmit the information to cloud-based servers using various types of communication networks, and use controllers and actuators to automatically control other modules of the production line. The systems perform computations (data analytics) and provide information to the user for monitoring and control. The students and industry employees that will be trained will use various components of the learning factory based on the purpose of the educational component, training or applied research.

The key elements described in the Design Framework have been implemented through a series of machine tools and specialized stations with focus on Industry 4.0, IoT and IIoT. Their main components are expected to address the educational, research, and training components of the SEPT Cyber Physical Systems Learning Centre. The SEPT Learning Factory includes an IoT learning station, cyber physical system stations with IIoT implementations, collaborative and mobile-intelligent robots, several additive manufacturing stations, and modern machine tools equipped with sensors and actuators that address the Industry 4.0 manufacturing digitization aspects. A series of sensors and actuators are active process participants in a manufacturing automation network. These stations, machine tools and sensors, shown in Fig. 1, are described below.

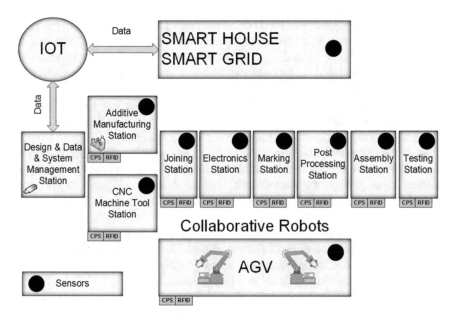

Fig. 1. Major components of the SEPT learning factory

The IoT Learning Station (not shown in Fig. 1) is a display platform designed in house that demonstrates the applications of IoT in various fields such as energy (smart home, smart grid, renewable energy), automation, transportation (automotive, aerospace), eHealth, and education.

Several Cyber-Physical Systems (CPS) Stations with IIoT Implementation have been designed and built in house for several manufacturing processes that include manufacturing mechatronics systems with mechanical and electronic components, marking, post-processing, assembly, and testing. Each of these stations is equipped with a Programmable Logic Controller (PLC) with standard inputs and outputs (I/O), communication modules (an IO-Link module for smart sensors, and industrial Ethernet communication interface, a managed switch for secure integration with the enterprise network), electric drives (an AC drive for motor control, a servo drive and a servo motor for motion control application), and smart sensors (photo switch, laser distance and object detection sensor).

Each CPS Station is designed to be equipped with a smart camera for inspection, correct orientation and placement, and a 2D bar code reader with industrial Ethernet capabilities that ensures integration with the PLC. Two of the current CPS stations are equipped with these devices.

It is frequently stated that Radio Frequency Identification (RFID) systems can make a major contribution to the realization of Industry 4.0. The new RFID technology can allow manufacturers to pursue this move to Industry 4.0 far more easily than they do now. In the SEPT Learning Factory each CPS Station is equipped with an RFID reader/writer connected to a software system that allows depositing or retrieving data directly to or from a database. Furthermore, each manufacturing item, component or assembly, is expected to be equipped with RFID chips holding all the data relevant to production of this item. The item itself is carried through different CPS Stations and processes. The plan for the near future is to deploy RFID scanners on machines or branches of a conveyor system that will read the data and issue corresponding commands to robots and logistics systems.

The SEPT Learning Factory also includes a Kanban Station that can be used in the classical Kanban mode or with RFID implementation to demonstrate the automated management of the supply chain of complex assembly systems.

All the stations and machine tools included in the SEPT Learning Factory are designed to include various sensors that can independently report errors and statuses to the control system. Some stations also include actuators that receive and process signals. The communication systems used by these sensors and actuators is IO-Link – an industrial network communication technology that offers new options for communication between the system control and field level. Each IO-Link device includes a sensor, an actuator, or a combination of both. The use of IO-Link devices is expected to result in cost and process optimization throughout the entire supply chain as these smart sensors are supplier of information for Industry 4.0 applications.

The major machine tools included in the SEPT learning Factory are a metal additive manufacturing station and a 5-axis CNC milling machine. They are used in the manufacturing of the mechanical components of the systems produced in the learning factory and for applied research. These machine tools are equipped with sensors that provide information about the manufacturing processes to the highest level of the software components of the learning factory – the manufacturing execution system (MES).

The complex machine tools described above are complemented by a series of plastic 3D printers that are expected to be used extensively in the production of some of the components of the mechatronic assemblies to be produced in the learning factory.

The SEPT Learning Factory also includes four collaborative and mobile-intelligent robots. One of the robots, mounted on a programmable Mobile-Intelligent Robot system that includes laser scanners and ultrasonic sensors, is designed for pick and place services in an autonomous fashion across the entire learning factory space. A similar robot is planned to be used on an Automated Guided Vehicle (AGV) for automatic loading and retrieval of parts form the 5-axis CNC machine. A third robot is mounted on a platform for collaborative assembly of parts. The fourth robot can be trained to assemble electronics components.

The capabilities on the CPS Stations, machine tools and robots allow for vertical and horizontal communications in the SEPT Learning Factory, one of the key requirements for Industry 4.0 implementation.

4   Planning Framework

The SEPT Learning Factory provides a series of modern capabilities that makes it attractive to undergraduate students for learning and development of capstone projects, for graduate students for research purposes, and for the manufacturing industry for training employees. The goals of the each of these groups are different. Furthermore, the machine tools, sensors and CPS station will vary from group to group. The multitude of undergraduate programs that will use the facilities alone will create a myriad of various needs between specializations. To address this issue, a planning framework has been developed.

The first approach to mitigate the issue described above is the existence of the SEPT Cyber-Physical Systems Learning Centre. This Centre will have the learning factory at its core but will include a series of specialized labs for various undergraduate programs, as described in [8]. The entire product development process, from an idea to a manufactured product, will take place in various specialized facilities of the Learning Centre. This approach is expected to balance the use of the learning factory between its major roles: learning and development of capstone projects; research; and industrial training.

Some Industry 4.0 technologies modules are already delivered in SEPT in various labs associated with the SEPT CPS Learning Centre. Examples of courses, teaching modules and capstone design projects include Industrial Networks and Smart Grid Networks, delivered in the Advanced Power Systems Lab as described in [9], as well as the Industrial Networks & Controllers and Advanced System Components & Integration courses [10, 11] delivered in the Automation Lab.

Due to space limitations, it is not possible to have in the production line all the CPS stations built for the SEPT Learning Factory. Furthermore, some projects require a certain set of CPS stations, while some other applications need other sets. To address this issue, all CPS stations have been built on movable carts. The users can select the configuration of stations that match their needs by bringing in the production line relevant for their manufacturing process.

The CPS stations that have been built use the Industry 4.0 aspects that are available at acceptable prices at the time of their construction. Considering the evolution of the Industry 4.0 technologies and the availability of the components and sensors, it is expected to build new CPS stations periodically. This is not only expected to create complications related to space, but will also generate challenges for planning purposes.

When the learning factory is fully functional, it is expected to identify a series of other constraints that will require further planning purposes. They will need to be addressed using a ranking system that has not yet been defined.

5   Anticipated Outcomes

The heart of the advanced manufacturing paradigm shift is the digitalization of the manufacturing process. A key aspect of this digitalization is the application of modern information and communication technologies involving mobile devices, additive manufacturing, CPS, IoT, IIoT, cloud computing and cognitive computing – the foundations of the Industry 4.0 technologies. Traditional engineering skills are no longer sufficient to meet the challenges posed by the Industry 4.0 revolution, therefore it is essential to train students and industry employees providing them with the specialized knowledge and hands-on skills required to prepare them for careers in industry, government, and academia. The SEPT Learning Factory will be used to develop a series of trainings for students and industry employees. The originality and novelty of the training programs developed in this learning factory lay primarily in the implementation of Industry 4.0 foundational elements in advanced manufacturing training, education, and technology adoption in the industry.

SEPT is an educational unit in the Faculty of Engineering at McMaster University that delivers its programs with a strong emphasis on student-centered learning. The School includes several undergraduate programs focused on engineering technologies and a number of specialized graduate programs with a concentration on engineering practice. The SEPT Learning Factory whose frameworks are described in this paper, is the major component of the SEPT Cyber-Physical Systems Learning Centre that is expected to compliment the students' and industry employees' qualifications and abilities by providing new technical skills that emphasize the inherent multidisciplinary nature of smart systems and advanced manufacturing.

5.1   Impact on the Undergraduate Curriculum

The SEPT Learning Factory is a major component of the SEPT Learning Centre that enriches the undergraduate program by replacing some existing laboratory experiments with developments of smart systems. Each undergraduate program is expected to develop smart applications related to their specialization in the learning factory (e.g. smart vehicle for the Automotive program, smart home and smart transportation for the Civil program, smart health for the Biotechnology program, smart manufacturing for the Manufacturing program, and so on).

Undergraduate students, in their desired programs of study will learn how to define a problem, build a prototype using modern Industry 4.0 concepts, write a business proposal, and present their solution during their capstone design course. These skills will be extremely valuable after graduation.

5.2   Impact on Research

One of the important aspects of the SEPT Learning Factory is the research on modern manufacturing approaches such as additive manufacturing. McMaster University and Mohawk College located in Hamilton, ON, Canada have partnered and have an agreement to combine efforts and use similar additive manufacturing metal 3D printers for undergraduate and graduate teaching, and for research.

Research activities involving graduate students and postdoctoral fellows will focus on: (i) extending the use of additive manufacturing technologies for the production of highly functional prototypes and custom final parts; (ii) addressing the issue of man-machine interface for advanced robotics; (iii) developing advanced sensors, real time data analytics and advanced controls for the manufacturing environment; and, (iv) extending and applying CPS and IoT to digital manufacturing processes supported by the SEPT Learning Factory.

5.3 Impact on Industry

The SEPT Learning Factory is also designed to train industry employees. The facilities are expected to provide trainees with tools for prototyping, manufacturing, and integration between information and operational technologies. The industry employees will acquire specific skills that will allow them to implement and improve the adoption of emerging technologies such as advanced materials and additive manufacturing in the context of the IoT/IIoT and Industry 4.0 technologies as applied to advanced manufacturing.

The learning factory is also meant to provide a university-industry/community partnership where students design and implement projects that allows collaboration between industry, community, faculty and students, thus benefiting themselves, their environment and the sponsors.

The learning factory is expected to attract a series of industrial partners that will play a significant role in guiding the training programs by identifying research opportunities, defining student projects, providing internship opportunities for the students, and providing input to the proposed training plans. This close collaboration will ensure the relevance of the research projects to industry, which is both essential in the manufacturing field and of major benefit to students. This collaboration will also provide an avenue for career opportunities for the students. Students will have the opportunity to present their results/findings and have regular interactions with industrial partners through various information sharing opportunities such as workshops, seminars, and conferences.

6 Summary

The paper presents an innovative framework for developing a learning factory at the W Booth School of Engineering Practice and Technology at McMaster University, Canada. The focus of the learning factory is the provide students, industry employees and researchers an experiential learning laboratory that allows its users to understand, learn and apply modern manufacturing approaches such as Industry 4.0, IoT, IIoT, and additive manufacturing.

The SEPT Learning Factory provides a university–industry/community partnership where students design and implement projects that cultivates industry/community-student-faculty collaborations, thus benefiting themselves, their environment and the sponsors. This unique collaboration in the Learning Factory aims to culminate in an environment that creates world-class engineers impacting the adoption of new technologies at a faster pace.

When fully functional, the SEPT Learning Factory will be a hands-on training facility that will allow small, medium and large companies to be competitive in the current market by using modern manufacturing approaches that implement various levels of digitization and cyber-physical systems. The facility will allow the development of applications that include elements of smart production, smart vehicles, smart energy, smart connectivity, smart home and city, and smart health.

References

1. Abele, E., Metternich, J., Tisch, M., Chryssolouris, G., Sihn, W., ElMaraghy, H., Hummel, V., Ranz, F.: Learning factories for research, education, and training. Procedia CIRP **32**, 1–6 (2015)
2. Wagner, U., AlGeddawy, T., ElMaraghy, H., Mueller, E.: The state-of-the-art and prospects of learning factories. Procedia CIRP **3**, 13–18 (2012)
3. Gräßler, I., Pöhler, A., Pottebaum, J.: Creation of a learning factory for cyber physical production systems. Procedia CIRP **54**, 107–112 (2016)
4. Kemeny, Z., Nacsa, J., Erdos, G., Glawar, R., Sihn, W., Monostory, L., Ilie-Zudor, E.: Complementary research and education opportunities—a comparison of learning factory facilities and methodologies at TU Wien and MTA SZTAKI. Procedia CIRP **54**, 47–52 (2016)
5. Kemeny, Z., Beregi, R.J., Erdos, G., Nacsa, J.: The MTA SZTAKI smart factory; platform for research and project-oriented skill development in higher education. Procedia CIRP **54**, 53–58 (2016)
6. Wank, A., Adolph, S., Anokhin, O., Arndt, A., Anderl, R., Metternich, J.: Using a learning factory approach to transfer Industrie 4.0 approaches to small- and medium-sized enterprises. Procedia CIRP **54**, 89–94 (2016)
7. Hadlock, H., Wells, S., Hall, J., Clifford, L., Winowich, N., Burns, L.: From practice to entrepreneurship: rethinking the learning factory approach. In: Proceedings of the 2008 IAJC-IJME International Conference, Paper 081, ENT P 401 (2008)
8. Centea, D., Singh, I., Elbestawi, M.: Framework for the development of a cyber-physical systems learning centre. In: Auer, M., Zutin, D. (eds.) Online Engineering & Internet of Things. Lecture Notes in Networks and Systems, vol. 22, pp. 919–930. Springer, Cham (2017)
9. Singh, I., Al-Mutawaly, N., Wanyama, T.: Teaching network technologies that support Industry 4.0. In: Proceedings of the Canadian Engineering Education Association, CEEA 2015. ISSN 2371-5243, https://doi.org/10.24908/pceea.v0i0.5712
10. Wanyama, T., Singh, I., Centea, D.: A practical approach to teaching Industry 4.0 technologies. In: Auer, M., Zutin, D. (eds.) Online Engineering and Internet of Things. Lecture Notes in Networks and Systems, vol. 22, pp. 794–808. Springer, Cham (2017)
11. Wanyama, T., Centea, D., Singh, I.: Teaching Industry 4.0 technologies and business modeling: case of energy efficiency and management. In: Proceedings of the Canadian Engineering Education Association Conference (CEEA 2016), Halifax, Canada, 19–22 June 2016 (2016)

Remote Structural Health Monitoring for Bridges

Mohammed Misbah Uddin, Nithin Devang, Abul K. M. Azad$^{(\boxtimes)}$, and Veysel Demir

College of Engineering and Engineering Technology, Northern Illinois University,
Dekalb, IL, USA
aazad@niu.edu

Abstract. Effective and efficient structural monitoring of bridges is an important factor in ensuring a safe transportation system. In this effort this paper presents the design development and implementation of a cloud based laboratory scale bridge monitoring system that can assist the management for real-time monitoring. Accelerometers are used to collect the vibration data and then passed to the cloud after some initial processing through an embedded system. The cloud server as well as a local server are used for further data analysis and web presentation for remoter user.

1   Introduction

US has about 600,000 bridges, almost 4 out of 10 are 50 years or older [1]. As per statistics 1 out of 4 of the bridges has received ratings of either structurally deficient or functionally obsolete. The deterioration in these structures is mainly due to prolonged traffic loads, environmental factors and extreme events like earthquakes [2]. It is important to monitor the structural changes of these critical infrastructure like bridges. The changes in structural properties can be in terms of stiffness, mass and dampness of the infrastructure. These changes can affect the way the bridge responds to ambient motion in terms of vibration behavior [2]. Through vibration response the modal properties of the bridges can be obtained and measured to detect any variations and anomalies. Modal parameters contain important characteristics of the structural dynamic response but are highly compressed compared to raw data, easing further analysis and storage. Operational modal analysis is typically used to identify the modal model in terms of modal parameters of the structure from the dynamic responses under operational conditions [3]. Structural Health Monitoring (SHM) involves many differentiated methods to analyze the health status of a bridge structure. Methods include but not limited to traditional visual inspections, ultrasonic testing to inspect the reinforcements within bridge components and smart sensors to measure displacements, acceleration/vibration responses on the bridge. Vibrational analysis using accelerometers is commonly used to evaluate the condition of bridges and other critical infrastructures. We consider bridge as a critical infrastructure mainly because of its structural complexity and its economic value, being a lifeline for effective functioning of a society. Structural health monitoring of bridges can be broadly classified into four categories: Traditional inspection, wired sensors, and wireless sensors, and other approaches.

© Springer International Publishing AG, part of Springer Nature 2019
M. E. Auer and R. Langmann (Eds.): REV 2018, LNNS 47, pp. 363–377, 2019.
https://doi.org/10.1007/978-3-319-95678-7_41

Visual Inspection comes under the category of traditional inspection. Where inspectors would visualize the cracks and keep record of the depth and size of the cracks. By tracking the change in the size of the crack over a period of time the structural integrity of the bridge could be assessed. However, this method did not allow insight about the internal properties of the bridge. The stiffness, condition of reinforcement bars etc. could not be evaluated. This is not a reliable system for bridge infrastructure monitoring [4]. The authors has conducted a comprehensive study of the reliability of visual inspection methods and have noted several inconsistencies within. In addition, they also claimed that the routine inspections are completed with significant variability.

The most common sensor types include but not limited to accelerometer, strain gauge, and inclinometer. Growing along with the popularity of sensors, the algorithms used to interpret the collected data has emerged steadily. Abudayyeh conducted a health monitoring of the Parkview bridge deck using strain gauges. Parkview bridge is located near Western Michigan University's Engineering Campus in Kalamazoo, Michigan [5]. Issue with these systems is the data loggers and the sensors have to be confined to a short distance. Also, because of low conductivity of the wires the data is some time erroneous and could misinterpret the readings.

Wireless sensors allow connectivity over a large area using a wireless network. Thus came the use of wireless sensor network (WSN) technology which allows connectivity of several sensors connected to each other using multi-hop network technology. Yu et al. used wireless sensors for vibration monitoring for Dongying Huanghe River Bridge [6]. The major issue found out was battery limitation. With the easiness in installing wireless sensors, multiple tests could be conducted at different locations. In this project three wireless accelerometers were placed on the mid-span of 3 girders while another three were placed on the quarter span of the same girders. Gangone et al. placed 20 wireless accelerometers between the quarter span and the mid-span on the five girders bridge [7]. For each girder, two wireless sensors were placed on the quarters spans while the other two were placed half way between the quarter spans to the mid-span. Lynch placed three wireless accelerometers near the quarter span of the girders and another four wireless sensors near the mid-span of other girders of the bridge [8].

Lin and Yang installed sensors on vehicles crossing the bridge and collected vibrational data from these vehicles for that duration [9]. Data was collected from the vehicles rather than the bridge itself, a mode of indirect SHM. The authors were able to obtain the fundamental frequency of the bridge from the vibrational data. Similar approach was also carried out by Toshinami [10]. However, past studies of such indirect-SHM concentrated mainly on identifying certain interaction properties and indexes, such as the fundamental frequency of the bridge, power spectral density magnitude variations from vehicle data. Thus, there is a need to complement these efforts with a more robust vibrational response analysis of the bridge itself which can take account of other characteristic effects as well.

Yin and Tang proposed a finite-element method to simulate the interaction of a vehicle and cable-stayed bridge [11]. The relative displacement of a passing vehicle of a bridge with known damaged conditions was used to generate a vector basis. The proper orthogonal decomposition on the relative displacement of a vehicle passing a bridge with an unknown damage condition is optimized with the known basis, and parameters of the unknown damaged bridges were reconstructed. Similarly Sirigoringo and Fujino proposed an approach to estimate the fundamental natural frequency of a bridge using the response of a passing instrumented vehicle [12]. The method was validated experimentally by fixing accelerometers on the vehicle and measuring the response with different speeds with which the vehicle travelled on the bridge.

In 2008 Maryland State Highway along with University of Maryland conducted a survey of wireless sensor systems connected to the Internet for structural health monitoring purposes [13]. They concluded that wireless sensors are easy to install, read and capture data and it's a very actively developing field. Also in 2011 Texas Department of Transportation along with Texas A&M University conducted a remote monitoring setup on two bridges: US59 over Guadalupe River Bridge and SH80 over San Antonio River Bridge [14]. In June 2017 Arcadius et al. did a survey and reviewed the use of Internet of Things (IoT) enabled systems for structural health monitoring [15]. They demonstrated the importance of remote monitoring in today's time.

The paper reports the design and development of an IoT enabled SHM system to provide an automated real-time diagnosis of the structural health of an infrastructure. For this study a laboratory scale suspended-bridge was used along with accelerometers mounted for data collected. Figure 1 shows the system diagram. The sensors are connected with an embedded processor system for data collection and pre-processing. Processed data are simultaneously passed to the cloud as well as a remote server for client access. The paper is composed of six sections including this one. The second section details for construction of a laboratory scale model bridge along with the instrumentation and embedded processor system for data collection and pre-processing. The third section presents the cloud integration for data analysis as well as the server deployment. The fourth section describes the graphical user interface for client access. The fifth and last section provides concluding remarks highlighting the major achievements as well as the issues that need to be considered for deployment of such a system for a full scale structure.

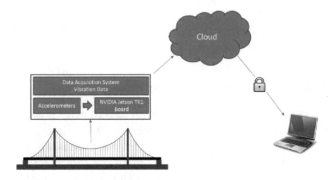

Fig. 1. System diagram of the developed monitoring system

2   Laboratory Scale Model Bridge

To implement the project a laboratory scale physical model bridge has been designed and built. Vibration data was collected using digital accelerometers along with the use of embedded processor board for initial processing and signal conditioning. This section will provide the details of the model bridge construction along with the instrumentation processor integration.

2.1   Bridge Structure

To implement the project idea a small scale suspended bridge was designed and developed. As construction material wood, strings, and polyethylene are used. There are five main components of this construction: wooden platform, wooden edges, strings, wooden pillars, and steel L-plates. Figure 2 shows the model of the bridge with all the main components. The bridge is six feet in length, five inches in width, and eighteen inches in height.

Fig. 2. Suspended bridge model

The pillars and base are made of wood, while the deck is made of polyethylene sheet. The strings are used as cables for the designed bridge. They are fixed in place by use of nails and screws and steel L-clamps, making them firmly affixed to the base. The bridge is stationed on a wooden platform. The pillars are placed at approximately 2 feet apart and are fixed to the wooden base using L-plates and screws.

2.2   Instrumentation and Embedded Processor System

The goal of the project is to collect the vibrational data from the model bridge and use an embedded processor to pre-process before passing to the cloud and the server. To have an effective structural health analysis the location of sensors is of vital importance. Mostly mid-span region of the bridge is considered for accelerometer placement as most of vibrations will occur around that region [2]. With this in mind one accelerometer is placed around the middle of platform/deck where maximum vibration occurs. Another accelerometer is mounted on a pillar to capture the movements due to high winds. A third accelerometer is placed on the wooden base and serves as a reference for other accelerometers. The third one will cancel any vibration on the ground. Figure 3 shows the locations of the accelerometers on the bridge and are marked with X.

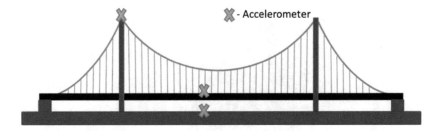

Fig. 3. Placement of accelerometers on the bridge

Figure 4 shows two different 3-Axis accelerometers that were used for this project. Figure 4a shows the accelerometer from Adafruit – LIS3DH, which has +/−2G to +/−16G sensitivity (Dimensions: 20.62 mm × 20.32 mm × 2.6 mm/0.8″ × 0.8″ × 0.1″ Weight: 1.5 g). Figure 4b shows another accelerometer from Sparkfun Technologies – MPU 9150, which also has +/−2G to +/−16G. The accelerometers are directly connected to an embedded processor board Jetson TK1, developed by NVIDIA [16]. Details of a TK1 board is shown in Fig. 5.

(a) Accelerometer from Adafruit

(b) Accelerometer from Sparkfun

Fig. 4. Two different accelerometers

The accelerometers allow Inter-Integrated Circuit (I2C) and Serial Peripheral Interface (SPI) communication interfaces provide suitable interface with the TK1 board. A python script pulls data from the accelerometers and stores it onto a SQL database on the TK1 board. The data collection rate can be adjusted by the user as needed for a given application. I2C communication allows communication of a single master with multiple slaves. In this case the TK1 board is be the master and accelerometers are as slaves. It does so by using separate address bits for the slave lines. The address bits in such a configuration should be unique, which allows the master to separately pull data from each of the slave modules connected to this bus line. Figure 6 shows how the master and slaves are connected on the I2C bus. SPI is another mode of communication between the modules; however, it does not allow multiple slave connections as this is needed for this bridge project. Hence I2C was preferred in the project.

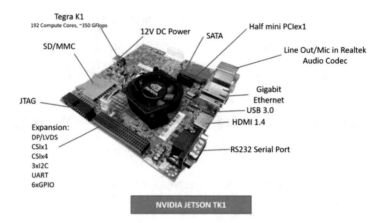

Fig. 5. NVIDIA Jetson TK1 - components and I/O terminals

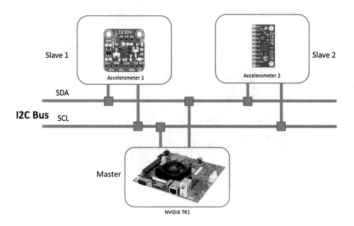

Fig. 6. Connection diagram utilizing I2C bus line

The TK1 board is used as the processing module to pull data from the accelerometers. There are three accelerometers connected to the same board, sending data at the same time. The accelerometers are connected to the TK1 board using I2C communication using the Serial Clock (SCL) and Serial Data Line (SDA). The board is connected to the Internet using an Ethernet cable, allowing the data to be passed to the cloud/server as required for client access. The board also allows provision to connect other types of sensors, like temperature, pressure, wind and process to display their data as well. The design of the system is made in a universal format allowing others to connect any sensor, network and use data for various other analysis.

2.3   Signal Conditioning and Analysis

To have an effective frequency analysis it is important to filter the data to remove unwanted frequencies. A cell-phone's vibration feature is used to excite the bridge structure. The frequency characteristics of the excitation strength is almost flat throughout the frequency range of interest. The vibrations generated in the bridge were then captured by the accelerometers and passed to the TK1 board. A filtering scheme is implemented within the board. The filtering involves a high-pass filter with a cutoff at 2 Hz and a low-pass filter with a cutoff at 220 Hz.

The filtered data is then processed to transform from time domain to frequency domain utilizing Fast Fourier Transform (FFT). Scipy library is utilized to perform filtering and frequency transform in Python before pushing it to the Datastore. Scipy is an open source Python library used for scientific computing, it has modules for linear algebra, FFT, signal processing, and interpolation [17]. Both the filtering and FFT is performed on the TK1 board. A diagram showing the signal conditioning and FFT process is provided in Fig. 7. For the data collection the bridge is excited by a cell phone vibrating with a frequency range from 0–150 Hz. Figure 8 shows the FFT analysis of the cell phone vibration data.

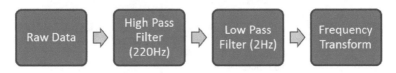

Fig. 7.   Steps for filtering and analysis

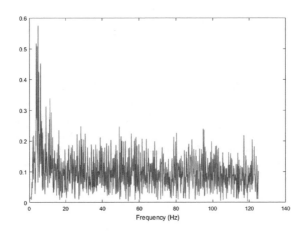

Fig. 8.   Spectral distribution of cell phone vibration data

3   Cloud Integration

The center of the cloud integration is the NodeJS server program running Heroku cloud. This server is connected with the NVIDIA TK-1 board as well as a local web server and Datastore (Fig. 9). The Heroku cloud accepts/requests data from all the sensors, TK-1 board's, supporting programs, and other users [18, 19]. Sensor data are stored in Elasticsearch (Lucene) which is running in the local Datastror. Elasticsearch is a Non-SQL

Database, which provides faster indexing, fuzzy searching, and analytics of very large data. It is highly scalable compared to the SQL [20]. The NodeJS server program in cloud provides Representational State Transfer (REST) end point to TK-1 board and push the data to the Datastore. This architecture makes it highly scalable to connect multiple monitoring sensors across multiple infrastructures. The collected data is pushed to Elasticsearch by the NodeJS program after receiving it from Python program running on TK-1 board. The central NodeJS server perform tasks which require high availability and computation such as serving multiple devices, clients such as web users, triggering alert, pushing the data to Datastore, and spectral analysis.

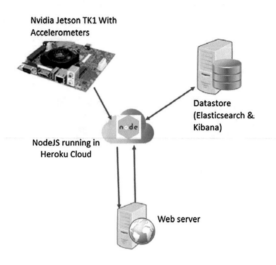

Fig. 9. System interaction using cloud

Considering limited storage capacity on TK-1 board, data is never stored locally but is pushed to the external server where large volume storage is possible through central NodeJS server. In addition to transferring data between the servers there is an additional service running in the web server, which trigger alerts such as e-mail, when there is critical breach in set threshold value. NodeJS in Heroku cloud automatically checks this threshold in real-time.

3.1   Data Collection and Storage

Data collection is done through python script running in TX-1 board. Multiple scripts run in parallel in multi-threading mechanism for each sensor (Fig. 10).

There are multiple services running in NodeJS. It fetches and pushes data from Elasticsearch, perform spectral analysis, provide service to running program to trigger threshold alert, support web application such as reading and updating threshold. Node.js is a JavaScript runtime that uses an event-driven, non-blocking I/O model that makes it lightweight and efficient. NodeJS runs on a single thread but its event driven approach makes it highly efficient with increased throughput [18].

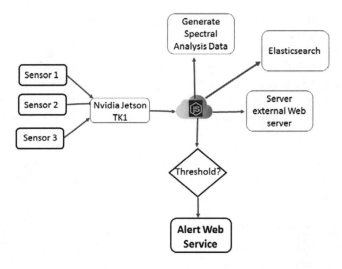

Fig. 10. Data collection threads

3.2 Data Visualization

There are three accelerometers fitted with the bridge, each providing data for three axes (x, y, and z). Data from each accelerometer will have two graphs, one is the time series and the other is the spectral analysis. Graphs are plotted using Kibana, which is an open source analytics and visualization platform designed to work with Elasticsearch [21]. Kibana graphs are embedded with the web page using 'IFrames'. Kibana also provides

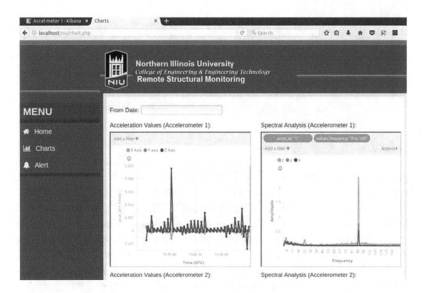

Fig. 11. Accelerometer data (time series and spectral analysis)

advanced filtering options in the graph. In addition, it can run Elasticsearch Query DSL within the graph to apply custom filter.

Figure 11 shows two graphs for one accelerometer that will be displayed within a web browser. The left-hand side graph is showing time series for one axes and the right-hand side graph is showing its spectral analysis. Within the web page the users can choose the start date for plotting. The time series graphs are dynamic and fetches new data from the Elasticsearch Datastore every 5 s. Spectral analysis is done by NodeJS server running in the cloud and pushed into Elasticsearch every minute. Each time NodeJS takes recent 2048 data points from time-series data to perform spectral analysis and pushes them to Elasticsearch to be plotted by Kibana via the graphical user interface (GUI). Figure 12 shows spectral analysis of three axis of a single accelerometer along with Kibana user interface.

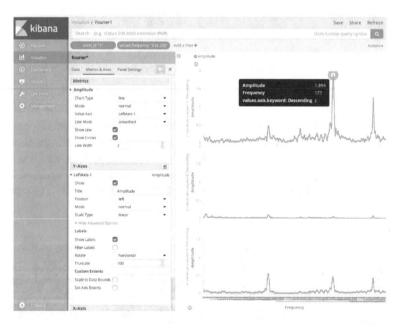

Fig. 12. Spectral graphs along with Kibana user interface.

3.3 Validation

It is important to validate the FFT analysis done over the web. The validation is done by comparing the results projected through web server via GUI and from offline FFT analysis. The procedure involves the collection of vibration data, which is used for both avenue of FFT analysis. For the data collection the bridge is excited by a cell phone vibrating with a frequency range from 0–150 Hz. Figure 8 shows the FFT analysis of the vibration data from the cell phone. The vibration data are collected by an accelerometer and then transferred to the web server, where the FFT analysis is performed and presented over the GUI. The same data is also used for offline FFT analysis using

MATLAB. While using MATLAB for FFT analysis, the data is passed through a band pass filter (combination of high pass and low pass filter) with the same cut off frequencies set for the web server. Following the filtering the FFT was computed using the MATLAB's inherent *fft()* function. It was found that the offline FFT analysis computed by MATLAB provides the same response as the web server's FFT analysis. Figures 13 and 14 are showing the frequency spectrum graphs obtained by offline analysis and web analysis, respectively.

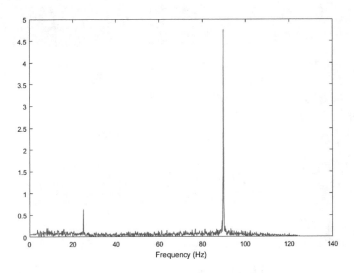

Fig. 13. Spectral distribution of vibration data with offline analysis.

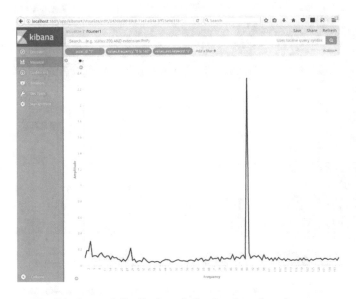

Fig. 14. Spectral distribution of vibration data via web server.

4   Graphical User Interface

The Fig. 15 shows an image of the main GUI with project information as well as a menu on the left for navigation. In addition to the Home there are two other menu choices, Chart and Alert. The Chart item provides access to the graphs when the Alert item allows users to set a threshold level for alert.

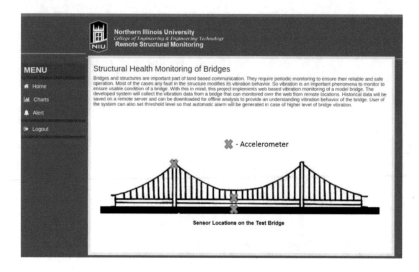

Fig. 15.   An image of the home page.

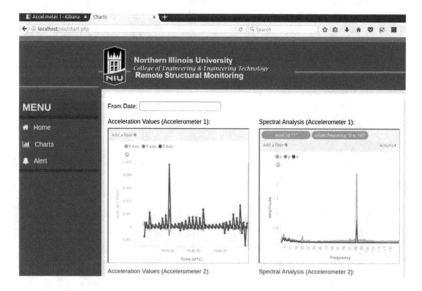

Fig. 16.   Real time graphs for one accelerometer within the web page.

The Chart menu item provide graphs for all three accelerometer data for all three axis (x, y, and z) both in time and frequency domains. Figure 16 shows graphs from one accelerometer.

Alert menu item open a web application provide a feature for configuring alert threshold. This allows users to set threshold for various axis of the accelerometers. Figure 17 shows an image of the web application to set up alert threshold level.

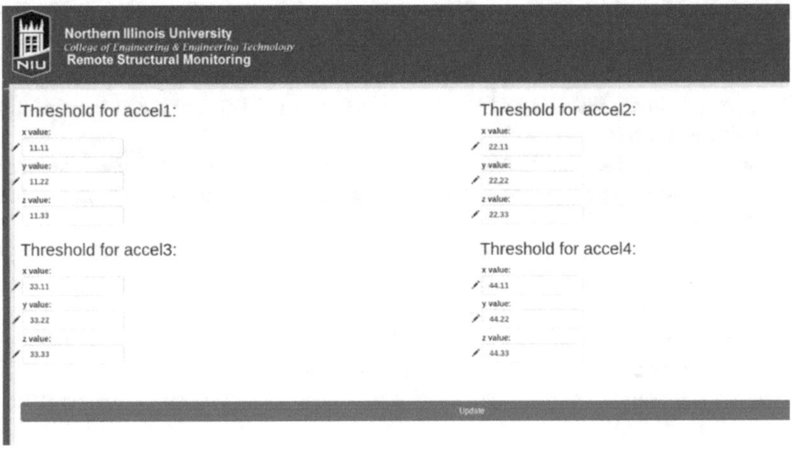

Fig. 17. Image of user interface for setting the threshold.

5 Conclusion

The paper describes the design, development, and implementation of a laboratory scale bridge monitoring system so that management can be assisted in terms of preventive maintenance in a proactive manner. The collected bridge data are presented as a vibration characteristics of the bridge and can be utilized to identify any deviation from standard vibration characteristics of the bridge structure. The web based data analysis implementation is also validated through off line analysis. Three accelerometers are placed on three key locations and are connected to a NVIDIA embedded processor board. After initial processing within the board the data then pushed to the cloud server for data analysis and web presentation. The system is password controlled and in addition to the near real-time graphs, the system can also produce an alert email to the management to notify any immediate threat that may need quick attention. The user can also change the alert threshold level using an application within the web page. The developed system demonstrates that the cloud based structural monitoring have potential to be incorporated for full scale bridge monitoring and will provide an additional tool for the management to ensure a safer transportation.

References

1. Infrastructure Report Card (2017). https://www.infrastructurereportcard.org/
2. Teng, C.-K.: Structural health monitoring of a bridge structure using wireless sensor network. Master's theses, Department of Civil and Construction Engineering, Western Michigan University, US (2012)
3. Guan, H., Karbhari, V.M.: Vibration-based structural health monitoring of highway. Final Report, Department of Structural Engineering, University of California, San Diego, California, US, December 2008
4. Moore, M., Phares, B., Graybeal, B., Rolander, D., Washer, G.: Reliability of visual inspection for highway bridges. Final Report, vol. 1, US Department of Transportation, Federal Highway Administration (2001)
5. Abudayyeh, O.Y., Barbera, J., Abdel-Qader, I., Cao, H., Almaita, E.: Towards sensor-based health monitoring systems for bridge decks: a full-depth precast deck panels case study. Adv. Civ. Eng. (2010)
6. Yu, Y., Xie, H., Ou, J.: Vibration monitoring using wireless sensor networks on Dongying Huanghe River bridge. In: Earth and Space 2010: Engineering, Science, Construction, and Operations in Challenging Environments. ASCE (2010)
7. Gangone, M.V., Whelan, M.J., Janoyan, K.D., Minnetyan, L., Qiu, T.: Wireless sensor performance monitoring of an innovative bridge design in New York State. In: Bridge Maintenance, Safety, Management and Life-Cycle Optimization, Clarkson University Department of Civil and Environmental Engineering, Potsdam, New York (2010)
8. Lynch, J.P.: An overview of wireless structural health monitoring for civil structures. Philos. Trans. R. Soc. A: Math. Phys. Eng. Sci. **365**(1851), 345–372 (2007)
9. Lin, C.W., Yang, Y.B.: Use of a passing vehicle to scan the fundamental bridge frequencies. An experimental verification. Eng. Struct. **27**(13), 1865–1878 (2005)
10. Toshinami, T., Kawatani, M., Kim, C.W.: Feasibility investigation for identifying bridge's fundamental frequencies from vehicle vibrations. In: Bridge Maintenance, Safety, Management and Life-Cycle Optimization: Proceedings of the 5th International Conference on Bridge Maintenance, Safety and Management, pp. 329–334 (2010)
11. Yin, S.-H., Tang, C.-Y.: Identifying cable tension loss and deck damage in a cable-stayed bridge using a moving vehicle. J. Vib. Acoust. **133**, 021007 (2011)
12. Siringoringo, D.M., Fujino, T.: Estimating bridge fundamental frequency from vibration response of instrumented passing vehicle: analytical and experimental study. Adv. Struct. Eng. **15**(3), 417–433 (2012)
13. Fu, C.C., Jaradat, Y.: Survey and investigation of the state-of-the-art remote wireless bridge monitoring system. State Highway Administration, Research Report, University of Maryland, US (2008)
14. Lin, T.-K., Lin, Y.-B., Chang, K.-C., Wu, S.-H.: Remote bridge monitoring system with optical fiber sensors. In: 13th World Conference on Earthquake Engineering, Vancouver, Canada, 1–6 August 2004
15. Tokognon, C.A., Gao, B., Tian, G.Y., Yan, Y.: Structural health monitoring framework based on internet of things: a survey. IEEE Internet Things J. **4**(3), 619–635 (2017)
16. NVIDIA: Jetson TK1 (2017). http://www.nvidia.com/object/jetson-tk1-embedded-dev-kit.html
17. SciPy (2017). https://www.scipy.org
18. NodeJs: About Node.js (2017). https://nodejs.org/en/about/

19. Heroku (2017). https://www.heroku.com/
20. Elastic (2017). https://www.elastic.co/
21. Elastic: Kibana User Guide (2017). https://www.elastic.co/guide/en/kibana/current/introduction.html

XOR

A Web Application Framework for Automated Security Analysis of Firmware Images of Embedded Devices

Christoph Vorhauer and Klaus Gebeshuber[(✉)]

Institute of Internet Technologies and Applications,
FH JOANNEUM University of Applied Sciences, Kapfenberg, Austria
klaus.gebeshuber@fh-joanneum.at

Abstract. Embedded Linux devices are very popular today and due their simple implementation they have become ubiquitous in our daily lives. Manufactures provide firmware updates online to enhance software security and quality of their products. In this context, software updates and firmware versions represent an enormous potential for security analysts to perform vulnerability analyses, because no real device is needed to gain valuable insights into systems. Previous solutions show the enormous advantages of automatically performing vulnerability research, but do not correlate program versions with 'Common Vulnerabilities and Exposures' (CVE) entries. Our solution provides a remote debugging interface to analyse extracted programs.

Therefore, in this paper we present *XOR*, an expandable web application framework, which supports the manual reverse engineering process of embedded Linux firmware images. In addition, an internal correlation database offers valuable insights for further research. *XOR* features detect system services, provides a remote debugging interface and allow the correlation between program version and CVE entries. We analysed 47 firmware images of 20 different vendors and found 487 related CVE entries.

Keywords: IoT · Reverse engineering · Embedded devices · CVE · Firmware

1 Introduction

Reverse engineering of firmware images is a common way to find vulnerabilities in embedded Linux devices. Security analysts can download images from manufacture webpages for free. This is a huge benefit, because no real device is needed. An analyst normally follows a strict process during reverse engineering of embedded firmware images. First, static approaches are used to gather information about the firmware and its software without executing any binaries. Normally, this involves reading source code and searching for security lapses, which can be exploited. Second, in dynamic analyses the whole firmware image itself or single programs are executed to find vulnerabilities and to proof that exploits are working.

Every step is manually executed and many repeating tasks are performed. Automatic analyses show an enormous potential, because a huge amount of firmware images can

© Springer International Publishing AG, part of Springer Nature 2019
M. E. Auer and R. Langmann (Eds.): REV 2018, LNNS 47, pp. 378–385, 2019.
https://doi.org/10.1007/978-3-319-95678-7_42

be analyzed and the extracted data such as password hashes can be correlated. Previous research on this concentrate on emulation, automation and finding pre-defined vulnerabilities in images such as backdoors or exploits topic [1–3]. XOR, where the name relates to the binary operation 'xor', instead is a single web application framework, which is based on the experience of manually reverse engineering of embedded firmware images and deals with repeating tasks. In detail, XOR correlates a programs version with a CVE [6] database, which shows potential threats of a device before further research is being made. In addition, XOR uses Docker [5] to run extracted programs in a safe environment and provides a remote debugging interface. In summary, we make the following contributions:

- XOR shows advantages of automatic extracting firmware images and static analyses designed to spare time during reverse engineering of embedded Linux firmware images.
- XOR highlights the modular expandability of its architecture and main components.
- The features of XOR allow correlation of static data, such as extracted certificates and passwords and remote debugging of emulated programs.
- XOR's web services provide interfaces to hash cracking servers and to an adapted CVE-search [7] database server to get information about possible vulnerabilities of a program.

This paper is organized as follows: Sect. 2 presents related work on automated firmware analyses. Section 3 gives an overview of repeating tasks during manual firmware analyses, where Sect. 4 presents the XOR web application framework and its outcome is discussed in Sect. 5. The last section presents the conclusion and future work.

2 Related Work

XOR is not the first automated framework for analysing firmware images. In 2014, the paper 'A Large-Scale Analysis of the Security of Embedded Firmwares' introduced a cloud framework that analyses 32.356 sample firmware images [1]. The researchers discovered 58 password hashes, which occurred in 538 distinct devices and found 38 new vulnerabilities. One year later, a publication about dynamic analyses of web applications - which are in firmware images - was released [2]. In detail, the authors run about 246 web servers and determined about 225 web vulnerabilities. FIRMADYNE framework instead concentrates on emulating complete firmware images and automatically tests different exploits of the Metasploit [4] framework [3]. This is the first open source framework, which was published in 2016. In general, XOR approach differs from existing projects in the following ways:

- XOR uses Docker to run extracted programs in a safe environment
- XOR correlates CVE entries with determined program versions
- XOR provides a debugging interface, which can be triggered online.

3   Repeating Tasks

The main advantage of XOR is time saving during the first analyses steps by automated recurring activities. Therefore, this section of the paper describes the basic principles of recurring activities during reverse engineering of embedded device firmware images.

Firmware Information Gathering
The first activity is to analyse the firmware image itself in order to get an impression, which components are included in the file. In this context, the firmware image can contain some of the following information:

- Linux Kernel
- Bootloader
- Root file system
- Certificates
- Signatures
- Files such as applications, pictures and webserver files
- Processor architecture

The availability of those parts differs among vendors, where firmware images can be classified by complexity [11]. Sometimes these images are also encrypted and entropy analyses have to be performed [10].

Unpacking and Mounting the Firmware
In order to get a deep view and to perform further analyses, a firmware image has to be unpacked to get access to its root file system. There are many different tools to achieve this goal e.g. binwalk [20], which was invented by Graig Heffner. Binwalk automatically tries to mount examined file systems, but sometimes this fails and mounting has to be manually performed with build-in Linux tools [12].

File Analyses
Once a firmware image is extracted, the included files are inspected. In detail, information about the file type, instruction set, passwords, configuration files, software versions and certificates is gathered. One golden rule while determining files is to never rely on a filename extension in order to diagnose the identity of a file [13]. Linux tools such as strings, file, readelf and hexdump are very powerful and recommended in this stage [12].

In the paper "Embedded Systems Security" published by Papp, Ma and Buttyan, the authors provide a deep view into existing vulnerabilities for embedded devices in CVE database [14]. For example, CVE-2010-0597 [21] allows remotely authenticated users to read or modify the device configuration, to gain privileges or cause a denial of service. In detail, firmware images often rely on many third-party software and libraries, which are often outdated, because of bad release management. In general, the version of the program is used to find vulnerabilities in vulnerability databases like CVE.

Emulate the Target
There are two different possibilities to emulate a firmware image. The firmware could be executed in a change rooted (chroot) environment with QEMU [15, 23]. In this

context, Linux changes the root directory to the root directory of the extracted root file system of the firmware and QEMU is executed in this directory with a given program. This is a huge benefit for embedded device testers, because with this technique several vulnerabilities easily can be found. For example, it is possible to start webservers and test websites for common vulnerabilities like remote code executions or cross side scripting. But this approach has also a few drawbacks, for example it is not always possible to fully emulate the peripherals and specific kernel extensions of the embedded devices, which can lead to false positives [2].

For this reason, QEMU should be prepared to run the complete firmware. Often the kernel image is missing in extracted firmware image and therefore, an own kernel must be built to get the virtual machine running [11, 16].

Penetration Testing

Penetration testing is the major activity during reversing firmware images with the intention to find vulnerable and exploitable code. In this context, analysts search for buffer overflows, web application vulnerabilities, authentication bypasses, unintended bugs and code injections [3, 9, 17–19].

Additionally, discovered weaknesses are used to exploit the target and the firmware in QEMU proofs that the exploit is working.

4 XOR

In this section, we describe the core architecture, features and interfaces of the XOR firmware analyze framework. Figure 1 illustrates the four core components of XOR, which run on virtual machines during testing. In detail, the web server is implemented in Python Flask [24] and it runs on a virtual Ubuntu 14.04 [25] machine. Additionally, Docker and the database run on the same machine for convenience. CVE-search database server runs on a separate virtual machine with Cent OS 7 [26].

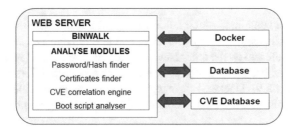

Fig. 1. XOR main architecture

Web Server

The web server consists of several main functionalities. The first function provides an interface to a mongodb [27] database. Second, web services provide an interface to hash cracking servers and storing the corresponding plain text passwords. Furthermore, a web service interacts with an adopted CVE-search database server, which receives common vulnerabilities queried by a specific program version.

The last function provides analyze modules, which are extendable. The extraction module of the web server uses binwalk to find and unpack the root file system of the given firmware image. Similar to Craig Smith's firmwalker [22], XOR's analyze modules search for password hashes, certificates, web server and executable files. These files help to determine the instruction set of the firmware image, which is mandatory to run programs with different CPU types. Finally, these modules extract the programs' version, which is mandatory for CVE correlation.

The boot script analyzer determines processes such as telnetd or sshd, which are executed during system boot of the device. Figure 2 illustrates the outcome of the dynamic analyze phase, where "RC Files" indicates processes, which are started during boot. In this example, XOR also automatically identifies web servers and its root directory.

Fig. 2. XOR software identifier

Docker
Docker is a virtual container, which holds a minimal Ubuntu bash image [28] that can run QEMU in a changed root environment [1]. On one hand Docker is very easy to handle and terminate and on the other hand, Docker provides a strict separation between the host system and firmware related programs. This approach differs much from other existing solutions [3, 9]. Docker is primary used to run any executable in a secure environment to determine the programs version and to start debugging process.

Database
The database stores all information about the firmware image and the outcome of the analyse module. Therefore, a detailed research and correlation between different manufactures are possible.

CVE-Search Database
This external database provides detailed information about CVE entries by querying the programs version and is based on the CVE-search project. For this project, we had to

adopt CVE-search and added our own request method. For example, Fig. 3 illustrates a vulnerability in OpenVPN version 2.0.9 located in an embedded firmware image.

Fig. 3. XOR CVE correlation engine

5   Outcome

At the time of writing this article, *XOR* analysed 47 firmware images of different vendors, wherein total 55 password hashes of ten different vendors and 256 certificates of seven vendors could be automatically discovered. Furthermore, *XOR* discovered 487 related CVE entries for 34 different deployed programs, but by now it has not been proven that these vulnerabilities can be exploited.

For vendor-only correlation, we analysed 30 Ubiquity[1] firmware images, where 23 programs are listed in CVE database and 15 programs have several vulnerabilities. Lighttpd 1.4.30 has five CVE entries, which are found in six different firmware images. Dropbear 0.51 server lists two CVE entries found in seven firmware images. For example, CVE-2013-4434 states that different time delays on Dropbear's login error messages allow remote attackers to discover valid usernames. We also extracted 27 'ubnt' and three 'root' hashes of Linux users, where 25 hashes are equal. In total, 920 Ubiquity programs are determined and 809 are running in QEMU.

6   Conclusion and Future Work

Basically, the main intention of implementing such a framework is the lack of security in firmware images of different vendors [14]. Often their images are available for free and have no proper encryption, which makes automation valuable. In addition, XOR was implemented in few hours' work, which exemplifies how fast a simple framework could be set up. The modules in XOR architecture provide a fast way to improve the overall analyzing tasks. Furthermore, XOR tries to examine CVE entries in early stages of firmware analyses, which is used to find existing vulnerabilities and exploits. Moreover, XOR has a debugging functionality, where debugging easily can be started in the web interface displayed in Fig. 4. In this context, debuggers such as IDA Pro [29] can be attached to a specific port, which is a huge benefit to the existing solutions.

---

[1] https://www.ubnt.com/.

Fig. 4. XOR debugging interface

The algorithm for identifying the version number of a program was a challenging task, because some binaries have different commands to show the version information. For example, to identify the version of busybox the command 'busybox –help' is used and for OpenSSL 'openssl version' is used. Therefore, an appropriate solution should be implemented.

Furthermore, other databases like Metasploit or exploit-db [30] should automatically be searched for existing vulnerabilities of running services. Moreover, XOR already identifies the web applications and corresponding root file systems. Therefore, it should be possible to run an automatically web application vulnerability research similar to the paper of Costin et al. [1] too. Frameworks like XOR could be used by security companies for penetration testing of embedded firmware images or by vendors to enhance the quality of their products.

References

1. Costin, A., Zaddach, J., Francillon, A., Balzarotti, D.: A large-scale analysis of the security of embedded firmwares. In: Proceedings of the 23rd USENIX Security Symposium 2014 (2014). https://www.usenix.org/system/files/conference/usenixsecurity14/-sec14-paper-costin.pdf
2. Costin, A., Zarras, A., Francillon, A.: Automated dynamic firmware analysis at scale: a case study on embedded web interfaces. In: Proceedings of the US Blackhat 2015 (2015). http://adsabs.harvard.edu/abs/2015arXiv151103609C
3. Chen, D., Egele, M., Woo, M., Brumley, D.: Towards Automated Dynamic Analysis for Linux-based Embedded Firmware. Boston University (2016). https://github.com/firmadyne
4. Metasploit. https://www.metasploit.com
5. Docker. https://www.docker.com/
6. CVE. https://cve.mitre.org/
7. CVE-search. https://github.com/cve-search/cve-search
8. Heffner, C.: Exploiting embedded systems part 3. In: /DEV/TTYS0 Embedded Device Hacking, blog post, 25 September 2011. http://www.devttys0.com/2011/09/exploiting-embedded-systems-part-3/
9. Costin, A., Zarras, A., Francillon, A.: Automated dynamic firmware analysis at scale: a case study on embedded web interfaces. In: Proceedings of the US Blackhat 2015 (2015). http://adsabs.harvard.edu/abs/2015arXiv151103609C
10. Heffner, C.: Differentiate encryption from compression using math. In: /DEV/TTYS0 Embedded Device Hacking, blog post, 12 June 2013. http://www.devttys0.com/2013/06/differentiate-encryption-from-compression-using-math/
11. Zaddach, J., Costin, A.: Embedded devices security and firmware reverse engineering. In: Proceeding of the US Blackhat 2013 (2013). http://media.blackhat.com

12. Siever, E., Figgins, S., Love, R., Robbins, A.: Linux in a Nutshell. O'Reilly Media Inc., Sebastopol (2009)
13. Eagle, C.: The IDA Pro Book. No Starch Press, San Francisco (2011)
14. Papp, D., Ma, Z., Buttyan, L.: Embedded systems security: threads, vulnerabilities, and attack taxonomy. In: Proceeding of the 14th Annual Conference of Privacy, Security and Trust (PST) (2015). http://www.cse.psu.edu/~pdm12/cse597g-f15/readings/cse597g-embedded_systems.pdf
15. Heffner, C.: 'Exploiting Embedded systems part 3', /DEV/TTYS0 Embedded Device Hacking, blog post, 25 September 2011. http://www.devttys0.com/2011/09/exploiting-embedded-systems-part-3/
16. Sally, G.: Pro Linux Embedded Systems. Springer, New York (2010)
17. Strackx, R., Younan, Y., Philippaerts, P., Piessens, F.: Efficient and Effective Buffer Overflow Protection on ARM Processors. University of Leuven (2010). https://lirias.kuleuven.be/bitstream/123456789/266377/1/paper.pdf
18. OWASP: OWASP Top 10, wiki (2013). https://www.owasp.org/index.php/OWASP_Top_Ten_Cheat_Sheet
19. Shoshitaishvili, Y., Wang, R., Hauser, C., Kruegel, C., Vigna, G.: Firmalice - Automatic Detection of Authentication Bypass Vulnerabilities in Binary Firmware. UC Santa Barbara (2015). http://angr.io/
20. Binwalk. http://binwalk.org
21. CVE-2010-0597. http://cve.mitre.org/cgi-bin/cvename.cgi?name=2010-0597
22. firmwalker. https://github.com/craigz28/firmwalker
23. QEMU. http://wiki.qemu.org/Main_Page
24. Flask. http://flask.pocoo.org/
25. Ubuntu 14.04. http://releases.ubuntu.com/14.04/
26. CentOS. http://isoredirect.centos.org/centos/7/isos/x86_64/CentOS-7-x86_64-Minimal-1511.iso
27. mongodb. https://www.mongodb.com/
28. Ubuntu bash image. https://github.com/Blitznote/docker-ubuntu-debootstrap
29. IDA PRO. https://www.hex-rays.com/products/ida/index.shtml
30. Exploit database. https://www.exploit-db.com

Development Models and Intelligent Algorithms for Improving the Quality of Service and Security of Multi-cloud Platforms

Irina Bolodurina and Denis Parfenov[✉]

Orenburg State University, Orenburg, Russia
{prmat, fdot_it}@mail.osu.ru

Abstract. The novelty of the presented research is the combination of two modern breakthrough technologies in the field of network organization and virtualization of its components for managing resources and data flows in software-defined networks based on real-time virtualization of network functions in order to provide the required level of security and maintain a specified quality of service for applications and services of multi-cloud platforms located in virtual data centers. The task of describing firewall rules and the rules for selecting nodes of the physical infrastructure for the placement of security elements built on the basis of virtualization technology for network functions in the software-driven infrastructure of the multi-cloud platform will be solved using Data Mining methods. For algorithmic implementation of the data flow management functionalities, the system approach, methods of data mining and cluster approach are used.

Keywords: Multi-cloud platforms · Software-defined networks
Software-defined infrastructure · Virtual data center · Virtual network functions

1 Introduction

Currently, significant amounts of convergent traffic are circulating within the virtual data center. The creation of effective mechanisms for managing and protecting such traffic flows in a data center network requires an understanding of its structure, as well as the definition of the load on its computing and network nodes when it is serviced. To solve this problem, within the framework of the research, the task was solved to develop an algorithm for firewalling and to ensure the quality of servicing of data flows in a network environment of a virtual data center. The solution proposed by us is based on an approach that allows aggregating all transmitted flows in a single analytical system for collecting SIEM data. The basic element of the proposed approach is a distributed network consisting of virtualized modules that perform processing and analysis of traffic passing through them. Due to decentralization, the proposed solution allows for unlimited scalability. All collected and analyzed information is submitted with aggregated and compressed form to a single network management center, where they are converted into the corresponding controller rules according to the OpenFlow standard.

© Springer International Publishing AG, part of Springer Nature 2019
M. E. Auer and R. Langmann (Eds.): REV 2018, LNNS 47, pp. 386–394, 2019.
https://doi.org/10.1007/978-3-319-95678-7_43

The presented research is aimed at solving problems related to improving the efficiency of quality of service provision, as well as the security of the software-defined infrastructure of multi-cloud platforms located in virtual data centers.

2 Problem Formulation

The presented research developed the architecture and the algorithmic solutions for an autonomous self-organizing security system, quality of service, detection of flows and protection from cyber attacks in the software-defined infrastructure of a multi-cloud platform located in a virtual data center. The ensemble of models allows us to describe its structure, as well as the principles of its operation. The originality of the developed solutions consists in using the technology of containerization of basic services and services, as well as the use of software-defined networks and virtualization network functions. The mathematical models presented in the article allow us to describe the self-organization of data flow control and firewall rules and the choice of physical infrastructure nodes for the placement of security elements to ensure quality of service and network security in the virtual data center. This approach allows you to fine tune the rules for filtering and classifying packets.

One of the most important tasks for increasing the efficiency of the network is to analyze the performance of all the components, which support the work of the data center [11, 12]. A significant difference of the virtual data center is the use of virtual networks for routing data flows of cloud applications and services over the existing physical infrastructure. Virtualization at the network level allows us to run simultaneously many different types of applications with different performance requirements for compute nodes, network objects, and QoS. Therefore, we should be aware of the flows circulating in the network at present time to provide the information about the quality of service at the network of the virtual data center. Most modern methods of traffic classification define the flow as a group of packets, which have the same destination IP address and use the same transport protocol and port numbers. This definition allows considering bi-directional flows. Therefore, the packets that transfer requests from the user and answers from compute nodes are a part of the same flow in a network of the virtual data center. The approach to classification based on selected characteristics is a fairly simple task; however, the accuracy of such solutions for modern cloud applications and services are not quite high. In terms of architecture, modern cloud applications are comprehensive objects consisting of many clustered and distributed services. In turn, each cloud service used by the cloud application has its own set of requirements for QoS and dependencies. In the virtual data center, the procedure of user request processing can be represented as a multi-phase queuing system, where data flows are described by different laws of distribution. Thus, the problem of analysis and classification of the traffic flows of cloud applications and services in the virtual data center becomes non-trivial. Another feature of using a software-defined network is the use of dynamic port numbers for routing flows for the same type of cloud application. This feature makes it difficult to classify the traffic of applications by this attribute. The existing solutions based on deep packet inspection (DPI) work slowly and require a lot of processing power.

In this investigation, we propose an approach for accelerating the learning process and improving the accuracy for traffic flow classification of cloud applications and services in a network environment of the virtual data center at the initial stage of data analysis.

3 Model of Self-organization of Data Flow Control

The model of self-organization of data flow control is created on the basis of the decomposition method and approaches used in queuing theory. To determine the management objects at the input of each virtual network node, sparse traffic outflows from other nodes aggregate. The solution of systems of linear algebraic equations allows determining the intensities and average values of time intervals between neighboring packets for each flow. The flow of packets coming from an external source to the network is described by the law of distribution of time intervals between neighboring packets. This means decomposition of the network into separate nodes, which allows you to identify all the main traffic flow indicators and provide the required quality of service within the virtual data center network.

Among the most important quality-of-service (Q) targets are usually the following set of characteristics:

$$Q = \{AvgAvl, MinUAvl, AvgRespT, MaxURT, AvgB, IReport\}, \tag{1}$$

where AvgAvl – average availability, expressed as the average number of failures during the service provision period; MinUAvl – minimum availability for each user; AvgRT – average service response time; MaxURT – maximum response time for each user; AvgB – average throughput; IReport – methodology for calculating metrics and frequency of collection of reports.

To improve the quality of services and the security of multi-cloud platforms, we need to detail the main object, which should give a prediction about the architecture of the network of the virtual data center. The network of the virtual data center can be described as

$$VDCnetwork = (V, e, u, FE), \tag{2}$$

where V – vertices of the graph (network nodes), e – arc graphs (network connections), $FE : E(G) \rightarrow R^+$ - flow of requests in a virtual data center network; $u : E(G) \rightarrow R^+$ - the cost of implementing a network connection in a data center network.

To maintain the required quality of service (QoS) and a given level of uninterrupted operation, we will write the balance equations:

$$ex_f(u, v) := \sum_{e \in \delta^-(U,V)} FE(e) - \sum_{e \in \delta^+(U,V)} FE(e), \forall e = (u, v) \in E(G) \tag{3}$$

The presented description is necessary for transition to modeling at the level of flows of transmitted and analyzed data. Each record of the flow $FE_{kij} = FE_{ki}(t)$ is dynamic and changes at times t. Each thread has a set of characteristics that uniquely identify it. In the traffic management system, the flow can be represent as form

$$FE_{kij} = \left(Match_{kij}, Actions_{kij}, TimeOut_{kij}, Flow_{kij}, Counters_{kij}(t) \right), \qquad (4)$$

where $Match k_{ij}$ this is a set of fields to check for matches with the headers of the package; $Actions_{kij}$ – A set of actions performed on the package, if its headers match $Match_{kij}$; $TimeOut_{kij}$ – time of fixing the flow in the system; $Flow_{kij}$ – the thread to which this OpenFlow rule applies; Counters – statistical counters OpenFlow.

In this case, for more efficient traffic analysis, as a rule, a signature approach is used in networks. Within the framework of this research, this approach is supplemented by methods of data mining, which allow reducing the set of characteristics, which significantly speeds up the processing of information. This is important when analyzing Big data flows that occur in networks of virtual data centers because of multiple intersections of communication channels. The flow analysis scheme at the point of connection to the network, on the node of the data center, can be formalized as follows:

$$Analyzer_{ki} = (Node_{ki}, CurrentFlows_{ki}(t), SuspiciousFlows_{ki}(t)), \qquad (5)$$

where $Node_{ki}$ – the network node on which the signature traffic analyzer works, $CurrentFlows_{ki}(t) = \{CurrentFlows_{kiJ}\}$ and $SuspiciousFlows_{ki}(t) = \{Suspi ciousFlow_{kiJ}\}$ – respectively, the set of current and the set of suspicious flows detected by the analyzer at time t.

For the model presented, firewall rules have been developed separately for the L2 and L3-L4 levels of the OSI model. All rules have two parts - the headers that identify the packages, and the action. The headers for L2-level rules include the source port of the packet, the source switch, the MAC addresses of the sender and receiver of the packet, the type of protocol, etc. The headings for the L3-L4 layer rules contain the IP addresses and ports of the source and destination of the packet, the type of the encapsulated protocol, etc. Possible actions - the removal or resolution of the package.

Such rules can be combined into chains if necessary. Each chain is an ordered list of rules that has an identifier. The entire chain of rules is checked in order of priority, until there is a rule suitable for the parsed packet. In this case, the action specified in this rule is executed, and the subsequent execution of the chain is terminated. In order to transfer the rule to another chain, the corresponding action with its identifier can be specified. The originality of this model lies in the fact that within the rules all Open-Flow headers are supported up to version 1.5, inclusive.

4   The Algorithm of Adaptive Firewall and Ensuring the QoS

From an algorithmic point of view, we can represent classification or clustering of a traffic flow as the function $f : X \rightarrow C$, which puts the label $c_j \in C$ in correspondence with each object $x_i \in X$. The set of C is defined in advance. In the task of clustering, neither the set of C nor its dimensionalities is determined. In this research, we have used the classification and clustering of traffic flows of the cloud applications and services to improve the efficiency of QoS inside the virtual data center.

A classification traffic flows of cloud applications and services located in the virtual data center can be divided into the following elements: classification, clustering and identification of association rules. Let's consider these elements of the model separately.

For the effective classification of traffic flows in the virtual data center, we need to determine the set of all applications. Suppose that we have set of cloud applications and services defined as $X = \{x_1, \ldots x_n\}$. Each cloud application and service is characterized by a set of attributes $X_j = \{a_1, \ldots a_m, y\}$, where a_i is the observed attributes, whose values represent characteristics of the traffic of cloud application or service; y is the target attribute that identifies the class of a cloud application or service. Each attribute a_i takes a value from some set $A_i = \{a_{i,1}, a_{i,2}, \ldots\}$, which describes valid values of characteristics of the attributes in the subject area under study. In the framework of solving the tasks of traffic classification, suppose a limited number of classes applications $y \in C$ circulate in the network of the virtual data center, where $C = \{c_1, \ldots, c_k\}$. With regard to our task, we will explore the traffic flows of the cloud applications and services according to the communication scheme of their interaction with the network objects within the virtual data center.

The algorithm of the adaptive firewall for software-defined infrastructure with multi cloud platform, taking into account the context (state) of the data transmission. The main idea of this algorithm is the following: if the packet arrives at the OpenFlow controller, and then, if inside IP protocol is encapsulated, then for L3-L4. The algorithm supports sessions (status monitoring) for TCP and UDP protocols. The beginning of the session for the TCP protocol is monitored by the first step of the three-step handshake (the packet with the SYN flag), and the end by the timeout in 300 s or when receiving packets with FIN and RST (in this case, the timeout is adjusted to 60 s). For the UDP protocol, the beginning is for the first packet, the end is only for a timeout of 300 s.

Each OpenFlow rule that is implemented in the OpenFlow is disabled in the OpenFlow, which allows optimizing the limited sizes of the thread tables OpenFlow switches.

The main advantage of the algorithm is the closest to the packet receiver, and then on the switch closest to the packet sender. In general, the algorithm can lock on any OpenFlow switch, any analyzing inter-node traffic. However, with this approach, the performance of the corporate network may decrease, therefore, within the framework of this research, it is proposed to use the developed firewall. The generalized algorithm for adaptive firewall and ensuring the QoS in virtual data center can be represented as the following sequence of steps:

Step 1. To identify the traffic flows, we use the data received from the controller of a software-defined network, which controls the placement of cloud applications and services in the virtual data center.

Step 2. Basing on the obtained data, we formed a graph in compliance with the communication schemes of interaction between cloud applications and services.

Step 3. The obtained schemes are overlaid on the current topology of the software-defined network to evaluate the network bandwidth, the usage of virtual channels and the analysis of primary transmitted thereon data basing on moments of the packets reception time distribution.

Step 4. The data obtained are ranked in descending order by the load of the communication channel and by the priority of the traffic flows of cloud applications. To adjust the routes, 20% of the most loaded channels are selected.

Step 5. For the selected virtual channels, the traffic flows of cloud applications are classified more thoroughly to identify the degree of channel usage by particular types of applications.

Step 6. For the applications identified in the previous step, we analyze the traffic, get the parameters of state from physical network devices and identify the most loaded objects.

Step 7. High loaded devices are excluded from the current route and the traffic is redistributed between less loaded nodes by using association rules for routing. These changes are also updated in the communication schemes of interaction between applications and services.

Step 8. The results of the analysis are used to apply QoS policies on the controller of a software-defined network in the virtual data center as well as in a retrospective analysis of data for error correction.

Thus, the main goal of the developed algorithm is to find the optimal solution and to maximize the performance of the physical network taking into account the existing flows of applications and services and their demands for delays in the work of the virtual data center.

5 Experimental Results

To assess the effectiveness of the developed algorithm for optimizing the adaptive routing of data flow balancing in the applications and services in the virtual data center, we have conducted a pilot study. We have chosen the Openstack cloud as the basic platform. For comparison, we have applied the algorithms used in the OpenFlow version 1.4, for route control of the software-defined network in the experiment. For the experimental research, a prototype has been created, including basic nodes, as well as software modules for the developed algorithms that redistribute data flows and applications. To verify the developed algorithm of optimal routing and traffic balancing in case of dynamic changes channels in a software-defined network of the virtual data center, several experimental networks consisting of 10, 50, 100, 500 and 1000 objects have been deployed. All generated requests were played consequently on two pilot sites: the traditional routing technology (Platform 1, NW) and the technology of the software-defined networks (Platform 2, SDN). This restriction is caused by the need to compare the results to a traditional network infrastructure, which is not capable of dynamic reconfiguration. The experiment time was one hour, which corresponds to the most prolonged period of peak demand, recorded in a real traffic network of a heterogeneous cloud platform. We have chosen response time of applications and services that work in a cloud platform as a basic metrics to assess the efficiency of the proposed solutions. The results of the experiment are provided in Table 1.

Table 1. Computational experiment result

Number of switches in the network		10	50	100	500	1000
Dijkstra Algorithm	NW	4.029	11.59	19.71	63.078	153.03
	SDN	0.029	5.19	10.533	33.078	53.007
Algorithm Jarri	NW	3.183	8.615	15.992	50.105	145.542
	SDN	0.033	4.237	8.592	30.105	45.542
Adaptive Routing Algorithm	NW	2.022	3.175	5.520	15.13	20.31
	SDN	0.016	0.078	0.460	0.503	10.538

6 Discussion

In article [1], the problem of responding to an incident in the cloud is considered. An incident is an approach to eliminating and control the consequences of a security breach or a network attack. As a key criterion of the model authors select the minimization of the incident processing time is considered due to the introduction of security controllers and security domain in the cloud infrastructure for the analysis of network threats.

To counter network cyber attacks, technologies began to appear to provide continuous monitoring on any device connected to the network in order to increase fault tolerance by detecting and mitigating targeted threats. The authors of the research [2] describe the Gestalt security architecture, which is based on the principles of strong isolation, the policy of least privileges, the concept of providing information security (defense-in-depth), cryptographic authentication, encryption and self-healing. Remote monitoring is carried out through an organized workflow through a multitude of components connected by a specialized secure communication protocol that together provide secure and sustainable access.

Conventional firewalls are used to enforce network security policies on the boundaries within the network. In [3], the authors take advantage of the SDN to turn the network infrastructure into a virtual firewall, thereby improving network security. The virtual firewall as ACLSwitch is presented, which uses the OpenFlow protocol to filter network traffic between OpenFlow switches. The authors also introduce domain policies that allow the use of different filtering configurations for different network switches.

The optimal use of IDS and IPS systems is also presented in [4]. The authors investigated the open source snort system. The implementation, tuning, installation of the system and the problems arising during the research are studied.

Firewalls often have vulnerabilities that can be exploited by cybercriminals. The publication [5] is devoted to the investigation of some possible fingerprinting methods (fingerprint, identification) of the firewall, which turned out to be quite accurate. The authors also studied DoF attacks (Denial of Firewalling) in which attackers use carefully processed traffic to overload the firewall. In [6] the implementation of a third-level firewall with a full-mesh topology with 1 controller and 6 switches is presented. Modification of the learning switch code for the POX controller is performed for the full-mesh topology inside the Mininet network emulator. In this case, the packet flow between hosts is controlled in accordance with the rules recorded in the learning switch through the OpenFlow controller.

In the context of the joint use of SDN and NFV, the NetFATE architecture is presented in [7], a platform designed for placing virtual network functions on the network boundary. This architecture is based on free open-source software on the provider's nodes and client equipment, which leads to a simpler deployment of functions and lower management costs.

An important part of any network architecture is application identification, which contains information about the activity of network applications, and their use of bandwidth. The article [8] focuses on the promotion of open source technology to identify applications. This technology is seen as the future of firewalls, where the administrator can view more detailed information about network traffic compared to current approaches.

The issues of achieving elasticity for network firewalls are discussed in detail in [9]. Elasticity here means the ability to adapt to network load changes by releasing and allocating resources in an autonomous manner. The elasticity of cloud firewalls is aimed at satisfying a consistent performance evaluation using the minimum number of instances of the firewall. The author's contribution of the publication is to determine the number of instances that must be dynamically adjusted in accordance with the load of incoming traffic and the base of firewall rules.

Due to software processing of network functions, the performance of VNF significantly decreases depending on the types of VNF and the configuration of VNF applications. In the publication [10], the authors pay special attention to the analysis of the virtual firewall as a representative of VNF. The paper proposes a method for estimating the latent load of a virtual firewall using rules in the ACL and the amount of traffic for each rule.

7 Conclusion

The developed methods and algorithms allow to increase the level of research in the field of information security and quality of service in multi-platform platforms in the virtual data centers. The application of the proposed approaches based on the joint use of virtualization network functions and software-defined networks allows for more efficient planning of data flows and providing the required quality of service and a given level of security in the data center network. Developed models and algorithms have a high innovative potential and can serve as a basis for developing an industrial system for detecting threats and protecting against cyber attacks in virtual data centers.

Acknowledgments. The research work was funded by Russian Foundation for Basic Research, according to the research projects No. 16-37-60086 mol_a_dk and 16-07-01004, and the President of the Russian Federation within the grant for state support of young Russian scientists (MK- 1624.2017.9).

References

1. Adamov, A., Carlsson, A.: Cloud incident response model. In: Proceedings of 2016 IEEE East-West Design and Test Symposium (EWDTS), pp. 1–3 (2016)
2. Atighetchi, M., Adler, A.: A framework for resilient remote monitoring. In: Proceedings of 2014 7th International Symposium on Resilient Control Systems (ISRCS), pp. 1–8 (2014)
3. Bakker, J.N., Welch, I., Seah, W.K.G.: Network-wide virtual firewall using SDN/OpenFlow. In: Proceedings of 2016 IEEE Conference on Network Function Virtualization and Software Defined Networks (NFV-SDN), pp. 1–7 (2016)
4. Kashif, M., Zahoor-ul-haq: An optimal use of intrusion detection and prevention system (IDPS). In: Proceedings of 2015 European Intelligence and Security Informatics Conference, pp. 190 (2015)
5. Liu, A.X., Khakpour, A.R., Hulst, J.W., Ge, Z., Pei, D., Wang, J.: Firewall fingerprinting and denial of firewalling attacks. IEEE Trans. Inf. Forensics Secur. J. **12**(7), 1699–1712 (2017)
6. Kumar, A., Srinath, N.K.: Implementing a firewall functionality for mesh networks using SDN controller. In: Proceedings of 2016 International Conference on Computational Systems and Information Systems for Sustainable Solutions, pp. 168–173 (2016)
7. Lombardo, A., Manzalini, A., Schembra, G., Faraci, G., Rametta, C., Riccobene, V.: An open framework to enable NetFATE (network functions at the edge). In: Proceedings of the 2015 1st IEEE Conference on Network Softwarization (NetSoft), pp. 1–6 (2015)
8. Patel, N.V., Patel, N.M., Kleopa, C.: OpenAppID – application identification framework. In: Proceedings of 2016 Online International Conference on Green Engineering and Technologies (IC-GET), pp. 1–5 (2016)
9. Salah, K., Calyam, P., Boutaba, R.: Analytical model for elastic scaling of cloud-based firewalls. IEEE Trans. Netw. Serv. Manag. J. **14**(1), 136–146 (2016)
10. Suzuki, D., Imai, S., Katagiri, T.: A new index of hidden workload for firewall rule processing on virtual machine. In: Proceedings of 2017 International Conference on Computing, Networking and Communications (ICNC): Communications QoS and System Modeling, pp. 632–637 (2017)
11. Parfenov, D., Bolodurina, I.: Development and research of models of organization storages based on the software-defined infrastructure. In: Proceedings of 40th International Conference on Telecommunications and Signal Processing, pp. 1–5 (2017)
12. Parfenov, D., Bolodurina, I., Shukhman, A.: Approach to the effective controlling cloud computing resources in data centers for providing multimedia services. In: Proceedings of International Siberian Conference on Control and Communications – 2015, pp. 1–5 (2015). https://doi.org/10.1109/SIBCON.2015.7147170
13. Bolodurina, I., Parfenov, D.: Development and research of models of organization storages based on the software-defined infrastructure. In: Proceedings of 40th International Conference on Telecommunications and Signal Processing, Barcelona, pp. 1–5. IEEE Press, Barcelona (2017)

The Application of the Remote Lab for Studying the Issues of Smart House Systems Power Efficiency, Safety and Cybersecurity

Anzhelika Parkhomenko[✉], Artem Tulenkov, Aleksandr Sokolyanskii,
Yaroslav Zalyubovskiy, Andriy Parkhomenko, and Aleksandr Stepanenko

Zaporizhzhia National Technical University, Zaporizhzhia, Ukraine
parhom@zntu.edu.ua

Abstract. The rapid development of cloud and mobile technologies, as well as the Internet of things and telematics, poses new challenges in the field of safety and cybersecurity of Smart House Systems. Of course, such technologies help to organize data storage, convenient interaction with Smart House services, remote surveillance and device monitoring effectively. At the same time, the usage of such solutions can become a factor that significantly reduces the safety and cybersecurity of the Smart House systems.

Therefore, the vulnerabilities of modern Smart House systems, that can lead to serious problems, both financial and related to people's health and life are presented in this work. As well as, the methods of the level of safety and cybersecurity increasing for such systems are discussed.

The usage of Remote Lab Smart House & IoT for investigation of the issues of Smart House systems power efficiency, safety and cybersecurity was offered. The application of FMEA method in the case study of vulnerabilities and faults of Remote Laboratory Smart House & IoT hardware and software was shown. The architecture of Remote Lab Smart House & IoT cyber security was proposed, that can be used for real project of Smart House realization.

Keywords: Internet of Things · Remote Laboratory · Embedded systems
Smart House · Power Efficiency · Safety · Cybersecurity · FMEA

1 Introduction

The concept of such systems as Smart City and Smart House creating is popular all over the world. The main regions that are actively implementing Smart House technologies are North America, the Asia-Pacific region and Western Europe [1]. Nevertheless, according to a recent studying by Gartner, making decisions about Smart House system usage is still at the early stage. An online survey of nearly 10,000 respondents from the United States, the United Kingdom and Australia have shown that only about 10% of families currently use Smart House connected solutions [2].

The realization of the concept of Smart House System (SHS) is connected with the problem of simultaneously executing of several critical requirements: using a wide nomenclature of control and executive devices, ensuring a high degree of data protection

© Springer International Publishing AG, part of Springer Nature 2019
M. E. Auer and R. Langmann (Eds.): REV 2018, LNNS 47, pp. 395–402, 2019.
https://doi.org/10.1007/978-3-319-95678-7_44

and minimization of the costs associated with resource consumption [3–6]. Depending on the customer's requirements, the architecture of the home automation system, as well as the set of hardware and software solutions, can differ significantly.

On the one hand, there is a huge number of devices, platforms and technologies for creation of home automation systems on the market [7]. The main requirements for them are interoperability, flexibility, security, simplicity of installation and exploitation, etc. Some of the companies offer the integration of their smart assistants into existing devices (Alexa, Google Home). The others are working on the new smart home devices (Nest, Walmart, ABB, British Gas). There are a lot of wired and wireless technologies (KNX, Ethernet ZigBee, Z-Wave, Wi-Fi, Bluetooth, etc.) for serving smart home needs [8]. There are smart home platforms (Wink, SmartThings from Samsung, HomeKit from Apple, Iris from Lowe) and solutions from security providers (for example, Tor Home Assistant) [9].

On the other hand, the issues of devices unifying and certifying, as well as of their integration into a single, easily configurable system are still open. The consequence for the industry generally is the slowing of the market pace, because consumers doubt the efficiency and safety of such systems and don't rush to install them.

In Ukraine, the market of intelligent home automation is developing not for a long period of time and the most difficult task remains to convince the mass market of the value of such technologies and systems. However, in the last few years there has been made significant progress in the area of the creation of smart houses and apartments. Many people have already begun to build smart houses from scratch and equip existing houses and apartments with home automation systems. The most popular are the solutions for "smart" resources usage (electricity, water, various types of fuel) and home security. The concept of security concerns not only the penetration of strangers into the house, but also the occurrence of fires, water leaks, short circuits, carbon monoxide, smoke or fire. Another important component of security is the cybersecurity and smart house's information security. And if the issues of energy efficiency and safety of such systems are paid great attention, the cybersecurity issues remain in the shadow, although cyberattacks of recent years have shown the danger of such an attitude.

Thus, the development of efficient and reliable SHS is an actual scientific and practical task that requires the application of a system approach and consideration of the entire set of requirements at all stages of its solution [10–12].

The goal of this work is to investigate the issues of energy efficiency, safety and cybersecurity of home automation systems based on the architecture and the set of experiments of Remote Lab Smart House & IoT.

2   Analysis of the Problems of SHS Cybersecurity

Today, the main trends of the SHS development are the Internet of things and devices, mobile and cloud technologies, as well as telematics. Technologies of the Internet of Things (IoT) are actively used for the control of smart devices, transmission and processing of information, control of the energy system, water supply system, etc.

Telematics provides the correct interaction between wireless and wired systems, as well as it allows remote monitoring of smart home devices. That's why users increasingly prefer to use mobile devices for remote monitoring and control.

The usage of cloud technologies and services for convenient interaction of smart devices and software, as well as storing of the various information (images from videocameras, data from various sensors, etc.) become more popular.

However, the main problem of the information security is precisely the permanent connection of intelligent home systems to the Internet. As the results there are risks associated with storing personal data in the network and the possibility of access to them by strangers. Nowadays, home automation systems do not have the proper level of protection to repel attacks of intruders. The most vulnerable smart devices are Wi-Fi or 3G-4G-routers and video cameras.

Hackers can not only crack the house access control subsystem, but also deactivate certain sensors, blind video-cameras, switch off the fire safety system, remotely control the heating/ventilation systems and even destroy the data and the house control software. The consequences of these actions are fraught with great material damage and are dangerous for people's health and life. Moreover, the access to the control system of a large number of people (several family members) is significantly weakening the protection of the SHS.

Therefore, professionals give clear recommendations on the level of house security improvement and minimizing the risks of cyber-hacking [13]. As long as most smart devices do not practically resist cyber-attacks, also it is necessary to adhere to certain rules related to their using [14].

In this way, the task of ensuring the reliable functioning of the home automation system should be solved at all levels, starting with requirements analysis, system architecture development and ending with system's operation. At the same time, it is necessary to apply methods of qualitative and quantitative assessment of the level of functional and information security at all stages of the system life cycle in order to timely identification of the vulnerabilities and the development of measures to eliminate weaknesses and reduce the risks of attacks.

3   Remote Experiments for Issues of SHS Power Efficiency, Safety and Cybersecurity Studying

The Remote Laboratory (RL) Smart House & IoT is the part of the integrated complex REIoT, which also includes the RL RELDES (REmote Laboratory for Development of Embedded Systems) [15, 16]. The RL Smart House & IoT can be effectively used for IoT technologies studying and investigations of the features of SHS development (Fig. 1).

Fig. 1. Web interface of the RL Smart House & IoT

For studying of the issues of resource conservation and energy efficiency, the following experiments are possible, which allow to control the parameters of lighting, heating, air conditioning, ventilation and recuperation systems as well as solar energy production. Solar Station – the basis of obtaining and accumulating renewable energy from solar panels; Climate control – control of temperature, humidity and air quality in different areas of sensors installation; Presence control – control of presence in the zone for effective lighting control; Ventilation – principles of operation of the air ventilation system with heating and control of the flow rate parameters, temperature and humidity; Lighting control and Illumination control – studying of various lighting systems usage and control of the level of illumination in different zones.

The following experiments for the safety issues studying can be used and they allow to monitor access to controlled zones and objects. Access control – principles of authorization and access control systems; Safety control – the basics of security systems with motion sensors and intrusion detection systems; Zone control – principles of security control of the perimeter of the zone.

The following experiments for cybersecurity issues studying can be performed, which enable the usage of wireless, cloud and mobile technologies for system monitoring and control, as well as for data collection and visualization. Wireless control – principles of connecting devices via Wi-Fi protocol, checking the stability and security of the connection, as well as the demonstration of exceptional situations; OpenHAB control – additional features of the OpenHAB platform (my.openhab, MailControl and Telegram), which allow to control SHS through the Internet, collect device statistics, receive notifications on users mobile devices.

So, the RL Smart House & IoT can be effectively used for investigation of the issues of energy efficiency, safety and cybersecurity of the SHS, as well as for solving the task of its comprehensive analysis, qualitative and quantitative assessment of the system indicators.

4   Case Study: FMEA of Remote Lab Smart House & IoT

As known, a lot of methods are successfully used for analysis of the systems' reliability, such as FMEA (Failure Mode and Effects Analysis), FMECA (Failure Mode, Effects and Criticality Analysis), FTA (Fault Tree Analysis), HAZOP (Hazard and Operability Analysis), RBD (Reliability Block Diagram), MM (Markov's models) [17, 18].

FMEA method was chosen and applied for identification of the most critical failures of the RL Smart House & IoT, with the aim of their eliminating or mitigating the aftereffect, as well as making recommendations for improvement of the system design. Architecture of the system was divided into three levels: management level, communication level and automation level (Fig. 2).

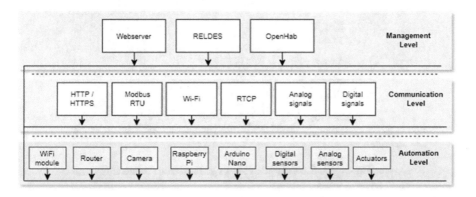

Fig. 2.   Tree-level architecture of RL Smart House & IoT

The Management Level includes components that provide system control: Webserver – software and hardware needed for RELDES functioning; RELDES – a web-oriented system, that provides control functions for the RL Smart House & IoT; OpenHab – integration platform of the RL Smart House & IoT, which also provides users interface.

The Communication Level includes all used protocols, interfaces for communication and data transfer, which can be implemented on the basis of wired and wireless technologies: HTTP/HTTPS and RTCP are used with the interaction of RELDES with the RL Smart House & IoT for sending control commands and video broadcast; Modbus RTU and Wi-Fi are used for data transfer between RL Smart House & IoT experiments and Raspberry Pi platform, which is the server of RL Smart House & IoT; Analog signals and Digital signals are the methods of data transfer from sensors to the Arduino controllers in each experiment. The data of these components are of particular importance, as the response of these experiments depends on this data.

The Automation Level includes hardware and software platforms (Arduino, Raspberry Pi), smart devices (web camera, router, Wi-Fi module), components (sensors and actuators) that make the basis of the system and provide basic functions.

The results of FMEA analysis includes information about possible failures and aftereffects and can be used at the design stage, as well as during the modernization and operation of the system. Based on this data, the architecture of RL Smart House & IoT cybersecurity was proposed (Fig. 3).

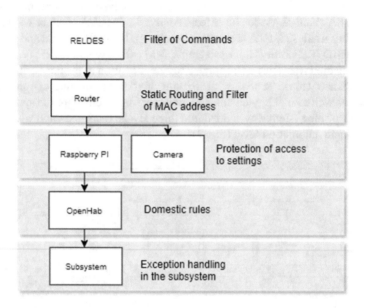

Fig. 3. RL Smart House & IoT cybersecurity architecture

The component RELDES protects the RL by filtering and overriding commands that are called by users to conduct experiments, excluding the sending of unauthorized commands. At the same time, control is available only to authorized laboratory users. The component Router performs command's routing and denies access if an extraneous device is invaded into the local network with a MAC address that was not reserved. In addition, the component is protected by a username and password to access the routing settings. Raspberry Pi is a component, that only accessible by a secure protocol SSH (Secure SHell). The component Camera requires authorization to prevent unauthorized resetting of settings. The component OpenHAB uses predefined rules that exclude the execution of unauthorized actions with actuators.

The component Subsystem corresponds to the RL Smart House & IoT experiment.

All subsystems are created on the basis of a typical structure, which allows integrating them into a common system effectively. A typical structure includes a communication module and Arduino controller for receiving, processing data from sensors and controlling actuators. In order to upgrade and improve the protection of the laboratory, the Serial interface that had been used previously was replaced by RS485 with the usage of Modbus protocol. This protocol guarantees the delivery of information packets, and if the data integrity is broken, the delivery is not performed. In the case of signal absence or master on the communication bus failure, each Subsystem has its own algorithm for correct completion of actions, which is implemented on the main controller of each subsystem. This protection concerns potentially dangerous laboratory devices that can disable the system and cause damage (for example, heating elements).

In this way, the RL Smart House & IoT is protected at the RELDES web interface level by redefining and filtering control commands, it is protected on the local network by the MAC address filtering with hard IP routing. Also, the lab is protected by the rules of the OpenHAB platform, which allow to exclude erroneous device control. At the controllers' level, simple conditions are realized to solve an exceptional situation, or to return the experiment to its original state where it is necessary.

From the point of view of RL reliability improvement, it was decided to develop an additional communication shield, the usage of which would reduce the number of connecting wires, thereby reduce the number of failures. The structure of the communication shield and the scheme of its interaction with the system components are shown in Fig. 4. Communication Shield is a platform for the placement of the Arduino controller, communication module RS485-TTL, as well as screw connectors, which provide a quality connection to the sensors and actuators of the subsystem.

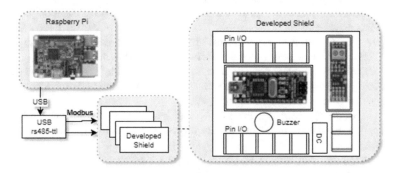

Fig. 4. Communication shield structure and interaction with RL components

5   Conclusions

The development of efficient and reliable SHS is an actual scientific and practical task, which requires a comprehensive solution based on modern methods and tools.

The RL can be effectively used as a prototype for studying of the issues of energy efficiency, safety and cybersecurity of SHS. The application of FMEA method for the SHS investigation can provide useful information about possible failures and their aftereffects, which can be used at the different stages of the system's life cycle.

References

1. Rucinski, A., Garbos, R., Jeffords, J., Chowdbury, S.: Disruptive innovation in the era of global cyber-society: with focus on smart city efforts. In: The 9th IEEE International Conference on Intelligent Data Acquisition and Advanced Computing Systems: Technology and Applications, pp. 1102–1104. University Politehnica of Bucharest, Romania (2017)
2. Gartner Survey Shows Connected Home Solutions Adoption Remains Limited to Early Adopters. https://www.gartner.com/newsroom/id/3629117

3. Haller, H., Nguyen, V.-B., Debizet, G., Laurillau, Y., Coutaz, J., Calvary, G.: Energy consumption in smarthome: persuasive interaction respecting user's values. In: The 9th IEEE International Conference on Intelligent Data Acquisition and Advanced Computing Systems: Technology and Applications, pp. 804–809. University Politehnica of Bucharest, Romania (2017)

4. Zhang, L., Xiong, X.: Optimization of the power flow in a smart home. In: Online Engineering & Internet of Things. Lecture Notes in Network and Systems, vol. 22, pp. 721–730 (2017)

5. Levshun, D., Chechulin, A., Kotenko, I.: Design lifecycle for secure cyber-physical systems based on embedded devices. In: The 9th IEEE International Conference on Intelligent Data Acquisition and Advanced Computing Systems: Technology and Applications, University Politehnica of Bucharest, Romania, pp. 277–282 (2017)

6. Hiromoto, R.E., Haney, M., Vakanski, A.: A Secure architecture for IoT with supply chain risk management. In: The 9th IEEE International Conference on Intelligent Data Acquisition and Advanced Computing Systems: Technology and Applications, pp. 431–435. University Politehnica of Bucharest, Romania (2017)

7. Best smart home system. https://www.techhive.com/article/3206310/connected-home/best-smart-home-system.html

8. Zhogov, N.: Communication protocols for the "smart house" (in Russian). https://www.ferra.ru/ru/digihome/review/SmartHome-Protocols/#1-Wire

9. TOR HOME ASSISTANT: How to protect Smart House with "anonymous" network? (in Russian). https://moy-domovoy.ru/news/20161214/tor_home_assistant_kak_zashchitit_umnyy_dom_s_pomoshchju_anonimnoy_seti/

10. Teslyuk, V., Beregovskyi, V., Pukach, A.: Automation of the smart house system-level design. Informatyka, Automatyka, Pomiary w Gospodarce i Ochronie Środowiska **4**, 81–84 (2013)

11. Tabunshchyk, G., Van Merode, D., Arras, P., Henke, K.: Remote experiments for reliability studies of embedded systems. In: International Conference on Remote Engineering and Virtual Instrumentation, Madrid, Spain, pp. 68–71 (2016)

12. Tabunshchyk, G., Van Merode, D., Arras, P., Henke, K. Okhmak, V.: Interactive platform for embedded software development study. In: Online Engineering & Internet of Things. Lecture Notes in Network and Systems, vol. 22, pp. 315–321 (2017)

13. Cybersecurity of Smart House (in Russian). http://www.bestron.ru/news/kiberbezopasnost_umnogo_doma

14. Smart home and cybersecurity: how to protect your data (in Russian). http://aquagroup.ru/news/umnyy-dom-i-kiberbezopasnost-kak-zashchitit-svoi-dannye.html

15. Parkhomenko, A., Tulenkov, A., Sokolyanskii, A., Zalyubovskiy, Y., Parkhomenko, A.: Integrated complex for IoT technologies study. In: Online Engineering & Internet of Things. Lecture Notes in Network and Systems, vol. 22, pp. 322–330 (2017)

16. Parkhomenko, A., Gladkova, O., Ivanov, E., Sokolyanskii, A., Kurson, S.: Development and application of remote laboratory for embedded systems design. iJOE **11**(3), 27–31 (2015)

17. Abdulmunem, A.-S.M.Q., Al-Khafaji, A.W., Kharchenko, V.S.: The method of IMECA-based security assessment: case study for building automation system. Syst. Inf. Process. **1**(138), 138–144 (2016)

18. Povolotskaya, E., Mach, P.: FMEA and FTA analyses of the adhesive joining process using electrically conductive adhesives. Acta Polytech. **52**(2), 48–55 (2012)

Human Machine Interaction
and Usability

Human-Computer Interaction in Remote Laboratories with the Leap Motion Controller

Ian Grout[(✉)]

Department of Electronic and Computer Engineering, University of Limerick,
Limerick, Ireland
`Ian.Grout@ul.ie`

Abstract. In this paper, the role of the human-computer interface for remote, or online, laboratories is considered. The traditional equipment used in such remotely accessed laboratories is based on a desktop or laptop computer with keyboard, mouse and visual display unit (VDU). This is typical for both the client and server side computing. In recent years, tablet PCs (personal computers) and smart phones have enabled remote and mobile access using device touch screen based communications, changing the manner in which the user interacts with the laboratory services. The remote laboratory therefore needs to be set-up in order to allow for different user interaction (input-output) requirements. However, other forms of user-laboratory interaction via alternative human-computer interaction (HCI) devices and approaches could also be accommodated. For example, hand position/motion/gesture control and voice activation are modes of HCI that are of increasing interest, driven in many cases by computer gaming and home automation requirements. To illustrate an alternative user control of the remote experiment, this paper will present an example remote experiment arrangement that uses the Leap Motion controller for hand position, motion and gesture control of laboratory test and measurement equipment.

Keywords: Remote laboratory · Human-computer interaction · Accessibility

1   Introduction

With the move towards an increase in developing robust whilst flexible access to software and hardware resources via the internet, such as those used in remote laboratories [1, 2], the end-user requirements now take on a greater emphasis. When the laboratory resources are used as on-line services, the needs and wishes of the end-user take on ever more importance. The user therefore must be considered in all activities from remote laboratory design, through development, to deployment. In addition, the manner in which the user can interact with the computing platform(s) can vary, based on aspects such as system requirements and user needs or preferences. Hence, the available forms of HCI and devices used to achieve the right level of interaction need to be considered. The purpose of this paper is to consider and discuss the role of the HCI [3] devices in remote laboratory access. With the different interaction devices possible, customization of the user interface for different HCI devices must now be considered. The paper will

© Springer International Publishing AG, part of Springer Nature 2019
M. E. Auer and R. Langmann (Eds.): REV 2018, LNNS 47, pp. 405–414, 2019.
https://doi.org/10.1007/978-3-319-95678-7_45

discuss different interaction devices, such as typical computing devices and alternative devices that are found in applications such as gaming and home automation.

To consider the alternative forms of remote laboratory access, a demonstrator system used to develop an approach using the Leap Motion controller [4] will be presented. This will use hand position, motion and gesture control of remote laboratory equipment and will be elaborated by using the Leap Motion controller to control laboratory test and measurement equipment via the web server. It would therefore be seen as a complement to the keyboard and mouse control of an experiment. The remote experiment presented will be based on a semiconductor device characterization experiment. Specifically, the diode will be considered in its different regions of bias (zero, forward and reverse). The user will control the experiment by moving their left hand over commands displayed on the user computer screen and a downward motion of the index finger will send the appropriate command to the remote laboratory. Figure 1 shows the experiment control principle. Results will then be automatically returned to the user and displayed for further analysis.

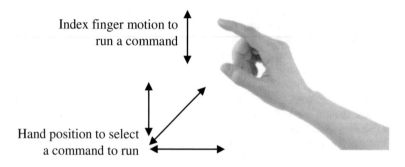

Index finger motion to run a command

Hand position to select a command to run

Fig. 1. Remote experiment control principle

The paper will be based on considering two key aspects. Firstly, HCI and its role in remote laboratories will be discussed and the different ways in which a user can interact with the remote laboratory using different approaches. Traditional keyboard and mouse, along with other forms such as non-contact approaches based on non-verbal position, motion and gestures, along with (verbal) voice control, will be identified as useful interaction approaches with the remote laboratory. Such different approaches allow for a personalized, or customized, approach that can be based on the individual user requirements. For example, HCI devices are used in a range of assistive technology (AT) [5–7] solutions enabling individuals with disabilities to interact with computing platforms and hence support personal and professional activities. The remote laboratory, if suitably designed, would be an aid to individuals to access laboratory equipment and hence, in an educational environment, can act as an assistive technology supporting access to education for all. Secondly, hand position, motion and gesture control of remote laboratory equipment will then be discussed with reference to an example design based on the use of the Leap Motion controller and a software interface based on the Processing software [8, 9]. An example system used to control of test and measurement equipment will be the focus of this example design.

The paper will be presented as follows. Section 2 will consider HCI and its use in accessing remote laboratories. Different forms of HCI will be identified and how they can interact with the laboratory experiment considered. Section 3 will consider HCI as a communications channel with an experiment will be discussed. Section 4 will consider a demonstrator experiment based on the use of the Leap Motion controller and the Processing software. Section 5 will conclude the paper.

2 HCI and Its Uses in a Remote Laboratory

HCI (also referred to as human-machine interaction (HMI)) considers the design and use of computer technology with a focus on the interfaces between people (the users) and computers [10, 11]. It considers a wide range of aspects in the creation of a computer based system that requires effective and efficient interaction with people. Such an approach is required for the overall system (people and machines) to work together in an optimal manner and to be intuitive for the person to use. The field of HCI is embedded in computer science, although this is just one aspect for consideration. With reference to Fig. 2 which shows different ways in which a person can interact with a computer, HCI considers the fields of engineering, design, computer science, ergonomics & human factors, ethnography, semiotics and branding, sociology, psychology and language.

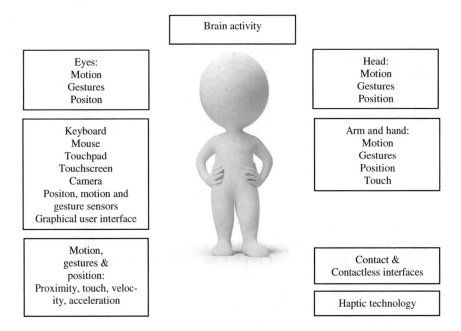

Fig. 2. User inputs and output

In Fig. 2, different parts of the human body can be used to provide laboratory control. A typical approach would be to use the hands and fingers to operate a keyboard and mouse or touchpad arrangement or, if using mobile devices, a touchscreen. Such approaches would require physical contact of the input device(s). Alternatively, prox-imity, position, motion and gesture control using a contactless interface could be used. Such approaches could use devices such as cameras, proximity sensors (e.g., magnetic Hall effect sensors), position/motion/gesture sensors (such as accelerometers and cameras with embedded digital signal processing (DSP)) and haptic technology. There-fore, it is possible to develop an interface that considers all parts of the human body that can be sensed as inputs to control the operation of a remote laboratory. Such an approach could be used to enable access to people with specific needs (e.g., [12]).

3   HCI Devices as a Communications Channel Between the User and the Remote Laboratory

If the control of a remote laboratory was to consider user input devices other than the keyboard, mouse, touchpad or touchscreen, it would be possible to incorporate suitable other forms of HCI into the user experience. It would depend on a range of needs that include those of the user, the intended outcomes from undertaking a remote experiment, the laboratory infrastructure and the experiment to perform. It is therefore possible to develop a laboratory system of the form shown in Fig. 3 that incorporates the necessary devices.

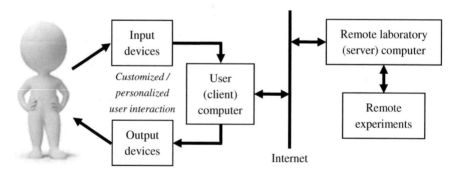

Fig. 3. User interaction with the laboratory experiment

In this generalized form, the choice of device can be made based on the user require-ments as to what types of input/output (I/O) devices would be most appropriate based on an assessment of the user needs and wishes. Each user could therefore have a custom-ized interface. The user (client) computer would deal with the customization aspects of the user experience in that it would be capable of working with various forms of I/O device without the need for any user intervention. The customization operations (set-up and use) would be transparent to the user.

4   Example Interaction Using the Leap Motion Controller

This section will consider how the Leap Motion controller is used and its application to a remote laboratory. This controller is a device that connects to the USB port of a computer and allows for a way to interact with a computer without the use of a traditional keyboard, mouse or touchscreen. Figure 4 shows the controller hardware. By using hand positions, motions and gestures, for example by moving the hands up or down, left or right, or by pinching the fingers, this data can be interpreted in order to control a computer software application.

Fig. 4. Leap Motion controller

This hardware is a form of NUI (Natural User Interface) in that it allows the user to interface with a computer using a form of interaction that is direct and consistent with the natural behavior of the user. Touch, gesture and voice based user interfaces are commonly referred to as NUIs. This form of user interaction can provide a useful way in which the user can interact with the computer without the need for any physical contact. It can also take into account situations where a user cannot use a keyboard/mouse arrangement. Instead of keyboard button presses or mouse movements, it essentially uses a non-verbal language to enter data and control values. With the Leap Motion controller, this is achieved by taking frames of the controller camera images and using software algorithms to interpret the image data. If the user interaction is correctly set-up, the user can readily learn the use of the system to be controlled without the need for a long learning path or reading of extensive documentation. However, whilst the user interface might be natural, there should be limits set as to what interaction can be supported in order to prevent unintended behavior of the system. Figure 5 shows the controller placed on a flat surface and connected to a laptop computer. Either or both hands are positioned above the controller and the hand positions, motions and gestures can then be captured. For example, a simple mode of interaction would be to place one

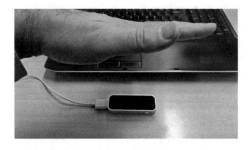

Fig. 5. Leap motion controller use

hand over the controller, to use the hand position in the x-y-z directions for location to select a command and then to use the fingers to run the required command.

In the design study considered in the paper, the remote experiment is an electronic circuit experiment where the user would investigate the current-voltage (*IV*) characteristics of three diodes (a silicon diode, a Schottky diode and a Zener diode). The diode operation under zero bias, forward bias and reverse bias conditions are controlled by varying a power supply voltage in the range 0 V to +5 V, selectable in 1 V steps. The user can select a bias condition, run the experiment and plot the results on a graph within the client side display. The controller is therefore used to enable or disable the power supply output voltage and to set and vary the power supply output voltages. The user operates the experiment by setting a hand position in the x-y-z directions for selecting the command and then use the index finger position to run the command. The Processing software was used in this study for interaction between the user, the controller and the remote laboratory experiment (both the web server and the hardware via a COM port connection).

The interface between the user (client) and remote laboratory (server) can be created using different software applications and approaches to how the user would interact with the experiment. In a typical arrangement, the user accesses the laboratory via an internet browser such as Microsoft *Internet Explorer*, Microsoft *Edge*, Google *Chrome*, Mozilla *Firefox*, Apple *Safari* and Opera Software *Opera*. The remote laboratory computer runs a web server software application and has a unique IP address to which it responds. Typically, an Apache HTTP Server (HTTPD) [13] is used and in many cases, the XAMPP [14] Apache Distribution is used. Alternative approaches use the LabVIEW Web Services from National Instruments™ and allow a user who is familiar with LabVIEW to create a LabVIEW application and deploy as a Web Service [15]. These are not the only possible approaches to developing the required software code. In this paper, Processing was used for client and server side software development. Figure 6 shows the overall system arrangement.

Figure 7 shows a photograph of the experiment hardware (front) and the power supply (back). The circuity on the prototyping board consists of the Arduino UNO board, the experiment circuit (six diodes and six resistors) and a MAX3232 "3 V to 5.5-V Multichannel RS-232 Line Driver/Receiver" integrated circuit (IC). By selecting the Agilent E3631A power supply [16] remote operation mode, the Arduino UNO can control the voltage to apply to the experiment. The diode experiment consists of the three diode types connected in series with a resistor to enable a current range (both forward and reverse bias) of 0 mA to +6 mA. This is achieved by using two circuits, one for forward bias and one for reverse bias as the +6 V output of the power supply is used. In reverse bias, only the 2.7 V Zener diode would show a current flow characteristic. To achieve the required current range, in forward (and zero) bias the resistor value is 1 kΩ. In reverse bias, the resistor value is 390 Ω. As the input voltage is only a positive voltage, to achieve reverse bias condition then the diode connections were reversed and the software was coded to differentiate between the different bias conditions. Figure 8 shows the client side (user) interface created as a Processing display. This is simple 2D layout with a control panel on the left side and a results display (graph

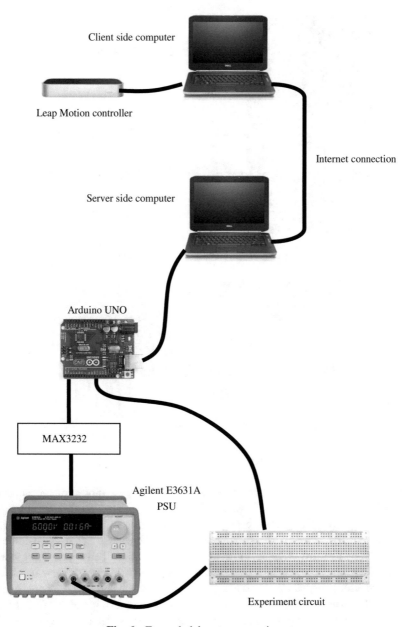

Fig. 6. Example laboratory experiment

on the right side). As the user moves their hand over a particular command on the control panel, it changes color (from grey to green).

Once selected, the index finger is gestured downwards to run the command. As a precaution, the command is run only once and the hand must be moved away from the selected command to enable it again.

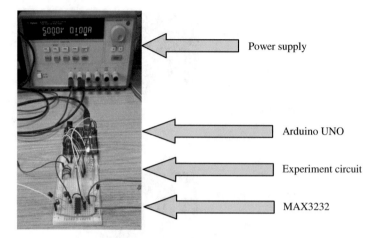

Fig. 7. Experiment prototype realization

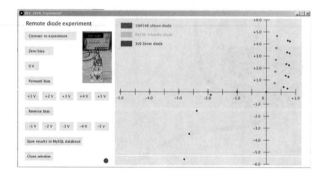

Fig. 8. Processing sketch display: client side

The graph in Fig. 8 shows the three diodes in forward bias (diode voltage and current are positive), zero bias (zero voltage and current) and reverse bias (diode voltage and current are negative). In forward bias, the user can see the *IV* characteristic as an exponential increase of current with a small increase in voltage as expected. Each diode used has a different forward voltage. In reverse bias, the diodes behave as expected. The silicon and Schottky diodes show near zero current whilst the Zener diode with a selected breakdown voltage of 2.7 V shows the current increasing with voltage, again as expected. The user can view the results on the graph and also identify values by positioning the hand over a reading and selecting the reading as well as saving the results to a MySQL database. In this arrangement, a number of specific user actions have been omitted such as a login/password access mechanism as well as selecting the values to the be stored in the MySQL database. It is assumed here that these operations have either been completed or have been automatically undertaken.

To run the remote experiment, assuming all prerequisites have been successfully completed, the user places their hand above the controller and the finger coordinates are identified. The system was designed for a distance of between 10 cm and 30 cm between the hand and the controller to select the command to run and a height difference of 5 cm between the middle and index fingers to run the command. The position of the middle finger is highlighted on the display and the user moves their hand left/right (x) and up/down (y) to position the middle finger over the command to run. To run the command, the index finger is motioned in a downwards direction and the height difference (of 5 cm) between the middle and index fingers indicates that the command is to be sent to the server. All hand based actions are also replicated by mouse motions and left button clicks. Figure 9 shows the server side (user) interface created as a Processing display. This is simple 2D layout that just provides the administrator with text messages on the experiment status and usage. As the administrator does not require a visual display with graphics, the visual appearance was designed to be simple and provide the basic information for monitoring purposes. In addition, the administrator can use a traditional keyboard/mouse arrangement or another form of HCI. Although in this design a web cam was not incorporated into the system, it would be straightforward to incorporate such a visual feedback mechanism to the server and client side displays. The experiment hardware is connected to an available USB (universal serial bus) port and written to/read from whenever a command is received from the user. Time delays were built into the experiment operation to allow operations to fully complete before committing to the next operation. A control panel is positioned on the bottom edge of the display window and the middle right portion of the display indicates server activity (*System log*) and client input (*User command*). For testing purposes, the client and server side Processing sketches were run on the same laptop computer using the local server IP address (i.e., *127.0.0.1*).

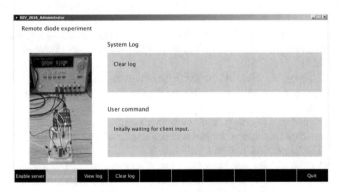

Fig. 9. Processing sketch display: server side

5   Conclusions

This paper has presented and discussed a remote laboratory prototype arrangement where a contactless interface is used to interact with a remote experiment. The Leap Motion controller was used in conjunction with Processing sketches to provide user interaction between the client and server computers. A prototype demonstrator design for a semiconductor diode experiment was developed and presented. The experiment was undertaken, via a remote laboratory arrangement, using a client/server arrangement. The demonstrator showed that remote laboratory control using hand motions and gestures is possible provided that the laboratory is suitably set up.

References

1. Henke, K., Vietzke, T., Hutschenreuter, R., Wuttke, H.-D.: The remote lab cloud "GOLDi-labs.net". In: 2016 13th International Conference on Remote Engineering and Virtual Instrumentation (REV), pp. 37–42 (2016). https://doi.org/10.1109/rev.2016.7444437
2. Zutin, D.G., Auer, M.: A simple LabVIEW based framework to facilitate the deployment of iLab batch lab servers. In: 11th International Conference on Remote Engineering and Virtual Instrumentation (REV), pp. 328–331 (2014). https://doi.org/10.1109/rev.2014.6784182
3. Microsoft: Being Human - Human-Computer Interaction in the Year 2020. http://research.microsoft.com/en-us/um/cambridge/projects/hci2020/downloads/BeingHuman_A4.pdf. Accessed 4 Nov 2017
4. Leap Motion: https://www.leapmotion.com/. Accessed 4 Nov 2017
5. The Institution of Engineering and Technology (IET), Engineering & Technology. Assistive Technology. https://eandt.theiet.org/tags/assistive-technology. Accessed 5 Nov 2017
6. Institute of Electrical and Electronics Engineers, IEEE, the institute. Special Report: Assistive Tech, 19 June 2017. http://theinstitute.ieee.org/static/special-report-assistive-tech. Accessed 5 Nov 2017
7. Harris, N.: The design and development of assistive technology. IEEE Potentials **36**(1), 24–28 (2017). https://doi.org/10.1109/mpot.2016.2615107
8. Processing: https://processing.org/. Accessed 4 Nov 2017
9. Reas, C., Fry, B.: Make: Getting Started with Processing, 2nd edn., Maker Media Inc., (2015). ISBN: 978-1-4571-8708-7
10. Artal-Sevil, J.S., Montañés, J.L.: Development of a robotic arm and implementation of a control strategy for gesture recognition through leap motion device. Technol. Appl. Electron. Teach. (TAEE) 1–9 (2016)
11. Nuamah, J., Seong, Y.: Human machine interface in the Internet of Things (IoT). In: 12th System of Systems Engineering Conference (SoSE) (2017). https://doi.org/10.1109/sysose.2017.7994979
12. Murray. I., Armstrong, H.: Remote laboratory access for students with vision impairment. In: Fifth International Conference on Networking and Services, pp. 566–571 (2009). https://doi.org/10.1109/icns.2009.107
13. Apache: HTTP Server Project. https://httpd.apache.org/. Accessed 4 Nov 2017
14. Apache Friends: XAMPP Apache + MariaDB + PHP + Perl. https://www.apachefriends.org/index.html. MySQL. Accessed 4 Nov 2017
15. National Instruments: LabVIEW Web Services FAQ. http://www.ni.com/white-paper/7747/en/. Accessed 4 Nov 2017
16. Agilent: Agilent E3631A, Triple Output, DC Power Supply User's Guide. Part Number: E3631-90002, 8th edn., 25 October 2013

Visual Tools for Aiding Remote Control Systems Experiments with Embedded Controllers

Ananda Maiti[✉], Andrew D. Maxwell, and Alexander A. Kist

School of Mechanical and Electrical Engineering, USQ, Toowoomba, Australia
anandamaiti@live.com, andrew.maxwell@usq.edu.au, kist@ieee.org

Abstract. Remote Laboratories are used to teach a variety of experiment in an online mode. One of the fields of RAL is control systems where various embedded controller or similar devices such as FPGA, PLC and MicroController Units (MCU) are the main focus of teaching and learning. For these experiments, the students upload a program to the device. Once the program is uploaded, it runs on the hardware and the students must verify that the program is running correctly by providing further inputs. This is confirmed is the sensors return the right data according to the proper time response. This paper proposes a visual tool in the form a coloring scheme for a virtual model in a 3D virtual environment to aid the students in understanding the changes in the rig with a virtual "twin" of the real rig in the virtual world. The sensor data is streamed to the client user interface from the experiment site and the client-side virtual rig model aligns with its real remote counterpart. The main challenges are the timing issues when the rate of change in data is fast such that it is difficult to determine the changing values. The sensor data is proceeded to find their derivatives and is checked for a minimum threshold. The resulting system allows for greater visualization and ease of understanding of the data.

Keywords: E-learning · Remote laboratories · Virtual reality
Programmable logic controller · Embedded controller · Visualization

1 Introductions

Remote Access Laboratories (RAL) provide online teaching technologies for students to run experiments on physical hardware through the Internet for educational purposes. There are different types of RAL experiments which require various architectures for the experimental setup and methods of integration into the teaching system. Common elements of the RAL experiments include a User Interface (UI) for input and output modules to control and observe the experiments. Experiments usually have a video feedback to show the actual of the experimental rig.

One type of RAL experiment is of a control system. These experiments often involve embedded controllers such as MicroController Units (MCUs) [1], Programmable Logic Controllers (PLCs) or Field Programmable Gate Array (FPGAs) [2, 3]. Typical architectures for such experiments consist of an embedded controller which connects to sensors and actuators of a linear, or nonlinear control system. Students are required to create a program in a particular programming language corresponding to the controller

© Springer International Publishing AG, part of Springer Nature 2019
M. E. Auer and R. Langmann (Eds.): REV 2018, LNNS 47, pp. 415–424, 2019.
https://doi.org/10.1007/978-3-319-95678-7_46

to solve a given control problem. Users have to upload the program to the embedded controller according to the experiment requirements. The program is then executed on the embedded controller and validated by a protection unit attached to the controller. As students typically program the controllers directly, it is usual to implement protection units to verify outputs of the embedded controller before they are executed. This ensures that unsuitable controller commands cannot damage the hardware. The embedded controller is interfaced directly to the Internet or through a separate dedicated server as a part of a remote laboratory system. The users typically observe the system states through camera feedback in real-time as well as output from other sensors.

For electro-mechanical models the structure is of a complex nature with potentially non-linear movement of rig components. Additionally, sensor locations and modelling can be complex. Most sensors are also not directly observable where there is no way to visually understand their values. Such information is lost in video streams.

In this paper a "visualization scheme" containing an animated virtual representation of the physical model is proposed to be used to represent the sensors data in real time. It can overcome the limitation of students not being able to handle and measure the values hands-on. This paper discusses a generic model for physical experiments and their virtual twin in an attempt to make control systems experiments that involve complex movements of various parts more accessible. This enables greater visualization [4] of the apparatus which will assist students in obtaining a better understanding of the underlying control task. They can easily relate their programs to the actual experimental setup and identify errors.

The rest of the paper is organized as follows: Sect. 2 describes the experiments with embedded controller in a RAL as well as visualization tools used in RAL as well as educational aspects. Section 3 presents the proposed architecture with embed controller, *coloring scheme*, virtual model, and learning goals. Section 4 provides an example of a lift using PLC and Arduino Mega. Section 5 outlines the advantages and disadvantages of the proposed system.

2   Related Work

This section presents the ways Embedded controllers are used in RAL as well as virtual reality tools used in RAL and education.

2.1   Embedded Controllers RAL

Embedded Controllers (EC) are devices with reduced architecture compared to a computer. These devices are programmable and widely used for industrial, embedded electronics and prototyping purposes. This class of devices are important in education for control systems and automation practical experiments. Such experiments have been implemented in RAL as well. FPGA has been introduced in context of automatic control and augmented reality before [7]. The MCU is a low-cost limited functionality alternative used for various designs of experiments [1]. Programmable Logic Controller (PLC) are routinely used in industrial settings for reliable control and automation.

The common properties between all these are that the students upload a program to the device. Students then view the remote hardware through visual camera feedback. The experiments have a validator component to ensure that the devices are not damaged during use. Once a program is running, the students can then provide inputs to the program. An experiment rig consists of many components connected to the actual controller which is also connected to the internet with a computer or directly in case of MCUs. Operating such an experiment requires the students to monitor several sensors to ensure that the experimental rig is working as expected for the experiments objective.

2.2 Virtual Reality

Virtual Reality (VR) is a new way of interaction between users and computers. In the context of desktops, this type of interface presents 3D objects generated by the user interface such as the web browsers. The 3D objects are positioned in a virtual 3D world. Such 3D worlds provide a camera viewport from where the objects can be viewed. VR is widely used for desktop gaming purposes. VR is now being used for training as well. It has also been shown to improve reaction time for users [6]. There are several metrics for measuring reaction time and in context of this paper the stop and signal reaction time of 0.5 s is considered. Reaction time is the average time for which an event needs to happen in VR world for it to be registered by the users. This value is obviously dependent on the context of the experiments. There are several libraries e.g. ThreeJS that enables to create VR objects or load existing models.

3   The System Architecture

This section describes the system architecture involving the embedded controller, color scheme and the virtual models.

3.1 The Embedded Controller Properties

In order to generalize the embedded controllers such as the PLC, MCU or FPGA across several experiments such that they follow a uniform *color coding*, the following properties are analyzed:

- In terms of the MCU, there is a set of ports or I/O connection points. These connection points are numbered and can be referred to in a program. The value of the port can be altered to affect the state of the rig. Alternatively, the port may be read to get the current status of the rig.
- The device can be programmed i.e. a program can be written involving common programming paradigms utilizing the ports in the hardware. The program is compiled into an executable and uploaded into the device. Every device will have limitations with regard to the amount of memory for program storage.
- MCU limitations may also relate to the amount of power consumed which is directly proportional to the amount of computation and the number of peripheral units e.g. sensors and actuators attached to the embedded controller.

Apart from the EC the experimental rig also has the following properties:

- The state space of the rig is limited. There can be only a finite number of states and a finite number relationship between the states. The program must adhere to the state space of the rig and not break the state change rules.
- Sometimes the state change must also be accompanied with strict time constraints. This means that the state changes must be completed within a definitive time frame for the proper output to result.

Figure 1 depicts the overall system model. The apparatus of rig includes a number of sensors and actuators that are connected to the protection unit. The protection unit is connected to the embedded controller (here a PLC). The unit is programmed by students via the Internet to execute the control task in question. The protection unit also provides feature vectors that describe the current state of the system.

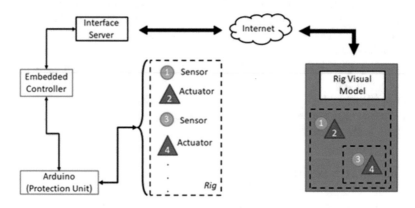

Fig. 1. The system architecture

The interface server integrates the experiment to the RAL system and connect to the client side. The client side contains the libraries to generate the 3D object in the virtual twin of the actual rig based on the data generated and sent by the EC. For creating an online experimental setup, developers must address the following:

Identifying Object Hierarchy
The experimental setup has many mechanical parts i.e. Sensors (S) and Actuators (A) pairings $R = \{s \in S, a \in A\}$ that can work independently and in relation with each other. The hierarchy and dependencies of objects must be implemented in the virtual model and uniquely identifiable.

Color Coding
A system of color coding is used to identify the dependent and independent objects. The color shades on the virtual objects on the 3D virtual model must comply with each item $r \in R$. The color intensity on the virtual model changes reflecting the current state of the actual equipment. The most critical part of the virtual model uses the critical colors i.e.

red followed by orange, yellow, green, and blue. The rig itself is of grey color in a blank background.

Camera and Timing Issues

The rig contains multiple moving parts and immobile parts such as sensors. The virtual models must be realigned to the view port corresponding to the relevant parts of the virtual model of the equipment where the action is taking place. This is done by rotating the camera view on the virtual objects as well as moving the light source and camera distance.

Another issue is the timing of coloring of the objects. It is very important as certain values i.e. states of sensors may appear for a very short period of time and a simple color changes indication may be too short to notice. Thus, a proper scheme is needed for notification of changes in the equipment.

Coloring Scheme with Warning and Data

The scheme is dependent on a few variables that can be set according to their suitability. Figure 2(a) shows two cases of sensor data change with time. In this figure, the Sensor 1 value changes more rapidly compared to Sensor 2. Two sets of points are placed on the graph (p, q) and $(p\prime, q\prime)$ where $t_p < t_q$ and $t'_p > t'_q$. t_p is the point where, the sensor data change is measurable. t_p is also the point of time when the changing data points can be brought to the attention of the users with visible object modifications. t_q is a point in time when the rate of change becomes too high for interpretation and registration by the users. In the case where rate of increment is very high it is necessary to identify the point t_q by pre-processing old session data (and possibly in real time during the session). It is possible that t_p has to be set before it is possible to detect the change in the sensor value. Essentially the difference between t_q and t_p must be static and suitable for user's learning goals. Conversely, the t'_q and t'_p are determined when the rate of change stabilizes and when the sensors data change becomes non-detectable.

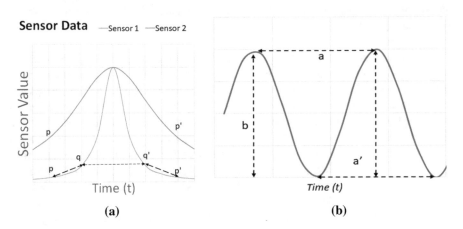

Fig. 2. (a) Sensor data change according to time (b) a cyclic data change from sensors

There is a minimum change that can be registered within a period of time interval for human interpretation in a virtual environment. This value is assumed to be 0.5 s [5]. During operation, sensor data is streamed to the user interface. The values are measures on the controller and sampled to generate and send at a rate of 100 Hz to 1 kHz. This means that there is a set of time T and corresponding value V for each sensor which may be obtained from running the experiment i.e. $v = f(t)$. This can be used to calculate the derivatives of all points. If the derivative for a time point $t \in T$ indicates that the rate of change of value is very high and will not meet the limit of reaction time (0.5 s), that point is determined by developers as the t_q. Correspondingly t'_q is chosen when the rate of change stabilizes. t_p is determined when enough change occurs that is noticeable for teaching and learning purposes. If no t_q or t'_q are determined, then they are set as $t_p = t_q$ and $t'_p = t'_q$.

Ideally, a value $d\left(\approx t_q - t_p \approx t'_p = t'_q \right)$ is set by the users which defines the *minimum warning time*. If the rate of change is slow then it is easier to pick time points and there is no need for t_q or t'_q to override the user settings for d. But if there is a sharp increase/decrease i.e. $d \approx 0$, then t_q and t'_q is determined. However, it may not be possible to honor the condition for d, in a *real-time session*, as t_q is too close to the peak value. In such as case, the warning time period can be held longer after t'_q even when the sensor has no change in value to satisfy at least $d = t'_p - t'_q$. If $d > t'_p - t'_q$, then the student is warned earlier or later by blinking the component before or after the actual change in its value. This allows the students to know that changes are occurring or about to occur at the component.

An extreme case of sensor data is where the change is binary i.e. the value goes from 0 to 1. The value of the t'_p is set according to the chosen d but t_p is irrelevant.

When the sensor value oscillates rapidly as shown in Fig. 2(b), the slope is steep or high and the value changes multiple time within a short time span. In this case, a' = a. This is dealt by rotating the color between {red, blue and green}. The changing color along with the intensity allows the user to identify the components changing value, but also follow which part of the cycle it is in. This is useful in depicting sensor values connected to motors for counting rotations and rotations will involve multiple 360-degree rotations.

Some components may have color coding even if they are not directly connected to sensors. This relation is determined by the *hierarchical* structure of the component relationship as described earlier. These components are connected to actuators and have a proportional relation to the main actuator and thus behaves with respect to the actuator. An example of this is for mechanical gears. While a main gear may be connected to a motor, the other gears still operate in relation to the main gear. Thus, while the main gear can be given a value corresponding to the sensor attached to it, the other gears may be color coded in the virtual model with a multiplying factor. The ratio of the gears determines the multiplication factor. For e.g. if Gear A with 30 teeth is connected to a motor and sensor, and Gear B has 20 teeth, then if Gear A has made a full 360-degree rotation, Gear B has moved 540-degrees. Thus, B's value is obtained by multiply 1.5 to A's value. The color coding reflects this accordingly.

3.2   Virtual Model of the Experiment

A virtual model of the experimental setup is presented in a 3D virtual world. Virtual models have the following components:

- *Static components*: These components are static throughout the experiment run and don't connected to any actuator.
- *Sensors component*: Position of the sensor with respect to immediate object is always static. However, the object on which it placed is an actuator, then the sensor moves. The sensors color changes as per the incoming data.
- *Actuator objects*: These objects consist of the actuator e.g. motor and its extensions e.g. shafts. These are the moving parts.
- Camera (Virtual): This provides a 3D point or viewpoint from where the user looks into the virtual rig.
- *Camera target*: This provides a specific 3D point in the virtual world usually a coordinate of an object of the experimental model.
- *Luminance*: Depending upon the experiment's nature the source of light and its vector towards the object may be set and altered during the experiment.

   A virtual object may not be an exact replica of the experimental setup, but must have a corresponding virtual object for each real object within the experimental rig. The virtual representation is rendered locally in the web browser window.

3.3   Learning Goals

The key visual and monitoring aspects for teaching these types of experiments based on embedded controller are:

- *Sensor/Actuator relationship*: Most sensors and actuators can be grouped in the equipment. An actuator is given a command and its position and orientation is measured with the corresponding sensor. In a RAL experiment utilizing a PLC (or similar controller), an actuator is almost always paired with at least one corresponding sensor. There can also be sensors that are independent of any actuator and return only data. The grouping can be one-to-one i.e. only one sensor measures one actuator as well as many to many. It may be possible that an actuator is not related to a sensor if it is used with a "timed constraint" to ensure discrete movements, in which case the actuator itself is colored separately based on its state.
- *Dependency and nesting of actuators*: An actuator can be dependent on another actuator in the state space i.e. a set of Sensor/Actuator $\{S_1, A_1\}$ can be dependent to another set $\{S_2, A_2\}$. This means that altering $\{S_1, A_1\}$ does not have any impact on $\{S_2, A_2\}$, but altering $\{S_2, A_2\}$ alters the portion of $\{S_1, A_1\}$.

4   A PLC Example with a Lift Model

This section describes PLC based "lift" example that uses the proposed visualization scheme.

4.1 Programmable Logic Controller

A PLC is used as a controller for a lift model as shown in Fig. 3(a). There are 4 levels of the lift and several hall effect sensors between each level. The PLC is programed using the 'Click' language. The lift is made of LEGO™ with motors connected to the PLC through an Arduino Mega which acts as a validator for commands from the PLC. The students are able program the lift box to be moved at different speeds with the aim of the program to ensure that the lift box can stop at different levels as required with perfect alignment. To achieve this, an accurate assessment of the speed of the lift must the determined by the students through different trials which prohibits the lift box from missing a level due to high speed. The Arduino utilizes WebSocket methods to stream data from the sensors to the client side.

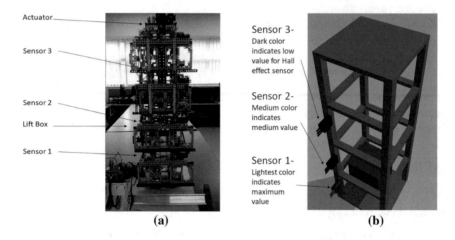

(a)   (b)

Fig. 3. (a) The lift built using a Lego parts, sensors, PLC and Arduino Mega, (b) 3D lift model

4.2 Lift Virtual Model

The lift model is created using a 3D object designer and in form of a PLY file as shown in Fig. 3(b) (*gray* color). The other objects contained in separate object files are the three hall effect sensors (*red* color) for each level to detect the current position of the lift box (*blue* color). Sensor data is used to determine the current level of the lift box. The lift has magnets attached to it and the value of the sensors change according to its current level. The range of the change of values is small.

The users provide their input to the program they upload to the PLC. The PLC then operates the lift according to further user inputs and determines where the lift box should go. One of the primary tasks of this exercise is to determine exact alignment of the "lift" with the "floor". This can only happen at certain points in time, and only when the value of the relevant sensor is at a peak i.e. the magnets are close to the hall-effect sensors. The corresponding 3D visual objects then change their color tone according to the current sensor value as streamed from the remote site, as shown in Fig. 3(b). If the speed of the lift is inaccurate, the values set for t_p and t'_p are not honored resulting in blinking of the visual component. The students can then identify problems with the speed and level to improve accuracy.

4.3 User Interface

The user interface contains several elements including the general description of the experiment, its aim, and instructions to perform the experiment. There is a platform to create and upload the program. The UI also enables the users to view camera feedback and the virtual twin model in two floating windows.

The virtual environment is built with ThreeJS library. The UI contains the WebSocket implementation that allows for streaming of sensor data from the remote site. The students can see the lift model and the camera feedback at the same time to determine the exact events occurring in the rig corresponding to the sensor data and the actuator movements. The user interface also has pan, tilt and zoom features to change the angle from which the virtual rig is viewed.

5 Advantages and Disadvantages

This section discusses the advantages and limitations of the proposed scheme.

5.1 Advantages

For the control systems experiments using embedded controllers, the user must compile and upload a program to the controller. The program then automatically runs and changes the state of rig. It is inevitable that the students would prepare a program that would be incorrect. The program validator will validate the program before allowing it to be uploaded to the controller. However, there may be states in the state space that can be passed by the validator as they will not break the rig. But these states may not be part of a correct program. With the proposed system, students will be able to identify the faults in their program. It simplifies the process of understanding the mechanical construction and state space of the equipment by enabling the users to view information not readily available through the simple camera view. Students can then identify any errors by observing the sensor values and altering their program accordingly. The students who program correctly will also understand the functionality of entire rig with respect to each sensor. In short, the system allows the students to visualize data that would be otherwise invisible or non-registrable to the camera.

The proposed system for determining the color scheme is generic for all embedded controller systems. The colors can be chosen according to the experiment nature, but all of them will implement similar techniques for choosing parameters for warning and visualization of sensor data.

5.2 Disadvantages

The main drawback of the current system is that the parameters (t_p and d) need to be set manually for each sensor based on trial session data. This is inefficient as each sensor will have different properties and ways it generates data within the rig. Another limitation is that the proposed scheme is designed for specific inflexible sample rates. Furthermore, the sampling rate must translate well into the communication rate. WebSockets are efficient in providing low latency and sufficient throughput for most experiments. But for experiments where the sensor data changes near instantaneously and reverts, the sampling rate can be supplemented with a notification system which identifies such events on the experiment site and send the information to the user side. The notification needs more manual setup from the developers of the experiment.

6 Conclusions

This paper proposes a new virtual color coding scheme what allows the users to identify the behavior of an experimental rig in response to the user inputs to their own program that is uploaded to the device. This is very useful within the context of control system experiments involving embedded controllers. This proposed scheme can be used uniformly across a wide range of experiments in a RAL Management System. This helps users to readily identify the sensors, actuators, and their relationship.

References

1. Maiti, A., Kist, A.A., Maxwell, A.D.: Real-time remote access laboratory with distributed and modular design. IEEE Trans. Industr. Electron. **62**(6), 3607–3618 (2015)
2. Santana, I., et al.: Remote laboratories for education and research purposes in automatic control systems. IEEE Trans. Ind. Inform. **9**, 547–556 (2013)
3. El-Medany, W.M.: FPGA remote laboratory for hardware e-learning courses. In: 2008 IEEE International Conference on Computational Technologies in Electrical and Electronics Engineering, pp. 106–109 (2008)
4. Gilbert, J.K.: Visualization: a metacognitive skill in science and science education. In: Visualization in Science Education, pp. 9–27. Springer, Dordrecht (2005)
5. Deleuze, J., et al.: Shoot at first sight! First person shooter players display reduced reaction time and compromised inhibitory control in comparison to other video game players. Comput. Hum. Behav. **72**, 570–576 (2017)
6. Dobrowolski, P., Hanusz, K., Sobczyk, B., Skorkom, M., Wiatrow, A.: Cognitive enhancement in video game players: the role of video game genre. Comput. Hum. Behav. **44**, 59–63 (2015)
7. Andujar, J.M., Mejias, A., Marquez, M.A.: Augmented reality for the improvement of remote laboratories: an augmented remote laboratory. IEEE Trans. Educ. **54**, 492–500 (2011)

Process Mining Applied to Player Interaction and Decision Taking Analysis in Educational Remote Games

Thiago Schaedler Uhlmann[(⊠)], Eduardo Alves Portela Santos,
and Luciano Antonio Mendes

Post-Graduate Program in Production and Systems Engineering,
Pontifical Catholic University of Parana, Curitiba, Brazil
tsu@tsu-it.com, {eduardo.portela, lmendes}@pucpr.br

Abstract. The usage of Game-Based Learning in engineering education, especially games and simulations to foster students interest and motivation, has been gaining attention by researchers. The present study has as main objective to explore possibilities of application of Process Mining techniques in the analysis of interactions among players, and the involved decision taking processes, in non-digital games. The analysis was applied to board games and to interactive remote experiments enhanced with ludic elements (remote games, as long as the game structure keeps similarity with board games). Activity logs collected during both board and remote game sessions were analyzed through process mining techniques available in ProM 5.2 software. From results, positive hypothesis were established regarding the applicability of process mining in finding behavior patterns presented by students playing educational games.

Keywords: Game-Based Learning · Process mining · Remote experimentation
Hybrid games · Game mechanics

1 Introduction

The use of active learning methods in engineering education is increasing, especially the use of games to simulate problem situations in different educational contexts across the engineering areas.

According to a systematic review conducted by Bodnar et al. (2016), there is evidence of the application of games in different modalities (digital, board or interaction-based) in production, manufacturing, mechanical, aeronautics, computing, among others.

Remote Experimentation (RE) is another major trend in engineering education. There is evidence of the application of RE with the use of game elements to teach STEM (Science, Technology, Engineering and Mathematics) areas, e.g. based upon the use of small and medium robots (Iturrate et al. 2013).

This exploratory work aims to help the development of game-based solutions that can contribute to the learning process, by introducing more immersive engineering practices both for the teacher (who proposes the activity and the problem) and for the

© Springer International Publishing AG, part of Springer Nature 2019
M. E. Auer and R. Langmann (Eds.): REV 2018, LNNS 47, pp. 425–434, 2019.
https://doi.org/10.1007/978-3-319-95678-7_47

student. The goal is to provide the latter a learning experience that is more pleasant, easier and based on situations that approximate those that he/she will experience in his/her future professional activities. It also aims to contribute to the improvement of the Remote Games concept (Uhlmann and Mendes 2016), which consists of RE's (physical experiments performed at a distance) enhanced with Game Based Learning elements.

That said, the research described in this paper has as main objective to explore possibilities of application of Process Mining (PM) for purposes of analyzing the interactions among players and the game elements in non-digital games, contemplating board games and the remote games whose structures have similarities with board games ones.

2   Methodology

The research described below has an exploratory character on possibilities of application of PM tools in order to analyze players interactions in GBL (Game-Based Learning) activities, considering aspects related to engagement, flow and immersion (Hamari et al. 2016). The analysis process specifically contemplated interactions that occurred in traditional board games and/or remote games: among participants; between participants and the game/learning mechanics, i.e., cooperation, collaboration, competition, exploration, experimentation, levels, feedback, rewards and penalties, as mapped by Arnab et al. (2015). The possible applications of PM tools in the analysis are then proposed in the form of hypotheses that can be tested in studies to be carried out in the future.

As to support the techniques of data collection and analysis used in this study, a literature search was first carried out in order to locate existing related research. The search was carried out on international databases in Engineering area, using a logical combination of search terms covering three axes: Process Mining, Weblabs, Game-Based Learning. The search terms have been structured so that located papers contain at least one term from each axis in Table 1.

Table 1. Research axes and search terms

Axis 1 Process Mining	Axis 2 Remote Experimentation	Axis 3 Game-Based Learning
"Process Mining"	"Weblab" "Remote Experimentation" Experiment*	"Game Design" "Serious Game" "Board Game"
SEARCH TERM ("Process Mining") AND (Weblab OR "Remote Experimentation" OR Experiment*) AND (Game OR "Serious Game" OR "Board Game")		

Searches were performed in August 2017. The following databases returned at least one related paper: ScienceDirect, Wiley, IEEE Xplore and SpringerLink. The number of publications located totaled 54. Next, publications were filtered with the criteria 'Relationship', 'Applicability' and 'Adherence'.

The criterion named 'Relationship' consisted of verifying the connection between title, abstract and/or keywords of each publication with the purposes of this study.

The criterion 'Adherence' consisted in reading the full text of the papers selected by the first filtering criterion, and discard those which did not address the aforementioned axes under the context considered in this work.

The third criterion, 'Pertinence', consisted of a second reading of the previously selected texts, in order to filter the papers whose scope of study was pertinent with the research purposes described in this paper.

The compiled results from the database searches are shown in Table 2.

Table 2. Papers by filtering criteria

	ScienceDirect	Wiley	IEEEXplore	SpringerLink
Initially returned	8	19	1	26
A – Relationship	4	11	1	4
B – Adherence	4	9	1	3
C – Pertinence	2	4	1	1

The second phase of this research consisted of a practical evaluation of the PM tools contained in ProM 5.2 software in order to verify the feasibility of these tools to be used in the evaluation of interactions and mechanics in Game Based Learning. For this purpose, records of board game sessions applied to groups of undergraduate students were collected. Also, records of actions were collected in a remote game.

Both records were submitted to PM using the aforementioned software, in order to evaluate the feasibility of its application. Once discovered, evidences of applicability were translated into hypotheses to be explored in future studies.

3   Related Work

Fu et al. (2014) describe a literature review with the objective of comparing and analyzing statistical methods for assessing the performance of players in Serious Games. PM is cited as one of these possible tools and is suitable for log files analysis of behavior patterns in the assessment after the game sessions.

Romero and Ventura (2013) describe the application of Educational Data Mining (EDM) in the analysis of interactions in educational games and in virtual reality environments. The application of PM tools, in this particular case, arguably promotes reflexions on the conformity of these educational techniques and the learning process.

A similar positioning is adopted by Reimann et al. (2014), using EDM with the objective of analyzing human performance regarding neurobiological processes and the environment with which it relates.

The use of EDM specifically for processes evaluation in the context of Game-Based Learning is addressed by Zhu et al. (2016). These authors used network schemes obtained through PM techniques - for example, Markov models and Petri nets - and empirical data collections, for the evaluation of SBT - Scenario-Based Tasks.

Harman et al. (2016) describes the use of S-BPM tools with the representation of characters in virtual worlds in order to improve the understanding of business processes. In this case, PM, as well as other methods described by the authors as robust, may result in an imprecise analysis of the results if they are not combined with supplementary methods that seek to contemplate the parties involved in these processes (stakeholders).

Furfaro et al. (2017) describe the process of developing a tool for evaluating security risks related to Internet-based devices of Things (IoT). PM was applied as one of the development activities of this platform, in the process of analyzing records of systems infected by malware, being the result used for the reverse engineering of malware that was applied as one of the simulated obstacles to be overcome in this virtual world.

From the above results, it can be stated that PM tools can be applied to the analysis and evaluation of player interactions in simulated environments under different aspects, as well as being used in the process of developing learning and game mechanics.

4   Data Collection Process

The systematic review described above did not find evidence of the application of PM processes in the evaluation of players behavior and game mechanics in Non-Digital Games or Remote Games.

ProM 5.2 software was selected for this study, due to its range of mining and analysis tools, and also because it is freely available for personal and academic use. This software relies on PM through the Alpha Algorithm (Van der Aalst et al. 2012).

The Alpha Algorithm is the base for the application of other methods used in ProM 5.2, including Markov Chain and Role Hierarchy Miner (Van der Aalst 2016).

The application of this software was carried out upon data samples took from the records of three game sessions, which were conducted under the supervision or participation of the researcher.

The first data record was made with the researcher's participation in a session of the board game PowerGrid (Friese 2004). This game simulates a chain of electricity supply and distribution, where players build, in cities located on a map, power plants that use coal, oil, rubbish, uranium or wind power resources, collecting financial results with correct decisions in the game. The game sessions are segmented into three phases, each phase being composed of shifts where players can perform three actions: buy a plant, buy raw material and build a network of distribution of energy in a specific place. The player that obtains the best financial result wins the game. Figure 1 illustrates the components of this game. In total, two players participated in the game session, besides the researcher.

Fig. 1. PowerGrid board game.

The second data record was performed with the application of a session of the game Container (Delonge and Ewert 2007) as part of a pedagogical activity in a Management course in a higher education institution. In this game, players simulate a production chain for a commodity, encompassing purchase, production, distribution and sale activities. The products, represented by containers in five different colors, had monetary values differentiated for each player. At the end of the game, the player that obtains the highest financial income wins. Figure 2 illustrates the game session.

Fig. 2. A session of container board game.

The third data record, unlike the previous ones, consisted of the collection of actions performed by the players in a remote collaborative game located in a production engineering laboratory in a Brazilian university. This game, called LyMIE Kanban Remote Game (Uhlmann and Mendes 2017), consists of an experiment that simulates the usage of warehouse inventory by a production system, requiring the player, according to his continuous observation of these stocks, to administer Kanban cards to avoid wasting resources. The utilization rate of these stocks is simulated by means of the light intensity of the environment where the experiment is located. Such intensity is collected by the system using an Arduino Mega board, connected to a photosensitive cell. The player, through access to a web page hosted in a cloud server, remotely observes the stocks of the warehouse and sends his actions to the game: to assign a

Kanban card (green, yellow or red) or to order a stock replenishment. If the card is wrong for the stock quantity of the warehouse or for each ordered replacement, the demand for the stock is also intensified. The actions are recorded in a MySQL database with a letter corresponding to each one: G (Green), Y (Yellow), R (Red) and F (Stock Reposition). The game ends when the stock in this simulated warehouse reaches a null value. The registration of the game sessions, in this case, was collected in the game databases located in the cloud server of this experiment, located on the Internet. Figure 3 illustrates the physical system as well as the website for this remote game.

Fig. 3. LyMIE Kanban remote game.

The following PM tools available in ProM 5.2 software were applied to the collected data: Alpha Algorithm, Role Hierarchy Miner and Markov Chain. The application process, as well as the establishment of hypotheses for future studies, are described next.

5 Data Analysis and Discussion

Data preparation for analysis using ProM 5.2 software was performed by structuring a table in XML format. Due to the nature of each game, and limitations encountered by the researcher at the time of data collection, the variety of data collected may not correspond to all types of data that may be collected in each game.

Fig. 4. Log summary applied to container board game.

A quantitative analysis was applied to the logs of game sessions through the Log Summary tool. It allows to verify the main events of the game session, as well as the most frequent occurrences of each action. In the case of the Container game, it was possible to verify the most frequent sequences of actions taken by the players, as well as the frequencies of interactions between players (Fig. 4).

The first mining process tested with the collected game records was Alpha Algorithm, resulting in a diagram which describes the logical sequences of actions adopted by the players. In the case of PowerGrid game, it was possible to visualize, graphically, the decision patterns adopted by players – e.g., buying/building power plants. From Alpha Algorithm, it was possible to detect, for example, that most of the decision sequences started from the decision to buy a specific plant. Thus, a positive hypothesis is established about the potential of Alpha Algorithm for the verification of the elements that influence or initiate decision-making processes in a game. Figure 5 shows the diagram obtained from the application of Alpha Algorithm to the aforementioned game.

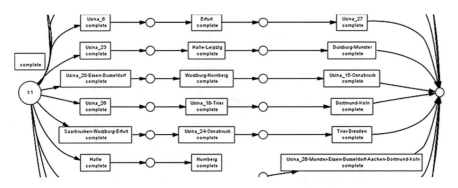

Fig. 5. Alpha algorithm applied to PowerGrid board game.

Additionally, it was found evidence of effectiveness of ProM 5.2 Role Hierarchy Miner method in the analysis of interactions between players, especially to the Container game, in which they occur constantly. Through the application of these method, it was possible to show which resources (in this case, players) interacted in a specific round, and what actions were performed in this interaction process. Thus, it can be stated a positive hypothesis on the applicability of the Role Hierarchy Miner method to discover and analyze behavioral patterns and possible tactical positions among players during a given game, and its results can be used as an input in assessment processes after the execution of the games (Fig. 6).

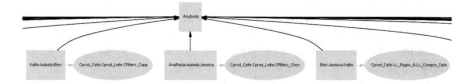

Fig. 6. Role hierarchy miner application for container board game.

In the case of games with no interaction between players, as in LyMIE Kanban Remote Game, the use of the Markov Chain method was useful to verify tendencies of actions adopted by players, as well as the frequencies of the decision patterns (shown with the action flow arrows). Thus, the hypothesis that this method can be used for the analysis of sequences of actions, beforehand to more in-depth methods, can be stated. Figure 7 illustrates the application of the Markov Chain method to the logs of the sessions of the analyzed games.

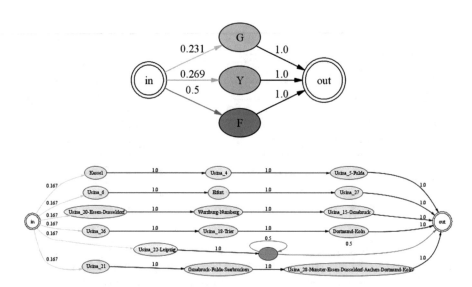

Fig. 7. Markov chain application for LyMIE and PowerGrid board games.

6   Final Considerations

The main purpose of this study was to explore possibilities of application of PM tools to non-digital games and remote games, in the analysis of player interactions, decision taking processes (as per the game mechanics).

From a systematic review, it was found that PM tools are used mainly in the analysis of players performances, and in studies related to EDM - Educational Data Mining. Studies directly addressing applications of these tools in the context of remote games or non-digital games were not found.

Results revealed several possibilities of applying PM methods to evaluate players performance in non-digital gaming sessions, including remote gaming and board games. Thus, it was possible to formulate positive hypotheses regarding the applicability of these methods - which can be further, throughly tested in future studies in order to accept or reject them – in finding behavior patterns presented by students playing educational games.

As for future work, the study may be extended to test other PM tools, with other board and remote games, under themes different from those presented in this study, broadening the basis for the generalization of conclusions.

Acknowledgment. The authors would like to thank the researched educational institutions, as well as the participants of the game sessions, for making the time and resources available to enable this research work.

References

Arnab, S., et al.: Mapping learning and game mechanics for serious games analysis. Br. J. Educ. Technol. **46**, 391–411 (2015)

Bodnar, C.A., Anastasio, D., Enszer, J.A., Burkey, D.D.: Engineers at play: games as teaching tools for undergraduate engineering students. J. Eng. Educ. **105**, 147–200 (2016)

Delonge, F.B., Ewert, T.: Container. Board Game. Valley Games Inc. (2007)

Friese, F.: Power Grid. Board Game. Rio Grande Games (2004)

Fu, J., Zapata, D., Mavronikolas, E.: Statistical methods for assessments in simulations and serious games. ETS Research Report Series (2014)

Furfaro, A., et al.: Using virtual environments for the assessment of cybersecurity issues in IoT scenarios. Simul. Model. Pract. Theory **73**, 43–54 (2017)

Hamari, J., Shernoff, D.J., Rowe, E., Coller, B., Asbell-Clarke, J., Edwards, T.: Challenging games help students learn: an empirical study on engagement, flow and immersion in game-based learning. Comput. Hum. Behav. **54**, 170–179 (2016)

Harman, J., et al.: Augmenting process elicitation with visual priming: an empirical exploration of user behaviour and modelling outcomes. Inf. Syst. **62**, 242–255 (2016)

Iturrate, I., et al.: A mobile robot platform for open learning based on serious games and remote laboratories. In: 1st International Conference of the Portuguese Society for Engineering Education (CISPEE) (2013)

Reimann, P., Markauskaite, L., Bannert, M.: e-research and learning theory: what do sequence and process mining methods contribute? Br. J. Educ. Technol. **45**(3), 528–540 (2014)

Romero, C., Ventura, S.: Data mining in education. WIREs Data Min. Knowl. Discov. **3**, 12–27 (2013)

Uhlmann, T.S., Mendes, L.A.: Jogos Remotos: perspectivas de aplicação conjunta de Aprendizagem Baseada em Jogos e Experimentação Remota no ensino de engenharia. In: COBENGE 2016 – XLIV Congresso Brasileiro de Educação em Engenharia, Natal, Brazil (2016)

Uhlmann, T.S., Mendes, L.A.: Proposta de arquitetura para um sistema de comunicação e o intercâmbio de dados em Experimentos Remotos voltados para a Aprendizagem Baseada em Jogos. Anais do Simpósio Ibero-Americano de Tecnologias Educacionais. 1(1), 278–286 (2017)

Van der Aalst, W.: Process Mining: Data Science in Action. Springer, Heidelberg (2016)

Van der Aalst, W., Adriansyah, A., Van Dongen, B.: Replaying history on process models for conformance checking and performance analysis. WIREs Data Min. Knowl. Discov. 182–192 (2012)

Zhu, M., Shu, Z., Davier, A.A.: Using networks to visualize and analyze process data for educational assessment. J. Educ. Measur. 53(2), 190–211 (2016)

An Approach to Teaching Blind Children of Geographic Topics Through Applying a Combined Multimodal User Interfaces

Dariusz Mikulowski[✉]

Faculty of Sciences, Siedlce University of Natural Sciences and Humanities,
Siedlce, Poland
dariusz.mikulowski@ii.uph.edu.pl

Abstract. Nowadays all pupils have easy access to different kinds of educational games that use images, sound and video. Despite the rapid development of computer technology the blind children meet a great restrictions with access to this type of content. The main reason may be that the producers do not perform accessibility requirements by their games. Otherwise, it is not cost-effective to create specifically tailored multimedia learning materials for the blind.

Then there is a great need to better sharing of multimedia educational content for this group of users. In this paper, an approach consisting in the replacement of images and video in usual educational game through a combination of audio recordings, synthetic speech, sound signals, electronic Braille, and convex graphics is proposed. An example of such interfaces was implemented as an educational audio game European Encyclopedia of Geography for the Blind that was developed for commercial sale.

Keywords: Multimodal interfaces · Blind children education
Alternative content

1 Introduction

Nowadays, most of the existing educational applications and in particular, computer games are not suitable to use for blind users. The reasons for that are different. However there are requirements regarding the accessibility of the web pages published by W3C [1], but without any guidelines for video games. Despite that, some of these requirements could be applied also in the case of creating games. More over, manufacturers often do not have motivations and funds to meet these requirements. In addition, the market sector of the blind is very small what causes that the creation specifically tailored multimedia learning materials for them is not cost-effective. On the other hand, requirements of blind user are sometimes opposite to the needs of low vision users. Despite this, there are few implementations of the classic board games i.e. chess or sapper and so-called paragraph games, but their scope is very limited.

© Springer International Publishing AG, part of Springer Nature 2019
M. E. Auer and R. Langmann (Eds.): REV 2018, LNNS 47, pp. 435–442, 2019.
https://doi.org/10.1007/978-3-319-95678-7_48

Taking it to account, it seems that there is a great need to take a step towards a better sharing of multimedia educational content for this group of users. In this paper, an approach consisting in the replacement of images and video in usual educational game through a combination of alternative interfaces in one time and place is proposed. Such interfaces are: audio recordings, synthetic speech, additional technical sound signals, electronic Braille, convex graphics embossed on the braille printers and even models made with 3d printer. The use of such interfaces set was implemented as the example application from the field of geography.

2   Related Works and Solutions

As we mentioned above, most of the existing computer games and other educational applications are not suitable to use for blind users. The only exception may be the implementations of the classic board games such as chess or sapper. But why the blind should be deprived of access to multimedia learning materials? Sanchez and Darin [2] have investigated the role of multimodal components in the development and evaluation of games and virtual environments targeting the enhancement of cognitive skills in people who are blind. They stated that multimodal games are a factor that positively stimulate the cognitive process of the blind children and help them to learn a new subjects. Although they found about seventeen multimodal games designed for the blind but there was only four proposals of models [3,4] explaining how to project and implement such applications. In turn, in the publication [4] a video game where cognitive map of knowledge is created from input data prepared by a teacher is proposed. This approach is very similar to the solution proposed in audio game untitled "Heart of the Winter 1812" [5]. In this production that was carried out on a large scale with the participation of professional actors and with music specifically wrote for the game.

A different studies also confirm that the blind pupils prefer the elements that are read by professional speakers than those that are read by a speech synthesizer [6]. The most games that are created for the blind have a graphical interface, but it is used rather to control the application than for content presentation. Some games have dual interface designed rather for blind and also for low vision users. But there are some (about 14%), that are dedicated only for the blind [2].

There are also attempts to use speech recognition technique, but it is very rarely method. The reason is that the blind are usually familiar with a keyboard and they dint need to speak to computer using voice. From other hand, that it is still difficult to implement speech recognition technique in some languages.

The most known portal that gathers information about games for the blind is "Audio games" [7]. It is worth noting, that out of more than 600 games that are stored in this collection only 9 was qualified in the category of educational games. Other several typical games on this portal such as logic games puzzles or the implementations of traditional games such as chess, or Chinese checkers can be classified as educational. But the most of the available productions that can be found there are entertainment games.

Therefore, there is the need to use educational multimodal applications for enhanced teaching blind students.

3   Applying an Aural User Interfaces for Educational Content

The multimedia games designed for sighted users have many UI elements, which make easy and attractive access to their main educational content. They are: solid graphics, videos, interactivity with the user, music and so one. Unfortunately, most of these items are not available for blind students. However, from one hand, each of these media can be replaced by another mechanism or interface that is available for them. In other cases, each inaccessible medium can be replaced by combination of several other interfaces or techniques. This can at least partially compensate no access this type of user to standard methods and media used in learning process. In the following few sections, a proposal to build such aural UIs for the blind pupil in sample educational game is described.

3.1   Using Speech Synthesizers and Screenreader

For many years the synthetic speech is used to provide access to computer features for the blind. But the speech itself is not merely as important as information that the computer says to the user. Therefore, any intermediate layer fulfill the gap between this what appears on the screen and what speech synthesizer really reads for the user is necessary. This layer is implemented as a software known as screenreader. There have already been a screenreaders for textual old systems such as DOS and different Unix family systems.

With the arrival of graphical systems i.e. Windows a new problem was appeared. How to read a graphical elements such as icons, widgets, frames, arrows, and pictures that are displayed on the screen. It was solved by describing them through the respective voice tag. Fortunately, there are screenreaders for almost all mobile and stationary operating systems. Due to this, each application that is programmed using standard controls is a well-articulated program for screenreader. However, in the case of video games the situation is even worse, because they are usually programmed using the nontypical graphical interface elements. Therefore, the accessibility mechanism of the game can be implemented natively with corresponding programming interface to the speech synthesizer that is built into the operating system. In the case of windows it is a SAPII interface. If the game is implemented using standard components, it features are readable by a screenreader. This approach was applied in sample educational game concerning geography described in this article.

3.2   Applying a Keyboard Shortcuts

Due to the nature of their disability a completely blind people usually works on a computer without a mouse. The cause is that most screenreaders does not support mouse movements in a satisfactory manner. In addition, it is very difficult for blind person to move accurately with the mouse on the Mat. Despite that blind users can currently use touch displays, but performing touch gestures on big staggering screen of a laptop or tablet is not convenient for them. Therefore, applications designed for the blind should be implemented for full using the keyboard. If the application is created using standard controls, such as dialog boxes, lists, menus etc., then navigation keys, letter keys, tab and shift keys can be used to move through the subsequent elements of the window or enter data to it. A game developers who don't always use standard components should take care of the full service level of the keyboard. In addition, all functions in such game must be accessible from a keyboard by intuitive hotkey e.g. pressing ctrl+s should lead to save the state of the game. The sample educational game implementing the method of aural UIs discussed in this article has full support for keyboard and a set of hot keys to facilitate user interaction with the program.

3.3   Enrichment of Video Material

Let us suppose, that in traditional educational application we have a material presented as a video sequence. So the question arises: how to share such material to a blind student? A solution may be to create the alternative content concerning with this video material and attach it to application. It can be an additional audio track so called "audio description" [8] which tells the user what is actually happening on the scene. During the creation of audio description is important that the soundtrack that describes the scene won't match the actors' dialogues. Moreover, the contents of narration should be brief and devoid of emotions but it should carrying as much information as possible, for example: "John turned away, raised his right hand and threw the stone."

3.4   Using an Auxiliary Sound Signals

Another element that supports the work of the blind user with the program can be auxiliary technical sound signals. This method lies in the fact that, for example, when the user opens the menu, he hears the sound of opening door. So, when he closes an application window, or menu options, he may hear also the corresponding sound signal e.g. sound of closing door. Another adequate sounds can be played when dialog box appears on the screen and also when application starts or finishes. However the auxiliary sounds cannot replace the information that is read via a speech synthesizer, but they are faster than speech synthesis and allows user to quick work with the program. Let us pay attention at this point to requirement that technical sounds chosen for the game must be the most intuitive as possible, however there is no common standard how to make such choice.

The sample educational game concerning a geography discussed in this article has been equipped with a set of technical sounds facilitating the students work. For example, when the student has resolved a quiz with questions about some country and its window has closed, then the sound of applause is played, which informs him on the correct solution he has made. But when the student has answered a question incorrectly, then prescribed sound, signifying disapproval is played. So let us notice, that with the help of such auxiliary sounds, the use of the game becomes more enjoyable and convenient.

3.5   Using Music and Oral Ilustrations

Another way to decorate educational program can be musical illustrations and narratives recorded by professional lecturer. It can describe the more important content items of the game.

The exemplary game discussed here has also such music and narrative elements. For example, if the student comes to information on any country, he can listen to its national anthem. He also can listen to information about this country, read by a professional speaker. So convenient to do this, the user must have access to a simple player, equipped with a few basic functions such as playback, stop playback, pause, fast forward or rewind, etc. All these functions should be accessible through shortcut keys. It is obvious that this method of presentation for the student is more interesting than listening to the usual textual information spoken by the speech synthesizer.

3.6   Applying a Convex Graphic and Description

The biggest barrier in access to different types of information for visually empaired users is the inability to view graphics especially the graphics presented on the computer screen. Therefore, in order to reach it for needs of a blind, the other solution should be used. One of them may be simultaneous synchronization of the following three user interface and physical elements:

1. text information supplied via a speech synthesizer,
2. tactile graphics printed on paper in the form of drawing, (Braille or on convex paper) along with the corresponding braille labels,
3. legend (printed in Braille and as an electronic text) where all symbols used on a drawing are explained.

Let us imagine, that in sample educational game - "Multimodal Encyclopedia of European Geography" a graphic explaining the outlines of any country i.e. Poland is placed. Then for needs of a blind student, a textual description explaining the shape of Poland with information about what other countries it borders is read by a speech synthesizer. In addition, the user is notified with a flash message, on which side in a paper braille book attached to the program the convex map of Poland and its legend can be found. An example drawing explaining map of Poland that is displayed on the screen in a training game discussed here is presented in Fig. 1.

Let us notice, that in such situation, blind user cannot see this map on his screen. But fortunately, due to message that he received from the program, he can find and touch paper braille drawing of a Poland map in attached book.

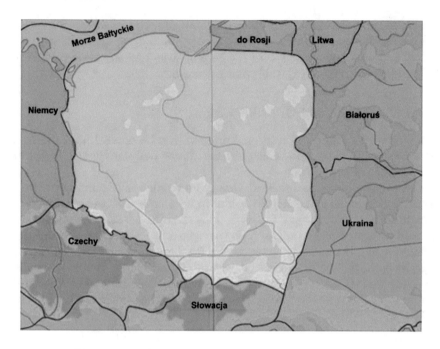

Fig. 1. Example map of Poland from sample application

An example drawing explaining the same map of Poland but in the form of a convex braille graphic printed on a paper and attached to a program is presented in Fig. 2.

However there is just another problem with braille graphics. Because the Braille graphics should be easily readable with touch it should not be too detailed and complicated. So only a small 2 3 letter symbols of the most important places, as well as small symbols, e.g., circles, squares or triangles that represent cities, rivers, lakes, etc. may be printed. Obviously, These symbols must be explained somewhere. For greater convenience, they can be explained in two places. At first in the application in electronic text form. Then user can obtain it by pressing, for example help key F1. A second place is a braille book attached to the program. In this book, a legend that explains the symbols used on the drawing is printed on the page next to the illustration. A example legend explaining braille symbols for drawing from Fig. 2 is presented in Fig. 3.

Let us assume, that only use all these three components in the same situation in the game can partially replace the graphic which the sighted user can see on the screen.

4   Summary

In this paper, an approach consisting in the replacement of the visual elements of user interfaces in the educational applications for the blind children by

Fig. 2. Example map of Poland in braille

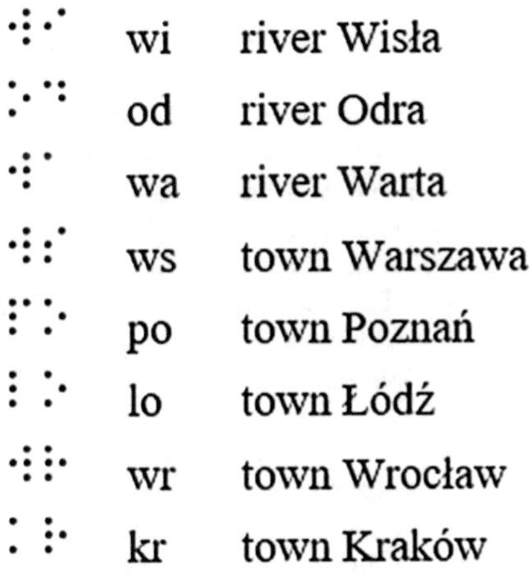

	wi	river Wisła
	od	river Odra
	wa	river Warta
	ws	town Warszawa
	po	town Poznań
	lo	town Łódź
	wr	town Wrocław
	kr	town Kraków

Fig. 3. Legend of a braille drawing explaining map of Poland

combination of alternative multimodal interfaces was presented. The use of alternative user interfaces described above was implemented as the example application from the field of geography. It has been tested and implemented in commercial sales.

However the exemplary implementation mentioned here relates to issues of geography, the method described above can be successfully applied to create solutions from various fields of education for blind pupil. An example of such other application may be the software created in the frame of PlatMat project [9], which dealt with issues related to learning mathematics of blind pupils.

The solution explained in this paper was implemented as an educational game untitled "The Multimedia European Encyclopedia of Geography for the Blind". The application was really developed and is currently sold as a commercial product by one of the companies involved in distribution of hardware and software for the blind in Poland [10]. The real deployment of the game has allowed us for collecting observations and suggestions from customers. The Elements that were well received by the users that were: an application of the simultaneous illustration of geographical questions using textual information and braille graphic, using of musical and narrative illustrations and the application of auxiliary sounds to aid in navigation across the game interface. The negative comments were: not quite intuitive search mechanism for information within all application and deactualization a substantial part of the material what necessitates the frequent creation of new versions of the program. Let us add that the project of encyclopedia was awarded the Golden medal at the exhibition of inventions "Invento" in Prague.

References

1. WCAG Overview—Web Accessibility Initiative (WAI)—W3C. https://www.w3.org/WAI/intro/wcag. Accessed Dec 2017
2. Sánchez J., Darin T., Andrade R.: Multimodal videogames for the cognition of people who are blind: trends and issues. In: International Conference on Universal Access in Human-Computer Interaction. Access to Learning, Health and Well-Being, UAHCI 2015. LNCS, vol. 9177, pp. 535–546 (2015)
3. Sánchez, J.: A model to design interactive learning environments for children with visual disabilities. Educ. Inf. Technol. **12**(3), 149–163 (2007)
4. Sánchez, J., Espinoza, M.: Designing serious videogames through concept maps. In: Kurosu, M. (ed.) HCII/HCI 2013, Part II. LNCS, vol. 8005, pp. 299–308. Springer, Heidelberg (2013)
5. Serce Zimy 1812 - Interaktywny Audiobook. http://www.sercezimy.pl/. Telekomunikacja Polska. Accessed Nov 2017
6. Pucher, M., Zillinger, B., Toman, M., et al.: Influence of speaker familiarity on blind and visually impaired children's and young adults' perception of synthetic voices. Comput. Speech Lang. **46**, 179–195 (2017)
7. Audio Games. https://www.audiogames.net/. Accessed Nov 2017
8. Lytvyn, V.V., Demchuk, A.B., Oborska, O.V.: Mathematical and software submission video content for visually impaired people. Radio Electr. Comput. Sci. Control **3**, 73–79 (2016)
9. Andrzej, S., Jolanta, B.-P.: Translation of MathML formulas to Polish text, example applications in teaching the blind. In: IEEE 2nd International Conference on Cybernetics (CYBCONF), Gdynia, Poland, 24–26 June, pp. 240–244 (2015)
10. P.H.U Impuls. http://www.phuimpuls.pl/?pobierz,25. Accessed Nov 2017

Development of a Virtual Environment for Environmental Monitoring Education

Jeremy Dylan Smith$^{(\boxtimes)}$ and Vinod K. Lohani

Virginia Tech, Blacksburg, VA 24060, USA
{smithjer,vlohani}@vt.edu
http://www.lewas.ictas.vt.edu/

Abstract. This paper presents details design and considerations to pilot an educational virtual environment (VE) development. The environment is designed for Freshman engineering students to achieve learning objectives regarding (1) engineering model understanding and prediction, and (2) experimental setup in the context of environmental education. Human computer interaction (HCI) methods are used to inform the design of the virtual environment of the existing physical site through user task analysis. Environmental data is displayed to exhibit the inverse relationship between two water quality parameters to the students. The virtual environment is provided through the use of a smart phone based head-worn display (HWD). Students reported being distracted, which are mostly due to hardware constraints (n = 14).

Keywords: Virtual environment · Education · Mediation interfaces
Usability · Fidelity

1 Introduction

The U.S. National Academy of Engineering identified providing access to clean water, advancing personalized learning, and enhancing virtual reality as three of their fourteen grand challenges for the 21st century [1]. We posit that these challenges may be addressed simultaneously, using technology to address educational needs. Our development is motivated by the democratization of science and engineering education to promote knowledgeable global citizens who are mindful of water-related issues concerning their communities.

The novel virtual environment (VE) is the next development in a long line of high-frequency environmental parameter learning developments. It builds upon an existing environmental monitoring system which has been implemented in at least twenty-nine courses at eight separate universities worldwide. The Learning Enhanced Watershed Assessment System (LEWAS) site is a high-frequency environmental monitoring system consisting of numerous sensors for water quality and quantity parameters [2].

© Springer International Publishing AG, part of Springer Nature 2019
M. E. Auer and R. Langmann (Eds.): REV 2018, LNNS 47, pp. 443–450, 2019.
https://doi.org/10.1007/978-3-319-95678-7_49

The user interface for LEWAS is the Online Watershed Learning System (OWLS); a cyberlearning system which can deliver integrated live and historical environmental monitoring data and imagery from the field to end users [3]. The OWLS features live video of the measurement site, an interactive live graph, local weather radar, background information on environmental monitoring, a glossary, and several case studies [4] (Fig. 1).

Fig. 1. Panorama image of the real world field site.

In this work-in-progress paper, we discuss the design and evaluation of a VE to improve the current LEWAS by providing higher fidelity, more immersive experiences which take place at a virtual outdoor field site.

2   Literature Review

Cyberlearning environments actively engage learners in the learning process and provide learners with personalized learning experiences [5]. Within the theoretical framework of situated learning [6], the existing system is designed to use data and imagery from the OWLS to cognitively situate users at the LEWAS site.

Virtual reality is a novel innovation which affords numerous strengths as an educational tool [7]. VEs have been a useful tool in training and simulation for decades, and previous studies suggest that knowledge and skills acquired in a VE can be transferred to real world situations [8–10]. There are numerous other labs integrating virtual realities and hydrological principles [11,12]. This design differs from those in our use of the video game engine Unity 3D. The previous are web-based implementations.

VEs are special in that they can provide high fidelity situations in a safe manner, globally. Pantelidis calls for considering virtual reality for teaching or training when using the real thing is dangerous, impossible, inconvenient, or difficult [13]. In our case we use a VE to conduct learning activities with students locally, and may expand the scope of our lab field site globally.

In past experiences, the quantity of information being provided was found to be overwhelming. VEs can stimulate learning and comprehension, due to the relationship between experiential and symbolic information [14].

Human-computer interaction techniques give an excellent evaluation method for a VE. Evaluating at different stages informs how we can measure distractions or constructs such as engagement [15]. To evaluate a VE it is critical to have a framework. In this case we utilized the powerful concept of fidelity, or realism, to describe our design decisions at early stages of development [16].

2.1 Fidelity

Fidelity is a way to define the realism of a VE component. The strength of this approach is that each component of the VE can be individually analyzed in terms of its fidelity [17]. Fidelity is comprised of three categories: interaction fidelity, scenario fidelity, and display fidelity [18]. Interaction fidelity, or naturalism, is the objective degree with which the actions used in the VE mimic real world interactions [16]. In this context we are implementing with low interaction fidelity for a number of reasons. The interactions are limited by the hardware, see Sect. 5, which relies on modest smart phone technology. Scenario fidelity is the objective degree of exactness with which behaviors, rules, and properties of objects are reproduced in a VE [19]. Our development prioritized scenario fidelity to mimic the outdoor field environment. Objects such as realistic buildings and roadways were included in the VE to improve fidelity in this sense. We also made sure the objects were all scaled properly, in a realistic way. The exceptions are the inclusion of a large three dimensional bargraph, and a hologram. This includes a representation of the triangular sensor suite emerging from the stream. Display fidelity, also known as immersion, is the objective degree of exactness sensory stimuli can mimic the real world [20]. This is largely a function of the available hardware, depending largely on the rendering software, output devices, and sensory stimuli produced [17]. In this case the limiting factor is the hardware available on the smart phone.

3 Context

The site of the study is a Freshman engineering course at a large research institution in the southeastern United States. We aim to address the following course learning objectives which are mapped to the same "Foundations of Engineering" course: (1) engineering model understanding and prediction, and (2) experimental setup, in the context of environmental education. While addressing these learning objectives, we explore the feasibility of a VE for environmental monitoring and engineering education. The results will further our understanding of VE effects on student learning with high-fidelity technology mediated educational material.

Our pilot study introduces the students to the field site and the sensors which are embedded there. Following this, an engineering model is introduced: the unintuitive, inverse relationship between water temperature and dissolved oxygen [21]. In the following section we address the design considerations and process using a Human-Computer Interaction (HCI) approach. Following

HCI considerations, the environment and interface were developed according to modern practices in VEs with consideration of fidelity concepts.

4   Human Computer Interaction Design

To develop the initial learning module, we adopt a HCI user-centered methodology to inform development [22]. The sequential user-centered design methodology allows for formative evaluation and iterative improvements throughout the lifetime of the solution. The methodology follows a series of sequential tasks including:

1. User task analysis
2. Expert guidelines-based evaluation
3. Formative user-centered evaluation
4. Summative comparative evaluations.

This paper presents the initial iteration of the VE including user tasks analysis, design minding expert guidelines, and formative user-centered evaluation.

User Task Analysis. The intended user of the VE is a freshman-level student enrolled in a general engineering program. In this pilot we focus on identifying which characteristics should be present in the VE. To address this we analyze a freshman engineering student's typical classroom tasks.

In a typical college lecture, the students sit at desks while the instructor is the front of the classroom. The users are not mobile, but confined to their chairs for the majority of the class. User mobility, or travel, is an aspect of interaction and scenario fidelity, and a major classifier of a VE. Student tasks inform the following development effort according to expert guidelines with formative user evaluation.

5   VE Development

A goal of the pilot is to create a virtual environmental monitoring field site which can be accessed remotely. This design exercise included six stakeholders, a number of concept maps, and storyboarding.

Hardware. There are two prominent methods for VE implementations: room-scale, externally tracked solutions and internally tracked solutions which require internal sensors such as inertial measurement units (IMUs) to track the user's head position [17]. Freshman engineering student task analysis reveals that mobility around the room is not required for the VE implementation, thus internal tracking will suffice. The hardware informed by user task analysis is a combination of a smart phone and a cardboard head worn display (HWD). Smart phones are internally tracked and the cardboard HWD allows for 3D viewing through stereoscopic rendering.

Software. To develop the high fidelity environment we utilized the Unity 3D game engine. This combined with a number of digital assets and three dimensional art pieces improve the fidelity of the field site. We also implemented a three dimensional graphing suite which allows for data visualization of the water quality parameters through standard 3D graphs.

Scenario Fidelity Considerations. In this development we make the water quality parameters visible and knowable to the student in the VE. Water quality and quantity parameters are mostly not observable without instrumentation. We introduced a 3D graph of water parameter values to show values of the parameters. This lowers the scenario fidelity of the VE, as free-floating graphs do not exist in the real world.

Interaction Fidelity Considerations. Interaction is limited in this design, as we utilize smart phones. There are no selectable or switchable parameters to interact with in the VE at this time. The only active technique is the changing their viewpoint from a seated position.

The students are led through the virtual lecture by an embodied voice that originates from a hologram representation of a senior lab member, S. The hologram guides the student through the site, giving an introduction to water quality parameters, explaining the sensors located at the field site, and describing the engineering model. A first person rendering of the VE can be seen in Fig. 2.

Fig. 2. First person view of the virtual LEWAS environment.

6   VE Evaluation

There are a number of methods which might be used to analyze educational VEs [23]. For the pilot, formative evaluation came in form of an online survey instrument of dichotomous and short response questions. We received IRB approval prior to the implementation.

The number of participants for the pilot was low (n = 14). Half of users reported being "distracted" by something within the VE, and half reported experiencing a lagging VE. Distractions were caused by the environment moving, the VE lagging, and audio issues. Some student users report the VE was rotating, requiring them to rotate their head to maintain a fixed view in the VE.

Limitations and Constraints of the Development. Limitations occurred due to the hardware which was used. The VE was in a constant, slow rotation. We believe this is a sensor erroneously detecting a constant rotational velocity. We anticipate newer smart phone technology and IMUs will not have this issue. While addressing the smart phone hardware, it may be worthwhile to consider numerous pieces to allow for parallel participation in the VE. Another limitation was the size of the graph, and number of datapoints it can handle. Future work will consider developing a large scatterplot which can represent numerous water quality parameters simultaneously in three dimensions.

7   Summary

This work-in-progress paper details an HCI approach to VE development to teach engineering concepts. The design considers student user tasks to inform the hardware decision. Design considerations stem from scenario and interaction fidelity components. The design was implemented with a small group of freshman engineering students, who reported distractions, mainly due to hardware issues. The formative evaluation will guide future developments of the VE.

8   Future Work

The VE was in a constant, slow rotation. We believe this is a sensor erroneously detecting a constant rotational velocity. We anticipate newer smart phone technology and IMUs will not have this issue. While addressing the smart phone hardware, it may be worthwhile to consider numerous pieces to allow for parallel participation in the VE.

We have also identified two upper level courses for further developing this work in spring 2018.

Acknowledgements. This material is based upon work supported by the National Science Foundation under Grant No. 479554. Any opinions, findings, and conclusions or recommendations expressed in this material are those of the author(s) and do not necessarily reflect the views of the National Science Foundation. In addition I would like to express gratitude to my colleagues Trevor Jones and Viktor Wahlquist and Yousef Jalali and Serena Emanuel for assisting in the VE development effort, and to my informal reviewers.

References

1. The National Academy of Engineering: Grand challenges - 14 grand challenges for engineering (2008). Accessed 30 Sept 2017
2. Basu, D.: Work-in-progress: high-frequency environmental monitoring using a raspberry pi-based system. Age **26**, 1 (2015)
3. Brogan, D.S., Basu, D., Lohani, V.K.: Insights gained from tracking users movements through a cyberlearning systems mediation interface. In: Online Engineering & Internet of Things, pp. 652–659 (2018)
4. Brogan, D.S., McDonald, W.M., Lohani, V.K., Dymond, R.L., Bradner, A.J.: Development and classroom implementation of an environmental data creation and sharing tool. Adv. Eng. Educ. **5**(2), n2 (2016)
5. Johri, A., Olds, B.M., OConnor, K.: Situative frameworks for engineering learning research. In: Cambridge Handbook of Engineering Education Research, pp. 47–66 (2014)
6. Johri, A., Olds, B.M.: Cambridge Handbook of Engineering Education Research. Cambridge University Press, Cambridge (2014)
7. Bell, J.T., Fogler, H.S.: The investigation and application of virtual reality as an educational tool. In: Proceedings of the American Society for Engineering Education Annual Conference, pp. 1718–1728 (1995)
8. Ahlberg, G., Enochsson, L., Gallagher, A.G., Hedman, L., Hogman, C., McClusky, D.A., Ramel, S., Smith, C.D., Arvidsson, D.: Proficiency-based virtual reality training significantly reduces the error rate for residents during their first 10 laparoscopic cholecystectomies. Am. J. Surg. **193**(6), 797–804 (2007)
9. Huang, Y., Churches, L., Reilly, B.: A case study on virtual reality American football training. In: Proceedings of the 2015 Virtual Reality International Conference, p. 6. ACM (2015)
10. Moskaliuk, J., Bertram, J., Cress, U.: Impact of virtual training environments on the acquisition and transfer of knowledge. Cyberpsychol. Behav. Soc. Netw. **16**(3), 210–214 (2013)
11. Ou, J., Dong, Yn., Yang, B.: Virtual reality technology in engineering hydrology education. In: International Symposium on Computer Network and Multimedia Technology, CNMT 2009, pp. 1–4. IEEE (2009)
12. Demir, I.: Interactive web-based hydrological simulation systems as an education platform using augmented and immersive reality (2014)
13. Pantelidis, V.S.: Suggestions on when to use and when not to use virtual reality in education. VR Sch. **2**(1), 18 (1996)
14. Bowman, D.A., Wineman, J., Hodges, L.F., Allison, D.: Designing animal habitats within an immersive VE. IEEE Comput. Graph. Appl. **18**(5), 9–13 (1998)
15. O'Brien, H.L., Toms, E.G.: What is user engagement? A conceptual framework for defining user engagement with technology. J. Assoc. Inf. Sci. Technol. **59**(6), 938–955 (2008)

16. Bowman, D.A., McMahan, R.P., Ragan, E.D.: Questioning naturalism in 3D user interfaces. Commun. ACM **55**(2), 78–88 (2012)
17. LaViola Jr., J.J., Kruijff, E., McMahan, R.P., Bowman, D., Poupyrev, I.P.: 3D User Interfaces: Theory and practice. Addison-Wesley Professional, Boston (2017)
18. McMahan, R., Ragan, E., Bowman, D., Tang, F., Lai, C.: FIFA: the framework for interaction fidelity analysis. University of Texas at Dallas, Dept. of Computer Science, Technical UTDCS-06-15 (2015)
19. Ragan, E.D., Bowman, D.A., Kopper, R., Stinson, C., Scerbo, S., McMahan, R.P.: Effects of field of view and visual complexity on virtual reality training effectiveness for a visual scanning task. IEEE Trans. Vis. Comput. Graph. **21**(7), 794–807 (2015)
20. McMahan, R.P., Bowman, D.A., Zielinski, D.J., Brady, R.B.: Evaluating display fidelity and interaction fidelity in a virtual reality game. IEEE Trans. Vis. Comput. Graph. **18**(4), 626–633 (2012)
21. LEWAS: Glossary - Dissolved oxygen (DO) (2014). Accessed 30 Sept 2017
22. Gabbard, J.L., Hix, D., Swan, J.E.: User-centered design and evaluation of virtual environments. IEEE Comput. Graph. Appl. **19**(6), 51–59 (1999)
23. Bowman, D.A., Gabbard, J.L., Hix, D.: A survey of usability evaluation in virtual environments: classification and comparison of methods. Presence Teleoperators Virtual Environ. **11**(4), 404–424 (2002)

School Without Walls - An Open Environment for the Achievement of Innovative Learning Loop

Carole Salis[✉], Marie Florence Wilson[✉], Franco Atzori[✉], Stefano Leone Monni[✉], Fabrizio Murgia[✉], and Giuliana Brunetti[✉]

CRS4 – Centre for Advanced Studies, Research and Development in Sardinia,
Pula, Italy
`{calis,marieflorence.wilson,fatzori,stefano.monni,fmurgia,`
`brunetti}@crs4.it`

Abstract. School Without Walls (SWW) is one of the novelties of the Iscol@ project - Line B2, conceived to address school disaffection in Sardinia, Italy. Lab. activities are based on, but not limited to, the use of mobile devices, Augmented Reality (AR) and georeferenced maps. Such technologies are used to create educational scenarios that connect what is being seen in real life to school curricula. All scenarios (geolocated data and related queries) are meant to be shared and reused through a Web platform. It will be possible to adapt existing content to different geographical sites by simply changing and/or setting the new coordinates, edit text and data to match the new location. The educational content and geopoints will be created by teachers and students on the Web platform, and scenarios will be accessed through a mobile App. Living technological enriched experiences at and outside school will contribute to a greater school-student retention. This paper describes the context and general objectives of this Lab. activity, which is to start at the beginning of 2018.

Keywords: Augmented reality & human machine interaction · Geolocation
Extracurricular labs · Technology enhanced learning activities
School dropout prevention

1 Introduction

The Iscol@ project-Line B2 is conceived to address school disaffection in Sardinia (IT). For academic year 2017–18 the Autonomous Region of Sardinia (RAS) allocated €2.518.000,00 [1] for Line B2, whereas €4,5 million were allocated for 2015–16 [2], and €3.500.000,00 for 2016–17 [3]. The reduction of the funds allocated is due to the fact that by the third years, schools are expected to have renewed their computing and technological equipment. Following the philosophy of Line B2, SWW activities require the involvement of 3 actors: **Schools** (teachers and students), **Institutions** (Regione Autonoma della Sardegna; Sardegna Ricerche: the Sardinian Regional R&TD (Research and Technology Development) Agency and CRS4 - Center for Advanced Studies, Research and Development in Sardinia - a Multidisciplinary Research Center), and the

© Springer International Publishing AG, part of Springer Nature 2019
M. E. Auer and R. Langmann (Eds.): REV 2018, LNNS 47, pp. 451–457, 2019.
https://doi.org/10.1007/978-3-319-95678-7_50

Economic Actors (experts from the many technological start-ups, cultural associations and SMEs that exist in Sardinia) [4–6].

In School Without Walls, students role is twofold: as contributors to the educational content they create and upload didactic digital scenarios and quizzes on the platform, and as end-users in non-school settings, they scan their surroundings through the camera of their mobile devices to find geo-tagged information which appears as an image (chosen by the scenario's author) overlaid on the camera view. Tapping the image, the scenario and queries will appear. Answering queries, students will either expand on what has been taught in the classroom, or be invited to think and direct their attention on a topic that has not yet been addressed by their teachers.

The Economic Actors will act as Technological Tutors. They will give technical support to teachers and students in creating the queries to be uploaded on the platform, but more important, they have the responsibility to transfer technological know-how to participating teachers, so that they will be able to use the technologies involved in this Laboratory to innovate their teaching methods.

Teachers get familiar with the involved technologies, create and upload their own educational scenarios, supervise and/or validate that of students.

1.1   Glossary

Scenario: geolocated educational content and related query and answers.
General Archive: all created scenarios that are shared with all registered users.
Map: set of scenarios belonging to the same learning unit consisting of geolocated queries semantically related to pedagogical criteria established by the teacher. Please note that a "map" in SWW corresponds to a "layer" in the Layar Web Platform.
Point Of Interest (POI): geographical coordinates linked to a specific scenario.

2   Motivation

One of the problems the Italian Educational System must face is that of modernizing school education, including the ageing teaching staff [7]. The unanswered need for the school system to offer a more modern didactics approach, more in line with the students expectations, has a negative impact on students desire to attend school activities. This contributes to school disaffection. Students commonly think that schools provide theoretical knowledge, that does not prepare them for the job market. Teaching methods are felt to be far from students interests and now-a-day reality. Learning inside and outside the classroom is one of the 11 promising Educational Technology Trends for 2017 [8].

The idea of SWW is not new: between 1960–70's there already existed a movement called Open Classrooms, based on informal and student-centered education in the Anglo Saxon world (UK and USA). What makes the concept modern is the use of technologies, such as AR and geo-tagged information, that broadens the learning experience to include outdoor activities and to use community facilities, as learning resources. This approach enables students to be at the center of their own learning.

This laboratory (School Without Walls) will give teachers and students the opportunity to combine relevant content that supplements school topics and a more innovative approach.

Teachers will experience a simple, useful way to use technologies for educational purposes. SWW will not only contribute to connecting classroom learning with real-world outcomes, but it will also help teachers better respond to student perspectives and concerns, modernize their didactics and teaching approach. Students, on the other hand, are encouraged to see and explore their surroundings, to make connections between what is being taught at school and real life, and contribute actively to the Web platform content thus building a bridge between the abstract school teaching of the curriculum and their surroundings. The learning process is no longer confined to the traditional face to face scholarly approach but becomes an ubiquitous learning experience that is expected to have a positive influence on behavioural outcomes, school and learning engagement [9].

3 The Choice of Augmented Reality and Geolocation

When used as part of the learning process, AR catches students attention, connecting reality and digital contents [10]. In SWW, POI, i.e. items of the surrounding real world, identified for having both a potential educational content, and a connection to the school curriculum, are enriched with overlaid extra digital information and related queries associated to the school curriculum. Overlaid features can be used either as the starting point of the theoretical knowledge to be covered at school or to supplement lessons with contents. This way, the teacher-centered lesson is enlarged to offer much more than is possible in frontal instruction. Moreover, students are more likely to be attracted by interactive activities than to complete traditional homeworks.

Geolocation can be used to locate historical events, to investigate the human impact on nature, to study demographic data, to apply correct math formulas, etc. It can help students develop spatial skills and improve geographical knowledge. In the context of this laboratory, students will be able to use geolocation to explore their surroundings and relate them to the school curriculum.

4 The Platform Architecture

CRS4 developed a user friendly Web content management platform for teachers and students who are not computer experts. The platform consists in a client-server architecture: the client-side is represented by the mobile application developed by Layar, that uses the GPS and AR technologies to show nearby POI [11, 12]. The server-side is represented by the Web platform developed by CRS4. The server front-end was implemented by using HTML5, Jquery and Bootstrap technology to provide responsive user interface. For back-end the Python language and the Tornado framework based on MVC model were used. To store user's answers, we opted for Neo4 J a Graph-Oriented database [13] instead of a traditional relational DB because meaningful patterns can easily emerge from examining the nodes (DB users, the students answers, the scenarios), edges and properties used. For example, this technical solution could help us better investigate the learning/teaching patterns, the most successful scenarios, etc. The platform server and the App client communicate thanks to a REST (REpresentational State Transfer) interface (Fig. 1).

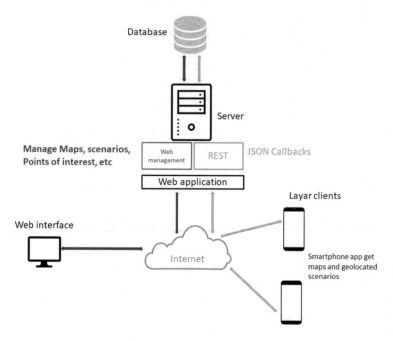

Fig. 1. Platform architecture

5   User Profile

Users will be required to register on the platform to create their own profile from which to manage the information. There are three possible profiles: "teacher", "tutor" (the Economic Actors) and "student". Of the three profiles, "teacher" is the most privileged user: he/she can create scenarios and maps, add scenarios to maps, share the latter. They can also authorize other teachers, tutors and students to access and edit their shared maps. Last but not least, they will validate the scenarios submitted by students prior to their publication on the platform. Tutors can create and submit new scenarios, either directly on the general archive or on the maps for which they have teacher's consent to do so.

Students can only create new scenarios for maps for which they are enabled by teachers. Such scenarios must be validated by their teacher prior to publication. Neither tutors nor students can create new maps, but tutors can edit maps previously created. Teachers and tutors will have the possibility to transfer and adapt another user's scenario to their own personal archive, act on its content and finally visualize statistics on the use of their scenario (Figs. 2 and 3).

Fig. 2. Access to my archive

Fig. 3. New scenario - entering a new geolocation of a scenario inside my archive

6 Approach

The project is based on the Open Teaching principle (free from common temporal and physical restrictions). The Open circuit guarantees the same fluidity and transparency found in up-front lessons that can be followed by question/answer and/or drills outside the classroom. In the Open paradigm, fluidity is given by the possibility for students to address or deepen topics independently, using the cues offered by the environment. Platform content will grow thanks to users collaborative approach: teachers, tutors and students will upload their AR geolocated contributions and make them available to other registered actors. Reuse of the uploaded georeferenced contents requires to change the coordinates of the selected scenario, to edit text and data to match the new location. This form of knowledge management is not limited to STEM but includes humanities.

Scenarios and queries can expand on what was taught in the classroom, or students could be asked to reflect on a topic that has not yet been addressed. This approach is as a flipped classroom strategy from the streets instead as from a home context. In a situation of flipped classroom strategy, part of the learning content is accessed by students outside the classroom. Since students can access material and explore topics before the lesson, they will come prepared with questions. This gives teachers more class time to review concepts that students found difficult to understand or dedicate time to more productive or creative activities.

7   Anticipated Outcomes

Bearing in mind that "Researchers find that extracting the full learning return from a technology investment requires much more than the mere introduction of technology with software and Web resources aligned with the curriculum. It requires the triangulation of content, sound principles of learning, and high-quality teaching - all of which must be aligned with assessment and accountability" [14]. We expect students to participate more actively, both in the classroom and outdoors, greater collaboration between teachers and a more appropriate representation of the abstract concepts students have to deal with. Finally, we hope that this mixed and opened learning strategy will promote the reuse of learning contents instead of, each time, starting from scratch. This should turn to teachers advantage, especially for those who are engaged with digital production.

8   Future Work

Our team would like to develop an ad-hoc mobile application to substitute Layar and specific tools based on the Graph-Oriented database versatility to enquire on teaching and learning patterns, in order to guarantee a long-term stability of the layer access system.

9   Conclusions

Since SWW Lab. activities will start early 2018, we are not yet able to evaluate its potential effects on students behaviour and on school drop-outs, but we expect the Lab. to promote ubiquitous and independent learning, encourage teachers to include technology in their lessons, and make students more aware of the existing links between the curriculum and their surroundings. Technology alone does not solve school disaffection but can help bringing schools and students closer.

Acknowledgment. The authors gratefully acknowledge the "Servizio Istruzione of Direzione Generale della Pubblica Istruzione of Assessorato della Pubblica Istruzione, Beni Culturali, Informazione, Spettacolo e Sport of RAS" and "Sardegna Ricerche".

References

1. Avviso "Tutti a Iscol@", ANNO SCOLASTICO 2017/2018, Allegato alla DD N. 284 PROT. N. 13682 DEL 06/10/2017 Integrazione Avviso per Piano straordinario di Rilancio del Nuorese (Deliberazione GR 46/5 del 03.10.2017). http://www.regione.sardegna.it/documenti/1_179_20171019090550.pdf
2. Salis, C., Wilson, M.F.H., Murgia, F. Monni, S.L., Atzori, F.: Public Investment on Education in Sardinia the "Tutti a Iscol@" Project, Introducing Innovative Technology in Didactics. In: Proceedings of 19th International Conference on Interactive Collaborative learning (ICL-2016), vol. 1, pp. 558–565, September 2016
3. Avviso "Tutti a Iscol@" Anno Scolastico 2016/2017, Assessorato della Pubblica Istruzione, Beni Culturali, Informazione, Spettacolo e Sport, Direzione Generale della Pubblica Istruzione Servizio Istruzione. https://www.regione.sardegna.it/documenti/1_38_20161028132434.pdf
4. Salis, C., Wilson, M.F., Murgia, F., Monni, S., Atzori, F.: Public Investment on Education in Sardinia—The "Tutti a Iscol@" Project. In: Auer, M.E., Guralnick, D., Uhomoibhi, J. (eds.) Proceedings of the 19th ICL Conference on Interactive Collaborative Learning, vol. 1, pp. 558–565
5. Salis, C., Wilson, M.F., Murgia, F., Monni, S.L.: Innovative didactic laboratories and school dropouts. In: Auer, M.E., Zutin, D.G. (eds.) Proceedings of the 14th International Conference on Remote Engineering and Virtual Instrumentation REV 2017, pp. 887–894. Columbia University, New York, 15–17 March 2017
6. Salis, C., Wilson, M.F., Murgia, F., Monni, S.L., Brunetti, G., Mameli, A., Atzori, F.: First Monitoring Results of the "Tutti a Iscol@" Project—A Technology Based Intervention to Keep Difficult-to-Motivate Pupils in the School Mainstream System, E-Learn. In: World Conference on E-Learning in Corporate, pp. 882–887. Government, Healthcare, and Higher Education 2016, 14 November 2016
7. Education and Training Monitor 2016—Italy annual Report released by European Union (2016). https://ec.europa.eu/education/sites/education/files/monitor2016-it_en.pdf
8. Kelly, R.: 11 Ed Tech Trends to Watch in 2017, Virtual Roundtable, Campus Technology (2017). https://campustechnology.com/articles/2017/01/18/11-ed-tech-trends-to-watch-in-2017.aspx
9. Burbules, N.C.: Ubiquitous learning and the future of teaching. In: Teacher Education in a Transnational World University of Toronto Press, pp. 177–187 (2014)
10. Li, R., Zhang, B., Sundar, S.S., Duh, H.BL.: Interacting with augmented reality: how does location-based AR enhance learning? In: Kotzé, P., Marsden, G., Lindgaard, G., Wesson, J., Winckler, M. (eds.) Human-Computer Interaction—INTERACT 2013. Lecture Notes in Computer Science, vol. 8118, pp. 616–623. Springer, Heidelberg (2013)
11. https://www.layar.com
12. Layar, from Wikipedia. https://en.wikipedia.org/wiki/Layar
13. Graph oriented DB. https://neo4j.com/
14. Lemke, C., Coughlin, E., Reifsneider, D.: Technology in schools: what the research says: an update. Commissioned by Cisco, Culver City, CA (2009). https://edtechtools.files.wordpress.com/2009/11/technology__in_schools_what_research__says.pdf

Low-Cost, Open-Source Automation System for Education, with Node-RED and Raspberry Pi

Phaedra Degreef[1]([⊠]), Dirk Van Merode[1], and Galyna Tabunshchyk[2]

[1] Thomas More University College, Sint Katelijne Waver, Belgium
{phaedra.degreef,dirk.vanmerode}@thomasmore.be
[2] Zaporizhzhya National Technical University, Zaporizhzhya, Ukraine
galina.tabunshchik@gmail.com

Abstract. In this article, the open-source Node-RED framework is considered as a low-cost, open-source automation system for education and research. It can be used not only to demonstrate and experiment with automation principles, but the open concept of the framework also allows customizations of the different components of the automation system.

Thanks to its shallow learning curve, good availability and a strong online community, this platform is very well suited for different education levels, from secondary school to universities. This paper will describe the different possible applications in education.

Keywords: Node-RED · Raspberry Pi · MQTT · IoT · Automation

1 The Node-RED Framework

1.1 Background

Automation once used to be the privileged domain of engineers, specialized in feedback systems, electronics and embedded systems programming. The learning curve was steep, and the system components required expertise to handle, configure, troubleshoot and maintain.

Arduino [1] and the likes and Internet of Things (IoT) brought a dramatic paradigm shift to the world of automation. It opened up the domain for non-expert users, hobbyists and educational institutes all over the world, with a dynamic and active online community to support them.

These microcontroller-based systems still require a minimal programming experience, though, inherently limiting the potential user-base and usage.

The Raspberry Pi [2] platform brought a new revolution when it came out, spawning yet another active community of enthusiasts, constantly looking for new challenging applications. Interfacing a Raspberry Pi with external sensors, however, again required some programming and electronics knowledge.

© Springer International Publishing AG, part of Springer Nature 2019
M. E. Auer and R. Langmann (Eds.): REV 2018, LNNS 47, pp. 458–465, 2019.
https://doi.org/10.1007/978-3-319-95678-7_51

That changed with the more recent appearance of the **node-RED** framework [3]. Node-RED is a visual web-based tool, which is even pre-installed on the latest Raspberry Pi distributions.

It allows a non-technical user, via a graphical drag-and-drop user interface, to assemble and connect different components into a working automation system, called a "flow". Interfacing with commercially available sensors and other hardware consists of downloading and installing custom node-RED "nodes" (best case) or programming your own custom nodes.

1.2 Sample Application

In the next figure, a "flow" was setup to display the temperatures measured by five DS18B20 sensors connected to a Raspberry Pi, store these values into a MySql database and display them via widgets on a "dashboard" web page, hosted on the internal web server.

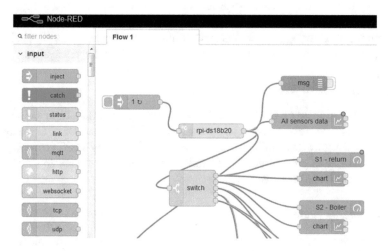

Fig. 1. Example flow to display DS18B20 temperature sensor data to a node-red dashboard

This web-based "dashboard" can be used to allow real-time interaction with the built automation system.

In the flow of the example above, (highly customizable) gauge and chart dashboard nodes were added to display the sensor data (Fig. 2):

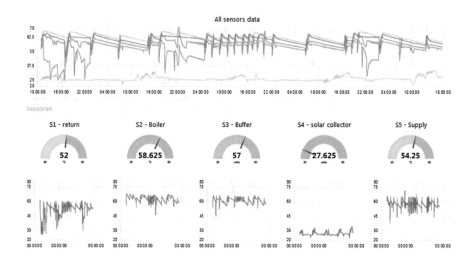

Fig. 2. Node-red dashboard page corresponding to the flow depicted in Fig. 1

Using this monitoring page for analysis of the system's behavior, a simulation flow was then created to experiment with new control algorithms for the system.

1.3 The Node-RED Framework and MQTT

The node-RED framework is a flow-based programming tool [4] that has its roots in the MQTT protocol [5]. It was first developed by Nick O'Leary and Dave Conway-Jones from IBM Emerging Technologies, as a means of visualizing MQTT messages (see below). It was soon enough developed further into the open-source tool it is today, reaching over 300.000 installs.

1.4 Node-RED Nodes

A node-red flow consists of so-called nodes, which can send and receive messages, and perform programmable logic with the message payload.

A node consists of three parts:

1. A package.json file describes the content of the node
2. The runtime behavior is defined by a JavaScript node.js file
3. The appearance inside the node-red editor is defined by a HTML file, accompanied by a custom icon (added as a .png file in a separate directory).

This structure makes it highly customizable, which makes it so interesting for educational purposes.

Nodes are drag-dropped in a visual, web-based editor, and inter-connected. These connections define the flow that the message will follow, as it travels from one node to another. Each node can then pass on the received message, or modify it first, or generate a new message and send it to other nodes. Message payload is structured in a JSON [6] format.

The visual flow editor can be used via a web browser and navigating to http://< ip address or name of host>:1880.

In a standard node-RED installation, 4 types of nodes are provided:

- Input nodes: input of messages via either a manual (inject) node, MQTT, http, etc.
- Output nodes: output of messages via debug nodes, MQTT, http and others
- Function nodes: allowing processing of messages
- Social nodes: interaction with social media

More nodes can be downloaded an installed via npm [7], the JavaScript package manager software registry, from a vast library of contributions to the node-red project.

If the required functionality cannot be found in that library, the end-user can create custom nodes, which can be achieved via the "npm init" command which will generate the 3 node files described above (package.json, the node.js and the HTML file).

Once the flow's design is completed, it can be deployed to the node-red runtime by a single click on the "Deploy" button.

A node type deserving special attention in this article, is the dashboard node. Using these dashboard nodes requires installation of the corresponding npm package.

Dashboard nodes allow user interaction via a separate web page, http://<ip address or name of host>:1880/**ui**.

Again, a number of standard nodes are available, like gauges, charts, switches, et cetera. But custom nodes can also be developed using a so-called template node, in which json data can be embedded in custom HTML code.

1.5   Message Queueing Telemetry Transport (MQTT)

MQTT is a lightweight messaging protocol that has gained popularity in the Internet of Things (IoT) world because it is well suited for low-bandwidth, low-reliability networks. This makes it very well adapted to mobile devices that do not have guaranteed connectivity all the time.

MQTT uses a publish-subscribe pattern, based on so-called Topics. Topics allow information contained in messages to be organized according to the topic. Subscribers can thus only receive messages from topics they subscribed to.

Node-RED fully implements this model in its flows.

1.6   Performance

With the overhead of the framework itself, the use of an interpreted JavaScript scripting language and the limited Raspberry Pi hardware, performance cannot be expected to be very good. Performance isn't the key point for node-RED. It is meant to be a prototyping and experimentation framework, and its performance will never be comparable to dedicated embedded software.

A new branch of node-RED for distributed processes was introduced by Mike Blackstock in 2014 [9]. It allows flows to be distributed over different processing nodes (Fig. 3):

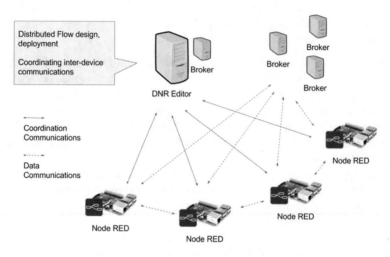

Fig. 3. Distributed node-RED architecture

A new editor, the dnr-editor [9] allows control over the flow distribution, using (Fig. 4):

- Capability/property (storage, memory or geographic location) of a device
- Constraints on nodes, which have to match the capability/property on the device it will be placed on

Fig. 4. dnr-editor example

2   Applications in Education

The main advantages of using the node-RED framework on Raspberry Pi in education are:

- It is a low-cost solution: a Pi can be bought for prices around 30 Euro.
- The open-source nature of node-RED and the customization possibilities makes the node-RED framework very interesting for educational purposes:
- Because it uses a graphical programming environment, its learning curve is shallow enough for secondary school students, to engage in elementary automation projects and experimentation.
- Its open nature and use of widespread standards allow for advanced customization by more ICT-oriented students (bachelor or master level);
- Engineering students can use this platform to actually design and build advanced automation projects, working with the concepts of feedback systems, test their stability, hysteresis, distributed processing (via MQTT), database interfacing, et cetera.

Usage is not limited to simulation or demonstration of their project, but could also extend to fast prototyping for embedded systems to be developed later.

2.1   Implementation in Elementary School

A set of Raspberry Pi computers with node-RED could be used to sparkle interest in designing and programming for automation. Inspired by commercial initiatives like Lego Mindstorms [8], the creativity of children at this young age can be used to make use of predefined and prebuilt "building blocks" to turn it into working projects or games.

2.2   Usage Examples in Secondary School

For secondary school students, the bar can be set a little higher. They can be challenged to implement a system according to a functional analysis. With the shallow learning curve of node-RED, not requiring a solid basis of electronics or programming, they can go about and process live or simulated data from physical sensors and actuate electrical devices via relays or other means.

External sensors and other devices can be connected for them in advance, not exposing them to the physical interfacing details if that is deemed unnecessary or too risky. But at the same time, customized hardware interfacing boards can be developed by the school staff to allow student some minimal physical interfacing experience.

Prior development of custom node-RED nodes by the school staff allows students to focus on the logic of the flow to be designed and the automation project requirements, and leave the technological details behind node-RED abstract. Development of custom dashboard and other nodes could even limit the user experience to simulation of the project, without even physically interfacing it with external components. Examples of these nodes could be:

– Injection of simulated data into the flow, in the absence of real sensors.
– Visualization of outputs, like for example LEDS that would depict GPIO output levels or simulated relay state.
– Integrate school or other logos into dashboard nodes, making the result more visually attractive and thus providing means for promotional activities.

Apart from implementing pre-designed systems, students could also be challenged to design simple automation systems, and thus learn the basic concepts of feedback systems and stability. Instead of focusing only on the theoretical background, they can discover the properties of feedback systems by experimenting.

2.3   Advanced Cases in Higher Education

At a higher level of education, the possibilities of this platform offer more potential.

Challenges in ICT-Oriented Education
Students can be challenged to design, develop and test custom nodes, exposing them to

– JSON
– JavaScript
– SQL, databases in general
– Angular (development of dashboard nodes)
– HTML
– Distributed processing and microservices (distributed node-RED)

A first test was conducted at Denayer Campus of Thomas More University college in January 2018, during a winter course within the context of project APPLE – "Applied curricula in space exploration and intelligent robotic systems". Sixteen participants from different countries used node-RED to simultaneously publish own temperature sensor data and subscribe to data from the other participants, displaying it on their dashboard. One Raspberry Pi was used as an MQTT broker and served as a WiFi access point for the MQTT data. For demonstration purposes, a number of additional MQTT publishers (on the same MQTT broker) were activated during the workshop:

• an ESP8266 microcontroller publishing temperature and humidity data every second.
• At the same time, 32 "dummy" node-RED MQTT nodes were setup as a stress-test, to simultaneously publish random messages every 3 s.
• Two additional separate Raspberry Pi demo systems published temperature and humidity sensor data every second.

The system proved robust enough for this kind of load as no adverse effects could be observed during a 3 h period of testing.

Node-RED Applications in Automation-Oriented Education
To experiment with interfacing physical sensors or other devices, custom nodes can be developed by the students to implement the required interface protocol(s) and to generate or process the corresponding JSON messages.

Engineering students could use node-RED as an experimentation platform for feedback and automation systems: learning the concepts of feedback systems, test their stability, implementation and the effects of hysteresis, et cetera.

Experiments in Data Communication and Protocols
The Node-RED framework can also be regarded as a testing ground for all sorts of existing or experimental protocols.

Take for instance MQTT, which is a single-hop communication protocol. A number of interesting experiments could be envisioned:

- Make MQTT messages routable, by encrypting them and using another protocol to communicate them between some sort of gateways connected to the Internet
- Test any wired or wireless communication technique or channel for transport of MQTT messages
- MQTT test message generator for functional or integration testing of MQTT-based automation systems.

3   Conclusion

The node-RED framework on the Raspberry Pi offers a low cost experimentation and prototyping platform that can be used on a broad range of educational levels.

It does not offer a replacement or competition for dedicated embedded software when performance or memory footprint is considered, but it can serve as a good starting base for the latter (prototyping).

Its shallow learning curve makes it interesting for younger students, while its open nature and extensibility make it particularly suited for use in higher education.

A first test proved the system to be robust enough for a workshop with 16 participants.

References

1. Arduino platform. https://www.arduino.cc/
2. Raspberry Pi. https://www.raspberrypi.org/
3. Sarkar, A.: Distributed Control System Technologies—NODERED, CODESYS, 4DIAC, DOME (2015). https://doi.org/10.13140/rg.2.1.3901.9609
4. Paul Morrison, J.: Flow-based programming. J. Appl. Dev. News. https://ersaconf.org/ersa-adn/papers/adn003.pdf
5. Shinde, S.A., Nimkar, P.A., Singh, S.P., Jadhav, Y.R.: MQTT—Message Queueing Telemetry Transport protocol. Int. J. Res. **3**(3). ISSN: 2348-6848. International Conference on Research and Recent Trends in Engineering and Technology, 27 Jan 2016
6. JSON JavaScript Object Notation. http://www.json.org/
7. npm, the software registry for JavaScript. https://www.npmjs.com/
8. Lego education. https://education.lego.com/en-us
9. Distributed node-RED. https://www.npmjs.com/package/dnr-editor

Demonstration: Face Emotion Recognition (FER) with Deep Learning – Web Based Interface

Andreas Pester[(✉)] and Kevin Galler

Carinthia University of Applied Sciences, Villach, Austria
`a.pester@fh-kaernten.at, k.galler@gmx.at`

Abstract. In this project, a pre-trained, python based deep learning algorithm for recognition of the emotional expression of a face on an image is used, to be accessed and executed in an online experiment. Therefore, newest web technologies are used, to get access to the front camera of the used device, to extract a picture of the face out of a continuous video stream. When the deep learning algorithm has finished its operation of detecting the emotions of the face, the results will then be displayed on the website.

Keywords: Deep learning · Face emotion recognition
Convolutional neural network · Adaptive web interface

1 Introduction

Deep Learning is an important part in the field of computer science and engineering, with the benefit of offering predictions and classifications for big data sets with a high level of accuracy, especially in the field of image processing. One of the examples to demonstrate the power of Deep Learning is emotion recognition of face images. It is an intensive discussed field in human-computer interaction and AI [1–3].

In this application, it is possible to localize faces on an image and based on a pre-trained deep learning algorithm the emotional expression of a face can be determined [6]. An already tested algorithm was used to develop an online experimentation bench for face emotion recognition for two reasons: to demonstrate the approach of Deep Learning on an example, which is very easy to understand and to have a demonstrator for multi-class classification with convolutional networks in TensorFlow [4]. The architecture is a usual 4-layer convolutional network, as activation functions the ReLu and Softmax functions are applied. The two convolutional layers are followed by two fully-connected layers. The code is described in [5]. This model is implemented in Python [7], trained and tested on standard data sets.

The objective of this demonstrator is to access the model via web-client in a client-server architecture. An implementation in a cloud architecture could be a next step.

As the given algorithm is written in Python programming language, a web-site was created which is able to process an actually taken photo of a person and compute as a multi-class classification the emotional expression of the inputted face for demonstration purpose.

© Springer International Publishing AG, part of Springer Nature 2019
M. E. Auer and R. Langmann (Eds.): REV 2018, LNNS 47, pp. 466–470, 2019.
https://doi.org/10.1007/978-3-319-95678-7_52

Therefore, it is crucial to have an input image, whose shape and location is close to the reference data set the deep learning algorithm is trained for.

With the help of WebAssembly implemented in a Javascript [11], the face of the user will be detected and extracted from an image, taken by a webcam, encoded and used as input on the server side. This will guarantee the best results for the emotional feature detection. To be sure to protect privacy, the user is going to be able to delete the taken image and the computed results after reviewing.

This demonstrator is mainly intended to be show the abilities of machine learning on an easy to understand example. The aim was less to develop a learning object, but to show the possibilities and the limitations of deep learning with convolutional networks.

2   Software

The given part of this project is the emotional recognition algorithm that is written in Python, using Deep Learning algorithms for object location and multi-class classification. To be able to implement a website, that is capable of executing the algorithm and capture an image, while being state of the art, the following software are used:

- HTML
  - CSS
  - JavaScript
- Python [7]
- Operating System – Core functions
- OpenCV [10].

3   Website

Most websites nowadays are developed either pure for desktop or mobile experience. This application works for both input devices. Therefore, the server must check if there is a mobile device accessing the website, to forward the mobile view, instead of the original one. This leads in many cases to an unsatisfying experience, regarding a lack of functions and unintuitive design, because of having two independent websites containing different abilities and features.

Furthermore, some companies rely on old adaptive web design, containing thousands of code lines using JavaScript to adapt everything on the website for every screen size.

With the help of CSS grid, a new and smart way of creating responsive web design is feasible, to provide full functionality and intuitive experience to the customers. Therefore, the website of this project is going to be CSS grid included, to be able to get the best state of the art experience on every device, regardless of the screen size.

3.1   Design of the Webpage

The expected design on the website and functionality is shown in Fig. 1. With dedicated buttons to delete the picture, take or retake a picture and start the emotional recognition, the user has full control of what he wants to do.

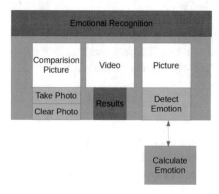

Fig. 1.   Expected design

3.2   Face Localization

To apply the FER (Face Emotion Recognition) algorithm it is necessary first to localize the part of the image or video, which reproduces the face in good shape, under an applicable angle of view and with the necessary resolution, so that the main algorithm can be applied. WebAssembly technology is used to access the device-build-in-camera, and the user can see his face on the screen and position it to get the best result for further progress. Because of the present lack of information about coding in [11], JavaScript is presently used for image capturing.

With the help of JavaScript functions, it is then possible to include face recognition and display the captured face beside the video stream, which will be the input of the FER algorithm.

4   FER Algorithm and Data Set

The used FER algorithm is a multi-class classification algorithm, based on convolutional networks. The classes, used, are:

anger, disgust, fear, happy, sad, surprise and neutral

The structure of the algorithm follows the scheme:

Input layer → 1. Convolutional layer → 1. Maxpooling layer → 2. Convolutional layer → 2. Maxpooling layer → Full connected layer → Output layer.

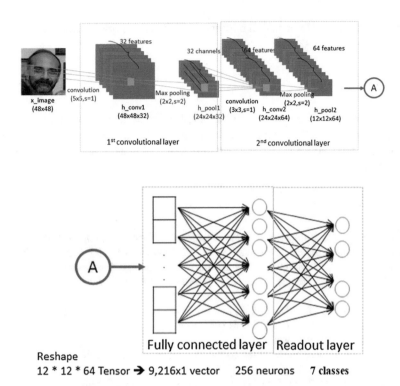

Reshape
12 * 12 * 64 Tensor ➜ 9,216x1 vector   256 neurons   7 classes

Fig. 2. Architecture of the FER algorithm [9]

In more details this looks so, as described below:

As can be seen from Fig. 2, the input is a $48 \times 48 \times 1$ grey scale image. If the device camera grabs a coloured RGB-image, it should be converted to a grey scale image. The filter used for the first convolution layer, is a $5 \times 5 \times 1$ filter, with stride 1 and no padding, repeated 32 times. The first Maxpooling layer uses a 2×2 filter with stride 2, the second Convolution layer uses a 3×3 filter with stride 1, followed by the second Maxpooling layer with the same dimensions as the first. The resulting $12 \times 12 \times 64$ tensor is flattened to a 9216×1 tensor and a fully connected layer with 256 neurons is applied. Finally, the output is given in the already described 7 classes. As a dataset is used the standard data set from Kaggle [8], with the following splitting in train data size = 3721, validation data size = 417 and test data size = 1312. The used learning rate is 1e−3 and 1000 iterations for training with a batch size of 128 are used.

To apply the trained FER algorithm to the taken image from the device-built-in-camera, this image should fulfil the requirements for the used train and test data. All faces should be pointed exactly at the camera and the emotional expressions should be exaggerated. Only in these cases reasonable results can be expected. It should be made sure that that there is no text overlaid on the face, the emotion is recognizable, and the face is pointed mostly at the camera. The results are reasonable, but improvements of the architecture of the used algorithm are under discussion. In [1] you can find a comparison of the implementation of different FER algorithms, but the described one is quite often used, see [9].

5   Conclusion

The aim of the project is to visualize the possibilities of deep learning algorithms and image processing. The finished project is going to be available on the open days of the CUAS to potential students, to get a first impression of what they can learn in this field of education.

References

1. Pramerdorfer, C., Kampel, M.: Facial expression recognition using convolutional neural networks: state of the art. Preprint arXiv:1612.02903v1 [cs.CV], 9 December 2016
2. Xia, X.-L., Xu, C., Nan, B.: Facial expression recognition based on TensorFlow platform. In: ITM Web of Conference, vol. 12, p. 01005 (2017). https://doi.org/10.1051/itmconf/20171201005
3. Xie, W., Shen, L., Yang, M., Lai, Z.: Active AU based patch weighting for facial expression recognition. Sensors **17**(2), E275 (2017)
4. https://www.tensorflow.org/
5. Zaccone, GMd, Karim, R., Menshawy, A.: Deep Learning with TensorFlow. Packt, Birmingham (2017). Chap. 4
6. Correa, E., Jonker, A., Ozo, M., Stolk, R.: Emotion Recognition using Deep Convolutional Neural Networks (2016). https://macalu.so/uploads/research/emotion_recognition.pdf
7. https://www.python.org/downloads/release/python-360/
8. https://www.kaggle.com/c/facial-keypoints-detector/data
9. https://exploreai.org/p/tensorflow-cnn
10. https://opencv.org/
11. http://webassembly.org

Poster: An Approach for Supporting of Navigation of Blind People in Public Building Based on Hierarchical Map Ontology

Dariusz Mikulowski$^{(\boxtimes)}$ and Marek Pilski$^{(\boxtimes)}$

Faculty of Sciences, Siedlce University of Natural Sciences and Humanities,
Siedlce, Poland
{dariusz.mikulowski,marek.pilski}@ii.uph.edu.pl

Abstract. The blind people encounters great difficulties with independent movement in unknown environment. However, navigation in open outside space is partially supported due to their ability to hear of sound reflected from obstacles, therefore movement across the unknown areas of public buildings is even harder. There are also various helping systems i.e. inbuilt physical items, such as the metal paths leading to important places or round protrusions are applied. But the installation of such support is usually too expensive. Therefore it is a great need of easier electronic supporting of navigation of blind people in public buildings. In this paper, a concept of specially developed hierarchical object map ontology is proposed. All objects in this ontology can be physically labeled by radio labels (i.e. bluetooth) and have description of their characteristic features such as their attributes and relationships. Due to this, the routes between two any objects will be easy to create.

Keywords: Navigation of the blind · Object map ontology
Relief daily life

1 Introduction

According to a World Health Organizations report [1], the total number of blind people in the world is estimated at 36 million. In Polish National Organization for the Blind is registered about 31000 totally blind members. It is a couple number of people who encounters great difficulties with independent movement, both in outdoor environment and also in closed area of public buildings. They are able to move around an open environment a little easier, due to ability to hear sounds reflected from obstacles (echolocation), sounds of passing cars, etc. But independent movement in open areas of public buildings is even more complex for them because it has not certain streets, pavements, passing cars, which they could hear.

© Springer International Publishing AG, part of Springer Nature 2019
M. E. Auer and R. Langmann (Eds.): REV 2018, LNNS 47, pp. 471–478, 2019.
https://doi.org/10.1007/978-3-319-95678-7_53

There are already many solutions for helping in navigation inside a building. They use some e.g. GPS replicators, radio waves, tags, acoustic or magnetic sensors, gyroscopes, compasses, strength signal meters, motion detectors or other sensory information collected by mobile devices. Also distance data provided by the sensors and e.g. building plans are used with a software associated with this equipment. More over, inside buildings a physical items, such as the metal paths leading to important places or round protrusions, informing, for example, on the stairs beginning are applied. However, very few buildings are set such services, and in addition, their installation is usually too expensive. Although there are a few commercial solutions, but there is no standard for building such navigation systems.

Taking it into account we can notice that there is a great need of electronic supporting of navigation of blind people in public buildings. Because the solution should be possibly easy accessible to blind user, popular mobile devices such as, smartphones and stickers on various objects in the building in such system, should be used. The navigation steering mechanism should be smart enough, for example, to allow making definition of safe routes between any two points in the building.

It seems that a good idea for a solution to this problem would be the use of semantic technology i.e. specially developed ontology. This ontology could be a specially designed hierarchical map which collects information about the objects placed physically in the building and relationships between them. All objects in this ontology must have described features as their attributes. Physically, they can be labeled based on the radio of the near range, such as Bluetooth so that the user is approaching the object on his smartphone device. Also the necessary relations between these objects and user i.e., the door is on the left must be designed. On the basis of such map the system that supports navigation can be implemented. In next few sections, a concept of such object map ontology and main properties of system that can be built on the basis of this ontology is briefly described.

2   Related Works

Blind people face a number of challenges when interacting with their environment because so much information is encoded visually. In the area of blind help, it is natural that attention is paid to compensating for the lack of sight. As one of the basic senses, sight is used every day for many activities e.g. recognizing objects, people and above all to movement in the real space and especially inside a building. There are many interesting researches in this area. In spite of the specific direction of research, we can distinguish several paths that researchers follow. But each path has a common feature - the blind help system uses different devices coming to the Internet of Things [2].

The first direction of research is to help the blind people **find a object or place**. A system that directly supports this type of interaction is presented by Bigham et al. [3]. It is based on a two-stage algorithm that uses information to guide users to the appropriate object interactively straight from their phone. The blind user must only send a photo of the object and ask for assistance in finding the object. The request is sent to remote workers who give outline the object, enabling efficient and accurate automatic computer vision to guide users interactively from their smartphones.

The next direction of research is **recognize indoor objects**. As an example we may point the work [4] where the authors present the prototype of system, which beyond the capabilities to move autonomously is offered to recognize multiple objects in public indoor environments. The system uses hardware components such as: camera, IMU and laser sensors that the blind person wears on his chest. For proper operation of the system, the data about indoor environment must first be loaded. The interaction between the user and the system is performed through speech recognition and speech synthesis modules.

The next approach is to help blind people in **navigation inside buildings** using various location technologies and other solutions supporting them in this task. The indoor navigation is more specific then outside one and it requires use much more diverse technologies, because there is no GPS signal and the path to the destination can follow many floors of the building. For this kind of navigation indoor positioning systems (IPS) [5] can be used. IPS uses two main technologies: *non-radio technologies* and *wireless technologies*. Non-radio technologies can be used for positioning without using the existing wireless infrastructure. A good examples of such approach are: magnetic positioning [6,7], inertial measurements [8], positioning based on visual markers or location recognition using known visual features [9]. Any wireless technology and any existing wireless infrastructure can be used for supporting indoor location recognition. Positioning accuracy is usually depending on the amount of infrastructure equipment and their cost. The architectures of these systems usually use: Wi-Fi-based positioning system [10], Bluetooth [11,12], Grid concepts and Choke point concepts [13], Long range sensor concepts [14], Received signal strength indication [15], Time of arrival [16] and others. Sometimes hybrid solutions connecting these technologies are applied.

The last approach we will describe represents **identification people and states of objects in known environment**. Hudec and Smutny [17] present an ambient intelligence system for blind people which provides full assistance at home environment. It also helps with various situations and roles in which blind people may find themselves involved. The system consists of several modules that mainly support or ensure recognition of approaching people, alerting to other household members movement in the flat, informs about the events that occur in the house, supports cooperation with a sighted person during his work on a computer, helps supervision of (sighted) children, allows control of heating and zonal regulation by a blind person. The modular application uses a lot of hardware such as microphones, an audio system, a server, temperature sensors, door switches, a gas boiler, motion detectors, etc.

Taking it into account it seems that work on the branch of development of indoor navigation is important. It opens the possibilities for a great number of directions and approaches which require further creation and integration of such systems into the environment of intelligent buildings in general.

3 Supporting Navigation of Blind in Unknown Public Building

The idea of supporting navigation of blind inside buildings proposed here is based on architecture which consists of several main components.

1. Application installed on the server in the building that manages the entire process of routing and the guidance of user to the selected point.
2. Mobile application installed on the user's phone, which is an interface of user to whole system. It is designed for routing the user by connecting with radio labels located in different places of a building.
3. Radio labels located in various areas of the building that allows for guidance of the user.

The core of the application that is installed on the server there is a special **hierarchical object map.** To create this map, a fundamental ontology with concepts such as object type and attribute is used. These attributes that means specific properties of objects are added to this ontology together with relationships that allows to connect objects and describe its properties in much more detailed manner. In order to make possible to determine routes between two any places in the space of the building a hierarchy relationship between object is very important.

The object map constructed in this way allows to describe the structural fragment of the world and planning of safety routes for the blind user in the unknown building space. A more detailed description of this object map is explained in Sect. 3.2.

3.1 System Functioning Scenario

Let us to describe functioning of a system based on object map ontology telling a following story:

Please imagine, that a blind user John comes to completely unknown shopping centre John wants to buy shoes in the NIKI store, which is located on the 2nd floor of this shopping gallery. However, he doesn't know how to find way to the shop.

Fortunately, he has an access to a system based on object map ontology installed in the building and the control app installed on his phone that is the part of this system. But how they will help him to get to the store and back? Upon entering the building, John stops and launches his smartphone application. The application connects to a local server that is installed in the building, and retrieves object map of the whole gallery. Then John receives a list of commercial facilities that are available in the gallery displayed on his screen. A Screenreader software that is installed on his smartphone is able to read this list for him. So John chooses so, from the this list and press the "Find a way" button.

At this moment, the mobile application connects to the server and sends a request for a route to the selected object. Then the server is routing the object map stored in the ontology to find a target object. Because John is located at the entrance on the ground floor of the shopping gallery and the target store is located on the 2nd floor, at first parent objects for starting and target points should be determined. Then the comparison should be made are two found objects are the same. For the transition in which John is standing, a parent object is the object named *floor0*. In turn, the store with shoes, a parent object is a, *floor2*. Because they are different objects, hence the parent object for both must be determined. It turns out that the parent object for the object *floor2* is a *gallery*. For the object *floor0* the parent object is also *gallery*. Thus, both objects *floor0* and *floor2* are on the same level (second) of hierarchy in the object map. Therefore, an application that analyzes the description of these objects in the ontology and checks which other objects they are related by the relationship neighborhood. It turns out that both *floor0* and *floor2* are connected by relation, *is a neighbor* with the object *escalators1*. So general list of objects describing the route, consisting of 3 objects: *floor0, escalators1* and *floor2* is created.

Now the route from the place where was John standing to the beginning of *escalators1* consists of objects belonging to *floor0* is defined. The route is calculated based on the relationship *is neighbour to* which connected objects belonging to the object *floor0*. This route consists of object *entrance, corridor,* and the *beginning of escalator1*. The same operation is performed inside the object *floor2* where object *end of escalator1, corridor2* and *store11* are combined into next fragment of a route. This creates the entire route from *entrance0* on *floor0* to *store11* leading through *escalator1*. The resulting object list is now updating with additional information such as distances between objects and directions from one to next object in a list. It is possible due to fact that these properties are stored as attributes of objects in the map ontology. The route obtained through the application was presented schematically in Fig. 1.

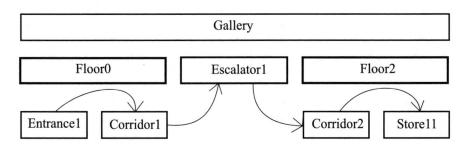

Fig. 1. A schematic view of a route.

At the end of the process, the auxiliary messages are added into a whole route to help John transit from *entrance1* into a destination NIKI store. Detailed description of this route that is displayed on the screen of John's smartphone can be as follows:

- Go straight 20 m from entrance to the start of corridor 1.
- Turn 90° right and go along corridor 1 30m.
- On the left you will find start of escalators.
- Turn 90° left and climb up the escalator 2 times.
- Then turn left and go up to 40 m along pass 2.
- The entrance to the Store, Nicky, you will find on the left.

At this moment, John can hit the road. Moreover, once he will be near the entrance to escalator, his application will receive the radio signal from the label on this site and will notify him about this situation.

Let us summarize, that due to use of such hierarchical map ontology, a routing mechanism and radio labels the blind user is able to run independently inside the building. He can use the system in similar way as ordinary sighted user currently can drive with the family navigation app in his car.

3.2   Concept of Hierarchical Map Ontology

As we mentioned above, the core of the system for supporting navigation of a blind user inside buildings is the specially constructed objects hierarchical map. It is derived from properly developed fundamental ontology that defines concepts such as object kind, object attribute, object type, and the concept of object itself. The object attribute is just its property that can be described by a name and a value. For example, object of type *door* can have the attribute "way to open" with values "swing", "slight" or "rotary". The type of the object determines a default set of attributes. For example, the *door* type of any object states that it should have an attributes such as: "way to open", "handler type", "width", "opening method", (with values to myself/from myself), etc.. There can be two kinds of objects: atomic or complex. An example of atomic object can be the *door* and complex object can be *room*. The difference between these kinds of objects is that an atomic object cannot have any child objects while the complex object usually has a child object. In the ontology describing object map, also relationships are defined. The main are the parenthood and neighborhood relations, but also can be others, such as: "is on left", "is on right", "is above", etc. For example, an atomic object *doors* can be the child object for an *room* object. Although the *door* object as atomic cannot have any child element, but it can be in relations of neighborhood with another atomic object, e.g. *wall*. The ontology also defines the notion of user. This is necessary in order to determine user who currently navigates through the building so his representative object is virtually in a specific point of the map. The user representative object can also be in relation of neighborhood with atomic objects. These relations, however, need to dynamically be changing at a time when the user moves in the building, what can help him in achieving his goal.

So this ontology we briefly explained above, can be a good schema for creating an instance of object map of any particular building. The construction of such an instance is made by creating the schema of the necessary types of objects, linking them by proper relations, assigning them identifiers, and assign values to all attributes. Let us add here that object types should not be mandatory. i.e. a specific object "entrance doors" into a shopping centre which is the object of the *door* type, with the specified list of attributes, can have additional attributes, such as color or weight. The sample structure consists of two atomic objects: *door1*, *window1* and complex parent object *room1* is shown in Fig. 2. Let us notice that although *door1* and *window1* objects have not child objects they are in relationship of neighborhood each to other.

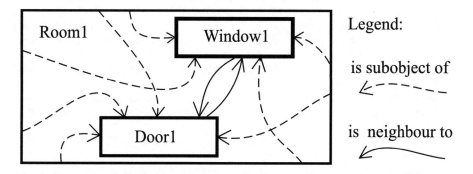

Fig. 2. The hierarchy of three simple objects.

To summarize, we can notice that with such a description of objects through the concepts of attributes, types and relationships, a creation of a map of any building is possible. As shown in Sect. 3.1, this map allows for calculation of routes between any two locations in this building and, as a consequence, the safe guidance of the user between them.

4 Summary

Nowadays, almost every publicly available buildings offer an infrastructure with functionalities such as access to a Wi-Fi network or monitoring system. In this paper, a system supporting navigation of the blind in the building based on the map ontology has been proposed. It seems that installation of such system on the existing infrastructure of the building would be quite inexpensive and relatively easy. The only cost would be the installation of software on the server and placing radio RFID or Bluetooth labels in the building.

Actually, the first version of the classes expressed in OWL language for hierarchical object map ontology was implemented. Also a conceptual version of the database to gather information about objects was established. In next steps, an implementation of prototype user interfaces to fulfill a database with testing data set will be provided.

References

1. Visual Impairment and Blindness, October 2017. http://www.who.int/mediacentre/factsheets/fs282/en/
2. Lee, G.M., Crespi, N., Choi, J.K., Boussard, M.: Internet of Things. In: Bertin E., Crespi N., Magedanz T. (eds) Evolution of Telecommunication Services. Lecture Notes in Computer Science, vol. 7768. Springer, Heidelberg (2013)
3. Bigham, J.P., Jayant, C., Miller, A., White, B., Yeh, T.: VizWiz::LocateItEnabling blind people to locate objects in their environment. In: Proceedings of the IEEE Computer Society Conference on Computer Vision and Pattern Recognition Workshops, San Francisco, CA, USA, pp. 65–72, June 2010
4. Mekhalfi, M.L., Melgani, F., Zeggada, A., De Natale, F.G.B., Salem, M.A.-M., Khamis, A.: Recovering the sight to blind people in indoor environments with smart technologies. Expert Syst. Appl. **46**, 129–138 (2016)
5. Curran, K., Furey, E., Lunney, T., Santos, J., Woods, D., McCaughey, A.: An evaluation of indoor location determination technologies. J. Locat. Based Serv. **5**(2), 61–78 (2011)
6. Sterling, G.: Magnetic Positioning the Arrival of Indoor GPS. Opus Research Report, San Francisco, June 2014
7. Subbu, K.P., Gozick, B., Dantu, R.: LocateMe: magnetic-fields-based indoor localization using smartphones. ACM Trans. Intell. Syst. Technol. (TIST) **4**(4), 73:1–73:27 (2013)
8. Foxlin, E.: Pedestrian tracking with shoe-mounted inertial sensors. IEEE Comput. Graph. Appl. **25**(6), 38–46 (2005)
9. Ecklbauer, B.L.: A Mobile Positioning System for Android based on Visual Markers. Mobile Computing, Austria, Hagenberg, 16 June 2014. http://www.academia.edu/7503224/A_Mobile_Positioning_System_for_Android_based_on_Visual_Markers
10. Petrou, L., Larkou, G., Laoudias, C., Zeinalipour-Yazti, D., Panayiotou, C.: Crowd-sourced indoor localization and navigation with anyplace. In: IPSN 2014 Proceedings of the 13th international symposium on Information processing in sensor networks, pp. 331–332, Berlin, 15–17 April 2014
11. Gilchrist, C.: Learning iBeacon. Packt Publishing Ltd., Birmingham (2014)
12. Lin, X.-Y., Ho, T.-W., Fang, C.-C., Yen, Z.-S., Yang, B.-J., Lai, F.: A mobile indoor positioning system based on iBeacon technology. In: 2015 37th Annual International Conference of the IEEE Engineering in Medicine and Biology Society (EMBC), pp. 4970–4973 (2015)
13. Reza, A.W., Geok, T.K.: Investigation of indoor location sensing via RFID reader network utilizing grid covering algorithm. Wireless Pers. Commun. **49**(1), 67–80 (2009)
14. Liu, H., Darabi, H., Banerjee, P., Liu, J.: Survey of wireless indoor positioning techniques and systems. IEEE Trans. Syst. Man Cybern. Part C (Appl. Rev.) **37**(6), 1067–1080 (2007)
15. Chang, N., Rashidzadeh, R., Ahmadi, M.: Robust indoor positioning using differential wi-fi access points. IEEE Trans. Consum. Electron. **56**(3), 1860–1867 (2010)
16. Zhou, Y., Law, C.L., Guan, Y.L., Chin, F.: Indoor elliptical localization based on asynchronous UWB range measurement. IEEE Trans. Instrum. Meas. **60**(1), 248–257 (2011)
17. Hudec, M., Smutny, Z.: RUDO: a home ambient intelligence system for blind people. SENSORS (Basel) **17**(8), Article Number: 1926 (2017)

Poster: A Mobile Application for Voice and Remote Control of Programmable Instruments

Burak Ece[1], Ayse Yayla[2], and Hayriye Korkmaz[3(✉)]

[1] Institute of Pure and Applied Sciences, Marmara University, Istanbul, Turkey
burakece9191@hotmail.com
[2] Vocational School of Technical Sciences, Marmara University, Istanbul, Turkey
acetinkaya@marmara.edu.tr
[3] Faculty of Technology, Marmara University, Istanbul, Turkey
hkorkmaz@marmara.edu.tr

Abstract. The purpose of this work is to add a new feature to bench-type conventional instruments used in Electrical and Electronics Engineering Laboratory which do not have any voice recognition and wireless communication technology. By this way, the user can control these instruments/devices remotely with voice commands and also monitor the results/values in graphical or numerical/text format as well over a mobile device screen. The only limitation is that such instruments should have a driver supported by any software such as NI LabVIEW and a PC connectivity interface such as USB, GPIB or LXI (LAN extensions for Instrumentation). Controlling the instruments (such as oscilloscope or signal generators which are frequently used for training purposes and whose functions are manually set) over a mobile device with voice commands will make life easier for disabled students who especially have difficulties in using their hands.

Keywords: LabVIEW · Android · Google Speech Recognition · Remote control
Internet of Things

1 Introduction

With the development of computers in the 1950s, the concept of internet has emerged, and wireless communication has become an indispensable part of our life with the recently added features on protocols. The presence of wireless communications and the concept of Internet of Things, launched by Kevin Ashton in 1999, allow users to control many devices remotely [1]. According to Gartner's research, it is estimated that in the year 2020, 26 billion products will be connected to the internet [2]. Thus, the ability to control electronic devices remotely will provide simplicity and comfort to users in many different application areas such as education. In human machine interaction, while the role of pressing the keys lost its importance; meanwhile voice-controlled operations gained popularity. With the successful recognition of spoken words and expressions and its conversion into a readable format by machines and also the existence of protocols enabling two or more devices communicate wirelessly with each other has led to the idea of managing electronic devices with remote voice commands. Thanks to the voice

© Springer International Publishing AG, part of Springer Nature 2019
M. E. Auer and R. Langmann (Eds.): REV 2018, LNNS 47, pp. 479–486, 2019.
https://doi.org/10.1007/978-3-319-95678-7_54

recognition and wireless communication technology of the mobile phones that people uses all the time, lots of products appeared that can be controlled by voice commands. This popularity and high demand for such mobile devices by people indicates that voice-controlled devices facilitate the human life.

With the development of wireless communication and voice recognition technologies and IoT concepts and their widespread and high impact throughout the world, it has become inevitable for laboratory instruments/devices to keep pace with this technology. A basic laboratory setup contains at least two instruments: one for supply or excitation of the system and the other for observation and measurements of the outputs. Distance control of experimental setup with voice commands and then monitoring the results on a mobile device screen by location-independently are very essential and useful features especially for remote lab applications or distributed instrumentation concepts. In this context, this work focuses the instruments that already supported by NI LabVIEW to convert them into voiced-controllable.

Paper organization is given as follows: In Sect. 2, the existing studies on remote laboratory and voice controlled applications over mobile devices are examined and similar works related to this paper are discussed. Section 3 describes the hardware and software components used in the application. In addition, the steps in developing remote instrument control by using speech recognition (SR) technology over a mobile device are explained in the same section. In the last section, the contribution of the application to the education technologies is emphasized. In addition, safety precautions taken to protect the computer and remote experimental setup from various dangers are discussed.

2   Similar Works

Students have the opportunity of doing real experiments after the theoretical learning in the traditional laboratories in engineering education. However, the traditional laboratories have many limitations: group working due to large number of students, lack of the number of instruments, high costs devices. This sometimes causes the students to attend the experiment sessions only as audiences. On the other hand, another important limitation of the traditional laboratories is the accessibility problem of disabled students. In the last decades, remote laboratories or e-learning platforms had been widely used and many studies were proposed for the solution of above problems.

With the rapid development of voice recognition technologies, it has become inevitable to add new features to traditional remote laboratories. While desktop computers are still widely used, but it seems that they will be replaced by mobile devices in the development of such applications. This new concept -by integrating voice recognition technology into traditional remote laboratories and controlling an instrument remotely over a mobile device by voice- seems to be a remarkable solution to eliminate deficiencies for handicapped students especially for the ones who cannot use their hands effectively. Besides the educational applications, mobile devices are widely used for home automation systems [3–6] or other control applications [7, 8], too.

Two main topics were considered in literature review: Remotely accessible laboratory applications and SR centric control applications in mobile devices.

2.1 Remotely Accessible Laboratory Applications

In [9], a small web-based remote laboratory framework is available for e-learning activities in the control systems for undergraduate students (Mindlab 2). The system supports hands-on e-learning experiences for robotics and automation content via internet.

The results of feedbacks for a remote laboratory used in the education of control engineering students are stated in [10]. In this paper, Stewart platform which is a ball-and-plate system with six degrees of freedom, driven by six servo motors, creating six possible motions which are three linear and three rotational and quadruple water tank system were addressed. These practical examples include challenging control concepts such as multiple-input, multiple-output control, decoupling, non-minimum phase systems, open-loop unstable systems, and PID control design. According to the results of the questionnaire applied to junior students, the remote laboratory needs further development to counter its limitations. And also, there is a keen interest on the remote laboratory to provide practical experiences in control engineering education.

An example for remote laboratories accessible via mobile device is stated in [11]. In this paper, mobile-optimized application architecture using Ionic framework is proposed for mobile learning (M-Learning). A control experiment according to innovative remote networked proportional–integral–derivative has been successfully implemented based on this new application architecture to demonstrate the effectiveness of the proposed new architecture for M-Learning. Students can connect to the real PID motor speed control experiment setup through the M-Learning environment anywhere and anytime.

A LabVIEW-based RF communications package to illustrate the different modulation techniques is presented in [12]. All the modulation parameters can be set by the user from the front panel. AM experiment is tested. The virtual and real time is accomplished via data-acquisition cards, installed on PC. NI USRP RIO2920 is used as transmitter in remote application. The user can specify the parameters such as amplitude, frequency and phase. The user can also add the noise to the signal.

An electro pneumatic system control designed by using LabVIEW is presented in [13]. This remote laboratory enables users to control equipment and monitor the results, allowing engineering students to perform experiments in real time, at their own pace from anywhere and whenever they want.

The speed control of DC motor based on PWM technique was achieved in [14] through SR technology supplied by Microsoft Speech SDK in LabVIEW. This application enables to control the motor by using vocal commands in English speaker independent. On the other hand, the motor can be controlled both manually and by voice. Speech processing procedures and recognition was also developed in LabVIEW.

A traffic light intersection application designed in LabVIEW was presented in [15]. NI USB 6008 module was used in order to simulate the system. Through remote panel, users can observe the application in real time via the Internet from a client machine. In addition, the durations of traffic lights' colors can be set on the remote panel.

2.2   Speech Recognition Centric Control Applications in Mobile Devices

Speech recognition technology provides an additional input to the mouse and keyboard. The usage of this technology has increased with mobile devices.

Voice controlled robotic arm vehicle using Android was developed in [16]. Hardware components consist of the low power 8-bit microcontroller ATMEGA328P, HC-05 Bluetooth module. Android application captures and transmits the voice commands to Bluetooth receiver and the controller. Command instructions written in the Arduino programming environment was used to program the controller. The robotic vehicle then operates as per the command received via android application.

A remote control system using SR technologies for blind and physically disabled individuals and also other individuals was presented in [17]. The main contribution of this system is to fulfill the most of the functions of a computer via speech. The configuration of the system consists of a mobile device such as a smartphone, a PC server, and a Google server. All these devices must be connected to each other. Users can perform the following tasks by voice commands: write emails and documents, calculate numbers, check the weather forecast, or manage the schedule. The proposed system also provides the text to speech conversion to announce the content of a written document saved in the computer for blind people by using Google Server.

In our study, in a similar manner as mentioned above studies, it was aimed to control LabVIEW supported laboratory instruments through a mobile device via voice. The main contribution of the proposed system is to integrate the SR technology into a remotely accessible laboratory setup in order to gain new facilities for disabled students in engineering education. In addition, it was realized that an ordinary instrument with PC connectivity interface and software support can become compatible with IoT technology through the application that we developed.

3   Materials and Methods

In this paper, a mobile application as shown in Fig. 1 has been developed to provide voiced-based remote control of the basic instruments which are widely used in Electrical and Electronics Engineering laboratories. In this setup, there is an oscilloscope, a signal generator, and an experiment board which includes an inverting amplifier using OPAMPs that is one of the basic applications in the electronic circuits course content. In the application, the signal generator and the oscilloscope are the instruments which can be controlled remotely over a mobile device. Android Studio and LabVIEW graphical development platform were used during the software development of the mobile application.

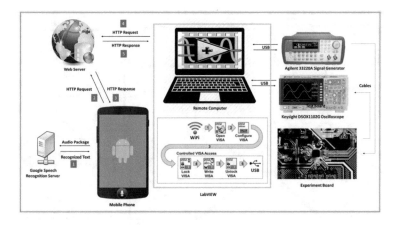

Fig. 1. General block diagram of the mobile application

3.1 Android Mobile Application

There are two main tasks in this application: the first is to convert the voice command into text format with the help of the Google Speech Recognition (GSR) Library, and the second is to transfer the text message to the remote server by using http post request as seen in Fig. 1. Our only criterion is that our mobile device must be open to internet access.

The application starts by converting the voice commands which is said in a predetermined format such as "generator volt 10", "generator function sinus", "oscilloscope autoset", "oscilloscope div set 1" into text by using GSR Service through the microphone of the mobile device.

Before transmitting the messages, detected commands were displayed on the mobile device screen in order to be sure that it is correct. A list of commands has been created depending on the features of the instruments supported by LabVIEW. If a detected command does not match to the commands in the list, it is not executed. Thus, sending of wrong messages to the instruments will be stopped at the beginning and the problems that may occur on the setup will be prevented. And finally, such a scenario the output signal of the generator is adjusted according to the commands and then applied to the experiment board. As a result, the expected output signal will be displayed on the oscilloscope screen. Remote control and monitoring of the experiment results on the mobile device will take place via the link provided to user by the Web Publishing Tool of the LabVIEW application running on the remote computer.

Figure 2a shows the flowchart of development steps in Android based mobile application.

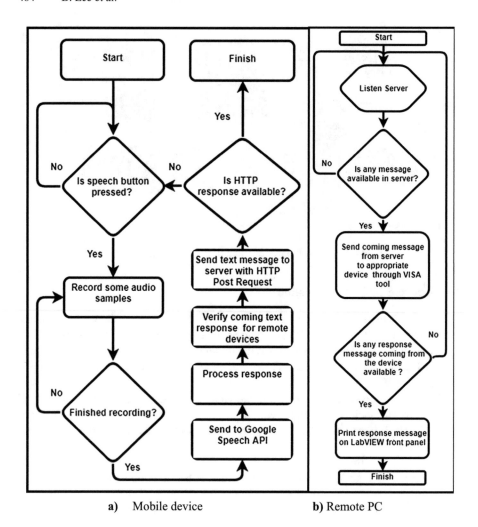

a) Mobile device   **b)** Remote PC

Fig. 2. Mobile application flowchart

3.2 Remote PC Application

As can be seen in Fig. 1, the Cloud continuously runs in the listening mode and the voice command for the desired action is sent to the corresponding instrument. After the drivers for signal generator and oscilloscope were installed on the remote computer, the LabVIEW VISA tool starts to communicate between those instruments and PC via USB interface.

Finally, the application is moved to the Internet using the LabVIEW Web Publishing tool. Thus, the user who sends a voice command over the mobile application can access the remote pc interface; change the input signal parameters to be applied to the experiment board, or observe the expected output signal for this input signal. Figure 2b shows the flowchart of the procedures conducted in LabVIEW.

4   Conclusion

Together with the IoT concept, the products compatible with this new technology are gradually being developed. In the near future, with the investments made on the products, most of the devices will be connected to the internet via Wi-Fi technology. Products developed under the concept of IoT often work with mobile devices and tablets. So, every object will have an application that can be accessed via a mobile device. As a result, those products can transfer their own notifications to their own mobile applications. In this work, it was realized that an ordinary instrument with PC connectivity interface and software support can become compatible with IoT technology through the application that we developed.

Measures are taken for security problems as the project uses remote http server. In this scenario, all students must be registered in order to use this application. In addition, the commands sent to the instruments have been encrypted with the AES algorithm. The encrypted commands will not be able to resolve without private key even if it is seized for 3rd party people. Also, since the commands were filtered out, each command cannot reach to the remote PC. Thus, the risks were tried to minimize and system safety has been improved.

Because of the lack of study on this subject, this paper has an innovative way that will provide important contributions to educational technologies. In addition, integrating voice recognition technology into the remote lab applications has a potential to provide important facilities for disabled students in engineering education. This will lead our future works.

References

1. Ashton, K.: That 'Internet of Things' thing. RFiD J. **22**(7), 97 (2011)
2. Gartner Says 8.4 Billion Connected "Things" Will Be in Use in 2017, up 31 Percent from 2016, Gartner Press Release. http://www.gartner.com/newsroom/id/3598917)
3. Soliman, M.S., Dwairi, M.O., Sulayman, I.I.A., Almalki, S.H.: Towards the design and implementation a smart home automation system based on internet of things approach. Int. J. Appl. Eng. Res. **12**(11), 2731–2737 (2017)
4. Chen, S.-C.,Wu, C.-M., Chen, Y.-J., Chin, J.-T., Chen, Y.-Y.: Smart home control for the people with severe disabilities. In: International Conference on Applied System Innovation (ICASI) (2017)
5. Wahab, M.H.A.: Iot-Based home automation system for people with disabilities. In: 5th International Conference on Reliability, Infocom Technologies and Optimization (Trends and Future Directions) (ICRITO) (2016)
6. Mittal, Y., Toshniwal, P., Sharma, S., Singhal, D., Gupta, R., Mittal, V.K.: A voice-controlled multi-functional smart home automation system. In: Annual IEEE India Conference (INDICON)

7. Leela, R.J., Joshi, A., Agasthiya, B., Aarthiee, U., Jameela, E., Varshitha, S.: Android based automated wheelchair control. In: Second International Conference on Recent Trends and Challenges in Computational Models (ICRTCCM) (2017)
8. Viswanath, N., Anbarasan, M., Jaisiva, S.: Android based automated wheel chair control for physically challenged person. Asian J. Appl. Sci. Technol. (AJAST) **1**(6), 26–29 (2017)
9. Di Giamberardino, P., Temperini, M.: Adaptive access to robotic learning experiences in a remote laboratory setting. In: 18th International Carpathian Control Conference (ICCC)
10. Chevalier, A., Copot, C., Ionescu, C., De Keyser, R.: A three-year feedback study of a remote laboratory used in control engineering studies. IEEE Trans. Educ. **60**(2), 127–133 (2017)
11. Wang, N., Chen, X., Song, G., Lan, Q., Parsaei, H.R.: Design of a new mobile-optimized remote laboratory application architecture for M-learning. IEEE Trans. Ind. Electron. **64**(3), 2382–2391 (2017)
12. Kandasamy, N., Telagam, N., Seshagiri Rao, V.R., Arulananth, T.: Simulation of analog modulation and demodulation techniques in virtual instrumentation and remote lab. Int. J. Online Eng. (IJOE) **13**(10), 140–147 (2017)
13. García-Guzmán, J., Villa-López, F.H., Vélez-Enríquez, J.A., García-Mathey, L.A., Ramírez-Ramírez, A.: Remote Laboratories for Teaching and Training in Engineering, in Design, Control and Applications of Mechatronic Systems in Engineering. InTech, London (2017)
14. D'Souza, C.R., D'Souza, C.D., Souza, S.D.D., D'Souza, S., Rodrigues, R.L.: Voice operated control of a motor using labview. Electr. Electron. Eng. **7**(2), 60–64 (2017)
15. Patrascoiu, N., Ioana, C.R., Barbu, C.: Virtual instrumentation for data acquisition and remote control. In: 2017 18th International Carpathian Control Conference (ICCC) (2017)
16. Alka, N., Salihu, A., Haruna, Y., Dalyop, I.: A voice controlled pick and place robotic arm vehicle using android application, Unpublished, Mechanical Engineering Mini Project, Abubakar Tafawa Balewa University, Bauchi-Nigeria (2017)
17. Jeong, H.-D.J., Ye, S.-K., Lim, J., You, I., Hyun, W., Song, H.-K.: A remote computer control system using speech recognition technologies of mobile devices. In: Seventh International Conference on Innovative Mobile and Internet Services in Ubiquitous Computing (IMIS) (2013)

Biomedical Engineering

Organic Compounds Integrated on Nanostructured Materials for Biomedical Applications

Cristian Ravariu[1(✉)], Elena Manea[2], Florin Babarada[1],
Doru Ursutiu[3], Dan Mihaiescu[4], and Alina Popescu[2]

[1] Faculty Electronics, Electronic Devices Circuits and Architectures,
BioNEC Group, University Politechnica of Bucharest,
Splaiul Independentei 313, Bucharest, Romania
cristian.ravariu@upb.ro
[2] National Institute for Research and Development of Microtechnology—
IMT Bucharest, Str. Erou Iancu Nicolae 126 A, Voluntari, Ilfov, Romania
[3] Transilvania University of Brasov, Brasov, Romania
[4] Faculty of Applied Chemistry,
Department of Organic Chemistry Costin Nenitescu,
University Politechnica of Bucharest, Bucharest, Romania

Abstract. This paper intends to present review aspects of our recent developments in the area of biodevices, nano-structures, organic compounds and integration techniques. Some simulation techniques accompany the hardware products or depict some virtual tools frequently exploited during the design stage. The main results reefer to TiO_2 nanostructures used to improve the enzyme adherence in biosensors, planar Nothing On Insulator nano-devices simulation, nano-particles application in biofabrication, besides to organic semiconductors in conjunction with bio-receptor integration for pesticide or glucose detection.

Keywords: Nanomaterials · Organic semiconductors · Integration techniques
Enzymes immobilization · Electronic biosensors

1 Introduction

The organic compounds are frequently encountered in recent nano-electronic devices, [1]. The integrated biosensors comprises a Field Effect Transistor (FET) or Organic Thin Film Transistor (OTFT), as transducer element, [1]. The biomaterials are incorporated near the transducer element, [2], sometimes using intermediate nano-structures for the adherence increasing. A key element during the biodevice design stage, is to allow a sensitive modulation of the current by the bio-chemical analyte, [3]. The successful biosensors are based on key enzyme entrapping. Novel technologies uses the organic semiconductors as buffer devices between the electronic circuitry and the biorecognition elements. Similar approaches are encountered for some bio-mimetic drugs that allow their electrochemical detection, combining Carbon nanotubes nanostructures, immobilized enzymes as acetylcholinesterase, light polymers and single strands of DNA as genetic material, [4].

© Springer International Publishing AG, part of Springer Nature 2019
M. E. Auer and R. Langmann (Eds.): REV 2018, LNNS 47, pp. 489–497, 2019.
https://doi.org/10.1007/978-3-319-95678-7_55

This effervescent state of the art, authorize us to approach a review paper, about our previous results of nano-structured materials and organic compounds that allow to integrate different materials and structures able for biodetection applications.

2 Nanostructured Materials Used in Biorecognition

2.1 Nanostructured TiO₂ Used in Biosensors

In this section, some intermediate nanostructured layers are briefly depicted to emphasize their usage within the biosensor. When the enzymes stand for the receptor element integrated onto a Si-wafer, porous materials starting by Si-nanoporous [5] are used for immobilization. The nano-porous materials synthesis onto at a silicon surface begins by a metal deposition (e.g. Al or Ti). The next technological step consists in metal conversion in its oxidic compounds, like Al_2O_3, [6] and finally, by anodization processes in a nano-porous layer. The pore sizes are strongly influenced by the chemical condition of the anodization bath.

The main role of this nano-structured material inside a biosensor is to increase the adherence of the enzyme solutions, assisted by capillary phenomena. On the other hand, this layer must be synthesized onto the Si-substrate, in agreement with the microtechnological processes and must ensure a robust anchoring to substrate. The TiO_2 can be converted into nano-tubes of TiO_2 by anodization, being more biocompatible than Si-porous with the living matter, as the majority of prostheses proves. For large organic molecules, the pore sizes are modified by the anodization reaction constrains or altering the electrolyte concentration.

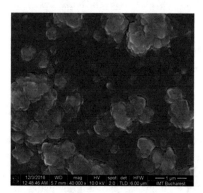

Fig. 1. Our nanostructured TiO_2 material characterized by SEM.

In our laboratories, thin films of nano-structured TiO_2 are synthesized in order to be subsequently used for enzymatic layers adsorption. The optimal TiO_2 properties are achieved if a constant pH < 7 is kept and an electrolytic solution of NH_4F, 40% for the anodization process, Fig. 1.

2.2 Fe_2O_3 Nano-Core-Shell - Used in Biofabrication

In the last decades, a special attention was paid to the synthesis of nanostructured composites, [7] or Fe_2O_3 nano-core-shell synthesis. One example is the silica meso-porous structure based on exceptional physical-chemical features, keeping cheap processes and green synthesis routes. These nano-composites can provide new hybrid products, if they are mixed by bioactive plant compounds from Achilea millefolium or Calendula officinalis, covered by polymer shells and MCM41 hexagonal silica np's cores, [8]. The product effect is antibacterial.

Another study of Ficai's group takes into account the collagen/hydroxyapatite-magnetite-cisplatin nanomaterial combined into a nanostructure able to control the drug delivery for the cytostatic agents, [9]. The co-existence of collagen and hydroxyapatite fosters a rapid healing of the injured tissue, besides to the antitumoral behavior of the magnetite under an appropriate electromagnetic field stimulation.

Instead Fe_2O_3, the Copper nanoparticles under some plant leaf extracts exhibit similar antimicrobial properties, after recent studies of Egziabher and colab., [10].

2.3 Planar Nothing on Insulator Nano-Structure Within a Pesticide Biosensor

This section briefly presents a pesticide biosensor fabricated in the Si-technology, using Acetyl-Choline-Esterase (AcHE) adsorbed onto Si-nanoporous as receptor, [11], designed to detect paraoxon that is a parasympathomimetic pesticide, which acts as an acetyl-cholinesterase inhibitor. Additionally, the transducer of this kind of biosensor consists in an interdigited capacitive electrodes, that recently were modeled by a special nano-structure - the planar Nothing On Insulator (p-NOI) device, included in the capacitive detection system. The biodetection is based on the paraoxon hydrolysis under the Acetyl-cholinesterase (AChE) enzyme action. The capacitive p-NOI configuration electrodes convert the Oxygen ions into an electrical capacity deviation.

Since 2005 the Nothing On Insulator (NOI) structure was conceptual proposed and time to time upgraded, [12–14]. Firstly, the succession n-Si/Vacuum nanocavity/n-Si (nVn) on insulator, with the device main body formed by the Vacuum cavity or Nothing space On Insulator was adopted. Later, Oxide (O) replaced the Vacuum (V) nano-cavity and additionally, a source metal replaced the source semiconductor island, providing the metal/Oxide/Si-n (mOn) succession. This structure still conserves the NOI tunneling mechanism. This rotated NOI structure by 90 degree, can be implemented by the Si-planar technology, known as planar-NOI (p-NOI). Other researchers experimentally proved similar tunneling conduction thru a metal/Insulator/metal (MIM) structure, [15]. The p-NOI and MIM structures are based on a tunneling principle thru a sub-5 nm thin insulator and benefit on the materials placement on the front plan of the Si-wafer. The p-NOI structure, which is integrated inside the biosensor transducer, has another facet in this work: one p-NOI structure still exists between top metal on insulator placed on silicon and a second p-NOI structure is present between two adjacent lateral metal fingers. The first one must accomplishes an isolation thru the bottom nanoporous material, Fig. 2a. But the liquid droplet that connects two adjacent fingers by an ionic conductor, offers a novel conduction route.

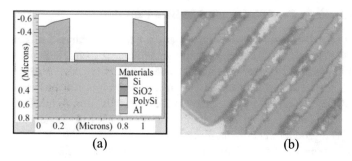

Fig. 2. (a) The p-NOI device description by Athena/Silvaco software; (b) The interdigited electrodes on the capacitive transducer of the pesticide biosensor.

The final interdigited biosensor transducer posses 98 metallic traces in a p-NOI configuration, which are starting from the central pillar for each electrode. This multiple traces optimum increases the active area, detail in Fig. 2b. The entire detailed technological process and the masks design for the capacitive electrodes are previously presented, [11].

3   Organic Layers Attached to Nanostructures

3.1   Organic Semiconductors for Integrated Electronics

The next Organic transistors correspond to the Thin-Film Transistors (TFT) placed on insulator with dual gate, [16]. The simulation of an Organic Thin Film Transistor (OTFT) is performed in Atlas/Silvaco, considering the pentacene as the p-type organic semiconductor, available in the Atlas library. The simulated device configuration keeps the agreement with the standard technology. Hence, a conductor ITO layer 10 … 20 nm thickness is deposited onto a conductor substrate covered by an organic insulator as polyimide of 20 nm. Above, the organic semiconductor layer was deposited as Pentacene doped by $3 \times 10^{17} cm^{-3}$ uniform concentration. The gold electrodes are deposited on the top of the structure. In this way, the simulated OTFT belongs to the organic transistors class named Bottom-Gate-Top-Electrodes, Fig. 3a.

From the static characteristics presented in Fig. 3b, the sub-threshold currents are visible for $V_G > 1$ V, in agreement with the distances among source, drain and both gates. Also a sub-threshold swing more than 200 mV/dec, besides to two firm ON, OFF states are visible, in agreement with the literature, [16]. One application could be the device transducer in those sensors that need the current variations without interfaces interferences. Another application concerns the synthesis pathways of the ferrite magnetic nanoparticles coated with polyethylene glycol, [17].

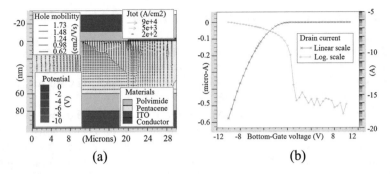

(a)   (b)

Fig. 3. (a) The distribution of the potential and holes mobility in the OTFT and current vectors inside the pentacene film; (b) static characteristics of the OTFT transistor biased at $V_D = -10$ V and variable bottom-gate.

3.2 Glucose-Oxidase on Nanostructured TiO_2

More complex organic layers, like enzyme films, can be integrated onto Silicon wafer. In biosensors with FETs, the receptor can be the Glucose-Oxidase enzyme, GOX, designed to detect the glucose. The optimal intermediate layer is TiO_2, [18, 19]. The GOX work principle is based on the oxido-reductase enzyme action to stimulate the β-D-glucose oxidation into D-glucono-δ-lactone plus hydrogen peroxide. In this way, the O- ions concentration increases. A Field Effect Transistor can be used as integrated transducer for biorecognition, (Bio-FET). We recently proposed this kind of Bio-FET with GOX and TiO_2, [20]. The Bio-FET's channel is modulated by the Oxygen concentration, after a remote concept previously developed in biosystems, [21].

Unfortunately, the SEM images with the upper GOX membrane entrapped in nafion, demonstrated a poor anchoring on the Si-substrate, even if an intermediate nanostructure of TiO_2 was utilized, [18]. If the nafion is replaced by glutaraldehyde as crosslink element, a better adherence of the GOX enzyme to substrate, was measured, [20]. Additionally, the enzymatic membrane has homogenous thicknesses, Fig. 4.

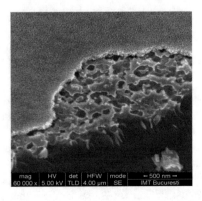

Fig. 4. SEM image of a cross-section thru the top layer onto TiO_2-nanostructure anchored to Si-substrate.

4 Simulations Techniques for a Virtual Analysis

During the design stage of our Glucose BioFET, different materials from the BioFET structure, were described by the material statement, in the Atlas simulator, [20]. For instance, the Al_2O_3 material can be defined in ATLAS as a general insulator, specifying its model parameters (e.g. $\varepsilon_r = 8.2$, [6]).

The Bio-FET simulations begin from a Si-substrate of 32 μm thickness and a p-type doping concentration of $7 \times 10^{15} cm^{-3}$. The channel has 10 μm and some source and drain zones as n^+-islands can be connected by the inversion channel so that the total length reaches 32 μm, Fig. 5a. The doping concentration in the source/drain regions as in the ring deep diffused zones reaches a Gaussian pick profile of $7 \times 10^{17} cm^{-3}$.

During the masks design stage, another simulation tool was used: LEDIT for the layout editing. Two FETs are simultaneously designed on the same chip: one is our active Bio-FET and the other one is a reference MOSFET fabricated by the same technological steps as the Bio-FET, for characterizations. The layers, which finally constitute the masks, are designed using the standard EDA tool LEDIT from Tanner. This double-transistor structure gets the MOSFET positioned in the left side of the structure and the BioFET positioned in the right side of the structure, Fig. 5b.

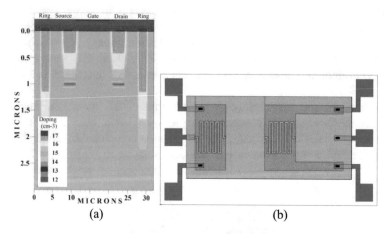

Fig. 5. The simulated FET part from our Bio-FET with 10 μm channel length; (b) The LEDIT simulation of the reference transistor at left and active biodevice at right.

In Atlas, the nano-structure of TiO_2 or GOX description are performed by the Ambient user-defined parameter. For instance, Ambient1 with a dielectric permittivity of 14 and a rare mesh is adopted for the TiO_2 roughnesses at interfaces. The option Ambient2 with a dielectric permittivity of 80, specific to aqueous solutions, is used for the GOX enzyme defining.

The simulator found a current of 0.8 nA for a gate insulator of 150 nm thickness [20], but this could be much thicker for a real BioFET that includes the nitride onto oxide and nano-structure of TiO_2, [20].

5 Biomedical Applications and Discussions

The biostructures fabricated with thin films of nano-structured TiO_2 are applied to produce glucose biosensors [20], able to monitor in vitro the glucose level in the blood stream. Other authors used the nano-structured TiO_2 in a total organic compounds sensing in aqueous media, [18], using excellent biocompatibility and environmentally friendly semiconductor properties of TiO_2. When the TiO_2 nanomaterial is combined by graphene into microspheres able to encapsulate haemoglobin, a mediator-free biosensor of high stability are produced, [19]. The biorecognition elements belong to a large spectrum that usual convert the biological analyte in H_2O_2 with a linear range between 0.1 to 360 μM and a low detection limit of 10 nM, [19].

The Nano-Core-Shell (NCS) are already used for biofabrication of antibacterial products: (i) the ferrite NCS were assembled into a thin layer with antibiotic effect on Staphylococcus aureus biofilm [8]; (ii) the gold NCS prepared by reduction of auric acid with sodium citrate were successfully tested for its toxic response, using Chlamydomonas reinhardtii (a green algae) and Vibrio fischeri (a bacteria). There is a different response for plant versus animal cells, under the gold NCS particles, in respect to the cell wall rigidity. It was found that the cytotoxicity of gold-NCS on microalgae is highly dependent on the binding sites to the membrane receptors. Further studies are also in progress.

Researchers are focused to find new drugs or to improve existing chemotherapy in the cancer therapy. In osteosarcoma a promising result was obtained by product based on the collagen/hydroxyapatite-magnetite-cisplatin nanomaterial able to targeted delivery of the cytostatic agents, [9]. The nano-compound has an antitumoral role, beside to a regenerative role stimulated by the collagen content.

The Organic Thin Film Transistor stands for another class of electronic devices that are recently applied in the field of biomedical products. The detection of the human Immuno-globulin IgG by organic transistors operated at low voltage under 3 V, was proved. By the threshold voltage shift, under the exposure of the device to biotinylated proteins (like IgG) was recognized up to a limit detection of 8 nM, [22].

The transistors with organic compounds represented by enzymes or antibodies are intensively developed as integrated biosensors. They cover a large palette of in vitro medical applications: (i) urea detection by the key urease enzyme deposition onto the gate space of an Enzyme- Field-Effect Transistor (En-FET) [23], (ii) aliments monitoring by the key antibody using a quartz microbalance [24] or Carbon-nanotubes (CNT) suitable for CNT-FETs construction, [25].

The use of Field Effect Transistors or Thin Film Transistors as transducer elements for biosensor applications offers real advantages as low noise, high sensitivity, low operation voltage and good long-term stability.

6   Conclusions

This paper proposed a review analysis about organic and biorecognition elements and their integration techniques with nano-structured materials. Firstly, the contributions in TiO_2 anchoring techniques, Fe_2O_3 nano core-shell biofabrication applications or planar NOI nanodevice, were presented. Secondly, the organic semiconductors used in OTFT simulations, besides to more sophisticated biomaterial compatibility with the semiconductor products, were investigated. Finally, some specific biomedical applications of the above discussed products are discussed in respect with the obtained results including comparisons and references to other sensors that are currently available.

Acknowledgments. This work was partially supported by a grant of Ministry of Research and Innovation, CNCS - UEFISCDI, project number PN-III-P4-ID-PCE-2016-0480, within PNCDI III, project number 4/2017 (TFTNANOEL) and also supported by a grant of the Romanian National Authority for Scientific Research and Innovation, CNCS/CCCDI UEFISCDI, project number PN-III-P2-2.1-PED-2016-0427, within PNCDI III, project number 205PED/2017 (DEMOTUN).

References

1. Liu, N., Hu, Y., Zhang, J., Cao, J., Liu, Y., Wang, J.: A label-free, organic transistor-based biosensor by introducing electric bias during DNA immobilization. Org. Electron. **13**, 2781–2785 (2012)
2. Bergveld, P.: Thirty years of ISFET-ology: what happened in the past 30 years and what may happen in the next 30 years. Sensor Actuat. B Chem. **88**, 1–20 (2003)
3. Umar, A., Ahmad, R., Kumar, R., Ibrahim, A.A., Baskoutas, S.: $Bi_2O_2CO_3$ nanoplates: fabrication and characterization of highly sensitive and selective cholesterol biosensor. J. Alloy. Compd. **683**, 433–438 (2016)
4. Viswanathan, S., Radecka, H., Radecki, J.: Electrochemical biosensor for pesticides based on acetylcholinesterase immobilized on polyaniline deposited on vertically assembled carbon nanotubes wrapped with ssDNA. Biosens. Bioelectron. **24**, 2772–2777 (2009)
5. Dhanekar, S., Jain, S.: Porous silicon biosensor: current status. Biosens. Bioelectron. **41**, 55–64 (2013)
6. Wen, Y.P., Wei, H.C., Jie, H.Y., Cheng, C.M., Hao, L.H., Shi, L.S., Ting, L.C.: A device design of an integrated CMOS poly-silicon biosensor-on-chip to enhance performance of biomolecular analytes in serum samples. Biosens. Bioelectron. **61**, 112–118 (2014)
7. Mollaamin, F.: Bioorganic adsorption on nano-clusters through disabling language cognition. Adv. Nano-Bio-Mater. Devices **1**(3), 135–145 (2017)
8. Grumezescu, A.M., Cristescu, R., Chifiriuc, M.C., Dorcioman, G., Socol, G., Mihailescu, I.N., Mihaiescu, D.E., Ficai, A., Vasile, O.R., Enculescu, M., Chrisey, D.B.: Fabrication of magnetite-based core–shell coated nanoparticles with antibacterial properties. Biofabrication **7**, 015014–015020 (2015)
9. Ficai, D., Sonmez, M., Albu, M.G., Mihaiescu, D.E., Ficai, A., Bleotu, C.: Antitumoral materials with regenerative function obtained using a layer-by-layer technique. Drug Des. Dev. Ther. **9**, 1269–1279 (2015)

10. Egziabher, H.M., Prakash Yadav, O., Kebede, T., Yadav, L.: Green synthesis of plant leaf extract mediated copper nanoparticles and their antimicrobial activity. Adv. Nano-Bio-Mater. Devices **1**(3), 161–171 (2017)

11. Syshchyk, O., Skryshevsky, V.A., Soldatkin, O.O., Soldatkin, A.P.: Enzyme biosensor systems based on porous silicon photoluminescence for detection of glucose, urea and heavy metals. Biosens. Bioelectron. **66**, 89–94 (2015)

12. Ravariu, C., Babarada, F.: Resizing and reshaping of the Nothing On Insulator NOI Transistor. Adv. Nano-Bio-Mater. Devices **1**(1), 18–23 (2017)

13. Han, J.-W., Meyyappan, M.: Introducing the vacuum transistor: a device made of Nothing. IEEE Spectr. **7**, 25–29 (2014)

14. Di, Y., Wang, Q., Zhang, X., Lei, W., Du, X., Yu, C.: A vacuum sealed high emission current and transmission efficiency carbon nanotube triode. AIP Adv. **4**(6), 045114:1–6 (2016)

15. Suzuki, M., Sagawa, M., Kusunoki, T., Nishimura, E., Ikeda, M., Tsuji, K.: Enhancing electron-emission efficiency of MIM tunneling cathodes by reducing insulator trap density. IEEE Trans. Electron Devices **59**(8), 2256–2262 (2015)

16. Kumar, B., Kaushik, B.K., Singh Negi, Y., Goswami, V.: Single and dual gate OTFT based robust organic digital design. Microelectron. Reliab. **54**(2), 100–109 (2014)

17. Augustin, M., Balu, T., Rejitha, S.G., Arockia Raj, A.A.: Synthesis and characterization of cobalt ferrite magnetic nanoparticles coated with polyethylene glycol. Adv. Nano-Bio-Mater. Devices **1**(1), 71–77 (2017)

18. Qiu, J., Zhang, S., Zhao, H.: Recent applications of TiO_2 nanomaterials in chemical sensing in aqueous media. Sens. Actuators B Chem. **160**, 875–890 (2011)

19. Liu, H., Guo, K., Duan, C., Dong, X., Gao, J.: Hollow TiO_2 modified reduced graphene oxide microspheres encapsulating hemoglobin for a mediator-free biosensor. Biosens. Bioelectron. **87**, 473–479 (2017)

20. Ravariu, C., Manea, E., Babarada, F.: Masks and metallic electrodes compounds for silicon biosensor integration. J. Alloy. Compd. **697**, 72–79 (2017)

21. Huimin, Z., Liu, Q., Li, Z., Liu, J., Jing, X., Zhang, H., Wang, J.: Synthesis of exfoliated titanium dioxide nanosheets/nickel-aluminum layered double hydroxide as a novel electrode for supercapacitors. Roy. Soc. Ch. Adv. **5**, 49204–49210 (2015)

22. Minamiki, T., Minami, T., Kurita, R., Niwa, O., Wakida, S.-I., Fukuda, K., Kumaki, D., Tokito, S.: Accurate and reproducible detection of proteins in water using an extended-gate type organic transistor biosensor. Appl. Phys. Lett. **104**, 243703:1–5 (2014)

23. Alizadeh, T., Akbari, A.: A capacitive biosensor for ultra-trace level urea determination based on nano-sized urea-imprinted polymer receptors coated on graphite electrode surface. Biosens. Bioelectron. **43**(5), 321–327 (2013)

24. Crosson, C., Rossi, C.: Quartz crystal microbalance immunosensor for the quantification of immunoglobulin G in bovine milk. Biosens. Bioelectron. **42**(4), 453–459 (2013)

25. Puertas, S., de Gracia Villa, M., Mendoza, E., Jorquera, C.J., dela Fuente, J.M., Fernandez-Sanchez, C., Grazu, V.: Improving immunosensor performance through oriented immobilization of antibodies on carbon nanotube composite surfaces. Biosens. Bioelectron. **43**(5), 274–280 (2013)

Towards an Automated Analysis of Forearm Thermal Images During Handgrip Exercise

Pedro Silva[1], Ricardo Vardasca[2(✉)], Joaquim Mendes[2],
and Maria Teresa Restivo[2]

[1] Faculdade de Engenharia, Universidade do Porto, Porto, Portugal
up201102738@fe.up.pt
[2] INEGI-LAETA, Universidade do Porto, Porto, Portugal
{rvardasca,jgabriel}@fe.up.pt,
trestivo@gcloud.fe.up.pt

Abstract. Infrared thermography is an imaging modality that is able to map the surface skin temperature and has been widely researched and applied in biomedical applications. It can be used as a monitoring method for upper limb condition support diagnosis and treatment assessment. The aim of this research is the development of an image processing method for rapid analysis of forearm thermographic images. The images were taken from 13 participants who undergone in a handgrip force exercise using a FLIR A325sc thermal camera. The obtained images were analyzed with FLIR ThermaCAM Researcher Pro 2.10 software package, and by a customer developed program in Matlab 2016b. Three regions of interest (ROI) were defined for both analysis. At the Matlab developed application, the forearm was segmented from the background and the ROIs placed automatically and average temperature and standard deviation were extracted. The results from both analysis were compared and showed good correlations for the examined 1365 images. The Matlab developed analysis is much faster than the time-consuming analysis performed using the standard camera manufacturer provided tool, however the first has to be used to export the temperature matrix from the proprietary image format in order to be used by the Matlab developed program. The development of a reliable tool for automated thermographic analysis of the forearms ROIs was successful implemented. For further work, it is proposed to read directly from the raw thermal images proprietary formats to hasten the analysis.

Keywords: Automated analysis · Forearm · Handgrip
Infrared thermal images

1 Introduction

The handgrip exercise using a dynamometer provides physiological parameters, which are relevant for frailty, pathological, rehabilitation assessments and nutrition status [1].

Musculoskeletal disorders may be caused due to exertion or prolonged exposure to physical factors, and can lead to body function impairment and to people's physical and psychosocial hardship [2]. Besides, in musculoskeletal disorders, the skin blood flow control is also an important factor to be taken into consideration, since it is necessary

© Springer International Publishing AG, part of Springer Nature 2019
M. E. Auer and R. Langmann (Eds.): REV 2018, LNNS 47, pp. 498–506, 2019.
https://doi.org/10.1007/978-3-319-95678-7_56

for the internal temperature and blood pressure maintenance, during exercise. The ageing process and various pathologies are often responsible for the impairment of such mechanisms that regulate skin blood flow [3]. There has been progress in the development of assessing tools for easy evaluation of disease severity or body function impairment for therapeutic purposes [4].

Infrared thermography (IRT) is an imaging technique that can be applied in the monitoring of the temperature distribution of the human skin surface. IRT is a remote, non-invasive, safe, relative low cost and rapid method that can be applied to analyze physiological activity in real-time [5–8]. Through the use of thermal imaging cameras, examiners are able to acquire the skin emitted thermal radiation in the form of a thermogram [9]. When applied to the upper limbs temperature monitoring, this technique provides a physiological imaging method that detects the temperature changes of the limb's skin, which may be a signal of a pathological state that can be related to peripheral vascular, neurological or musculoskeletal conditions [10, 11].

Currently, the available software for analyzing thermal images is generic, designed for general applications such as industrial, security or military. As this software uses proprietary files format, it makes difficult to design and develop customized software for specific applications, such as biomedical [12]. Despite this situation, several manufacturer software packages offer the feature of exporting the temperature matrix, so that it can be used as input to customized data analysis applications.

The aim of this research is to develop a method based in Matlab for automatically analyze thermal images of forearms, which will be useful on the optimization of future studies, in which there will be a vast number of upper limbs thermographic images to analyze for frailty, pathological and rehabilitation purposes, etc.

2   Methodology

2.1   Forearm Thermographic Examination

This study consists on the analysis of temperature behavior of the forearm muscles and main vascular areas, when subjected to mechanical stimulus of handgrip. The study involved 13 healthy participants (9 men and 4 women) with ages of 26 ± 5 years old, with a 26.0 ± 4.5 of body mass index (BMI), to which the procedure was explained and signed an inform consent. The examinations were divided in three main phases: acclimatization, griping and resting phases. The first phase consisted of 10 minutes' period of thermal acclimatization to promote the thermal equilibrium with the surrounding environment. The griping phase consisted of an exercise of 10 consecutive grips of 5 s each. The final phase, resting, involved a period of 60 s in which the forearm remained still holding the handgrip device.

The thermal images were recorded, one image per second, by the thermal camera FLIR A325sc (focal plane array sensor size of 320×240, a Noise Equivalent Temperature Difference of <50 mK at 30 °C and a measurement uncertainty of $\pm 2\%$ of overall reading), which was turned on 15 min before the first capture to enforce electronics thermal stability. For facilitating the stabilization of the forearm, a support was built (Fig. 1).

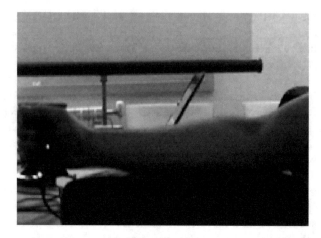

Fig. 1. Forearm resting at the support and holding the handgrip dynamometer

The images were taken in a room with stable environment conditions, no air conditioning neither force air were in use and the windows were closed, being the temperature 21.5 ± 0.5 °C and humidity around $47.3 \pm 2.1\%$, measured by a Testo 175H1 temperature and humidity data logger. There was also an absence of artificial illumination in the examination room. In order to cope with internationally accepted thermal image capture protocols [13, 14] and avoid interference of participant related variables, he/she was instructed to do not have any heavy meal, coffee, tea, alcohol or drugs and smoking in the 2 hours' prior the examination and to do not participate in any sports of physiotherapy in the day of the appointment as well as not having any oil or ointment in the forearm.

2.2   FLIR ThermaCAM Researcher Pro 2.10 Image Analysis

The recorded thermographic images were opened and displayed on the FLIR ThermaCAM Researcher Pro 2.10 program and three Regions of Interest (ROI) were defined and used (Fig. 2) to extract temperature parameters (mean temperature and standard deviation). The ROIs used were a circle over the digital flexor muscle (ROI1), a little square over the ulnar artery wrist region (ROI2) and another little square over the radial artery wrist region (ROI3). Once the three ROIs were placed on the desired locations (Fig. 3a), the mean temperature and the standard deviation of these sites were extracted from the thermographic images of the forearms. The ROIs size were the same for the whole images.

2.3   Matlab Image Processing Developed Program

A Matlab algorithm was developed in order to automatically analyze the previously captured thermal images of the forearms, which were exported from the FLIR ThermaCAM Researcher Pro 2.10 software as temperature matrices and loaded into Matlab 2016b as grayscale images (Fig. 3b).

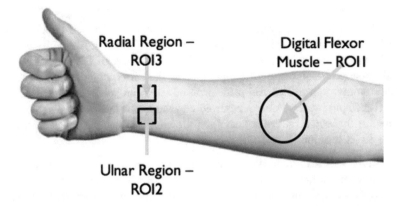

Fig. 2. The definition of the Regions of Interest (ROI) used

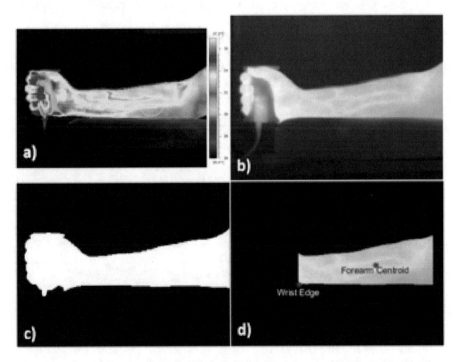

Fig. 3. Images of the developed methodology: (a) An example of a forearm thermal image loaded into the FLIR ThermaCAM Researcher Pro 2.10 software with the ROIs used; (b) Example of a temperature image loaded into Matlab; (c) Example of the binarized image from the histogram thresholding; (d) The forearm segmentation and the location of the reference points for the placement of the ROIs

The first step in image processing was to isolate the upper limb from the background, using the Otsu binarization method (Fig. 3c). The next step was to isolate the forearm from the rest of the upper limb, through a rectangle mask. The width of this rectangle is calculated based on the number of horizontal white pixels of the forearm region, which is calculated by detecting the discontinuities of the wrist and elbow were the vertical height of white pixels is minor. The rectangle height is the inside rectangle mask higher number of vertical white pixels that are part of the forearm. With the isolated forearm by subtracting the inner area of the surrounding rectangle, two main reference points coordinates for placement of ROIs are obtained with the Matlab function *regionprops*, the forearm centroid and the lower left wrist edge (Fig. 3d).

Based in these two reference sites the following calculations are needed to determine the ROIs location (Fig. 4):

- The ROI1 radius corresponds to the difference between the y coordinate of the forearm centroid and the y coordinate of the lower left wrist edge.
- The ROI1 x coordinate results from the x coordinate of the forearm centroid added with the ROI1 radius.
- The ROI1 y coordinate is obtained by subtracting the y coordinate of the lower left wrist edge with the division by 2 of the height of the forearm white pixels at the x coordinate of the ROI1.
- The ROI2 and ROI3 square side corresponds to the height of the forearm white pixels at the x coordinate of the lower left wrist edge divided by 4.
- The ROI2 and ROI3 x coordinates result from the addition of the x coordinate of the lower left wrist edge with the ROI2 and ROI3 square side.
- The ROI2 y coordinate is obtained by subtracting the y coordinate of the lower left wrist edge with 3.5 times the ROI2 square side.
- The ROI3 y coordinate corresponds to subtracting the y coordinate of the lower left wrist edge with 1.5 times the ROI3 square side.

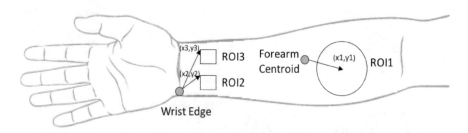

Fig. 4. Diagram of the forearm ROI's placement in Matlab

This is the procedure for the right forearm, for the left forearm the correspondent method can be used, the mean temperature is obtained through verifying if the point belongs to the ROI, adding its temperature value, which will be later divided by the number of pixels of the ROI.

These ROIs were then utilized as masks to calculate the mean temperatures (T) and standard deviation (Std) from the original temperature image. This program automatically processes and analyzes the same forearm thermal images that were also analyzed in the FLIR ThermaCAM Researcher Pro 2.10.

2.4 Results Comparison

In order to assess the viability of automatic thermographic analysis of the forearms, during a handgrip exercise, the mean temperature values of the three ROIs obtained through FLIR ThermaCAM Researcher Pro 2.10 were compared to those obtained by the Matlab program. The Matlab software processed all temperature images of the participants making the calculation of the mean temperature of ROI1, ROI2 and ROI3. These values were then compared to the ones acquired FLIR ThermaCAM Researcher Pro 2.10 and Pearson correlation and the Intra Class Correlation (ICC) coefficient that provides the measure of the variability of the results and the alpha Cronbach coefficient gives the consistency and thus the quality of the data. The ICCs were performed between the values of both applications to assess the validity of the developed approach. A correlation value (in a scale of 0 to 1) was determined in the correspondent values for the three ROIs. When the correlations value between different image analysis programs is closer to 1, it means that the obtained results are similar.

3 Results

Using the Matlab customer program, the images were loaded and processed automatically, obtaining the mean temperature (T) and standard deviation (sd) of all ROIs (Fig. 5) and export them to a Microsoft Excel spreadsheet.

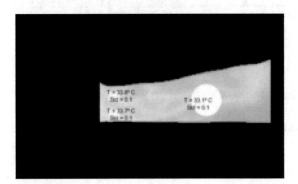

Fig. 5. Forearm image with three ROIs, their mean temperature and standard deviation, obtained from Matlab developed program

The Pearson correlation between the Matlab program and FLIR ThermaCAM Researcher Pro 2.10 software was of 0.8572 for ROI1, 0.8681 for ROI2 and 0.9124 for ROI3 (Fig. 6). The ICC statistical analysis showed good data consistency (alpha Cronbach coefficient) for all ROIs and good correlations among the used methods, the traditional one and the one developed (Table 1).

It is important to note that any single image analyzed in the manufacturer program takes 1 min and exporting it to a temperature array format takes around 10 s, analyzing it in the developed Matlab program, consumes in average 5 s.

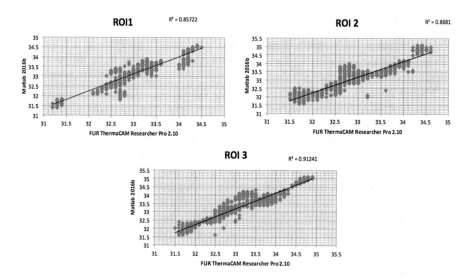

Fig. 6. Correlation between the Matlab developed program and FLIR ThermaCAM Researcher Pro 2.10 analysis values for the three ROIs used

Table 1. The ICC coefficient statistical analysis with the data consistency and ICC interval

ROIs	Alpha Cronbach coefficient	ICC	ICC interval (95% c.i.)
ROI1	.961	.925	.916 to .932
ROI2	.965	.932	.924 to .938
ROI3	.977	.955	.950 to .960

4 Discussion

Throughout this study, the forearm thermal images were acquired in a very controlled environment and using international accepted guidelines for thermal images capture [13, 14]. Not only were the room temperature and humidity adequate for the experiment, but they also remained almost the same, during the entire examination of the 13 participants. Thus, it was possible to capture thermal images with a lesser degree of external influences and intrinsic variables.

During the thermal image analysis through FLIR ThermaCAM Researcher Pro 2.10 software, the acquisition of the means and standard deviations values of the ROIs was a slow and tedious process. Not only there was a big number of images to analyze (more than a thousand), but also, due to the upper limbs movements of the participants (which sometimes could not be totally avoided during the image acquisition), there was a frequent necessity of adjusting the ROIs positions.

Moreover, the Matlab developed program was successful in the processing of the forearm thermal images, obtaining the ROIs and respective mean temperatures and standard deviations automatically and exporting the values to a Microsoft Excel spreadsheet, which was much faster, about 45 s less per image, than doing it manually with the provided manufacturer software. In the present sample, 1365×45 s = total saving time of 17 h. This was possible due to the fact that there was the ability of determining reference points (forearm center of mass and wrist edge) to respond to the constantly changing of the forearm's position and shape, which lead to an automatic adaptation of the ROIs sites in the forearm.

The statistical results showed good data consistency at the studied ROIs and correlation between the values obtained from both software's, being generally higher those providing from the Matlab solution. It is worth to mention that the drawback of this approach is the need of exporting the temperature matrix images from the proprietary provided software, which did not allow a fully automatic processing of the raw thermal images.

At the time that the application was tested, only data from 13 subjects was available and it provides enough power for the statistical assessment, there are 1365 images with data about three ROIs, being a significant sample size.

The developed tool is characterized by its simplicity. In order to be used, the exported images from the manufacturer provided tool in a matrix of temperature values are loaded and it will automatically detect and extract information of the defined ROIs exporting them to a Microsoft Excel spreadsheet. No specialized technical skills or knowledge are required to use it when compared with the traditional solution.

The added values are to minimize the human error in placing manually the ROIs and to standardize the images analysis. Applications for highlight phenomena or to classify images can be built over this solution, but they will require enough previous knowledge to make the decisions based on it.

The proposed method can be applied to any research involving IRT imaging of human anatomical areas, the modifications needed will vary with the number, size and location of required ROIs, being manly positioning.

5   Conclusion

Over the course of this research, it can be concluded that it was possible to develop a reliable method for fast thermal image analysis of the forearm using the Matlab program, since this software provided results with an acceptable correlation to the traditional method (FLIR ThermaCAM Researcher Pro 2.10) results. This indicates that the Matlab algorithm can provide trustworthy results, and be a viable method for future studies. Moreover, it is examiner independent avoiding common human mistakes.

For further research, it is proposed to study the proprietary manufacturer's raw images format in order to avoid the tedious exporting of the temperatures matrix from the manufacturer software and load them directly into Matlab.

Acknowledgment. The authors gratefully acknowledge the funding of project NORTE-01-0145-FEDER-000022 - SciTech - Science and Technology for Competitive and Sustainable Industries, co financed by Programa Operacional Regional do Norte (NORTE2020), through Fundo Europeu de Desenvolvimento Regional (FEDER) and of project LAETA - UID/EMS/50022/2013.

References

1. Guerra, R.S., Amaral, T.F., Sousa, A.S., Fonseca, I., Pichel, F., Restivo, M.T.: Comparison of jamar and bodygrip dynamometers for handgrip strength measurement. J. Strength Cond. Res. **31**(7), 1931–1940 (2017)
2. Piper, S., Shearer, H.M., Côté, P., Wong, J.J., Yu, H., Varatharajan, S., et al.: The effectiveness of soft-tissue therapy for the management of musculoskeletal disorders and injuries of the upper and lower extremities: a systematic review by the Ontario Protocol for Traffic Injury management (OPTIMa) collaboration. Manual Ther. **21**, 18–34 (2016)
3. Brothers, R.M., Wingo, J.E., Hubing, K.A., Crandall, C.G.: Methodological assessment of skin and limb blood flows in the human forearm during thermal and baroreceptor provocations. J. Appl. Physiol. **109**(3), 895–900 (2010)
4. Bos, I., Wynia, K., Drost, G., Almansa, J., Kuks, J.B.: The extremity function index (EFI), a disability severity measure for neuromuscular diseases: psychometric evaluation. Disabil. Rehabil. **40**, 1–8 (2017)
5. Hildebrandt, C., Raschner, C., Ammer, K.: An overview of recent application of medical infrared thermography in sports medicine in Austria. Sensors **10**(5), 4700–4715 (2010)
6. Ring, E.F.J., Ammer, K.: Infrared thermal imaging in medicine. Physiol. Meas. **33**(3), R33–R46 (2012)
7. Vardasca, R., Simoes, R.: Current issues in medical thermography. In: Topics in Medical Image Processing and Computational Vision, pp. 223–237. Springer, Dordrecht (2013)
8. Vardasca, R., Vaz, L., Mendes, J.: Classification and decision making with infrared medical thermal imaging. In: Dey, N., Ashour, A.S., Borra, S. (eds.) Classification in BioApps: Automation of Decision Making. Studies in Lecture Notes in Computational Vision and Biomechanics Book Series, pp. 79–104. Springer, Dordrecht (2018)
9. Skala Kavanagh, H., Dubravić, A., Lipić, T., Sović, I., Grazio, S.: Computer supported thermography monitoring of hand strength evaluation by electronic dynamometer in rheumatoid arthritis–a pilot study. Periodicum Biologorum **113**(4), 433–437 (2011)
10. Gatt, A., Formosa, C., Cassar, K., Camilleri, K.P., De Raffaele, C., Mizzi, A., et al.: Thermographic patterns of the upper and lower limbs: baseline data. Int. J. Vasc. Med. (2015)
11. Sanchis-Sánchez, E., Vergara-Hernández, C., Cibrián, R.M., Salvador, R., Sanchis, E., Codoñer-Franch, P.: Infrared thermal imaging in the diagnosis of musculoskeletal injuries: a systematic review and meta-analysis. Am. J. Roentgenol. **203**(4), 875–882 (2014)
12. Vardasca, R., Plassmann, P., Gabriel, J., Ring, E.F.J.: Towards a medical imaging standard capture and analysis software. In: 12th International Conference on Quantitative InfraRed Thermography, Bordeaux, France, pp. 162–168 (2014)
13. Ring, E.F.J., Ammer, K.: The technique of infrared imaging in medicine. Thermol. Int. **10**(1), 7–14 (2000)
14. Ammer, K.: The Glamorgan Protocol for recording and evaluation of thermal images of the human body. Thermol. Int. **18**(4), 125–144 (2008)

Handgrip Evaluation: Endurance and Handedness Dominance

Ricardo Vardasca[✉], Paulo Abreu, Joaquim Mendes, and Maria Teresa Restivo

INEGI-LAETA, Universidade do Porto, Porto, Portugal
{rvardasca, pabreu, jgabriel}@fe.up.pt,
trestivo@gcloud.fe.up.pt

Abstract. This work contributes to the development of a repeatable and objective methodology for relating the physiological energy spent during a handgrip exercise, identified through the variation of skin temperature, with the average grip force, and evaluate its influence on exercise endurance and handedness dominance. For that purpose, a special handgrip dynamometer is used as well as an Infrared Thermal Imaging (IRT) to map large areas of skin surface temperature. Results suggest that at least a 10-grips test with the dynamometer is required to produce reliable thermal results and the dominant hand should be used. In the future, relationship between the thermal variables and mechanical work involved during the handgrip should be addressed. The developed methodology should be applied to populations at health risk conditions to which the use of the handgrip dynamometer can provide information for diagnose and treatment assessment.

Keywords: Grip endurance · Handedness dominance · Handgrip
Infrared thermography · Skin temperature

1 Introduction

In many nutritional status or status alterations, the voluntary hand/arm force (Handgrip strength (HGS)) has been shown to be an easy and relevant indicator to assess muscle function (MF). Therefore, HGS is used as an indicator of overall muscle strength in routine clinical practice [1–3]. HGS is also one of the five established characteristics within the Criteria to Define Frailty [4] and is being used in multiple studies [5–10]. The devices for HGS evaluation are dynamometers offering an inexpensive and portable method for easy assessment.

This work uses a dynamometer prototype [11] developed at LAETA-INEGI, University of Porto, with patent application P300.3WO (September 2016). It offers a fully instrumented dynamometer with wireless communication for reliable data recording and fast processing. It allows the analysis of grip force over time and associated parameters that can be added to the traditional maximal grip strength for extended studies.

The prototype has demonstrated a very good performance when compared with Jamar dynamometer, considered as a reference in clinical use [11]. It has a parallelepiped

© Springer International Publishing AG, part of Springer Nature 2019
M. E. Auer and R. Langmann (Eds.): REV 2018, LNNS 47, pp. 507–516, 2019.
https://doi.org/10.1007/978-3-319-95678-7_57

shape with reduced dimensions. These features make it very convenient for frailty tests in ageing, recovering, rehabilitation and with children. Despite being a dynamometer, its features are not limited to that of a typical dynamometer, it can also provide different information by post-processing the handgrip force over time evolution as, e.g., estimation of relevant part of the user expended energy, applied mechanical work, rate of rising and dropping of grip force and association with user parameters to be explored as, e.g., endurance, resilience.

The Infrared Thermal Imaging (IRT) is able to map large areas of skin surface temperature distribution, which is linked to and an indirect method to estimate the peripheral blood flow [12, 13]. It is a remote, non-ionizing and safe imaging method that can provide physiological information in response to mechanical stimuli [14].

IRT has been used in the past to identify chronic forearm pain in patients when exposed to a mechanical stimulus, typing on a keyboard, though monitoring temperature changes in the forearm [15].

The used dynamometer measures the muscular mechanical force (N) and the elastic energy applied to the device during the handgrip test. The physiological energy the user spends during the grip test is then accessed in terms of the effect on the change on the skin temperature resulting from the muscle activity and on the mechanical energy and applied force measured with the dynamometer.

It is aim of this research to develop a repeatable and objective methodology for quantifying the physiological energy spent at a handgrip test by infrared thermal imaging and to evaluate its influence on estimation of the endurance and the handedness dominance within the test.

2 Methodology

The designed procedure involves the execution of consecutive handgrip exercises using a dynamometer while recording the grip force and the thermal images for identifying skin temperature. For enforcing the correct positioning, handgrip exercise and avoid unwanted movements of the forearm, a support was designed and developed. Figure 1 shows a data capture setup.

Fig. 1. The data capture setup

There were made several variations of the exercise, first with only 1 grip, then with 5 consecutive grips and finally with 10 consecutive grips. After initial tests, it was decided to proceed using 10 grips since they provide major thermal amplitude. Thermal images were taken at a rate of one per second during the exercise and up to 60 s counted from the last performed grip. In this additional period the user is still holding the dynamometer, with his arm resting over the support. An acclimatization period of ten minutes with the forearm unclothed to promote thermal equilibrium with the surrounding environment was adopted before starting the data collection.

For measuring and recording grip force over time it was used a special handgrip dynamometer prototype (BodyGrip) that was developed at the authors research group. This device allows the measurement of isometric gripping or puling forces associated with multiple muscle groups, such as the ones from the hand. The BodyGrip device uses onboard electronic circuits for wireless communication and processing data, being powered by a chargeable battery. An application, running on a personal computer, allows the continuous registry of the force data during assessment time, apart from enabling to configure test conditions such as the test duration. The design of the BodyGrip using special force measurement load cells instrumented with strain gauges' sensors, allows a compact ($0.144 \times 0.022 \times 0.045$ [m]) and low weight device (0.25 kg), offering force range of ± 900 N, and resolution of 1 N. Considering working details within the specially designed load cells it is expected a maximum displacement of each load cell free end for the full load, with a value of 0.001 m (exhibiting an elastic constant of 1140 N/mm), allowing to determine the elastic energy applied to the device. From the dynamometer, the force (N) and time (ms) is obtained and recorded in a CSV file. The maximum grip forces per handgrip were retrieved and later associated with the thermal data parameters using the same time interval.

To capture the infrared thermal images a laptop attached thermal camera FLIR A325sc (sensor Focal Plane Array of 320×240, Noise-equivalent temperature difference of <50 mK at 30 °C and a measurement uncertainty of $\pm 2\%$ of the overall reading) was used.

A total of 21 participants (13 men and 8 women), all right-handed, with ages 35 ± 14 years old and Body Mass Index of 25.7 ± 3.8, participated in the study. To all, the test procedure was explained and an informed consent was obtained. All experiments followed the ethical direction of the declaration of Helsinki. In order to avoid metabolic influencing variables, the participants prior to the experiments refrained from having a heavy meal, alcohol, coffee, tea and drugs and smoking up to 2 h before the appointment. They were also instructed to avoid any sport or physiotherapy and not using oils or ointments in the forearms.

The data was collected in a controlled environment room, in line with the international guidelines [14, 16], with average temperature 23.8 ± 0.4 °C and relative humidity (RH) of $47.4 \pm 5.3\%$, values that were monitored using a Testo 175H1 temperature and humidity data logger (range of -20 to $+55$ °C and 0 to 100%RH; accuracy of ± 0.4 °C and $\pm 2\%$RH; and resolution of 0.1 °C and 0.1%RH). There was an absence of airflow and incident illumination over the participants to prevent thermal reflections.

The data analysis to study the thermal effects of the handgrip exercise, involved the definition of three Regions of Interest (ROI), Fig. 2, being the ROI1 over the digital flexor muscle, the ROI2 over the wrist ulnar artery and the ROI3 over the wrist radial artery. The ROI1 is upon the main muscle responsible for the handgrip test and ROI2 and ROI3 are over the main arteries that feed the hand. The geometry of the ROIs was chosen because of the facility in placing them at the appropriate location, the circle for ROI1 over the muscle and the squares for ROI2 and ROI3 at the wrist over the arteries.

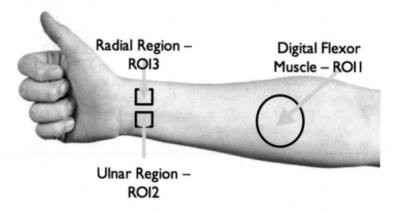

Fig. 2. The definition of the Regions of Interest (ROI) used

For all the three ROI and per thermal image, at the FLIR ThermaCAM researcher pro 2.10 software, the value of mean temperature is obtained and registered in a Microsoft Excel spreadsheet.

A baseline image holding the dynamometer was then taken and the handgrip at maximum force for 5 s was exerted.

Three time moments are considered: "B" - baseline (before the starting of the exercise), "A" - immediately after the exercise (1 s after for 1-grip, or 15 s after the last exercise for the other test types), and "F" - final recorded image. The middle point was spotted after analyzing the temperature evolution pattern over the exercises and corresponds to the moment where the maximum decrease in temperature occurs (one example is provided at Fig. 5). At the three moments, gradients between the ROIs were also calculated (ROI1–ROI2, ROI1–ROI3 and ROI2–ROI3). Other thermal parameters per ROI were DT1 (the difference between B and A) and DT2 (the difference between F and A). The mean temperatures for the ROIs and the calculated thermal gradients were statistically evaluated to verify the influence of handedness dominance, age, body mass index (BMI), sex and correlation with the average force applied in the handgrip test.

The collected data was later imported to the SPSS v24 statistical analysis software package. Every variable was statistically verified if it follows the normal distribution using the Shapiro-Wilk test. If the values follow the normal distribution, the parametric tests student t-test and Pearson correlation are used, if not the non-parametric tests Kruskal-Wallis and Spearman Correlation are used instead. Influence of factors such as sex, age and BMI are assessed. The ANOVA test is used to analyze the differences among the three-different endurance tests means. The statistical significance value used is $p < 0.05$.

3 Preliminary Results

As the study involved a group of subjects and not a single subject, the Table 1 presents the average grip force measured per grip in the 10-grips test and the final average grip force that was used for the correlation with the thermal variables. The maximum recorded force in a subject was 708.6 N for the right limb and 703.2 N for the left limb. It can be verified that the grip force decreased for the upcoming grips when compared with the first.

Table 1. The average grip force measured with the dynamometer per handgrip test and limb.

Grip	Right limb (N)	Left limb (N)
1	329.0 ± 135.1	299.8 ± 145.5
2	303.0 ± 124.5	303.3 ± 134.5
3	288.0 ± 120.4	279.4 ± 120.1
4	270.6 ± 113.1	266.7 ± 109.5
5	273.1 ± 89.6	257.6 ± 90.7
6	265.7 ± 97.4	247.3 ± 85.3
7	259.7 ± 91.3	244.2 ± 81.2
8	250.7 ± 80.3	232.5 ± 88.9
9	244.7 ± 76.2	228.4 ± 74.6
10	236.2 ± 70.7	221.7 ± 70.4
Average	272.1 ± 95.5	258.1 ± 96.7

An example of a thermal image obtained from the forearm is shown in Fig. 3 with the ROIs superimposed on the picture.

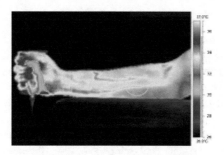

Fig. 3. An example of IRT image with the defined ROIs positioned.

In all the three different tests with both hands it was verified that the temperature decreased during and immediately after the workout, increasing thereafter. In the Fig. 4 it can be seen that comparing the relative gradients, the maximum variation was obtained in the 10-grips test (±0.3 °C) and it increased proportionally with exercise, being minimal with 1-grip test (±0.1 °C) and 5 grips test (±0.2 °C). This, despites the lack of statistical evidence (ANOVA p > 0.05) of independence between the results of the three types of tests, shows that the endurance matters in terms of thermal energy generated.

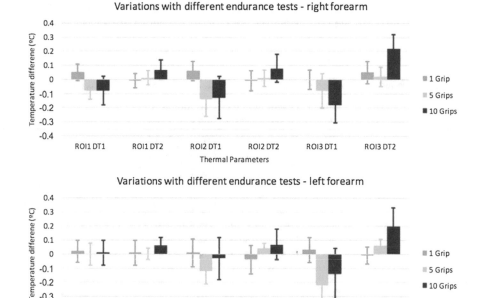

Fig. 4. The thermal variation at the thermal parameters on the three endurance tests for the two limbs (top – right forearm, bottom – left forearm; green – 1 grip, yellow – 5 grips and red – 10 grips)

The Fig. 5 present the average evolution of the mean temperature difference from baseline on the 3 ROIs of the participants during the 10-grips test. It can be seen that the temperature decreases during the test, reaching the minimum 15 s after the last grip and rising thereafter. The ROI3 showed a response with larger temperature amplitude, followed by ROI2 and ROI1, that presented the minimal amplitude.

From the statistical analysis, using non-parametric methods, it was verified that not all the variables followed the normal distribution. In the single grip and 5-grips tests, for both hands, there was no statistical evidence of the influence of age, BMI and sex in the obtained results. In the 10-grips test, it was found evidence (Kruskal-Wallis p < 0.05) of age in the gradients ROI1–ROI3 at A and F, ROI1–ROI2 at A, ROI1 DT1

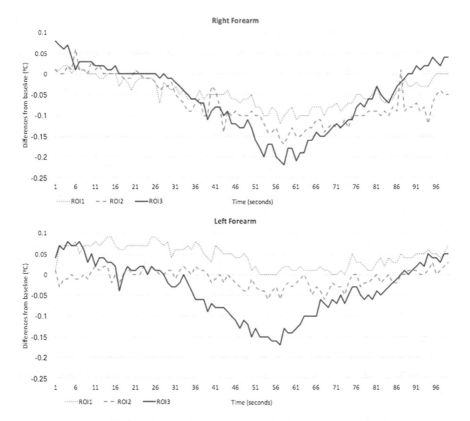

Fig. 5. The evolution of the temperature differences from baseline per ROI over the 10-grips exercise at the forearms (top – right forearm, bottom – left forearm)

and ROI2 DT2, and of sex in ROI DT2. There was no statistical evidence of the influence of BMI.

In terms of correlation between the force measured with the dynamometer and the thermographic studied variables, it was found statistical evidence (Spearman $p < 0.05$) on ROI2 F, ROI3 A, ROI3 F and ROI1–ROI3 F, of R = -0.305, -0.328, -0.291 and 0.327 respectively.

4   Discussion

To the authors knowledge, this was the first attempt to measure thermal energy from handgrip exercise and relating it to grip force measured with dynamometer. The developed methodology follows the international standards of dynamometer monitoring [1–4] and IRT imaging [14, 16] of human subjects.

It was verified that during exercising and immediately after the skin temperature decreased recovering thereafter, which is in line with previous related literature [15].

This may be justified by the demand of blood supply by the muscles during exercise and posterior release of the generated energy after the exercise.

This study allowed to evaluate the endurance of an individual exercise with the BodyGrip. Although there is no statistical evidence to support the assumption that more repetitions of the exercise generate more heat, empirically this has been demonstrated and at least ten repetitions are required to come out of the uncertainty range of the thermographic camera.

The handedness dominance question was not statistically demonstrated, however the responses to right-handed individuals were more pronounced in the right limb, which may result from a better endurance capacity than is expected to be the dominant.

Correlations were found between strength and thermal variables. However, the value of R is low, which may mean that the thermal energy expended may be different from the related work, being complementary to the mechanical elastic energy exerted on the device.

In order to corroborate the elastic energy, it is suggested the use of chemical markers for measurement of metabolic equivalents (METs) or explore the measurement of the mechanical tension of the muscular fibers through ultrasound.

The limitations of this experimental setup are: the need of a controlled environment, the need of collaboration of the subject, both measurements have to be taken separately and join together for global quantification of the energy spent by performing the exercise. Although, this data can be used to aid diagnosis and treatments assessment.

Different applications using the dynamometer combined with IRT may include among others the prediction of nutrition status, and of falls and risk of limited mobility in elders.

There are temperature variations at the different studied sites, but in this particular research, different endurance duration tests were considered along with the effect of hand dominance.

5   Conclusion

The aim of developing a repeatable and objective methodology for relating the physiological energy spent during a handgrip exercise, identified through IRT, with the average grip force, and evaluate its influence on exercise endurance and handedness dominance was fulfilled.

There is a measurable correlation between the handgrip force and the thermal variables (ROI2 F, ROI3 A, ROI3 F and ROI1–ROI3 F). With the developed procedure, it was found that the exercise involving ten consecutive grips achieves a significant thermal variation on the skin temperature. It can be concluded that, despite not having statistical significance, there is a higher amplitude in the thermal variables, which is in line with the correlation between force and those variables.

For further work, it is suggested to research the relationship between work produced and thermal variables and apply the developed methodology to populations at risk of health conditions to which the dynamometer can provided information for diagnostic and treatment assessment.

Acknowledgment. The authors gratefully acknowledge the funding of project NORTE-01-0145-FEDER-000022 - SciTech - Science and Technology for Competitive and Sustainable Industries, co financed by Programa Operacional Regional do Norte (NORTE2020), through Fundo Europeu de Desenvolvimento Regional (FEDER) and of project LAETA - UID/EMS/50022/2013.

References

1. Bautmans, I., Gorus, E., Njemini, R., Mets, T.: Handgrip performance in relation to self-perceived fatigue, physical functioning and circulating IL-6 in elderly persons without inflammation. BMC Geriatr. **7**(1), 5 (2007)
2. Leal, V.O., Stockler-Pinto, M.B., Farage, N.E., Aranha, L.N., Fouque, D., Anjos, L.A., Mafra, D.: Handgrip strength and its dialysis determinants in hemodialysis patients. Nutrition **27**(11), 1125–1129 (2011)
3. Guerra, R.S., Fonseca, I., Pichel, F., Restivo, M.T., Amaral, T.F.: Handgrip strength and associated factors in hospitalized patients. J. Parenter. Enter. Nutr. **39**(3), 322–330 (2015)
4. White, J.V., Guenter, P., Jensen, G., Malone, A., Schofield, M., Academy of Nutrition and Dietetics Malnutrition Work Group, et al.: Consensus Statement of the Academy of Nutrition and Dietetics/American Society for Parenteral and Enteral Nutrition: characteristics recommended for the identification and documentation of adult malnutrition (undernutrition). J. Acad. Nutr. Diet. **112**(5), 730–738 (2012)
5. AGS Geriatric Evaluation and Management Tools (Geriatrics E&M Tools) support clinicians and systems that are caring for older adults with common geriatric conditions. http://familymed.uthscsa.edu/gerifellowship/redirect/articles/CLC/Geriatrics%20Eval%20Management%20Tool%20for%20Frailty.pdf. Accessed Dec 2017
6. Leong, D.P., Teo, K.K., Rangarajan, S., Lopez-Jaramillo, P., Avezum, A., Orlandini, A., et al.: Prognostic value of grip strength: findings from the Prospective Urban Rural Epidemiology (PURE) study. The Lancet **386**(9990), 266–273 (2015)
7. Dudzińska-Griszek, J., Szuster, K., Szewieczek, J.: Grip strength as a frailty diagnostic component in geriatric inpatients. Clin. Interv. Aging **12**, 1151 (2017)
8. Dodds, R.M., Syddall, H.E., Cooper, R., Kuh, D., Cooper, C., Sayer, A.A.: Global variation in grip strength: a systematic review and meta-analysis of normative data. Age Ageing **45**(2), 209–216 (2016)
9. Yates, T., Zaccardi, F., Dhalwani, N.N., Davies, M.J., Bakrania, K., Celis-Morales, C.A., et al.: Association of walking pace and handgrip strength with all-cause, cardiovascular, and cancer mortality: a UK Biobank observational study. Eur. Heart J. **38**(43), 3232–3240 (2017)
10. Nofuji, Y., Shinkai, S., Taniguchi, Y., Amano, H., Nishi, M., Murayama, H., et al.: Associations of walking speed, grip strength, and standing balance with total and cause-specific mortality in a general population of Japanese elders. J. Am. Med. Directors Assoc. **17**(2), 184-e1 (2016)
11. Guerra, R.S., Amaral, T.F., Sousa, A.S., Fonseca, I., Pichel, F., Restivo, M.T.: Comparison of jamar and bodygrip dynamometers for handgrip strength measurement. J. Strength Cond. Res. **31**(7), 1931–1940 (2017). https://doi.org/10.1519/JSC.0000000000001666
12. Vardasca, R., Vaz, L., Mendes, J.: Classification and decision making with infrared medical thermal imaging. In: Dey, N., Ashour, A.S., Borra, S. (eds.) Classification in BioApps: Automation of Decision Making. Studies in Lecture Notes in Computational Vision and Biomechanics Book Series, pp. 79–104. Springer (2018)

13. Ring, E.F.J., Ammer, K.: Infrared thermal imaging in medicine. Physiol. Meas. **33**(3), R33–R46 (2012)
14. Ring, E.F.J., Ammer, K.: The technique of infrared imaging in medicine. Thermol. Int. **10** (1), 7–14 (2000)
15. Sharma, S.D., Smith, E.M., Hazleman, B.L., Jenner, J.R.: Thermographic changes in keyboard operators with chronic forearm pain. BMJ **314**(7074), 118 (1997)
16. Ammer, K.: The Glamorgan Protocol for recording and evaluation of thermal images of the human body. Thermol. Int. **18**(4), 125–144 (2008)

Digital Health for Computer Engineering Classes: An Experience

Lucia Vaira[✉] and Mario A. Bochicchio

Department of Engineering of Innovation, University of Salento, Lecce, Italy
{lucia.vaira,mario.bochicchio}@unisalento.it

Abstract. There are several teaching methods adopted in today's engineering faculties. The traditional teaching-learning style with frontal lessons has proven to be more useful for theoretical aspects. When practice comes to play, laboratories and hands-on sessions are more effective on learning outcomes. Considering the computer engineering faculty, the "*learning by doing*" paradigm can be supported by the "*learning by practice*" and "*learning by competing*" approaches, by stimulating innovation and creative thinking and by developing experience in teamwork and project execution. In this paper, in the context of the project work associated to our computer engineering master degree course, we propose three case studies, parts of eHealth research projects currently under development and testing at our University in Southern Italy, each one characterized by specific constraints and issues to be addressed in order to find innovative solutions to such problems. These research activities are aimed at demonstrating how "healthy competition", mainly based on team collaboration and cooperation, in project-based learning can be profitably deployed and exploited in computer engineering classes.

Keywords: Problem-based learning · Project-based learning
Student competition · University education

1 Introduction and Background

There are several teaching methods adopted in today's engineering faculties. The classical teaching-learning style with frontal lessons has proven to be more useful for theoretical aspects. When practice comes to play, laboratories and hands-on sessions are more effective on learning outcomes. *Problem-based learning* is an inductive learning method and it is considered to be the best method for engineering students [1]. The same is true for *project-based learning*, which is more or less the same as problem-based approach, that simulates the real-world problem and allows students to use their imagination to innovate, explore, analyze, synthesize and interpret the problem in hand [2]. In particular, considering a computer engineering school, the "*learning by doing*" paradigm, which allows students to achieve skills that otherwise would remain unexpressed or at a shallow, improving scientific enquiry and problem-solving attitudes [3], can be supported by the "*learning by practice*" and "*learning by competing*" approaches, by stimulating innovation and creative thinking and by developing experience in teamwork and project execution.

M. E. Auer and R. Langmann (Eds.): REV 2018, LNNS 47, pp. 517–527, 2019.
https://doi.org/10.1007/978-3-319-95678-7_58

Students competitions provide a valid opportunity for students to demonstrate and compare different design approaches and learn from each other. In electrical and computer engineering particularly, there are many regional, national, and international events that invite teams of students to compete in engineering activities. An example of competition is the "Student Hardware Competition" annually organized at the IEEE South East Conference [4].

It is often assumed that competitions are strong motivators that help students learn about real-world engineering problems [5, 6], which is consistent with the involvement required by any theory of education [7]. For the computer engineering background, the competition framework provides realistic deadlines, motivation and feedback by offering students the opportunity to conceive and design solutions which are afterwards developed and transformed into actions.

Students need now new teaching strategies due to their intense experience with technologies and digital tools (*digital natives*). They are, indeed, already full engaged on many different levels and are used to daily use digital devices in their life. In particular, the *mHealth and eHealth* arenas are under the focus of big companies like Apple, Samsung, etc., which provide every kind of health-oriented digital artifacts (Apps, sensor-enriched smart-watches, cloud services etc.) to support daily life habits (about hearth rate monitoring but also foods, workout activities, etc.).

In this paper, in the context of the project work (3 ECTS internship) associated to our computer engineering master degree course, we discuss three different research experiences, presented to students as case studies, each one characterized by specific constraints and issues to be addressed in order to find innovative solutions to given problems. Such case studies are part of a larger set of eHealth projects currently under development at our University, which is shared by three different courses, namely: Database, Software Engineering and Image Processing. The topics selected for each case study are respectively: smart health data monitoring; big data analytics for health and urgent needs of innovation in specific healthcare sectors.

Such activities are aimed at demonstrating how "healthy competition", mainly based on team collaboration and cooperation, in project-based learning can be profitably deployed and exploited in computer engineering faculties.

The remaining part of the paper is organized as follows. Section 2 summarizes the research method and the three research topics provided to students. In Sect. 3 we discuss achieved results. Finally, we draw conclusions in Sect. 4.

2   Research Method

Considering the class composition, students are partitioned into small groups and the partitioning is randomly performed. Here the configuration details:

- the class is composed by 33 attending students;
- we randomly divide students in 3 main groups (11 students per group) according to the 3 topics;
- each group secretly votes, with a valid motivation, 1 colleague (or himself/herself) to whom assign the "communication expert" title (in total, 3 communication experts,

1 for each topic). The most voted student will be the communication expert for a given topic;

- the remaining 10 students of each group are again randomly divided into 5 teams composed by 2 students.

Once the class has been subdivided into 15 teams, we illustrate to the class the 3 above mentioned topics. In particular, for each topic we describe the problem, its background, objective and constraints. Once explained all needed details for each topic, we describe to students the task assignment. In particular, we consider a two-steps process with two different levels of assignments: a theoretical proposal and a practical task. The theoretical proposal has to be considered as a team competition. Indeed, each team is asked to produce a valid theoretical solution within 1 week.

Each team formulates its own conceptual solution which is explained and discussed during a presentation session. Such presentation has to be held by the communication expert assigned to that specific topic with the support (when and if needed) of the two students belonging to the specific team. We evaluate each single presentation by judging on the basis of four main parameters (scale from 1 to 10):

1. value of the idea (in terms of strategies and software choices);
2. objective achievement;
3. trade-off between complexity (cost) and performance (effectiveness);
4. quality of presentation.

After our evaluation we calculate 3 rankings (1 for each topic) and the groups classified in the first 3 positions for each ranking are the winners of the first stage. Such 9 winner teams, each one composed by 2 students, have to deal with the development process of their own solution. The remaining 6 groups (2 teams for each topics) have to deal with the data management aspects (data design and data generation). The 4 students in charge to deal with data management aspects of a given topic, and hence for the 3 proposals of that topic, take the liberty to autonomously choose *"who makes what"* by means of a table of responsibilities (RACI table). This means that:

- for each topic: 1 communication expert, 6 developers and 4 data analysts are involved (as shown in Table 1);
- for each proposal: 1 communication expert, 2 developers and 4 data analysts must collaborate;
- each communication expert has to deal with 3 different proposals on the same topic;
- data analysts have to deal with 3 different proposals on the same topic;
- developers have to develop only their own proposal.

The duration of this second stage is set to 4 weeks. Students are free to manage their time, the shared communication experts and the shared data analysts. Given the educational purpose of the experiment, we offer students of each group the possibility to have 2 meetings with us (1 in the first week and 1 in the third one) in order to guide, support and advise them by means of prior and proper assessment/revision for the background, for a technical review, for conceptual aspects and for implementation choices. After the end of implementation and test phases, students are individually evaluated according to

Table 1. Class composition (33 students) and organization

	Communication Expert	Team 1	Team 2	Team 3	Data analysts
Topic 1	1 student (voted)	2 developers (winners)	2 developers (winners)	2 developers (winners)	4 students (self-organized)
Topic 2	1 student (voted)	2 developers (winners)	2 developers (winners)	2 developers (winners)	4 students (self-organized)
Topic 3	1 student (voted)	2 developers (winners)	2 developers (winners)	2 developers (winners)	4 students (self-organized)

their roles and responsibilities in each project for educational grade. Only the 3 best evaluated proposals (1 for each topic) are reported in the following subsections.

2.1 Smart Health Data Monitoring

Context Description and Problem Definition. Majority of healthcare systems are subject to several factors that negatively affect their success and adoption, but one of the major problem in health databases is the lack of data, which can be due to several (technical/social/political/legal) reasons: the local nature of traditional data collections, managed by bureaucratic units; the legitimate conflict of interest existing among physicians, patients and health administrations; the lack of adoption of proper data gathering techniques [8]. Health monitoring systems have the potential to change the way health care is delivered. Smart health monitoring systems empower patients letting them to automatically monitor their own health status. Even if smart health monitoring systems automate patient monitoring tasks improving the patient workflow management, while valid and interesting solutions have been proposed, their efficiency in clinical settings is still debatable and it is quite apparent that the problem is not just of a technical nature. A part of the problem stems from resistance to change, other obstacles are more structural in nature. In [8] we presented the overriding issues that are the stumbling blocks for the adoption of ICT in healthcare analyzing them from different point of views and trying to address some of the well-known challenges.

Objective Specification. To propose a smart data harvesting process, which refers to the possibility to concurrently adopt several strategies to obtain relevant data from all possible channels to monitor pregnant women during the pregnancy. Students can consider only one parameter to be monitored and they have to justify the choice.

Constraints Description. When dealing with personal and clinical data, security, privacy and confidentiality issues have to be considered. Although the technology that supports smart health monitoring systems is becoming more sophisticated, there are still concerns with quality of medical data, accuracy, usability, invasive devices, costs, acceptability by patients and physicians and the frequency of false alarms.

Students' Proposal. Students consider the hypertensive pregnant women as stakeholders, since high blood pressure, or hypertension, is a very common and serious concern for pregnant women. It is not always dangerous, but it can cause severe health complications for both mother and fetus. A daily monitoring for high-risk patients can be useful so that physicians are always informed whenever thresholds are exceeded and women feel more followed. Students propose a mobile smart health monitoring system mainly based on the possibility to use smartphones as the main processing station with the support of specific wearable devices or biosensors that can be worn by women in order to daily monitor pressure. In order to preserve privacy and confidentiality in the data gathering process, students imagine to provide all involved women written and oral information about the study (possible utilization of data) so that only patients who gave informed consent, authorize to store and manage their sensitive personal data. In Fig. 1 the conceptual architecture of students' proposal is showed.

Fig. 1. Smart health monitoring architecture

2.2 Analysis of Big Quantity of Health Data

Context Description and Problem Definition. Up to 50% of European adults search online for health information [9]. The need for widespread online access to accurate, updated and relevant information about diseases and therapies, as well as to personal health data, is essential. Access to a massive quantity of certified authoritative and constantly updated information, taking into account several parameters (environment, lifestyle, pathologies, etc.), could help to decrease the level of undue anxiety in patients [10]. It could act as a first-level support to patients and avoid unnecessary and expensive further investigation. The same is true on the research side, where access to a vast quantity of shared healthcare data for scientific purposes could reinforce, support and validate clinical studies to make progress. A Big Data repository, if organized hierarchically by chronic and acute diseases and within those by specific disease codes can offer the capability for real-time and continuous learning and can be useful to increase the ability to generate and test hypotheses that could lead to important insights.

Objective Specification. To propose a smart approach able to provide personalized analysis based on the specific features of a given patient, by considering the concept of Homogeneous Patient Groups [10], i.e. clusters of patients with similar genetic make-up (ethnicity, familial aspects, etc.) and similar environmental conditions (nutrition, smoking, drugs, etc.). This can be very useful for women during pregnancy to personalize the well-known fetal growth test according to the maternal and fetal parameters in order to take advantage from data and to try to apply historically archived data to a personal level.

Constraints Description. When dealing with personal and clinical data, security, privacy and confidentiality issues have to be considered. Moreover, when dealing with so high volume of data, traditional data analysis methods cannot be directly applied. Data are collected without discarding anything, so cleaning procedures and integration mechanisms are needed to guarantee data quality. It becomes essential to perform data clustering so that all records within a cluster are closer among them than with other records in an external cluster.

Students' Proposal. Students propose a simple but effective conceptual architecture mainly based on the possibility to empower pregnant women to adopt personal computers and smartphones as means to feed a big data repository containing a big amount of fetal data in order to perform significant analysis. In order to preserve privacy and confidentiality in the data gathering process, students consider anonymization techniques. In order to collect and integrate data coming from heterogeneous sources, students adopt a data warehouse system, which consists of a source layer (including multiple data pools), a reconciliation layer and a data warehouse layer (including both data warehouse and data marts). Students develop ETL (Extraction, Transformation and Loading) pipelines able to clean up and integrate the sources of input data to a common standardized form. To perform aggregation and multidimensional analysis in order to extract useful knowledge showed in a summarized but immediate manner, students develop a multidimensional cube representing the fetal growth data able to exploit the usual multidimensional operations (e.g. drill-down, roll-up, etc. [11]), looking for correlations among the available measures and dimensions. The conceptual model is represented by the Dimensional Fact Model (DFM) showed in Fig. 2. Each relevant and influential variable is modeled as a dimension of analysis, while each biometric parameter analyzed during the exam, is considered as a measure.

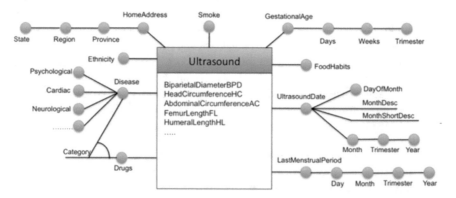

Fig. 2. Dimensional Fact Model of the ultrasound visit

With the precious help of data analysts, students have the possibility to work on a real big data repository made by randomly but realist generated data which consider 500.000 Italian pregnant women and about 4 millions of visits. Students analyze the "Head Circumference" biometric parameter evaluated during the fetal growth, considering different gestational ages, with the possibility to explore data from different points of views. In Fig. 3 the 50th percentile related to the Head Circumference in relation to the mother ethnic group is showed.

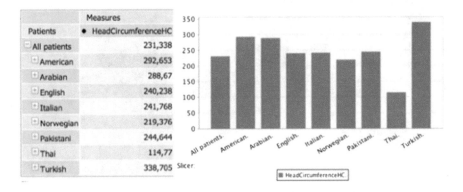

Fig. 3. Example of multidimensional analysis

2.3   Need for Innovation in Specific Healthcare Sectors

Context Description and Problem Definition.   The obstetric and gynecological sector has a huge potential for improvement at diagnosis, care planning and treatment level. Ultrasound has been the major development in the last three decades that allowed assessment and fetal management during pregnancy. The determination of the exact fetal head position is of paramount importance and can be very useful during labor in order to predict the success of labor itself, especially in case of malposition [12].

Objective Specification.   To propose a smart algorithm able to detect and reproduce the main geometrical parameters of a solid object by acquiring a series of ultrasound pictures and, in particular, the contouring images.

Constraints Description.   The intrapartum ultrasonography is currently an operator-dependent method that does not establish the exact position of the fetal head in the birth canal neither the degree of fetal pelvis position. Traditional sensors are unmanageable devices characterized by big dimensions and high costs.

Students' Proposal.   Students perform a detailed literature review about the available hardware solutions for an objective detection of the main geometrical parameters and find two main approaches: sensors and devices for the objective detection of the main geometrical parameters and of the degree of advancement of the fetal head in the birth canal; development of operator-independent techniques for the 3D representation of geometries. Once analyzed and compared the main techniques of volume reconstruction from 2D images, students chose the MultiPlanar Reconstruction algorithm as more appropriate. They develop a simple prototype able to reconstruct 3D surfaces from 2D ultrasound pictures. In particular, in order to obtain artificial ultrasound images, students adopt the Google SketchUp software: a 3D skull (Fig. 4a) is imported in the 3D environment and several sections (placed one every 5 mm) are created by using a section plane as shown in Fig. 4b. Such sections simulate what an ultrasound machine is able to detect when scanning a real object.

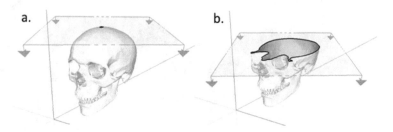

Fig. 4. 3D skull

Once extracted the various sections, students develop a Matlab algorithm that gets as input some images (the different sections) and returns a cloud of points. The algorithm handles images by converting them into data structures in the form of three-dimensional arrays. The resulting algorithm is composed by three main steps able to suitably manipulate the image: (a) edge detection and sampling (in Fig. 5 is showed a simplification of a real case); (b) contour analysis and conversion into points (Fig. 6); (c) location of points in 3D space to get the cloud of points (Fig. 7).

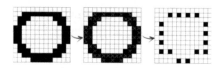

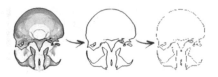

Fig. 5. Edge detection and sampling   **Fig. 6.** Edge analysis and conversion to points

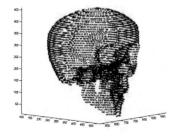

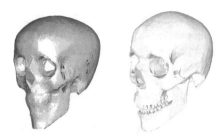

Fig. 7. Cloud of points in a 3D space   **Fig. 8.** 3D surface reconstruction

Once obtained the cloud of points, a meshing algorithm for 3D reconstruction is needed. Students adopt the Ball-Pivoting algorithm. In Fig. 8 is represented the final result with the comparison with the original 3D object.

3   Results and Discussions

In this section we will show the main benefits of our experiments and we will discuss about how potential disadvantages will be avoided. Competitions have a great motivational effect and generate interest from students who are highly motivated to produce the best possible solution in order to achieve a good performance. Students learned that often the simplest and least sophisticated solution can win with respect to complex ones. Students learned to manage the deadlines and pressure of releases, to work in groups and to compare each other's. Moreover, they learned more about colleagues (both from the personal side and the professional one).

Despite many educational psychologists claim that one of the major disadvantages of every competition is that there is typically one winner and several losers and that competitions can therefore generate disappointment, in our experience we state that this is not true because there are no official losers, every student participates to the project activities and gives its own contribution by knowing that its role is important for the accomplishment of the whole objective. Competition indeed has generated in the class interest and concentration.

The random subdivision has been chosen with respect to the classical approach we normally use to compose groups when assigning projects (each one chooses his/her colleagues) because we wanted to provide students with the realest situation they can find when working in a software house company.

The communication experts showed great responsibility and impartiality. One of the three voted students did not expect to be chosen by the companions for a similar role. From a psychological point of view, this resulted in a greater confidence in himself. The other two students, instead, seemed born to make it, so voting students have confirmed their nature.

Students profusely proved the self-assessment capacity and also the power of evaluation for their colleagues (e.g. for the choice of the communication expert).

The first step of the assignment (the theoretical proposal) tested their creativity and originality and gave birth to genuine ideas that have been transformed into very appropriate solution to the illustrated purposes. Success of students lied on the extent of being able to argue and explain, in a clarified manner (both verbally and by means of electronic presentations) to a third person outside the group (the communication expert) their own proposal, so that the communication expert could be able to present it in the best possible way. Some students (3 groups in total) initially showed some difficulties in the first step of the assignment due to their limited knowledge about the possible tools and methodologies to propose as potential solution, but after a "motivating meeting" they had the right charge to continue. The good and "healthy" competition makes the whole project work more dynamic and interesting. Moreover, a good competition should challenge the participants to give their best. Students showed indeed high interest for all proposals as alternative solutions for the same problem, trying to find positive and negative aspects for each proposal spontaneously taking their own work as a reference. This works as a valid lesson since students discover a variety of possible solutions that they had not imagined. We experimented also an increase in the awareness of the choice of the right tool depending on their own observations and experiences instead of passively accepting the assignments with no critical judgment and this is a valuable lesson.

The second step of the assignment considered by the data analysts point of view, showed also their astuteness in trying to hinder developers (encouraged by us) by generating dirty data, all characterized by the unfortunately common issues (incompleteness, ambiguous values, mistakes in unities of measurements, etc.). This allowed us to practically show students the role and importance of the data, how the way in which data are organized impacts on data management and quality; how the way in which data are presented affects the comprehension.

95% of them agreed that such approach to project work was beneficial in providing a more practical and impactful method to application design and development and a different perspective of the expected results (learning outcomes) for the method itself.

Students' feedback clearly indicates that they prefer learning on project work in a team even if the overall focus of the experiment is not on the objective's accomplishment, but on the process they go through, on the challenges they face and the knowledge, skills sand competences they gain by means of this experiences.

4   Conclusions

In this paper we presented a *"learning by competing"* approach in a computer engineering master class by involving students in real-world problems belonging to the healthcare domain integrating different engineering courses.

Such approach is demonstrated to be able to prepare students for an increasingly competitive job market. Students gain more confidence on tools and methodologies and being the competition "healthy", nobody feels looser since each student participates to the project activities and feels essential for the objective accomplishment. Every student has its own role but has to know everything on the project. Indeed, students are involved in several activities: problem definition, hypotheses formulation, experimentation design, data collection and interpretation, results evaluation and experiment result quality evaluation.

We plan to adopt the same approach in the next year in order to perform a second experiment on one or more different classes and to achieve also their feedback. Considering a larger number of students and other academic contexts can be very useful to validate and generalize the approach and its influence in students' training giving them a good glimpse of "real life" development in challenging domains.

References

1. Felder, R.M., Silverman, L.K.: Learning and teaching style in engineering education. Eng. Educ. **78**(7), 674–681 (1988)
2. Lemu, H.G.: On competition-driven teaching of multidisciplinary engineering education: implementation cases at University of Stavanger. Nordic J. STEM Educ. **1**(1), 278–286 (2017)
3. Clark, J., White, G.: Experiential learning: a definitive edge in the job market. Am. J. Bus. Educ. **3**(2), 115–118 (2010)
4. IEEE 2017 Student Hardware Competition. http://bit.ly/2nigjL1
5. Carroll, D.R., Hirtz, P.D.: Teaching multi-disciplinary design: solar car design. J. Eng. Educ. **91**(2), 245–248 (2002)
6. Meade, J.: Coach. PRISM **3**(3), 32–34 (1993)
7. Astin, A.W.: What Matters in College? Four Critical Years Revisited. Jossey-Bass, San Francisco (1993)
8. Vaira, L., Bochicchio, M.A., Navathe, S.B.: Perspectives in healthcare data management with application to maternal and fetal wellbeing. In: 24th Italian Symposium on Advanced Database Systems (SEBD), pp. 31–41 (2016)
9. EU Commission: Digital Single Market. Managing Health Data. http://bit.ly/2DHLNAp
10. Bochicchio, M.A., Vaira, L., Longo, A., Malvasi, A., Tinelli, A.: Multidimensional analysis of fetal growth curves. In: IEEE 2013 International Conference on Big Data. Bigdata in Bioinformatics and Health Care Informatics (BBH), Santa Clara (2013)
11. Elmasri, R., Navathe, S.B.: Fundamentals of Database Systems, 7th edn. Pearson Publishing (2015)
12. Malvasi, A., Bochicchio, M.A., Vaira, L., Longo, A., Pacella, E., Tinelli, A.: The fetal head evaluation during labour in the occiput posterior position: the ESA (evaluation by simulation algorithm) approach. J. Maternal-Fetal Neonatal Med. **27**(11), 1151–1157 (2014)

A Support System for Information Management Oriented for the Infant Neurodevelopment Study

Sergio Daniel Cano-Ortiz[1]([✉]), Yadisbel Martinez-Cañete[1], Lienys Lombardía-Legrá[1], Reinhard Langmann[2], and Harald Jacques[2]

[1] Universidad de Oriente, Santiago de Cuba, Cuba
{scano,ymartinez,lienys}@uo.edu.cu
[2] University of Applied Sciences Duesseldorf, Duesseldorf, Germany
R.Langmann@t-online.de, harald.jacques@hs-duesseldorf.de

Abstract. Since more fifteen years a multidisciplinary and scientific collaboration between Group of Speech Processing (GSP) from Universidad de Oriente and the Medical Consultation for Infant Neurodevelopment and Disability (CPNDI) of Santiago de Cuba have been successfully developed. Notwithstanding the successes and advances of the CPNDI, certain difficulties limit its action and impact as: non automation of the information flow related to the study of neurodevelopment and children's disability, the delayed detection of some cases of infants who then make their debut with deviations in their neurodevelopment and associated disabilities. Currently, a project is being developed to deploy a Management Computer System of the CPNDI (named PMSIND). This web-based system will need another support system that will allow it to incorporate software packages and diagnostic methodology both linked to the infant cry analysis, which will facilitate an early detection of neonates at risk to be treated later by the CPNDI. The main objective of the paper is the development and implementation of this support system made up of non-proprietary software tool. The support system developed by the authors should not only help to manage the clinical information of the neonate at risk, but also must guarantee the availability of all information with diagnostic value from the Infant Crying Analysis that facilitates the early detection of infants at risk to be followed up by the CPNDI.

Keywords: Telemedicine · Infant neuro-development outcome · Cry analysis

1 Introduction

The study of Infant Neuro-Development and Disabilities aims to determine in an early age those children with symptoms or in risk of debuting with a deviation of normal patterns in their cognitive development for the later treatment and follow up procedure. The Medical Consultation for Infant Neuro-development and Disability (CPNDI) in Santiago de Cuba province was founded in the 90 s and it is the one in charge to detect and control newborns with any neurological disease or with motor or functional disability in order to establish their diagnosis and prognosis. The study of infant cognitive development and its evaluation in newborns and young children is decisive for several aspects: an efficient differential diagnosis, a right early stimulation, to fix clinical risks

© Springer International Publishing AG, part of Springer Nature 2019
M. E. Auer and R. Langmann (Eds.): REV 2018, LNNS 47, pp. 528–535, 2019.
https://doi.org/10.1007/978-3-319-95678-7_59

and perspective treatment of disability. It is located in the Southern Infant Hospital and serves the entire child population of the province of Santiago de Cuba. CPNDI covers the following services:

- attention to children with risk factors to debut with some disability or deviation of their neuro-development
- early diagnosis of deviations in the neurodevelopment of neonates
- rehabilitation of children with physical, visual and auditory limitations.
- to facilitate the education of the child in a family environment
- to qualify professionals to care for these priority groups.
- development of local programs for attention and control of disabilities.
- to establish the integration between institutions to raise the quality and effectiveness of that attention.
- to elevate the psycho-biological-social orientation as the central nucleus of the program

The CPNDI has been also important scenery for research efforts developed by members of the Group of Speech Processing within Study Center for Neurosciences, Image and Signal Processing (CENPIS) from Universidad de Oriente, on the field of Infant Cry Oriented for Diagnosis. In this area the GPV has obtained important findings and information with potential for neonatal diagnosis from the digital processing of the crying signal [1, 2], clearly shown in the main result of the GPV: a neonatal diagnosis method based on Infant Crying Analysis (ICA) related to children suffering CNS diseases with a hypoxia background [3].

Despite the successes and advances of the CPNDI, certain difficulties persist that limit its actions and impact, such as: the non-automation of the flow of information related to the study of neurodevelopment and childhood disability, the early detection of some cases of infants that later debuted with deviations in their neuro-development and related disabilities. On the other hand there is no link between the diagnostic aid tools based on ICA with the information flow handled by the CPNDI. This link should guarantee: the introduction of the diagnosis aid methodology based on ICA [3], the handling of cybernetic and statistical information derived from the digital processing of crying, the help to the early detection of possible patients to be followed by the CPNDI as well as the multidisciplinary preparation of the CPNDI work team that will make use of the ICA-based diagnostic methodology.

At present, an Information Management System (named PMSIND) of the CPNDI is being developed in collaboration between the UO and the University of Applied Sciences Duesseldorf (UASD) that should significantly reduce the limitations previously addressed. For this purpose it is necessary first to implement a Support System to the PMSIND that guarantees the management of the information based on cry signal processing and the use of the ICA-based diagnostic methodology for an early detection of the neonates at risk to be treated and followed by the CPNDI staff. That means there is a lack of link between previous research results on Cry analysis (oriented for Infant neuro-development and diagnosis) and the PMSIND system. That is the problem should be solved the paper.

Thus the main objective of the paper is aimed to the development and implementation of a support system (SS) composed of non-proprietary software tools (open source) that contributes to the fulfillment of two specific aspects: (a) early detection of cases of newborns to be studied in its neurodevelopment by the CPNDI (according to multidisciplinary criteria that include analysis of infant crying) and (b) computerization of all the data related to the clinical profile and digital processing of the cry signal.

2 Materials and Methods

2.1 The ICA-Based Methodology of Aid to Neonatal Diagnosis

This methodology is the result of years of research work of the GPV [1–5] initially visualized in a doctoral thesis successfully defended in 2010 [3]. The methodology is in its initial phase of testing and introduction in medical practice. The methodology conceives the incorporation of cry signal classification into the traditional clinical diagnosis of the classification of childhood crying considering emergent techniques derived from soft computing and pattern recognition [4, 5]. At the output from ICA-based diagnostic methodology, a differential diagnosis must be given by the medical personnel staff that will determine the following-up or not of the neonate by the CPNDI. In Fig. 1 its operation is visualized.

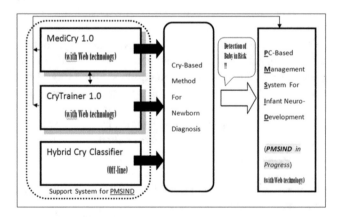

Fig. 1. Block diagram for the support system as a previous stage of the Information Management System (PMSIND) for CPNDI

As it is observed the support system provides to the multidisciplinary team who have the responsibility of applying the ICA-based diagnosis methodology, three types of information:

Set (1): all the initial clinical information and information obtained from the cry signal processing (including also wav files with cry recordings, the digital spectrogram and its corresponding pitch contour) supplied by MediCry v1.0.

Set (2): training of the multidisciplinary team members not familiar with the ICA, who are trained in acquiring abilities to find clues with diagnostic weight from the digital spectrogram of the baby's cry (provided by the Cry Trainer v1.0).

Set (3): the normal/pathological classification provided by a hybrid classifier.

As you see the first two sets of information are managed online via the web (the access links to PMSIND are provided). The classifier in set (3) delivers its information off-line.

When the methodology is applied all the information coming from the three sets plus the one from recognition exams of neonate (made by doctors) will yield a result that locates or not the baby in a situation of risk of deviation in his neurodevelopment (baby at risk), which is immediately sent to the CPNDI for study, assessment and possible follow-up process.

2.2   Management of Clinical and Cybernetic Information of the Newborn

Studies and research related to ICA show that the newborn cry can contribute to the early diagnosis of certain pathologies in a baby [1, 3, 6, 7]. The GPV of the CENPIS Universidad de Oriente has been developing the project "Infant Cry Analysis for the Diagnosis of Newborn" in close collaboration with clinical units in Santiago de Cuba province This project demands, among other things, the collection of a large number of cry samples, which implies a huge volume of information. Handling this volume of information would be very cumbersome if you do not have an adequate database (corpus).

Designed for this purpose MediCry 1.0 is software designed for consulting this database, which is accessed through a web environment. This consists of a master table with fields describing clinical cases of healthy and pathological children, with their respective crying signal, narrow and broadband spectrogram and pitch contour. In Fig. 2 a MediCry v1.0 screen is shown analytical and graphical information.

Fig. 2. A MediCry v1.0 screen

2.3   A Hybrid Classification of the Infant Cry Signal

One of the most important results obtained by GPV is the development of a hybrid classifier that makes use of soft computing and pattern recognition techniques for cry signal classification [5]. Several hybrid classifiers were developed and tested, combining connectionist models, evolutionary and fuzzy models with highest results to standard cry classifiers (based on neural networks alone, in threshold values) [4, 5].

The developed hybrid classifier will only indicate a possible deviation of normality in the acoustic pattern analyzed and will work off line within the support system. The output of the classifier is incorporated into the multidisciplinary analysis (derived from the methodology) performed by the medical team in charge of applying the ICA-based diagnostic methodology (see Fig. 1). In this decision (a definitive differential diagnosis) intervene several factors complementing each other: the result of the cry classifier, the evaluation of the neuro-physiological recognition applied to the newborn, the validation of information coming from the medical equipment and tests performed on the neonate, assessment of the statistical, visual and auditory derived information obtained from the ICA-based tools (Medicry). Here is where the convenience of preparing specialized medical personnel for an evaluation of that visual and auditory information present in the digital spectrograms of the recorded cries becomes evident.

2.4   Effective Use of Information by the CPNDI Medical Team and Specialists Using the ICA-Based Diagnostic Methodology

For this purpose, the CryTrainer v1.0 tool was implemented, which must guarantee through a friendly user interface the effective use of information from the digital processing of crying with diagnostic potential (digital spectrograms of crying and pitch contour). The CryTrainer is a software developed in Web environment able to train the user in associating the behavior of different parameters and acoustic phenomena of the crying signal, from visual and auditory information, with the presence of different neuro-physiological and pathological patterns present in newborn. The visual information consists of the digital spectrogram of the crying signal, informative text about the crying pattern; the auditory information is given by the audio file that contains the crying acoustic signal. The software can also be applied in the training and preparation of future specialists in phoniatrics and children's speech therapy.

Acoustic Attributes of Crying in the Digital Spectrogram

In general a lot of the acoustic attributes of crying can be reflected and induced by simple inspection in the digital spectrogram of crying. Figure 3 shows the schema of some of these attributes [3].

It is these attributes that, according to their behavior, presence or combination, can serve as clues or indicators to establish a possible neuro-physiological status in the neonate, who only has the cry emission to interact with the surrounding world (including the medical personnel who take care of him).

For example we see in Fig. 4 the digital spectrogram of the cry of a child with asphyxia problems (pathological crying).

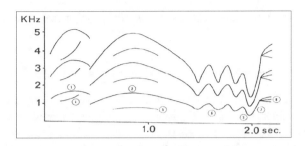

Fig. 3. Schematic diagram showing some characteristics or attributes in the signal of infant crying: 1 - Maximum displacement pitch; 2 - Maximum of the fundamental frequency; 3 - Minimum of the fundamental frequency; 4 - Biphonation; 5 - Rupture of Subharmonic; 6 - Vibrato; 7 - Sliding; 8 - Bifurcations

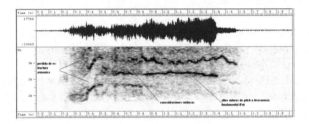

Fig. 4. Digital Spectrogram of a cry signal corresponding to a newborn with problems of hypoxia (asphyxia)

In the spectrogram some aspects related to the cry of asphyxia are observed as: weak cry (poor energy content), with loss of its harmonic structure, high values of fundamental frequency or pitch F0 (above 1000 Hz), as well as areas with Noisy concentrations (high energy levels given by blackness or level of gray but without established harmonic structure).

In Fig. 5a system output is shown. The software tool incorporates didactic methods that allow the student's progress through the step-by-step process of acquiring skills to locate these deviations with diagnostic potential, making use of text, visual (image) and

Fig. 5. Screen with the Evaluation section of CryTrainer v1.0

acoustic information (reproduction of the acoustic cry signal through multimedia components).

2.5   Use of Web Technology for the Development of Software Tools

The sub-systems (or sets) MediCry 1.0, CryTrainer 1.0 were developed using web technology: for the implementation of the sub-systems client-server architecture was used as well as the php programming language on the server side and Html and JavaScript in the client side. Moreover Apache was used as a web server, Dreamweaver as a development IDE, MySQL-Server system for database administration, and the content management system (CMS) JOOMLA [8–13].

3   Results and Discussion

Several control tests were developed in an environment limited to the UO intranet. The effective functioning of all functionalities for the online (online) components of the support system with web technology (MediCry 1.0 and CryTrainer 1.0) was verified: remote access, e-mail and chat services, security, access control by layers to database. For the off-line component (hybrid classifier) crying was pre-classified into 2 categories: healthy and pathological using the same data reported in the bibliographic references [4, 5]. The hybrid classifier tests were performed with the same database used in [5], as well as the same criteria were maintained for both sub-classifiers This information constitutes one of the inputs to the ICA-based diagnostic methodology where a multidisciplinary team will evaluate in a multidisciplinary way (as explained in Fig. 1) all the information obtained (clinical, pre-classification according to ICA, statistics, etc.).

Meanwhile the PMSIND is still in implementation, it was not possible to carry out an integral test with all its components assembled. But the expected benefits of each of the 3 subsystems (or sets) were properly verified. The tests with the ICA-based diagnostic methodology will begin when the initial prototype of the PMSIND be finished (it is scheduled for the end of 2018).

4   Conclusions

The proposed support system guarantees the management and control of the information prior to being supplied to the PMSIND of the CPNDI (currently under development). The developed software tools MediCry v1.0 and CryTrainer v1.0 (both with web technology) as well as the hybrid crying classifier (offline) facilitate the effective management of multidisciplinary information that should lead to an early detection of babies at risk that will be subsequently studied, evaluated and researched by the CPNDI. In the future it is recommended the online incorporation of the hybrid cry classifier as an integral part of the support system (also with web technology). The components of the support system successfully tested their functions on the CENPIS intranet under different Windows configurations. The authors considered the use of open source, standards of object-oriented programming and portability in the design as well.

References

1. Cano-Ortiz, S.D., Escobedo Beceiro, D.L., Socarrás Reyes, M.: The spectral analysis of infant cry: an initial approximation. In: Proceedings of EUROSPEECH'95 (sponsored by ESCA and IEEE). 4th European Conference on Speech Communication and Technology, Madrid, Spain, 18–21 September, vol. 3, pp. 1895–1898 (1995)
2. Cano-Ortiz, S.D.: Cry-based newborn diagnosis of CNS diseases and speech developmental aspects: software and hardware tools, cry databases, methodologies. In: Proceedings of the 7th International Workshop on Models and Analysis of Vocal Emissions for Biomedical Applications MAVEBA 2011, Firenze, Italia, August 2011
3. Escobedo Beceiro, D.I.: Análisis acústico del llanto del niño recién nacido orientado al diagnóstico de patología en su neurodesarrollo debido a Hipoxia. Ph.D. thesis (Registro: 1015–2010). Universidad de Oriente, Santiago de Cuba
4. Reyes Garcia, C., Reyes Galaviz, O. Cano Ortiz, S.D., Escobedo Beceiro, D.: Soft computing to the problem of infant cry classification with diagnostic purposes. In Melin, P. (ed.) Studies in Computational Intelligence, vol. 312, pp 3–18. Springer, Berlin, Heidelberg, ISSN 1860949X, ISBN: 978-364215110-1
5. Cano-Ortiz, S.D., Escobedo, D.I., Suaste, I., Ekkel, T., Reyes Garcia, C.A.: "A combined classifier of cry units with new acoustic attributes. In: Trinidad, M., et al. (eds.) Progress in Pattern Recognition, Image Analysis and Applications, CIARP 2006. Lecture Notes in Computer Science LNCS vol. 4225, pp. 416–425. Springer, Berlin, Heidelberg (2006)
6. Michelsson, K., Wasz-Höckert, O.: The value of cry analysis in neonatology and early infancy. In: Murry, T., Murray, J. (eds.) Infant Communication: Cry and Early Speech, pp. 152–182. College-Hill Press, Houston (1980)
7. Chittora, A., Patil, H.A.: Classification of normal and pathological infant cries using bispectrum features. In: Proceedings of 23rd European Signal Processing Conference (EUSIPCO). IEEE Xplore (2015)
8. Cano-Ortiz, S.D., Langmann, R., Martinez-Cañete, Y., Lombardía-Legrá, L., Herrero-Betancourt, F., Jacques, H.: A web-based tool for biomedical signal management. In: Auer M., Zutin D. (eds) Online Engineering and Internet of Things. Lecture Notes in Networks and Systems, vol 22. Springer, ISSN 2367–3370, ISSN 2367-3389 (electronic); ISBN 978-3-319-64351-9, ISBN 978-3-319-64352-6 (eBook). https://doi.org/10.1007/978-3-319-64352-6_71
9. Jacobson, I., Booch, G., Rumbaugh, J.: El Proceso Unificado de Desarrollo de Software. La Habana. Cuba. Editorial Félix Varela (2004)
10. Visual Paradigm: http://en.wikipedia.org/wiki/Visual_Paradigm_for_UML. Accessed 10 Feb 2010
11. https://developer.mozilla.org/es/docs/JavaScript/Acerca_de_JavaScript(12/04/2013)
12. Osmani, A.: Learning JavaScript Design Patterns, vol. 88. O'Reilly, Sebastapol (2012)
13. The HTML5 Creation Engine: http://www.pixijs.com/

Demonstration: Online Detection of Abnormalities in Blood Pressure Waveform: Bisfiriens and Alternans Pulse

Daniel Nogueira[1,2]([✉]), Rafael Tavares[3], Paulo Abreu[2],
and Maria Teresa Restivo[2]

[1] LAETA-INEGI, University of Porto, Porto, Portugal
dnogueira@inegi.up.pt
[2] Faculty of Engineering, University of Porto, Porto, Portugal
{daniel.nogueira,pabreu,trestivo}@fe.up.pt
[3] INEGI, University of Porto, Porto, Portugal
rtavares@inegi.up.pt

Abstract. Some cardiovascular diseases (CVD) can be characterized by abnormalities in the blood pressure (BP) waveform. This work explores the use of ML techniques in a screening system currently under development for detection of BP waveform abnormalities - Bisferiens and Alternans pulses. The system uses a tonometric probe for signal acquisition, signal processing involving period segmentation and image processing, and classification. The classification method used was the support vector machine (SVM) and it achieved an accuracy of 99.84%. Signal acquisition is done locally and sent to a remote server where the signal and image processing and classification is performed and the result prediction is sent back to the user.

Keywords: Blood pressure waveform
Bisferiens and Alternans pulse · Histogram of oriented gradients
Support vector machine

1 Introduction

According to the World Health Organization (WHO), cardiovascular disease (CVD) causes 17.7 million deaths each year, approximately 31% of all deaths worldwide. Of these deaths, more than 75% occur in low-income and middle-income countries [1].

The blood pressure waveform (BP) has characteristics that allow the detection of cardiac abnormalities in an individual. With direct analysis of the BP waveform, it is possible to determine criteria such as systolic blood pressure (SBP), diastolic blood pressure (DBP), second wave pressure (SWP), dicrotic notch pressure (DNP) and duration of each beat (Fig. 1), indirectly, heart rate (f) and pulse pressure (PP) [2], given by

$$PP = SBP - DBP \tag{1}$$

© Springer International Publishing AG, part of Springer Nature 2019
M. E. Auer and R. Langmann (Eds.): REV 2018, LNNS 47, pp. 536–545, 2019.
https://doi.org/10.1007/978-3-319-95678-7_60

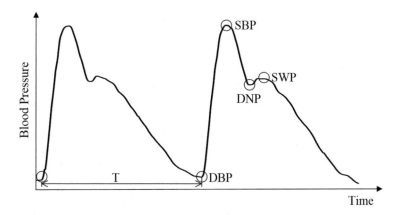

Fig. 1. Main points in BP waveform

However, the methods used to identify these points depend on a BP waveform without interference and noise and are used in combination with the monitoring of the ECG signals [3].

In an intensive care unit (ICU) environment, the monitoring of blood pressure waveform can make use of special equipment, even invasive measurement methods, operated by specialized staff. However, in regular check-up and clinical practice, it may be interesting to have non-invasive equipment, just for disease screening. In this way, such a system can be widely operated by supporting staff to identify some possible problems in cardiovascular function. Therefore, early detection of cardiac problems can be identified and patients referred to a cardiologist for further exams.

There have been many approaches to detect signal abnormalities in the BP waveforms that often rely on the actual comparison of the characteristics to their reference values. A signal abnormality algorithm that detects abnormal beats in BP waveforms through feature extraction of the pressure points from the signal and comparison with the abnormality criteria have been proposed in [2]. Other methods avoid using the BP values and focus mostly on features extracted exclusively from the waveform. An algorithm that determines the onset of arterial pressure pulses, through converting the waveform into a slope sum function signal and subsequent adaptive threshold for local search of pulse onset, shows that it is possible to extract rich information from the cardiovascular system abnormalities only using the waveform [4]. Algorithms to filter artefacts from BP signals have been proposed in order to detect Heart Rate Variability (HRV) which is closely related to Respiratory Sinus Arrhythmia (RSA) [5].

An approach that takes advantage of machine learning techniques for the pulse wave analysis signal collected with tonometric probes have been demonstrated using four different types of classifiers: Random Forest, probabilistic, decision tree and rule-based induction. An accuracy of 96.95% have been achieved when using the Random Forest classifier [6].

The use of tonometric probes has the disadvantage of being sensitive to sensor placement and to introduce artefacts due to small movements during the acquisition that may lead to signal distortion [7]. To minimize this sensitivity to sensor placement, interpreting the waveform signal as an image and its processing through contour recognition of each pulse wave would be an alternative that overcomes the need of having absolute pressure values for detecting pulse abnormalities.

In this article we present an automatic screening system, based on the analysis of BP waveform, using ML techniques. The system uses a sensor probe for data acquisition and the signal processing involves period segmentation, image processing and classification methods. For classification, support vector machine (SVM) and histogram of oriented gradients (HOG) were used.

The detection system considers two types of abnormalities in the BP: the Bisferiens and the Alternans. The Bisferiens pulse form (Fig. 2b) is characterized by a cardiac cycle with two peaks and indicates problems with the aortic valve, including aortic stenosis and aortic regurgitation, as well as hypertrophic cardiomyopathy-causing subaortic stenosis [8].

The Alternans pulse shape (Fig. 2c) is characterized by a beat-to-beat oscillation in the force of contraction of the heart muscle with constant heart rate. It is almost always indicative of left ventricular systolic impairment [9].

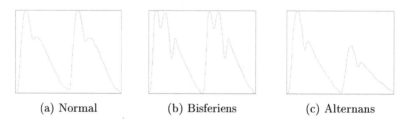

(a) Normal   (b) Bisferiens   (c) Alternans

Fig. 2. Normal blood pressure waveform and two abnormalities: Bisferiens and Alternans

The modeling of time series as an image has been used in association with machine learning (ML) techniques. Pereira et al. [13] propose a method for automatic identification of Parkinson's Disease (PD) using a smart pen, whose signal is converted to an image that is further processed with ML. With this approach, that uses shape analysis, contour and features extraction, an accuracy around 98% was reached in identifying PD.

For the identification of objects, the literature shows several applications using HOG [10] for contour detection and SVM as classifier [11].

HOG captures the edge or gradient structure that is the characteristic of the sought shape and does this in a representation with an easily controllable degree of invariance for local geometric and photometric transformations [12].

SVM is a supervised machine learning method of pattern recognition used for classification and linear regression. This process takes as input a set of data and predicts, for each given information, which of two possible classes the input is part of, which makes the SVM a non-probabilistic binary linear classifier.

The following sections of the paper describe the proposed methodology on how to identify the existence of abnormalities in the BP waveform and results obtained for our classification methods.

2   Methodology

The main purpose of the system is to allow a non-specialist to conduct a non-invasive screening test for detection of problems in the cardiovascular condition through the automatic analysis of the blood pressure waveform. If the system detects anomalies or irregularities in the BP waveform, the patient can be referred to a specialized physician. Furthermore, the system enables collecting and storing acquired data in order to build a database for mapping purposes. For the moment, the system is able to detect two common abnormalities in the blood pressure waveform, the Bisferiens and the Alternans pulses.

The operation of the system involves three main stages:

- Signal Acquisition;
- Signal Processing with Period Segmentation;
- Image Processing and Classification.

(a) Signal Acquisition
The signal acquisition was done using a custom developed tonometric probe. This sensor element consists of the force sensor coupled to an ergonomically designed handle and materialized by 3D printing, that is used to scan the signal strength and transmitted in real time to a computer. This implementation allows the signal acquisition of the waveform instead of the absolute values.

As a way to create data samples for the processing analysis, our tonometric probe was subject to a force produced by an actuator device, the generator of blood pressure profile. This force controlled device is able to produce force profiles with programmed patterns that mimic the blood pressure wave on an artery. The generator of blood pressure profile was used to produce three distinct BP waveforms: Normal (healthy), Bisferiens and Alternans (abnormalities), that were used as references signals. The setup used for signal acquisition of waveforms generated by the actuator device is shown in Fig. 3.

(b) Signal Processing with Period Segmentation
After collecting a sample, the signal is going through several phases of signal pre-processing, image processing and further image classification using machine learning.

The signal processing of the blood pressure waveform before turning into an image to be used in the machine learning classification requires several phases as shown in Fig. 4.

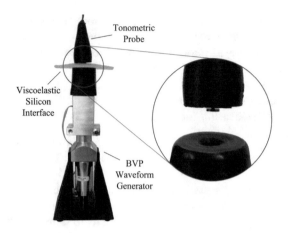

Fig. 3. Signal acquisition of the generated BP waveform using a tonometric probe

The first approach is to determine relevant inflection points in the signal, namely the high and low peaks and limits of the sample signal, through the analysis of derivative function.

The different pulse periods are identified through the crossing-points determination. The crossing-points, shown in Fig. 5c, result from the calculation of the mid-reference level instants of the first transition of each positive-polarity pulse and the next positive-going transition in the bilevel waveform [14].

The period estimation value is then used for a second waveform analysis to ensure that no periods are faulty detected or missed in the crossing-point calculation.

The last steps for the signal processing involves doing the period segmentation of our sample and normalize each period of the blood pressure waveform, that needs to be converted into an image to be suitable for the HOG.

The analysis of each period is enough to detect the bisfiriens abnormality but in order to also detect the alternans abnormality, a two period discretization is needed.

The images are generated using the crossing-points of two consecutive periods and are normalized in both axis in a way that only the waveform is important for the image processing analysis.

(c) Image Processing and Classification

A pre-processing is used to generate a binary from an indexed image, by converting the original figure to a binarize grayscale image. The data used to train the classifier are HOG resource vectors. The extraction of features through the HOG returns a variable amount of information regarding the size of the HOG cell. The cell size used was 8×8 which guarantees a sufficient amount of information for a good characterization of the waveform, as well as maintaining a short training time. The result of the HOG segmentation for a two period image is shown in Fig. 6. In practice, the HOG parameters must be selected according

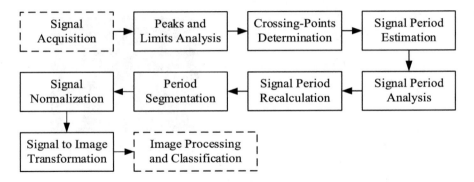

Fig. 4. Signal processing of the blood pressure waveform before image processing

to the training performance and classification test, repeated in order to identify the configuration parameters.

For the classification was used SVM that, from a set of training data, defines a plane separator, or hyperplane, of classes. The input data to train and test the SVM model were the contour vector resultant from the HOG segmentation. The new data is mapped in the same space and predicted as belonging to one of the two classes (normal or abnormal). This hyperplane seeks to maximize the distance between the closest points in relation to each of the classes. The support vectors are the data points closest to the separation hyperplane. In case of outlier the SVM looks for the best possible form of classification and, if necessary, disregards the outlier [15].

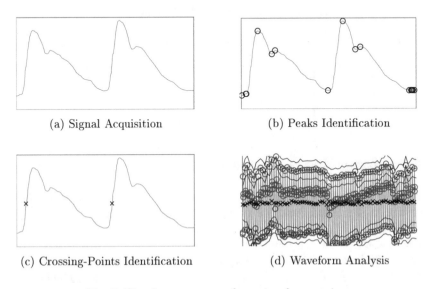

(a) Signal Acquisition

(b) Peaks Identification

(c) Crossing-Points Identification

(d) Waveform Analysis

Fig. 5. Blood pressure waveform signal processing

(a) Image for segmentation   (b) Result of segmentation

Fig. 6. HOG segmentation

To remove the necessity of having specific hardware for the signal and image processing and classification, the results of the signal acquisition are sent to a remote server. All the processing is done online in a remote server that is responsible for the most demanding calculations and processing, sending back only the result of the prediction.

3   Results

For the Alternans signals were acquired and analyzed 328 sets of two periods, for Bisferiens were 232 and for Normals were 192. Sixty percent of the total of these waves were used for training and 40% for prediction (Table 1).

Table 1. Number of images for training and test

Waveform	Number of images for training	Number of images to test	Total
Alternans	197	131	328
Bisferiens	139	93	232
Normal	115	77	192

Eight groups of different combinations were used to train and test the system (Table 2).

A mean of 19.03 s (standard deviation of 0.15 s) for training, 8.06 s for the classification (standard deviation of 0.04 s) and 99.05% of hits (standard deviation of 0.41%).

The Table 3 shows the result when using the training group with higher accuracy (99.67%) as the base, that is, there was variation only in the test group. Data obtained averaged 8.14 s for classification (standard deviation of 0.07 s) and accuracy of 99.84% (standard deviation of 0.18%).

Table 2. Data obtained from training and prediction for eight different groups of images

Group	Number of wrong predictions	Time to training (sec)	Time to predict (sec)	Accuracy (%)
1	5	18.98	8.11	98.34
2	2	18.76	8.07	99.34
3	3	18.93	8.03	99.00
4	2	19.21	8.04	99.34
5	4	19.20	8.13	98.70
6	3	19.09	8.08	99.00
7	3	19.07	8.03	99.00
8	1	19.01	8.01	99.67

Table 3. Data obtained from the prediction for eight different groups of images tested (keeping the same training group)

Group	Number of wrong predictions	Time to Predict (sec)	Accuracy (%)
1	0	8.09	100
2	0	8.05	100
3	0	8.16	100
4	0	8.19	100
5	1	8.04	99.67
6	1	8.24	99.67
7	1	8.19	99.67
8	1	8.18	99.67

4 Conclusions

The BP wave provides essential physiological data that can be used to identify CVD and, when detected during the anamnesis process, as one example, may shorten the length of hospital stay and increase the patient's chances of survival.

This online detection system under development can detect, through a non-invasive method, if the blood pressure wave has a profile of any of the two tested abnormalities (Bisfiriens and Alternans).

The efficiency of the classification approach of BP signal through SVM associated with HOG was demonstrated by the high accuracy of the proposed system that makes its use feasible in patient screening processes in hospitals, for example.

The online remote processing of signal and image decreases the necessity of having specific hardware in the local where the signal acquisition is done. This results in a less expensive dedicated tonometric probe that can send the acquired signals to the main server and allow the detection of CVD through abnormalities in the BP waveform in low-income and middle-income countries.

As future improvements of this approach, it is proposed the supervised application in patients to obtain the signals and their subsequent classification. In order to compare the reliability of our approach to detect abnormalities, it is suggested to use other classification algorithms and techniques (such as deep learning, for example).

Acknowledgment. Authors gratefully acknowledge the funding of Project NORTE-01-0145-FEDER-000022 - SciTech - Science and Technology for Competitive and Sustainable Industries, cofinanced by Programa Operacional Regional do Norte (NORTE2020), through Fundo Europeu de Desenvolvimento Regional (FEDER). This work was also funded by Project LAETA - UID/EMS/50022/2013.

References

1. WHO: Cardiovascular diseases (CVDs) May 2017. http://www.who.int/mediacentre/factsheets/fs317/en/
2. Sun, J., Reisner, A., Mark, R.: A signal abnormality index for arterial blood pressure waveforms. In: Computers in Cardiology 2006. IEEE, pp. 13–16 (2006)
3. Sun, J., Reisner, A., Saeed, M., Mark, R.: Estimating cardiac output from arterial blood pressurewaveforms: a critical evaluation using the mimic II database. In: Computers in Cardiology 2005. IEEE, pp. 295–298 (2005)
4. Zong, W., Heldt, T., Moody, G.B., Mark, R.G.: An open-source algorithm to detect onset of arterial blood pressure pulses. In: Computers in Cardiology 2003, pp. 259–262, Sept 2003
5. Luo, S., Zhou, J., Duh, H.B.-L., Chen, F.: BVP feature signal analysis for intelligent user interface. In: Proceedings of the 2017 CHI Conference Extended Abstracts on Human Factors in Computing Systems, ser. CHI EA 2017. ACM, pp. 1861–1868, New York (2017). https://doi.org/10.1145/3027063.3053121
6. Almeida, V.G., Vieira, J., Santos, P., Pereira, T., Pereira, H.C., Correia, C., Pego, M., Cardoso, J.: Machine learning techniques for arterial pressure waveform analysis. J. Personal. Med. **3**(2), 82–101 (2013). http://www.mdpi.com/2075-4426/3/2/82
7. Webster, J.G., Hendee, W.R.: Encyclopedia of medical devices and instrumentation, volumes 1–4, Physics Today, vol. 42, p. 76 (1989)
8. Riojas, C.M., Dodge, A., Gallo, D.R., White, P.W.: Aortic dissection as a cause of pulsus bisferiens: a case report and review. Ann. Vasc. Surg. **30**, 305-e1 (2016)
9. Euler, D.E.: Cardiac alternans: mechanisms and pathophysiological significance. Cardiovasc. Res. **42**(3), 583–590 (1999)
10. Mikolajczyk, K., Schmid, C., Zisserman, A.: Human detection based on a probabilistic assembly of robust part detectors. In: Computer Vision - ECCV 2004, pp. 69–82 (2004)
11. Ronfard, R., Schmid, C., Triggs, B.: Learning to parse pictures of people. In: European Conference on Computer Vision. Springer, Heidelberg, pp. 700–714 (2002)
12. Dalal, N., Triggs, B.: Histograms of oriented gradients for human detection. In: IEEE Computer Society Conference on Computer Vision and Pattern Recognition 2005, CVPR 2005, vol. 1. IEEE, pp. 886–893 (2005)

13. Pereira, C.R., Weber, S.A., Hook, C., Rosa, G.H., Papa, J.P.: Deep learning-aided Parkinson's disease diagnosis from handwritten dynamics. In: 2016 29th SIBGRAPI Conference on Graphics Patterns and Images (SIBGRAPI). IEEE, pp. 340–346 (2016)
14. Matlab signal processing toolbox 2017b
15. Tan, P.-N., Steinbach, M., Kumar, V.: Introduction to Data Mining. Addison-Wesley Longman Publishing Co., Inc., Boston (2005)

Augmented and Mixed Reality

The Effect of Augmented Reality in Solid Geometry Class on Students' Learning Performance and Attitudes

Enrui Liu[1], Yutan Li[1], Su Cai[1,2(✉)], and Xiaowen Li[3]

[1] VR/AR + Education Lab, School of Educational Technology,
Faculty of Education, Beijing Normal University, Beijing, China
caisu@bnu.edu.cn
[2] Beijing Advanced Innovation Center for Future Education,
Beijing Normal University, Beijing, China
[3] China Mobile Communications Corporation Government and Enterprise
Service Company, Beijing, China

Abstract. With the rapid development of mobile devices, Augmented Reality (AR) contents on the mobile are available and could be integrated with the course easily. Previous studies have shown that the integrating of Augmented Reality in the math courses would take more positive attitude of students, and the Augmented Reality could help students in learning some concepts which were not easy to understand in the courses. In this study, an Augmented Reality based mobile application running on tablet was developed and integrated into the solid geometry class in a junior high school to help students in three-dimensional geometry learning. Pre- and post-test were taken, the purpose of the current study is to learn about the effect of AR-based learning application and the students' attitudes and satisfaction to this application. The results show that Augmented Reality could make students' learning gains better in the traditional mathematics class, and students in junior high school are willing to study through this way.

Keywords: Augmented reality · Mathematics learning · Solid geometry

1 Introduction

The craze of AR-based mobile games like Pokémon GO in 2016 took the public attention to the AR, and the rapid development of mobile devices, the content with Augmented Reality (AR) could be got much easily, and learners could use the AR applications in the traditional classroom environment. In the elementary education field, Augmented Reality was a useful technology that could position the learner within a real-world context while interacting with the multiple modes of virtual learning contents, and it was primarily aligned with constructivist learning theory (Dunleavy and Dede 2014), also could be a valuable teaching tool in both elementary and high school classrooms (Billinghurst and Duenser 2012). The using of AR application in inquiry-based learning activities could bring higher motivations (Chiang et al. 2014a) and more engagement (Wang et al. 2014) in the learning approach, and the AR environment with inquiry-based learning would guide students to share knowledge in inquiry activities (Chiang et al. 2014b). Comparing with the computer-assisted learning tool, mobile-based AR learning applications provided significant supplemental (Cai et al. 2014). Moreover, AR could give a rise to better learning experiences for especially difficult courses (Kose et al. 2013). As for these reasons, the adoption of AR in the classroom and learning process were more and more common.

This study designed and developed an AR application for the junior high school students, helps them to explore and practice the three-view projection and drawing in the mathematic course. In China the junior high school students would learn and grasp the three-view projection and drawing in eighth or ninth grade in general, these contents were somehow difficult to students who were learning the solid geometry seriously for the first time. In learning activities in classroom, with the help of Augmented Reality application on mobile devices, students learned the basic conceptions of three-view projection and drawing and were practiced pairing the three-view drawing with the geometric shapes. The AR application would give the feedback of the result of pairing. Students could also interact with the solid geometries with the touchscreen, scaling, rotating and moving were implemented. Besides, students' achievements and attitudes towards the application were tested. The aim of the test was to reveal the effectiveness and influence of this AR application to real students.

2 Related Works

Augmented Reality could help students in a immersive learning experience in which the virtual and real contents were blended (Klopfer and Sheldon 2010), as imaging virtual space was a critical ability in the Geometry, the earliest AR application in education was about math and geometry (Kaufmann and Schmalstieg 2003; Piekarski and Thomas 2003). AR could build virtual 3D space easily, it was helpful in the geometry learning, especially in geometric constructions and spatial skills' improving, many AR applications in geometry were used to provide 3D geometry models to learners to help them (Banu 2012; Bergig et al. 2009).

In the mathematic education, previous studies have shown that the integrating of Augmented Reality in the mathematics courses would take more positive attitude of students (Billinghurst and Duenser 2012; Lin et al. 2015), and the response from the students was satisfying (Purnama et al. 2014). In addition, Lin et al. (2015) found that the students with low academic achievements demonstrated a positive attitude toward the AR learning contents. In short, AR in mathematics education could take more positive attitudes to students, that was a important advantage.

Liao et al. (2015) developed an Augmented Reality system to assist students in solving the Rubik's cube and learning the geometry concepts of volume and surface area, they found, beside the attitude in geometry learning, the spatial ability, geometry achievement, and attitudes toward whole mathematics learning could be improved with AR. Based on their study, AR was a good technology for improving the geometry learning and education.

Then, AR could also encourage motivation, comprehension and a higher involvement with the contents to be learned (Coimbra et al. 2015), they presented a state of the art mapped mainly by studies that focus AR in educational contexts, the creation of 3D contents in AR were described in their research.

As for the cognition, Bujak et al. (2013) presented that AR could provide scaffold in the process of learning and aid students' symbolic understanding, resulting in improving of understanding of abstract learning contents, that was very beneficial to mathematic learning.

Salinas et al. (2013) thought that AR could promote learners' visualization skills, they integrated AR with high school mathematics courses, to help students to learn equation, to serve students in the learning of a visual and tangible mathematics.

In conclusion, AR in the mathematic courses could bring good motivation, participation, involvement and students' attitude, it had potential in the cognition and understanding of learners. As for geometry learning, the need of good spatial ability of learners make AR as a proper technology in the courses. Solid geometry asks the learners to have a good level of spatial ability, that may be difficult for the beginner in solid geometry, AR could provide a good scaffold.

3 Methods

3.1 Participants

This study involved 75 students in the 7th Grade. The experiment of the AR application's impact was conducted in a junior high school in Beijing, China. All of them were from the same junior school and their ages ranged from 12 to 14. Before this study, these students had some prior knowledge of two-dimensional Euclidean geometry and spaces. In solid geometry, they only knew some basic conception of three-dimensional geometries (e.g., the conception of basic geometries like cube, cuboid, ball and so on), but they did not have conception of projection in mathematics and three-view drawing.

Fig. 1. The participants in the class, they took the pre-test and did learning activity with AR application

To make the AR application more suitable to use, the students were divided into groups of two. Two students in the same group would cooperate in the learning activity with AR application in the class using one Android tablet as shown in Fig. 1. And due to the large number of students, they were divided into two class, but the lecturers and content were the same.

3.2   Research Preparation and Procedure

In this study, a 40-min course was taken on the participants, the course mainly focused on the basic conception of three-view projection and three-view drawing, which was compulsory learning content in Chinese mathematics curriculum in junior high school. Before the experiment, researchers interviewed a mathematics teacher, learned some basic information about the difficult in the teaching and learning in the three-view drawing, the teacher expressed a wish to a simulation tool in the course to help students to construct the three-dimensional space. Then the application was used in a small group of five junior high school students, to make sure the application was easy to use, feedbacks were collected to revise and improve the application.

Before the course, a pre-test of some background knowledge of three-dimensional geometry was taken. Then in the 40-min class, researchers introduced the conception of projection and three-view drawing in easy-understanding words. The learning activity of pairing the geometries with their three-view drawing was the main part of this course, the AR application was used in this part.

After the learning activity, at the end of this course, a post-test about three-view drawing was taken to the participants. The difference between pre-test and post-test scores will represent the AR application's effect on students' learning performance. Additionally, a questionnaire about the AR application surveyed students' learning attitudes toward this application.

3.3   Research Questions

In this study, the researchers propose three research questions to be tested and examined by the experiment. These questions are listed as follow:

1. Whether there is a statistically significant improvement in students' scores on the pre- and post-test, after the using of AR application in the learning activity?
2. Whether there is a statistically significant difference in low-achieving and high-achieving students' learning gains in the AR assisted learning activity?
3. As mentioned above, AR technology in the courses can bring positive attitude in students, whether there is the same or similar effect on the AR application this research designed and developed?

3.4   Introduction of the AR Application

This study designed and developed an Android application called Three-View, which is an Augmented Reality based application with two types of cards to scan. In the learning activity, students were asked to pair two types of cards.

The first type which is called geometry name card, the name of basic three-dimensional geometries (e.g., cube, cuboid, ball) are written on these cards. Another type which is called three-view drawing card, a three-view drawing of a geometry based on the three-view projection was drawn on these cards.

Students in the class were asked to pair the cards in two types, and then they used the AR application to scan the paired-cards. If they paired correctly, a 3D model of the space geometry would appear, and they could interact with them on the touch screen and look at the model in different views by moving the tablet in the real world. Figure 2 showed the test of the AR application in the learning activity, students were looking at the geometry through the way of Augmented Reality.

The basic structure of the AR application and the way of using it are shown as Fig. 3.

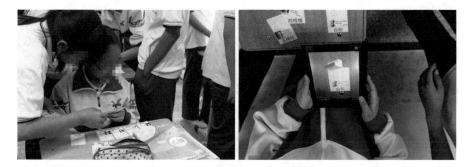

Fig. 2. AR application in using in learning activity

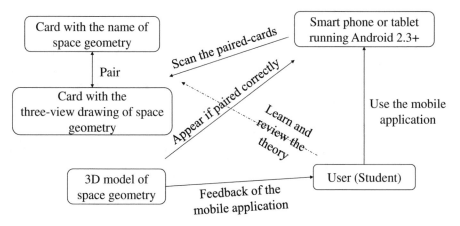

Fig. 3. The basic structure of the AR application

3.5   Tests and Questionnaire

The pre-test was mainly about the background knowledge of three-dimensional geometries, it was consisted by completing questions and drawing questions, the professional terms that students did not know about (e.g., projection, three-view drawing, front view, top view, left view) were changing to easy-understanding description in the pre-test. The full mark of pre-test was 25.

The post-test was based on the learning contents in the three-view drawing course, it also contained completing questions and drawing questions. It was used to test the effect of AR application in the learning performance. The full mark of post-test was 25.

In addition, a questionnaire was distributed to all participants, it adopts Likert scale with six options: a scale of 1, or "Strongly Disagree", to 6, or "Strongly Agree". The questionnaire consists of four constructs: (1) learning attitude, (2) satisfaction with the application, (3) cognitive validity and (4) cognitive accessibility, is revised on Cai et al. (2014)'s work.

The "learning attitude" construct includes seven items, which are revisions of items taken from Hwang and Chang (2011), it describe the attitude and motivation towards mathematics and geometry learning of students.

The "satisfaction with the application construct is covered by 14 items, which were revised and appended from Chu et al. (2010a, b), it describes students' satisfaction to the AR application in the learning activity.

The "cognitive validity" construct is covered by five items, and the "cognitive accessibility" construct is covered by four items, all of which are revisions of items taken from Chu et al. (2010a, b), they describe students' view of how the AR application could help them in learning and whether it is easy to use the AR application.

4   Results

4.1   Students' Learning Performance

Table 1 shows the descriptive statistics of the pre- and post-test in this study. The experiment produced 75 * 2 test copies (75 for the pre-test and 75 for the post-test), all of which are considered effective. The full mark of the test is 25 points.

Table 1. Descriptive statistics of pre- and post-test scores

	N	Mean	Std. deviation
Pre-test score	75	16.45	5.983
Post-test score	75	22.88	3.325

We conducted a paired t-test for the pre-test and post-test score variables. The tested variable is post-test score minus pre-test score, which stands for the difference yielded after using the AR application for each student. The results are shown in Table 2.

Table 2. Paired t-test for pre-test and post-test score variables

	Paired differences					t	df	Sig. (2-tailed)
	Mean	Std. deviation	Std. error mean	95% Confidence interval of the difference				
				Lower	Upper			
Post-test – Pre-test	6.427	6.288	.726	4.980	7.784	8.851	74	.000

Table 2 shows that the p-value (two-tailed) of the mean is close to zero ($t = 8.551$, p-value = 0.000). When the significance level is 0.05, we should reject the null hypothesis, which suggests that students' scores after using the AR application are significantly higher than those attained before the learning activity.

As a result, we conclude that with other unobserved variables controlled, the AR learning application has a statistically significant improvement on the score of the adopted learning performance test, and students' averages scores increased by 6.427 points.

Then the students are divided into three groups by their pre-test scores, the first 33% (25 students) as high-achieving students and the last 33% (25 students) as low-achieving students. The average learning gains (post-test scores minus pre-test scores) of both groups, as shown in Table 3.

Table 3. Average scores for low-achieving and high-achieving groups

Group	Pretest average	Posttest average	Gain average
Low-achieving	9.69	21.92	12.23
High-achieving	23.08	23.64	.56

The independent t-test for the low-achieving and high-achieving groups was conducted, the results are shown as Table 4. For the Levene's test for the equality of variances shown in Table 4, $F = 1.128$, $p = 0.294 > 0.05$, which suggests that we cannot reject the null hypothesis and should accept that the variance difference is not significant at the 0.05 significance level. In a word, the two groups students do not have significant difference in the learning gains.

Table 4. Independent t-test for low-achieving and high-achieving groups in learning gains

	Levene's Test for Equality of Variances		t-test for Equality of Means							
	F	Sig.	t	df	Sig. (2-tailed)	Mean difference	Std. error difference	95% Confidence interval of the difference		
								Lower	Upper	
Equal variances assumed	1.128	.294	12.451	48	.000	12.400	.996	10.398	14.402	
Equal variances not assumed			12.451	47.561	.000	12.400	.996	10.397	14.403	

4.2 Students' Learning Attitudes

In the questionnaire analysis, we calculated the score of each construct by averaging all the corresponding items within each construct. The descriptive statistics obtained are shown in Table 5. The mean values of all four construct are above 5 and close to the max value 6. That demonstrates that the students in junior high school have positive attitude and motivation in mathematics and geometry learning after the using of AR application, and they are satisfied with the AR application this study designed and developed, they think that the AR application can help them in the three-view drawing learning and it is easy to use in general.

Table 5. Descriptive Statistics of the questionnaire

	N	Min	Max	Mean	Std. deviation
Learning attitude	75	1	6	5.41	1.116
Satisfaction	74	1	6	5.18	1.343
Cognitive validity	69	1	6	5.28	1.196
Cognitive accessibility	69	1	6	5.61	1.060

5 Conclusion

Based on the results of tests and questionnaire, the Augmented Reality in solid geometry class has some positive effect on students' learning performance. After using the AR application in the learning activity in a 40-min course, students' scores in the pre- and post-test have significant difference, the using of AR could help students in the learning of three-view projection and three-view drawing, it is consistent with the results of some previous studies (Cai et al. 2014, 2016). As for the students in different achieving levels, the difference is not significant in this study, so the improving taken by AR is the same at different levels.

Students' attitude and motivation of geometry and mathematics learning are positive in general in the current study, that shows students in junior high school are willing to study through this way. It is consistent with the main stream idea of learning with AR and mobile devices (Banu 2012; Hwang and Hu 2013; Liao et al. 2015; Lin et al. 2015). The AR as a learning tool or scaffold in mathematics learning, especially in geometry learning, has great potential and good acceptance in learners. AR could become a useful technology in the learning in geometry and mathematics, especially at elementary level.

Further studies could design more strict experiment, compare the difference between AR and other multi-media tools, make AR technology correlated with well-designed inquiry-based learning. The more constructs in the questionnaire are also expected, just as students' learning conceptions, self-efficacy and so on. AR which helps learners in geometry learning in the more learning situations like informal learning and instruction in college is worth being tried.

Acknowledgement. This work is supported by the National Natural Science Foundation of China (Grant No. 61602043).

References

Banu, S.M.: Augmented Reality system based on sketches for geometry education. In: 2012 International Conference on Paper Presented at the e-Learning and e-Technologies in Education (ICEEE) (2012)

Bergig, O., Hagbi, N., El-Sana, J., Billinghurst, M.: In-place 3D sketching for authoring and augmenting mechanical systems. In: 8th IEEE International Symposium on Paper presented at the Mixed and Augmented Reality, 2009. ISMAR 2009 (2009)

Billinghurst, M., Duenser, A.: Augmented reality in the classroom. Computer **7**, 56–63 (2012)

Bujak, K.R., Radu, I., Catrambone, R., Macintyre, B., Zheng, R., Golubski, G.: A psychological perspective on augmented reality in the mathematics classroom. Comput. Educ. **68**, 536–544 (2013)

Cai, S., Chiang, F.-K., Sun, Y., Lin, C., Lee, J.J.: Applications of augmented reality-based natural interactive learning in magnetic field instruction. Interact. Learn. Environ. 1–14 (2016)

Cai, S., Wang, X., Chiang, F.-K.: A case study of Augmented Reality simulation system application in a chemistry course. Comput. Hum. Behav. **37**, 31–40 (2014)

Chiang, T.H., Yang, S.J., Hwang, G.-J.: An augmented reality-based mobile learning system to improve students' learning achievements and motivations in natural science inquiry activities. J. Educ. Technol. Soc. **17**(4), 352 (2014a)

Chiang, T.H., Yang, S.J., Hwang, G.-J.: Students' online interactive patterns in augmented reality-based inquiry activities. Comput. Educ. **78**, 97–108 (2014b)

Chu, H.-C., Hwang, G.-J., Tsai, C.-C.: A knowledge engineering approach to developing mindtools for context-aware ubiquitous learning. Comput. Educ. **54**(1), 289–297 (2010a)

Chu, H.-C., Hwang, G.-J., Tsai, C.-C., Tseng, J.C.: A two-tier test approach to developing location-aware mobile learning systems for natural science courses. Comput. Educ. **55**(4), 1618–1627 (2010b)

Coimbra, M.T., Cardoso, T., Mateus, A.: Augmented reality: an enhancer for higher education students in math's learning? Procedia Comput. Sci. **67**, 332–339 (2015)

Dunleavy, M., Dede, C.: Augmented reality teaching and learning. In: Handbook of Research on Educational Communications and Technology, pp. 735–745. Springer (2014)

Hwang, G.-J., Chang, H.-F.: A formative assessment-based mobile learning approach to improving the learning attitudes and achievements of students. Comput. Educ. **56**(4), 1023–1031 (2011)

Hwang, W.-Y., Hu, S.-S.: Analysis of peer learning behaviors using multiple representations in virtual reality and their impacts on geometry problem solving. Comput. Educ. **62**, 308–319 (2013)

Kaufmann, H., Schmalstieg, D.: Mathematics and geometry education with collaborative augmented reality. Comput. Graph. **27**(3), 339–345 (2003)

Klopfer, E., Sheldon, J.: Augmenting your own reality: student authoring of science-based augmented reality games. New Dir. Student Leadersh. **2010**(128), 85–94 (2010)

Kose, U., Koc, D., Yucesoy, S.A.: An augmented reality based mobile software to support learning experiences in computer science courses. Procedia Comput. Sci. **25**, 370–374 (2013)

Liao, Y.-T., Yu, C.-H., Wu, C.-C.: Learning geometry with augmented reality to enhance spatial ability. 2015. In: International Conference on Paper presented at the Learning and Teaching in Computing and Engineering (LaTiCE) (2015)

Lin, H.-C.K., Chen, M.-C., Chang, C.-K.: Assessing the effectiveness of learning solid geometry by using an augmented reality-assisted learning system. Interact. Learn. Environ. **23**(6), 799–810 (2015)

Piekarski, W., Thomas, B.H.: Interactive augmented reality techniques for construction at a distance of 3D geometry. Paper presented at the proceedings of the workshop on virtual environments (2003)

Purnama, J., Andrew, D., Galinium, M.: Geometry learning tool for elementary school using augmented reality. In: International Conference on Paper presented at the Industrial Automation, Information and Communications Technology (IAICT) (2014)

Salinas, P., González-Mendívil, E., Quintero, E., Ríos, H., Ramírez, H., Morales, S.: The development of a didactic prototype for the learning of mathematics through augmented reality. Procedia Comput. Sci. **25**, 62–70 (2013)

Wang, H.-Y., Duh, H.B.-L., Li, N., Lin, T.-J., Tsai, C.-C.: An investigation of university students' collaborative inquiry learning behaviors in an augmented reality simulation and a traditional simulation. J. Sci. Educ. Technol. **23**(5), 682–691 (2014)

Multimodal Data Representation Models for Virtual, Remote, and Mixed Laboratories Development

Yevgeniya Sulema$^{(\boxtimes)}$, Ivan Dychka, and Olga Sulema

Igor Sikorsky Kyiv Polytechnic Institute, Kyiv, Ukraine
{sulema,dychka,olga.sulema}@pzks.fpm.kpi.ua

Abstract. The main objective of the research presented in this paper is to provide developers of virtual, remote, and mixed laboratories with the powerful instrument for data representation. The data sets are supposed to have multimodal nature. Three models for multimodal data representation are presented and discussed in the paper. These models are the Muxel Model, the Multilevel Ontological Model, and the Spatio-Temporal Linked Model. The use of these models for implementation of different types of laboratories is discussed as well.

Keywords: Virtual and remote laboratories · Immersive technologies
Mulsemedia

1 Introduction

Virtual, remote, and mixed laboratories are powerful tools used in engineering education where the relation between theoretical material and physical objects, or processes to be studied should be evident and strong. Such laboratories give students an opportunity to investigate features of a subject (object, process, phenomena) of the study in a practical way – either through remote access to either a corresponding equipment or by means of a computer simulation.

Recently published bibliometric analyses [1–5] show that 'continued progress in computer graphics, virtual reality, and virtual worlds technologies can provide the opportunity to rapidly enlarge the use of virtual laboratory based systems applications, and can eventually reduce the need for real world laboratories altogether'. The future trends include 'the use of virtual worlds, immersive education, and other technologies, and to particularly target STE disciplines (particularly in engineering and robotics)'.

The main objective of the research presented in this paper is to provide developers of virtual, remote, and mixed laboratories with the powerful instrument for data representation. The data sets are supposed to have multimodal nature. This assumption looks natural because the more information of different types we can collect and provide to learners, the more completed 'picture' of the subject (object, process, phenomena) of the study they obtain and, therefore, the better results of the study process we can expect from the learners.

© Springer International Publishing AG, part of Springer Nature 2019
M. E. Auer and R. Langmann (Eds.): REV 2018, LNNS 47, pp. 559–569, 2019.
https://doi.org/10.1007/978-3-319-95678-7_62

Having in mind the necessity of multimodal data representation, processing, storing, and transmission, we focus our research on different use cases which depend on the type of a laboratory (virtual, remote, mixed, based on augmented reality, etc.). The separate use case we consider in our research relates to Digital Twin technology recently pointed out by Gartner research [6] as a promising emerging technology. Digital Twin is a virtual model of a process, product, or service. This pairing of the virtual and physical worlds allows analysis of data and monitoring of systems to head off problems before they even occur [7]. We consider Digital Twin as a possible use case for engineering education in future.

These different use cases set specific requirements to data representation: levels of detailing, data volumes, data sources, etc. Thus, the final goal of the research presented in this paper is to develop models for multimodal data representation that enable effective storage, processing, and retrieval of information in order to be used for the development of news virtual, remote, and mixed laboratories for effective engineering education.

2   Related Works

A critical overview of existing concepts and technologies in the field of fully-software-based virtual laboratories is presented in [1]. The analysis of the literature on virtual and remote labs is given in [2]. The investigation of creation and using virtual and remote laboratories for improving Science and Engineering teaching and learning is presented in [3]. Gravier et al. in [4] provide a literature review of modern remote laboratories as well as identify possible evolutions for the next generation of remote laboratories. The authors in [5] review a selection of the literature to contrast the value of physical and virtual investigations and offer recommendations for combining the two to strengthen science learning. A review of the different online delivery methods for virtual and remote laboratory development is presented in [8].

Sáenz et al. in [9] present an open course which offers several virtual and remote laboratories on automatic control, accessible to anyone. The authors in [10] present Embedded Systems' Hardware-Software CoDesign as well as an overview of the approach based on using ready platforms. The Smart Device specification to interface with remote labs as well as the extensible and platform-agnostic specification of the Smart Device services and internal functionalities are discussed in [11].

Kalúz et al. [12] present ArPi Lab remote laboratory for education in area of process control which is a cost-effective approach to remote experimentation. The authors in [13] describe a completely functional Android-based mobile Operational Amplifier iLab that will enable students all over the world perform experiments remotely from a mobile device using the Android platform. Tawfik et al. in [14] reports on a state-of-the-art remote laboratory project called 'Virtual Instrument Systems in Reality' which allows wiring and measuring of electronic circuits remotely on a virtual workbench that replicates physical circuit breadboards. The design and implementation of Networked Control System Laboratory 3D, which is a web-based 3D control laboratory for remote real-time experimentation, are introduced in [15].

The authors in [16] show how different universities have developed e-learning tools, their problems and advantages, and the necessity of integrate them in learning scenarios as well as investigate the use of virtual and remote lab in education. Andújar et al. in [17] propose a new concept in virtual and remote laboratories: the augmented remote laboratory which allows the student to experience sensations and explore learning experiences that, in some cases, may exceed those offered by traditional laboratory classes. Valera et al. in [18] describe Internet functions that include video file generation and real-time control as MWS features, which are useful for under-graduate courses. A remote laboratory NetLab, which has specially designed graphical user interface, is presented in [19]. Good examples of virtual labs and remote labs can be found in [20–22] and in [23, 24] respectively.

3   Approach

Virtual, remote, and mixed laboratories should provide visualization, description and/or reproduction of the subject of the study in the most informative way. It can be achieved by employing as many different modalities of data, which characterize the subject of the study, as possible. These modalities can include visual data, audio data, environmental data (temperature, pressure, atmosphere composition), data about physical properties and state of the object, process, or phenomena to be studied, etc. Multimodal data sets can be obtained by means of measurement, simulation, modeling, estimation, prediction, etc. Different data models should be considered in order to represent data in the most optimal way for their processing, storing, and transmitting while using in virtual, remote, and mixed laboratories.

3.1   Muxel Model

The Muxel Model (MM) of multimodal data representation enables the most detailed description of an object, process, or phenomena. The notion of a muxel (multimodal element) is the further advancement [25] of such well-known notions as a pixel (2D picture element), a voxel (3D volume element) [26], and a doxel (dynamic element) [27, 28]. A muxel can be described by heterogeneous data structure – a frame that keeps data of different modalities as slots (fields). The MM produces complex representation of different types of information about the subject of the study but it produces large amount of data. For completed description of a state of a matter in a certain point of a 3D scene in a certain moment of time, the muxel data structure (Fig. 1) can include the following data belonging to mulsemedia:

- Time data slot, it stores a moment when multimodal information is captured;
- Graphical data slot, it sets color and its transparency is a certain point;
- Audio data slot, it sets an amplitude value of acoustic signal in a certain time moment defined by time data slot;
- Olfactory data slot, it sets smell and its intensity in a certain point in a 3D scene;
- Taste data slot, it sets taste and its intensity in a certain point of a matter;

- Physical data slot, it sets both a type and a physical state of a matter in a certain point of a 3D scene; some examples of the matter type are air, water, wood, metal, glass; a physical state of the matter can be characterized by its temperature, pressure, humidity, density.

Every muxel can be in one of the following four states: informational, undefined, quasi-defined, and repeated.

The muxel in the informational state is defined by the frame that contains numerical values in every slot; the use of the label "by default" is allowed for some slots.

The muxel in the undefined state is defined by the frame that contains the label "unknown". This state means that data about muxel is either unknown or unimportant for some task, and therefore it isn't defined intentionally. For example, if the muxel model is used for describing the scene where an animate being is present, the internal constitution of this being is unimportant, because the monitoring is carried out only for its position in the scene.

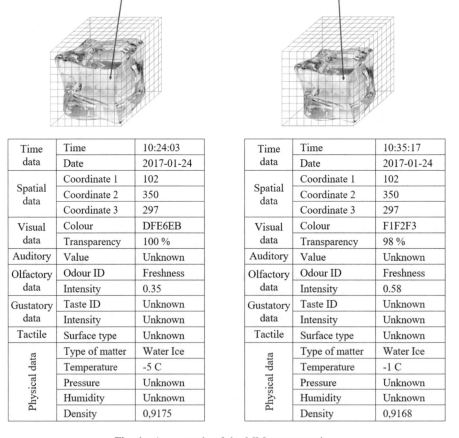

Time data	Time	10:24:03
	Date	2017-01-24
Spatial data	Coordinate 1	102
	Coordinate 2	350
	Coordinate 3	297
Visual data	Colour	DFE6EB
	Transparency	100 %
Auditory	Value	Unknown
Olfactory data	Odour ID	Freshness
	Intensity	0.35
Gustatory data	Taste ID	Unknown
	Intensity	Unknown
Tactile	Surface type	Unknown
Physical data	Type of matter	Water Ice
	Temperature	-5 C
	Pressure	Unknown
	Humidity	Unknown
	Density	0,9175

Time data	Time	10:35:17
	Date	2017-01-24
Spatial data	Coordinate 1	102
	Coordinate 2	350
	Coordinate 3	297
Visual data	Colour	F1F2F3
	Transparency	98 %
Auditory	Value	Unknown
Olfactory data	Odour ID	Freshness
	Intensity	0.58
Gustatory data	Taste ID	Unknown
	Intensity	Unknown
Tactile	Surface type	Unknown
Physical data	Type of matter	Water Ice
	Temperature	-1 C
	Pressure	Unknown
	Humidity	Unknown
	Density	0,9168

Fig. 1. An example of the MM representation.

The muxel in the quasi-defined state is defined by the frame that contains defined data in some slots only; all other slots keep a label "unknown". The case, when the muxel model is used for description of an object which odour is unimportant for certain task, can serve as an example of the quasi-defined state of muxels.

The muxel in the repeated state is defined by the frame that contains the same values in its slots (excepting time slot) as a neighboring muxel has. In this case such repeated values can be substituted by the label "repeated". The repeated state is necessary for the data compression of muxel sequence.

The rate of muxel data capturing depends on a rate of the most changeable data among used modalities. Usually an audio signal frequency defines this rate because an acoustic signal is the most changeable in time data sequence among other data sets.

The set of data slots depends on a specific task to be solved by using multimodal data representation. For example, in a chemical engineering virtual lab, the taste data slot is unnecessary, at the same time this data modality is of a special importance in a virtual lab for food manufacturing studies.

The advantage of the MM is a complete description of a real scene. Its disadvantage is that such full description produces large data volumes, thus, the MM can be used for description of small objects or scenes of a limited size.

3.2   Spatio-Temporal Linked Model

The Spatio-Temporal Linked Model (STLM) is aimed at the complex description of the subject (object, process, phenomena) of the study by composing multimodal data sets coming from different sources (local data storages, remote sensors, remote data bases, etc.). It requires data synchronization in terms of time and space.

The STLM of an object can be realized as a distributed database. The core of the data representation in the STLM is so called a 'primary database' which keep references (links) to the data sources depending on modality of data, their features, etc. (Fig. 2). Such sources can be external devices (cameras, mics, sensors) as well as other databases called 'reference databases'. A reference database includes typical data of a certain modality. Thus, a reference database of visual data can contain typical images of the object appearance. For example, if the object is a water tank, the reference database of visual data can contain pictures of water tanks of different models, size, color, etc.

<Object *i*>	
Data Modality: Visual	Data Source: Webcam in <Link 1>
Data Modality: Audio	Data Source: Audio file in <Link 2>
Data Modality: Environmental (temperature)	Data Source: Temperature sensor in <Link 3>
Data Modality: Environmental (humidity)	Data Source: Humidity sensor in <Link 4>

Fig. 2. An object description (record) in the STLM primary database.

The advantage of the STLM is the orientation to external sources of data like sensors, cameras, mics, etc. The disadvantage is that data synchronization is necessary what can require additional resources for data processing.

3.3 Multilevel Ontological Model

Since high detailing related to a separate muxel generates huge data volumes, it is reasonable to compose MMs (i.e. detailed models) of only relatively small objects and to describe a scene as a set of such detailed models. With this purpose we introduce a 'macro-level' and a 'micro-level' of detailing and distribute data between these levels by introducing the Multilevel Ontological Model (MOM).

The MOM is an ontology where every object (node) is described by data sets of different modalities (Fig. 3). This allows us to search similar objects by similarity of their characteristics.

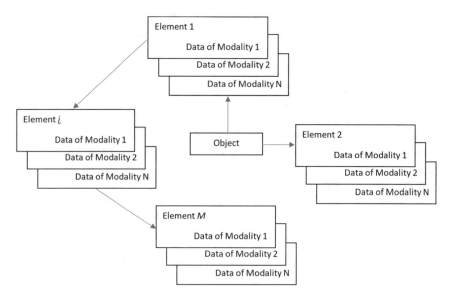

Fig. 3. The MOM of an object.

A macro-level of the MOM is a scene (which includes objects, processes, phenomena) in whole. A micro-level contains detailed data of each modality. The form of data representation on the macro-level is a graph of ontology with references to the tables containing data of the micro-levels. In some cases, the micro-level data can be represented by using either the MM or the STLM. The specific purpose of the MOM is data retrieval and analysis what can be useful for some specific tasks in virtual and mixed laboratories. The advantage of the MOM is its orientation onto complex search of multimodal data: it enables finding objects by a specific feature represented as a certain modality. The disadvantage is that depending on micro-level specification the MOM can be too complex.

4 Discussion

The proposed models can be used in different ways depending on the type of a laboratory to be developed (Fig. 4).

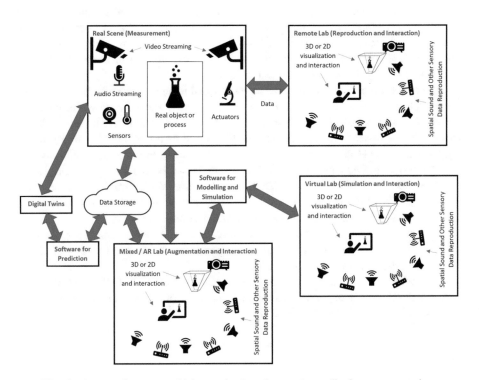

Fig. 4. A general concept of laboratories based on mulsemedia data representation.

Thus, the MM is useful when we need to collect and present very detailed information about a relatively small object. For example, the MM can be used for the full description of a certain matter (silicone, silver, wood, etc.) which features is a subject of learning. If the remote lab is equipped with a 3D display, a learner can 'fly' trough the matter sample and see the color, texture, density of this matter. If the lab equipment includes an olfactory display, a learner can also perceive odor of the matter. If the lab is equipped with a haptic glove, a learner can touch the matter sample and perceive the surface roughness or smoothness.

The MOM can be used when we would like to present a compound scene, an operating principle, or a concept in whole. For example, for effective learning of mechanical engineering principles a virtual lab can be developed with use of the MOM of an engine which can be demonstrated not only as the whole object but also as an aggregate which includes a number of components. Every component can be also demonstrated as the whole object as well as if it is a compound object it can be decomposed. In addition to visual demonstration, a learner can perceive information of other modalities, for example, he or she can hear the sound of the working engine, etc.

The STLM is ideal for remote and mixed laboratories because it allows developers to combine data sets coming from different sources. For example, a mixed lab for studying a thermonuclear reactor working principle can be developed where video and audio streams are supposed to come from the premises of a real nuclear station and these data are augmented by temperature graphs and other data about the process developing inside of the thermonuclear reactor.

The data presented by these models are supposed to be processed by using software for modeling, simulation, prediction, and other tasks. A special case is the representation of digital twins of real objects or processes. For example, we can model an aircraft engine, visualize it and its characteristics and then test this virtual object (the digital twin) instead of the real engine to predict its work in different conditions.

The programming language ASAMPL, which has been recently developed for effective processing of multimodal data [25], can be used for the implementation of the proposed models for different use cases. The distinctive features of ASAMPL language are orientation on storing mulsemedia (multimodal) data in external storages (e.g. in cloud storages) and optimization of multimodal data processing. The latter has been achieved by using special data structures–aggregates which are the carrier at the Algebraic System of Aggregates (ASA).

The following fragment of the program in ASAMPL shows how the STLM can be used for development of mixed laboratory in robotics (the general structure of a program in ASAMPL can be seen in [25]).

```
Program RoboticsMixedLab {
  Libraries { ... }
  Handlers { ... }
  Renderers { ... }
  Sources {
    VisualDataStream1 Is 'http:/robolab.edu.net/031605';
    HapticDataStream1 Is 'https://171.63.0.152:2096';
    SceneFile Is 'D:\Lab\task1.agg'; ... }
  Sets { ... }
  Elements {
    Duration is Time;
    StartTime = 09:30:00; ... }
  Tuples { ... }
  Aggregates {
    Object = [VisualDat, HapticDat];
    RoboticArm = [VisualDat, HapticDat, AudioDat]; ... }
  Actions { ...
    Timeline StartTime : Delta : (StartTime+Duration) {
      Download HapticDat From HapticDataStream
                     With default.HapticLib;
      Download VisualDat From VisualDataStream
                     With default.VisualLib;}
        ...
    Scene is [RoboticArm, Object];
    Upload Scene To SceneFile With default.all;
    Render Scene With [VisualRen, AudioRen, HapticRen];}
}
```

In this example the STLM of the scene, which includes a robotic arm and an object to be manipulated by this robotic arm, is composed by real data is being streamed from sensors of both the object and the robotic arm as well as augmented audio content with theoretical information, practical guidelines, comments, etc. The object is defined as an aggregate which unites data sets of 2 modalities: visual and haptic. The robotic arm is defined as an aggregate which unites data sets of 3 modalities: visual and haptic are obtained from external devices and audio is an augmented data set. The scene is rendered by using predefined libraries and devices as well as it is stored for further use.

Since the range of applications is wide (digital twins, virtual labs, remote labs), both the MM and the MOM can be also employed and represented by using ASAMPL.

5   Conclusions

The research presented in this paper is focused on the development of models of multimodal data representation for effective processing, storing, transmission, and reproduction (including visualisation) of synchronized data sets of different nature. These models can be a powerful instrument to be used for the development of virtual, remote, and mixed laboratories.

The proposed models for multimodal data representation are based on the recently introduced concept of a multi-image–an overall description of an object, process, or phenomena [25]. This concept uses the notion of a muxel (multimodal element) which is a least unit in the object, process, or phenomena features description.

The implementation of different types of laboratories based on the proposed models can be fulfilled by using ASAMPL programming language.

References

1. Potkonjak, V., et al.: Virtual Laboratories for education in science, technology, and engineering: a review. Comput. Educ. **95**, 309–327 (2016)
2. Heradio, R., et al.: Virtual and remote labs in education: a bibliometric analysis. Comput. Educ. **98**, 14–38 (2016)
3. Esquembre, F.: Facilitating the creation of virtual and remote laboratories for science and engineering education. ScienceDirect, IFAC-PapersOnLine **48**(29), 049–058 (2015)
4. Gravier, C., et al.: State of the art about remote laboratories paradigms—foundations of ongoing mutations. Int. J. Online Eng. **4**(1), 1–9 (2008)
5. de Jong, T., Linn, M.C., Zacharia, Z.C.: Physical and virtual laboratories in science and engineering education. Science **340**, 305–308 (2013)
6. Top Trends in the Gartner Hype Cycle for Emerging Technologies (2017). http://www.gartner.com/smarterwithgartner/top-trends-in-the-gartner-hype-cycle-for-emerging-technologies-2017/
7. What Is Digital Twin Technology and Why Is It So Important? https://www.forbes.com/sites/bernardmarr/2017/03/06/what-is-digital-twin-technology-and-why-is-it-so-important/#573fc3c2e2a7
8. Chen, X., Song, G., Zhang Y.: Virtual and remote laboratory development: a review. In: Proceeding of Earth and Space 2010: Engineering, Science, Construction, and Operations in Challenging Environments, pp. 3843–3852 (2010)
9. Sáenz, J., Chacón, J., de la Torre, L., Visioli, A., Dormido S.: Open and low-cost virtual and remote labs on control engineering. IEEE Access **3**, 805–814 (2015)
10. Parkhomenko, A., et al.: Development and application of remote laboratory for embedded systems design. Int. J. Online Eng. **11**(3), 27–31 (2015)
11. Salzmann, C., Govaerts, S., Halimi, W., Gillet, D.: The smart device specification for remote labs. Int. J. Online Eng. (2015)
12. Kalúz, M., Cirka, L., Valo, R., Fikar, M.: ArPi lab: a low-cost remote laboratory for control education. In: Proceedings of the 19th World Congress of the International Federation of Automatic Control, Cape Town, South Africa (2014)
13. Oyediran, S.O., Ayodele, K.P., Akinwale, O.B., Kehinde, L.O.: Development of an operational amplifier iLab using an android-based mobile platform: work in progress. In: Proceedings of 120th ASEE Annual Conference and Exposition (2013)

14. Tawfik, M., et al.: Virtual Instrument Systems in Reality (VISIR) for remote wiring and measurement of electronic circuits on breadboard. IEEE Trans. Learn. Technol. **6**(1), 60–72 (2013)
15. Hu, W., Liu, G.-P., Zhou, H.: Web-based 3-D control laboratory for remote real-time experimentation. IEEE Trans. Industr. Electron. **60**(10), 4673–4682 (2013)
16. Sancristobal, E., Martín, S., Gil, R., Orduña, P., Tawfik, M., Pesquera, A., Diaz, G., Colmenar, A., García-Zubia, J., Castro, M.: State of art, initiatives and new challenges for virtual and remote labs. In: Proceedings of 12th IEEE International Conference on Advanced Learning Technologies, pp. 714–715 (2012)
17. Andújar, J.M., et al.: Augmented reality for the improvement of remote laboratories: an augmented remote laboratory. IEEE Trans. Educ. **54**(3), 492–500 (2011)
18. Valera, A., Díez, J.L., Vallés, M., Albertos, P.: Virtual and remote control laboratory development. IEEE Control Syst. Mag. **25**, 35–39 (2005)
19. Nedic, Z., Machotka, J., Najhlski, A.: Remote laboratories versus virtual and real laboratories. In: Proceedings of the 33rd ASEE/IEEE Frontiers in Education Conference (2003)
20. Virtual Labs: Computer Science and Engineering. http://vlab.co.in/ba_labs_all.php?id=2
21. Learn Genetics: Virtual Labs: http://learn.genetics.utah.edu/content/labs/
22. BioInteractive. Virtual Labs: http://www.hhmi.org/biointeractive/explore-virtual-labs
23. Huawei Developer. Remote Lab: http://developer.huawei.com/ict/en/remotelab
24. Internet Remote Laboratory. http://remote-lab.fyzika.net
25. Sulema, Y.: ASAMPL: programming language for mulsemedia data processing based on algebraic system of aggregates. In: Advances in Intelligent Systems and Computing, vol. 725, pp. 431–442. Springer (2018)
26. Hill, D.L., et al.: Voxel similarity measures for automated image registration. In: SPIE Proceedings of the Visualization in Biomedical Computing, vol. 2359, no 205 (1994)
27. Carnero, J., Diaz-Pernil, D., Mari, J.L., Real, P.: Doxelo: towards a software for processing and visualizing topology computations in Doxel-based 3D + t images. In: Proceedings of the 16th International Conference on Applications of Computer Algebra, ACA 2010 (2010)
28. Gonzalez-Diaz, R., et al.: Algebraic topological analysis of time-sequence of digital images. Lecture Notes in Computer Science, vol. 3718, pp. 208–219 (2005)

Voice Driven Virtual Assistant Tutor in Virtual Reality for Electronic Engineering Remote Laboratories

Michael James Callaghan[(✉)], Gildas Bengloan, Julien Ferrer,
Léo Cherel, Mohamed Ali El Mostadi, Augusto Gomez Eguíluz,
and Niall McShane

Intelligent Systems Research Centre, Ulster University,
Derry, Northern Ireland, UK
mj.callaghan@ulster.ac.uk

Abstract. The first generation of affordable consumer virtual reality headsets and related peripherals are now available. Question-Answering (QA) systems and speech recognition/synthesis has improved dramatically over the last decade. Virtual assistants, based on speech-based services are growing in popularity and can be used in a range of diverse application areas. This paper explores the practical use of virtual reality, IOT and voice driven virtual assistants in remote laboratories to facilitate visualization of electrical phenomena and to tutor students; guiding them through each stage of an experiment; presenting supplementary teaching resources when requested; accessing, controlling and configuring instrumentation and hardware and providing feedback with summative and formative assessment. The re-purposing of existing teaching material for use in an immersive environment with a virtual assistant is shown and the limitations and opportunities offered by the approach taken and the technologies used are discussed. The process of integrating test instrumentation, the board under test, a switching matrix and additional teaching resources into virtual reality using IOT with the inclusion of virtual assistants is described. Two case studies of practical working examples of remote laboratories in virtual reality with a virtual assistant tutor are demonstrated and the viability and long-term opportunities for the use of virtual reality and virtual assistants in the context discussed.

Keywords: Virtual reality · IoT
Automatic Question-Answering (QA) systems · Speech recognition/synthesis
Virtual assistants · Voice user interfaces

1 Introduction

Affordable consumer virtual reality headsets and peripherals are increasingly common driven by advances in video game technologies and related hardware [1]. Automatic Question-Answering (QA) systems and speech recognition/synthesis functionality and accuracy has improved dramatically over the last decade allowing the use of voice interactions to automate and organize complicated tasks and directly answer domain

© Springer International Publishing AG, part of Springer Nature 2019
M. E. Auer and R. Langmann (Eds.): REV 2018, LNNS 47, pp. 570–580, 2019.
https://doi.org/10.1007/978-3-319-95678-7_63

specific questions using natural language and have the potential to revolutionize human interactions with devices and data [2, 3]. Virtual assistants, based on speech-based services are growing in popularity and are now entering the mainstream. These services come with a set of built-in capabilities and allow the creation and addition of new abilities e.g. the Cortana Intelligence suite provides a development environment and distribution ecosystem which includes IoT where developers can publish and distribute their voice based applications with functionality to connect devices to cloud based services [4]. These flexible, highly functional development environments and backend architectures allow the use of voice enabled services in a range of diverse application areas including education, industrial and home automation. Practical virtual and remote laboratories engineering laboratories for undergraduate students are evolving driven by the availability of more affordable instrumentation, hardware kit, internet growth, new types of interfaces/front ends and a move towards student-centered pedagogies [5].

This paper explores the use of virtual assistants in remote electronic and electrical engineering virtual reality laboratories to tutor students; guiding them through experiments; presenting supplementary teaching resources when requested; accessing, controlling and configuring test instrumentation and hardware and providing feedback through summative and formative assessment. Two case studies and practical examples of voice driven virtual assistants are demonstrated based on the enhancement of an existing remote laboratory and teaching resources for fundamental engineering circuits suitable for the first year of an undergraduate degree. The process of integrating test instrumentation, the board under test, a switching matrix, teaching resources, the IoT hub and virtual assistants into an immersive virtual reality environment are discussed.

Section 2 provides an overview of the Cortana Intelligence suite and the creation of voice user interfaces using the Microsoft Bot framework. Sections 3, 4 and 5 discusses challenges related to the re-purposing an existing laboratory for voice interactions/virtual reality and provide practical example of this process and explore deeper integrations. Section 6 presents the conclusion and possible future work.

2   Voice User Interfaces and Virtual Assistants

The Microsoft Bot framework is part of the Cortana intelligence range of cloud based services and provides a set of tools to create, deploy and publish conversational bots across a range of channels and interaction modes. The Cortana personal assistant is a speech-enabled channel in the framework that can send and receive voice messages using cloud based voice recognition and speech synthesis capabilities based on natural language processing (NLP) algorithms. It is integrated into Microsoft Windows 10, can be installed as an app on mobile devices, is available in smart speaker format and can be used as a home automation hub [6]. When operating in default mode Cortana continuously listens to all speech in its general vicinity and responds/becomes active when it detects the use of a "wake word". The voice command interactions detected after activation are sent to the cloud for processing through LUIS (Language Understanding Intelligent Service) and relevant responses generated (Fig. 1). Third party developers can create voice user interfaces that extend the capabilities of Cortana. These are called using an invocation name which is a key word used by the end user to initiate a set of voice interactions/responses with Cortana and the developed/connected services or hardware devices. Interactions and responses are defined by an interaction model which manages communications using an intent schema, sample utterances and entities [6]. Intents are the core functionality of your skill and are a list of common actions your skill can accept and process. Entities are parameters or values passed with an intent. Sample utterances specify the spoken words and phrases users can say to invoke intents.

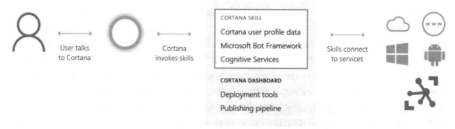

Action	Voice User Interaction (Interaction model)
Make a request	User says, "Ask Cortana to open practical engineering lab"
Collect more information from the user	Cortana replies "Which laboratory?" and then waits for a response.
Provide required information	User replies, "Series Parallel laboratory."
User request in completed	Cortana initializes Series Parallel laboratory and responses "Welcome to laboratory one, the Series Parallel…"

Fig. 1. Architecture/interactions of Cortana virtual assistant and related services

3 Re-purposing a Remote Laboratory for Voice/Virtual Reality

The architecture of a typical online laboratory allows the user, located in a separate geographical location, to access, control and conduct experiments remotely. The hardware control element is usually facilitated by the use of GPIB (General Purpose Interface Bus) or similar communication standards connected to a switching matrix allowing test instrumentation and experimental boards to be selected, connected and configured during experiments. The lab(s), related teaching resources, test instrumentation and the circuit board under test are accessed through the web using a client/server approach [7, 8]. The approach taken for this project was to reuse most of the existing local control protocols and add a Raspberry Pi running Microsoft Azure IoT (Fig. 2). This facilitated interactions and communications between the physical laboratory, the Cortana/LUIS/Bot framework and the client user interface which is an application created in the Unity games engine for the Oculus Rift/touch controllers and allows remote access to and control of the remote laboratory through Virtual Reality [9]. In practical terms this process involved creating a bot service in the Azure portal, enabling the WebChat/Cortana channels and extending the functionality/range of understanding of these channels using LUIS. Communications with the Unity Virtual Reality front end was through the Microsoft Azure IoT stack/Raspberry Pi using UDP.

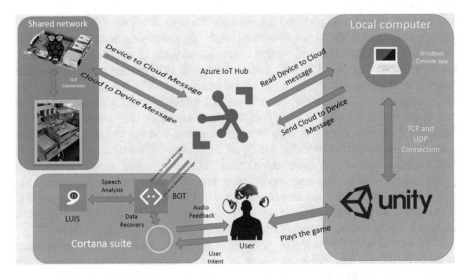

Fig. 2. Re-purposed laboratory with Cortana intelligence services

4 Series Parallel Experiment with Virtual Assistant Tutor in VR

The experiments in the existing online remote laboratories are focused on first year undergraduate level electronic and electrical engineering and cover topics ranging from series/parallel to oscillator circuits. The Series Parallel experiment was selected to develop a prototype using a virtual assistant tutor in virtual reality with circuit phenomena visualization to explore the viability and challenges of this type of approach (Table 1).

Table 1. Series Parallel game based laboratory for undergraduate engineering students

Laboratory	Objective/circuit	Theory	Learning outcomes
Lab 1 Series/parallel	Solve for R1 given Vi,R2,R3 to get value Vo	$V_o = \dfrac{R_{eq} \times V_{in}}{R_{eq} + R_1}$	Parallel and series circuits. Equivalent resistance. Circuits and current flow.

The Series Parallel experiment employs a game and time-based approach where using the formulas provided for equivalent resistance *(Req),* voltage out *(Vo)* and the current values of input voltage *(Vin)* and resistors *R2/R3* the students have to calculate the correct value of resistor *R1* to achieve the given target output voltage (*Vo*). A score is then awarded based on how close the calculated value was to the target value of *Vo* and the time taken to complete the calculations. When the laboratory is in progress, the game user interface/client communicates with the underlying hardware, switching matrix and instrumentation and connects the selected value of resistor *R1* to physically complete the circuit [10, 11]. Using this experiment as a starting point, a structured series of interactions suitable for a voice driven experience which included an overview of the laboratory, access to help, remote control and configuration of instrumentation and circuits and assessment with feedback to the student was created (Fig. 3 and Table 2). The web based user interface was removed and the teaching material and related resources re-purposed for use in an immersive 3D environment built in the Unity games engine and using the Oculus Rift/touch virtual reality headset and peripherals. A virtual laboratory environment was created to host the experiment and designed to facilitate the visualization of electrical phenomena (Fig. 4). The physical layout of the Series Parallel circuit was recreated at "room scale" to allow the student to explore individual circuit components and related voltage and current values.

The virtual laboratory contains a series of feedback panels which were synchronized with the Cortana virtual assistant to provide additional/complementary information to the student i.e. an overview of the circuit to solve, instructions on how to approach the experiment, an interface to select and connect individual components, written feedback from Cortana, a live streaming webcam showing the physical circuit and instrumentation in the real world lab and formative/summative when the lab was completed. The student starts the lab/interactions by launching the Virtual Reality Unity application on the PC and using the invocation name to begin the engagement process with Cortana. This initializes the remote hardware, circuits and instrumentation (Fig. 5). Cortana then welcomes the student to the laboratory, provides an overview of the experiment and sets out their objectives (Fig. 6). It guides the student through the laboratory with a series of voice prompts, underpinned and complemented by the supporting material on the feedback panel e.g. help provision and formulas for equivalent resistance and voltage out. The student is then asked to physically select and connect the correct resistor value using the Oculus Touch controllers to bias the circuit and achieve the required output voltage.

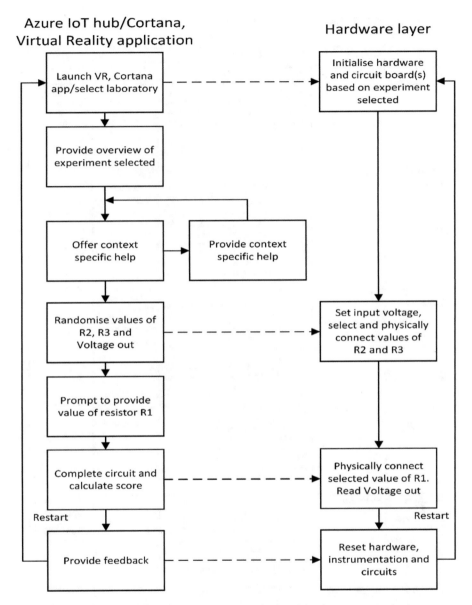

Fig. 3. High level overview of user interactions and hardware communication

Table 2. Interaction model for Cortana Series Parallel experiment

TASK	INTENT NAME	CORTANA PROMPT	UTTERANCES	HARDWARE
Launch Cortana app	app.launch	Welcome to laboratory one, the Series Parallel circuit. This session will help you to understand the operation of series parallel resistor networks. You will also learn about equivalent resistance. I have initialised the hardware and we are ready to begin. When you are ready, you can ask for your objectives, by saying, **what are my objectives**? Or you can say **leave** to exit the program.	What are my objectives. Leave.	Initialise hardware and circuit board(s) based on experiment selected
Provide overview of experiment selected	help	For this laboratory, you need to solve for the value of resistor R1 to achieve the required target output voltage Vout. I will provide the values of the input voltage Vin, resistor R2 and resistor R3 and the value of the output voltage Vout. From these values, you can calculate the required value of resistor R1. When you are ready to start, please say **ready to start**. Or you can ask for some formulas to help you solve the circuit by saying **formula please**. Or you can say **leave** to exit the program.	Ready to start. Formula please. Leave.	
Offer context specific help	selectFormula	Do you require the formula for the **voltage out** or the formula for the **equivalent resistance**?	Voltage out. Equivalent resistance.	
Provide context specific help	provideFormula *(equivalent resistance)*	Equivalent resistance is equal to the product of resistor R2 multiplied by resistor R3, divided by the sum of resistor R2 plus resistor R3. The value is in ohms. When you are ready to start, please say **ready to start**. Or you can ask for more formulas to help you solve the circuit by saying **formula please**.	Ready to start. Formula please.	
Provide context specific help	provideFormula *(voltage out)*	The value of the voltage out is calculated by multiplying the equivalent resistor, R equivalent by the value of voltage Vin. Then dividing the product by the sum of the values of R equivalent and resistor R1. When you are ready to start, please say **ready to start**. Or you can ask for more formulas to help you solve the circuit by saying **formula please**.	Ready to start. Formula please.	
Prompt to provide correct value of R1 to achieve target value of Vout	readytoStart	The value of the input voltage Vin, is 5 volts, the value of resistor R2 is {valueR2} ohms and the value of resistor R3 is {valueR3} ohms. Now please state the value of resistor R1 to achieve the target output voltage of {valueVo} volts. There are 4 options available for the value of resistor R1 which are 1000, 1500, 2200 and 3300 ohms.	{First} ohms. *(State value of R1 in ohms)*	Set input voltage, select and physically connect values of R2 and R3
Complete circuit and calculate score	circuitComplete	I have connected resistor R1 with the value of {resistance} ohms and completed the circuit. The output voltage is {voltage} volts and your score is {score}.	Restart. Leave.	Physically connect selected value of R1. Read Vout
Provide feedback	feedback	*Feedback on the score and on how to increase the score achieved is then given to the student based on the target value of voltage out and the calculated values of voltage out.*	Restart. Leave.	Reset hardware

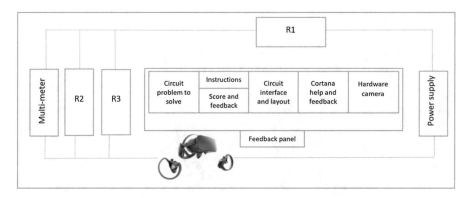

Fig. 4. High level overview of the virtual reality laboratory environment

Fig. 5. Remote hardware, circuits and instrumentation and virtual reality user interface

When the resistor is connected virtually it is also connected physically through the switching matrix. Visual feedback is provided on the hardware camera panel which shows a live feed of the multi-meter in the physical lab. Cortana provides the student with summative and formative feedback i.e. an overall score and a context related summary on how well they completed the experiment and areas of improvement if needed. The student can use the touch controllers to navigate and explore the virtual circuit further e.g. relative voltage drops and current flow through individual components.

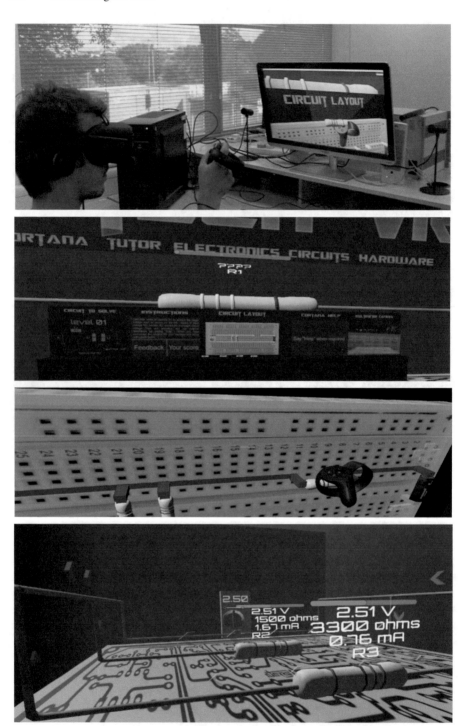

Fig. 6. Virtual reality/voice user interface and visualization of electrical phenomena

5   Deeper Integration with Unity Using Cognigy AI

The previous demonstration used the Microsoft Bot framework and Cortana as the virtual assistant. However, the current iteration of the Bot framework does not have a SDK for the Unity games engine. Cognigy AI offers a cross-channel voice and chat user interface with a deep Unity integration through its platform [12]. Figure 7 provides a high-level overview of the Cognigy integration in Unity. The plugin/extension works by using the computer microphone to record the students voice/input which is then parsed to text. The extracted text is sent to a Cognigy bot (as a flow) and associated with an intent. The intent processes the request and sends the response/action to complete back to Unity which then carries out any required actions and provides audio feedback to the users. The integration streamlines the development process and provides a better overall more cohesive user experience.

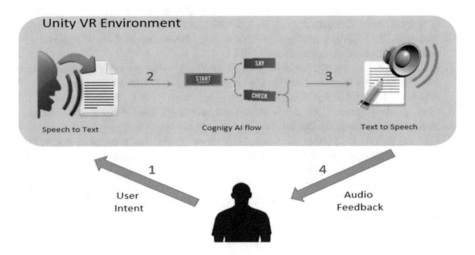

Fig. 7. Cognigy integration in Unity

6   Conclusion and Future Work

This paper explored the feasibility of using virtual assistants, voice user interfaces and virtual reality in campus based engineering laboratories to tutor and assess students. Practical case studies were presented based on the modification of an existing remote laboratory where a virtual reality front end with virtual assistants were added. The working examples shown demonstrated how this approach could be used to guide a student through an experiment, providing supplementary teaching resources and help when requested while accessing and controlling remote test instrumentation and hardware. It also demonstrated how voice user interfaces integrated into a virtual reality environment could be used for summative and formative assessment and to provide feedback to students. This area of research is set to grow rapidly as virtual assistants, consumer virtual reality and related peripheral devices become mainstream driven by low cost consumer hardware and cloud based services. Future work on this project will focus on developing formal, structured approaches to the creation of virtual assistant/voice user interfaces/virtual reality environments for engineering laboratories, investigating how this approach could be integrated with existing and widely used remote laboratory infrastructures and frameworks and exploring possible uses of these technologies to improve accessibility and access for students with disabilities.

References

1. De Floriani, L., Schmalstieg, D.: IEEE Tran. Visual. Comput. Graph. **23**(4), v–v (2017). IEEE Virtual Reality 2017 Special Issue
2. Bouziane, A., Bouchiha, V., Doumi, N., Malki, M.: Question answering systems: survey and trends. Procedia Comput. Sci. **73**, 366–375 (2015) ISSN: 1877-0509
3. Khillare, S.A., Pundge, A.M., Mahender, N.C.: Question answering system, approaches and techniques. Int. J. Comput. Appl. **141**(3), 34–39 (2016)
4. Microsoft. 2017 Cortana Intelligence services. https://azure.microsoft.com/en-gb/overview/cortana-intelligence. Accessed 03 Dec 2017
5. Halimi, W., Salzmann, C., Jamkojian, H., Gillet, D.: Enabling the automatic generation of user interfaces for remote laboratories. In: Auer, M., Zutin, D. (eds.) Online Engineering & Internet of Things. Lecture Notes in Networks and Systems, vol 22. Springer, Cham (2018)
6. Cortana Intelligence. https://www.microsoft.com/en-gb/cloud-platform/what-is-cortana-intelligence. Accessed 03 Dec 2017
7. Lindsay, E., Liu, D., Murray, S., Lowe, D.: Remote laboratories in engineering education: students' perceptions. In: Proceedings of the 18th Australian Association for Engineering Education (AaeE 2007) (2007)
8. Lowe, D., Murray, S., Lindsay, E., Liu, D.: Evolving remote laboratory architectures to leverage emerging internet technologies. IEEE Trans. Learn. Technol. **2**(4), 289–294 (2009). https://doi.org/10.1109/tlt.2009.33
9. Connect Raspberry Pi to Azure IoT Hub. https://docs.microsoft.com/en-us/azure/iot-hub/iot-hub-raspberry-pi-kit-node-get-started. Accessed 03 Dec 2017
10. Callaghan, M.J., Harkin, J., Prasad, G., McGinnity, T.M., Maguire, L.P.: Integrated architecture for remote experimentation. In: International Conference on Systems, Man and Cybernetics, vol. 5, pp. 4822–4827. IEEE (2003). https://doi.org/10.1109/icsmc.2003.1245746
11. Callaghan, M.J., Harkin, J.G., Scibilia, G., Sanfilippo, F., McCusker, K., Wilson, S.: Experiential based learning in 3D virtual worlds: visualization and data integration in second life. In: Remote Engineering and Virtual Instrumentation (REV 2008) Conference, Dusseldorf, Germany (2008)
12. Cognigy AI. https://www.cognigy.com/. Accessed 03 Dec 2017

Using Unity 3D as the Augmented Reality Framework for Remote Access Laboratories

Mark Smith, Ananda Maiti, Andrew D. Maxwell,
and Alexander A. Kist$^{(\boxtimes)}$

School of Mechanical and Electrical Engineering, USQ, Toowoomba, Australia
{mark.smith,andrew.maxwell}@usq.edu.au,
anandamaiti@live.com, kist@ieee.org

Abstract. Constructing augmented reality systems for remote access laboratory environments may seem daunting to many institutions. Utilizing open source tools may benefit from the large user and developer base, providing advice and support. Platforms such as Unity 3D provide comprehensive resources to developers when deciding how to construct a working framework, such as 3D graphic rendering, audio generation with built-in software tools. Comprehensive computer vision tools written with Unity 3D's C# or JavaScript compiler provide developers with the necessary Augmented Reality interfaces and feedback. This work discusses the Unity 3D framework, and the methods required to construct functional augmented reality support for remote access laboratories.

Keywords: Augmented reality · Computer vision · Remote access laboratories
STEM · Unity 3D

1 Introduction

Remote Access Laboratories (RAL) is a force multiplier for education facilities with underutilized and expensive laboratory equipment. RAL systems provide the mechanism for laboratory and workshop equipment to be accessed from a distance. Accessing equipment remotely allows for greater utilization and extends the reach of the practical aspects of the learning outcomes. Current development of RAL has occurred because of engineering departments interests in the technology and the enthusiasm of the student cohort [1]. Each RAL implementation looks feels and acts differently to other systems in the diverse range of remote experiments. Augmented Reality (AR) provides the experience of reality combined with computer generated virtual objects which enhances the sensory data to the user. Merging RAL and AR provides users of remote experiments with interactive sensory feedback. The sensory information may provide the user with additional data which enhances their experience, and creates an immersive experience.

Many RAL developers look to find a solution to the methods of implementation, not only for their RAL systems but also for including AR capabilities. Purpose built RAL applications are common, especially when the faculties proliferate with willing students ready to become involved. Regardless of the skill-set, the application of AR into the RAL environment is difficult. Continual growth of RAL systems requires the

M. E. Auer and R. Langmann (Eds.): REV 2018, LNNS 47, pp. 581–590, 2019.
https://doi.org/10.1007/978-3-319-95678-7_64

enthusiasm of both STEM and non-STEM fields. Attracting developer's means addressing not only the complexity of the design problem, but also engaging all other stakeholders. Augmented reality uptake for RAL systems has been slow with minimal concern in the final product. Even though AR is capable of engaging students at a deeper level, a lack of simple tools and skill forces AR to the shadows of RAL developers. A set of open source development tools are required which are accessible to all levels of user [2].

This work focuses on Unity 3D as an open source platform on which to build AR RAL systems. The object models of AR and RAL are discussed and catalogued to understand the requirements of any future supporting framework. Unity 3D's function and attributes are examined with a view for successful interfacing with the AR RAL model.

This paper is structured as follows. Section 2 provides a brief overview of remote access laboratories and augmented reality, with a summary of Unity 3D. Section 3 describes the AR RAL object model, while Sect. 4 examines Unity 3D's attributes. Section 5 defines an interface of AR RAL with Unity 3D, and Sect. 6 describes a simple sample configuration. Section 7 concludes this paper.

2   Related Work

While both remote access laboratories and augmented reality have rich histories, very little work exists combining the two. The first recognizable implementation of AR with RAL occurs in 2011, when Andujar [3] demonstrates a Field Programmable Gate Array (FPGA) and a microprocessor based control system. This RAL system exhibited the benefits of including AR with an elegant solution.

In recent years, surveys comparing real laboratories with remote or virtual laboratories have attempted to understand and measure the benefits and impact on the student cohort. An appreciation of AR systems for RAL has been difficult because there are too few configurations. Demonstration systems have been developed for assessment [4] and for concept demonstrations [5]. Straightforward physics or engineering experiments have so far been the leaders in AR RAL systems. Simple yet effective RAL experiments such as civil engineer projects focused on water seepage [6, 7], and electrical circuit simulation [8] have been very effective.

There are no current published works surrounding the open source package, Unity 3D. The product does have a very large developer base, supported by the Unity 3D site [9], and several other forums and web sites. Apart from using Unity 3D to introduce students to Object Orientated Programming [10], there has been no formal application of Unity 3D within the academic environment.

3   Augmented Reality and RAL Object Model

There is a lack of research regarding the functional models of AR and RAL. Some individual RAL installations have demonstrated an understanding of the experiments functional systems [11]. In general, an understanding of the simplified RAL functional model and the combined AR RAL functional model is necessary before building new experiments, or upgrading existing systems. Functional models identify the capabilities and activities of the systems, and provide an understanding of the interfaces necessary for integrating into secondary systems. A consistent approach to future enhanced experiments is possible.

3.1 Remote Access Laboratory

A simplified remote access laboratory functional model concerns its self with the two primary functions of sending commands to the equipment, and receiving measurement data. Many RAL installations rely on seeing the experiment unfolding, so the final functional requirement becomes live video streaming. Consider Fig. 1, in which the basic functions of a remote experiment can be summarized. From the user's point of view, there appears to be only the input (user inputs) and output (experiments results). Inputs are from the user or server system, which command the experiment equipment. User inputs are processed by the server prior to dispatch to the equipment. Outputs are the data received from the experiment. The video stream is not generally recognized as an output, but is an important output from the experiment. Video streams are the primary source of sensory feedback.

3.2 Augmented Reality

Augmented reality systems can become significantly more complex due to the senses engaged, and the critical timing required. From Fig. 2, the majority of AR processing centers on vision feedback; which provides the users with the greatest sense of immersion within the environment. Dedicated vision analysis systems are required to extract meaningful information from the video scene. Computer vision models perform a range of tasks such as image sharpening, edge detection, and image segmentation in an effort to locate and track objects within the video frame. An AR system functions through the generation of Virtual Objects (VO's). A VO generator creates sensory feedback, which may consist of virtual visual, aural, and physical data. Coordination and synchronization of all sensory feedback is critical and must be consistent with reality.

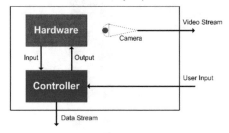

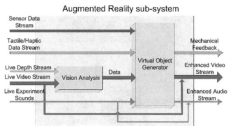

Fig. 1. Simplified RAL interface   **Fig. 2.** Simplified augmented reality interfaces

3.3   Combined Augmented Reality and Remote Access Laboratory

An augmented reality remote access laboratory functional model combines and merges key requirements of both systems. Simplifying the inputs to the AR RAL, as shown in Fig. 3, produces two important streams. The experiments data stream is the primary function of the RAL system. Data must be reliable and accessible to subsystems which may record the data or act upon it. Sensory data related to the experiment provides the second input. Video streaming tops the list of sensory inputs, but other inputs may include the sounds made by the experiment, and any feel or touch (such as temperature).

Outputs from the AR RAL functional model can be simplified as the command stream and the sensory stimulation stream. Instructions to the experiment devices propel the execution of the practical lesson. Sensory stimulation data consists of the enhanced streams such as real and virtual video data. Haptic and audio data may also be present on the sensory stimulation output.

Internally, an augmented reality remote access laboratory functional model is required to create information from the various input data streams. Command and control systems remain the same between RAL and AR RAL systems, with the exception that the controller parses data to the VO generator. This link helps to synchronize VO's with the commands sent to the equipment. Sensory data inputs consist of visual, aural and touch. Each may require preprocessing to extra meaning from the data stream. As shown in Fig. 2, the video feed is analyzed to extract knowledge about the scene. This may involve detecting objects within the scene and even tracking them. This information becomes important to not only generate computer generated feedback, but can become a valid measurement or data set for the experiment results.

Vision analysis processes can be decomposed into four main functions (see Fig. 4). There is a large resource of Computer Vision (CV) models developed over the decades, and the complexity of each function can become significant based on the level of accuracy and reliability required. Object detection is the principal focus for most CV work, where physical objects within the scene are isolated for future processing. The frame of reference of the video scene is equally important so as to ascertain the location of detected objects. The three-dimensional coordinates may also be necessary for future

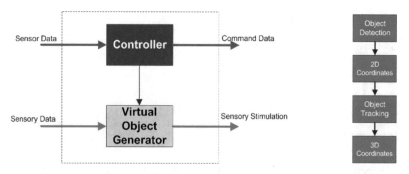

Fig. 3. Simplified AR RAL functional model   **Fig. 4.** Vision analysis processes

processing. Object tracking can assist in determining the 3D position of an object, if the frame of reference is known.

Augmented reality for RAL systems becomes actuality from the VO generator. Compiling the data from the available streams, computer generated enhancements are created by the virtual object generator. Any virtual objects must be properly incorporated into the sensory stream in a timely manner. Enhancements need to provide improved knowledge to the user without overwhelming or confusing the senses.

Unity 3D scripting language is suitable for the processing of internal augmented reality functionality.

4   Unity 3D Functions and Attributes

Unity is an open source cross-platform game development environment by Unity Technologies, providing degrees of complexity suitable for every level of developer. The graphical user interface supports drag-n-drop of game elements to create comprehensive two or three-dimensional worlds. The system also supplies scripting tools for greater control and distinctive interaction of the various elements [9]. The Unity system targets the multimedia libraries of the supported operating system platforms, allowing cross-platform development. The personal edition of Unity 3D was utilized for all testing. Initial assessments were with Unity version 5.2.1f1.

Summarized in Table 1 are relevant functions centered on visual processing. The Unity platform is based around the graphics engine. By default, Unity provides a 3D environment for developers. Figure 5 shows the X/Y/Z transform for every object which simplifies 3D calculations and motion. Two dimensional environments are also available and selected as an option when creating the environment. Both 2D and 3D environments are easily programmed and controlled through Unity's unique Game Object, Assets and Component model.

Table 1. Unity 3D features suitable for AR

2D Graphics	While not the normal state for object development, 2D world environments are provided by the graphics engine
3D Graphics	The primary option for developers is the 3D world environment. All object within Unity consist of 3D position, scale and rotation
PhysX	The popular NVidia's PhysX engine provides developers with access to objects that have attribute of mass, friction, gravity and velocity. This simplifies development as the engine maintains all attributes
SDK	Software Development Kits available for Unity allow developers to construct new features/capabilities which can be shared with other developers
Scripting	Several scripting languages are available. Mono is the current main language, exploiting the Microsoft .NET framework, and the large user base
Net Sockets	Access to the web sockets, UDP and TCP of Unity's transport layer, providing networking capabilities

Unity supports Mono [12] which is the open source version of the Microsoft .NET framework. Microsoft C# provides the object orientated programming (OOP) capabilities. Unity includes several namespaces to access the various supported engines and capabilities. Third party libraries are also accessible to both the scripting engine and Unity's project assets. Classes defined in the scripting language may access Unity Game Objects, influencing the objects attributes as the program executes. Unity allows defined C# classes to be added to objects as components. Public variables are visible to the developer on the Inspector panel. Shown in Fig. 5 is an example of the C# script mainButtons added as a component of the selected Game Object. Four public variables are also visible (Quit Now, Button, Icon, and My Skin) which allow the developer to set conditions or parse other Unity objects. More than one class can be added to the object at a time.

Each of Unity's objects consists of a collection of components as shown in Fig. 5. Components can be added or removed based on the needs of the project design. As objects are created and adjusted with additional components or attributes, they can be added to other objects, or become a set of resources for use by other objects. Important in OOP systems is the reuse of elements. Most Unity objects may become a parameter of another object, its operation changing based on the parent object.

5   Interface Model

Employing Unity 3D as a framework for augmented reality based remote access laboratories requires a clearly defined structure on which to develop. Once an understanding of the basic functional AR RAL model is formed, developers can focus on the methods to interface Unity 3D to their remote experiments.

Due to Unity's scripting capabilities, writing AR processes is greatly simplified. New or existing computer vision models will function just as well when written in C#. Interfacing the AR RAL functional model to the Unity framework requires mapping AR RAL data and processes to Unity features (see Fig. 6). Primary sensor data streams from the experiment, including video streams, will still be processed by C# classes and methods. Unity provides the infrastructure, and the means of generating virtual content.

Unity does not contain a specific video player object; however, there are methods available to display a video stream. Unity 5.6 includes a Movie Texture object, but it does not accept the most common video formats. This new feature was not tested. I have chosen another path. With a simple video streaming DLL library written for the purpose, each video frame triggers an event. The event handler within Unity's scripting environment receives a bitmap of the current frame. As shown in Fig. 7, the video frame is applied to a Texture object within the Unity scene. The frame is also sent to the computer vision analysis sub-routines for processing. After vision analysis, the extracted data is sent to the VO Generator.

Experimental data from the equipment sensors or other sources are passed unhindered to the experiments recording systems. The data is also supplied to the virtual object generator for two reasons. Firstly, the data helps to maintain synchronization between the VO's and the events occurring within the experimental environment. Secondly, key information may exist in the data that is necessary in the type of virtual objects presented to the user. For example, certain objects may appear or change when specific data measurements are made, such as virtual dials moving to indicate the measured values.

Other sensory data streams are processed the same, regardless of the Unity framework. The VO generator accepts and processes the data, or passes it through to the output. Unity includes an audio system, and can provide virtual audio objects if necessary. Programming the networking features of Unity allows the developer flexibility over the transfer of experimental data as well as the AR RAL functionality. Unity is well positioned to provide client/server capabilities as well as data streaming.

6   Building a Unity AR RAL System

A simple application was created to test the notion of Unity's suitability as an AR RAL framework. The interface had only three requirements; it must stream video, the video must be able to be analyzed by computer vision models, and it must pass user choices to the computer vision systems. Constructing a Unity based augmented reality remote access laboratory begins with the Unity 3D editor. All elements of the environment are created or accessed from the editor.

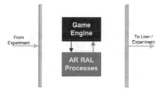

Fig. 6. Simplified relationship between Unity and AR RAL processes

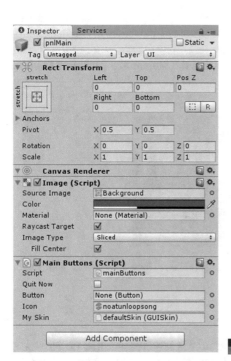

Fig. 5. Unity 3D Inspector panel showing a scripting class (Main Buttons)

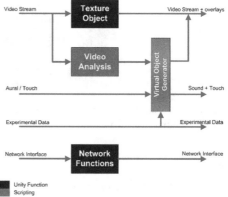

Fig. 7. Unity and augmented reality for remote access laboratory interfaces.

6.1   Video Streaming

Unity 3D version 5.2.1f1 does not support video streaming or popular image formats unless they have been previously imported into the Unity editor. Accessing the image files or video streams in real time during execution is not possible without a specific file loader. The file loader needs to convert an image into a byte array formatted for Unity's texture or sprite objects. The Unity community has constructed such utilities, and provides advice about such questions.

A small video player was locally developed to accept video streams. The purpose built library only provided a single public event to allow access to the current video frame. The frame was available as a bitmap image, which required conversion before Unity could utilize it. After reviewing historic queries within the Unity forums, the bitmap representation of the current video was successfully converted and applied to a Texture2D object.

6.2   Frame Analysis

Computer vision analysis of the incoming video stream accepts the bitmap image supplied by the video player library. The current frame is subjected to whichever analysis is required by the developers. For this simple implementation, tracking of the red pivot gear was performed using colour histogram colour zone segmentation and hotspot tensor tracking.

The RawImage object was placed on the canvas in the centre of the screen (see Fig. 8) and the mediaPlayer script attached. The script redrew the image on each Unity frame. The script also accepted the media player's new frame, sending the frames bitmap to the tracking routines and also converting the bitmap to a Texture2D object.

Fig. 8. Gear assembly remote experiment played within the Unity framework, with object tracking the red gear pivot

6.3   User Inputs

For the colour histogram tracking model, the user is required to click on an object within the video frame. The object's colour histogram is computed and for each frame, the image is segmented based on the histogram. Hotspot tensor tracking then follows the hotspot on each frame. Unity 3D has many options to receive user inputs. Simply tracking the mouse over the video frame and recording the position was enough to locate the users requested object.

Successful tracking of the selected object demonstrated that user inputs can be quickly and precisely activate actions within the augmented environment. A tracking box was drawn and followed the selected item through the video stream.

This sample environment, while simplistic, achieved success in the three requirements to validate Unity's suitability as an AR RAL platform. The tracking image overlay demonstrates the ease with which Unity can generate virtual objects. More comprehensive overlays and augmentation is apparent.

7 Conclusion

The very nature of using a gaming development platform as the framework for augmented reality systems means that numerous visual and audio capabilities will be available for use. Unity 3D provides a robust framework for generating images and sounds, suitable as virtual objects within an augmented environment.

Unity 3D application development occurs via a comprehensive graphical user interface that allows the user to immediately see the results of their actions and choices. The simple GUI allows all levels of developers to become involved with creating AR content. Drag and drop for most elements within the Unity development environment means that designs can be quickly trialed. Preconstruction of virtual objects, creating a library of useful images, animations and sounds permits the developer to rapidly introduce the objects to the environment. Libraries of C# code for generic computer vision models, such as object detection and tracking may then drop into video streams and initiate enhanced video feedback to the user.

Comprehensive video and audio processing systems are in-built to Unity. Video analysis and processing is historically processor intensive, yet Unity is tuned to achieving good results through its various engines. Required augmented reality enhancements are not expected to stretch the ample resources of the Unity systems.

The simple Unity AR RAL system developed has demonstrated the ease with which a generic framework can support future development. Video streaming, image overlaying and user interaction with the environment are the minimal key AR requirements, all tested and verified with minimal work. Current versions of Unity are reported to improve the image input capabilities, providing easier image handling mechanisms. Enhancing the simple video player would also create a robust library for further development.

This work has only touched on the suitability and basic capabilities of Unity 3D as a foundation for future augmented reality remote access laboratories. It is hoped that this initial work will inspire further research and the development of tools to expand the reach of augmented reality for remote access laboratories to many more institutions.

References

1. Maiti, A., Maxwell, A.D., Kist, A.A.: An overview of system architectures for remote laboratories. In: IEEE International Conference on Teaching, Assessment and Learning for Engineering (TALE), Kuta, Indonesia (2013)
2. Wu, H.-K., Lee, S.W.-Y., Chang, H.-Y., Liang, J.-C.: Current status, opportunities and challenges of augmented reality in education. Comput. Educ. **62**, 41–49 (2013)
3. Andujar, J.M., Mejías, A., Marquez, M.A.: Augmented reality for the improvement of remote laboratories: an augmented remote laboratory. IEEE Trans. Educ. **54**, 492–500 (2011)
4. Abu Shanab, S., Odeh, S., Hodrob, R., Anabtawi, M.: Augmented reality internet labs versus hands-on and virtual labs: a comparative study. In: 2012 International Conference on Interactive Mobile and Computer Aided Learning (IMCL), Amman, Jordan, pp. 17–21 (2012)
5. Benavides, X., Amores, J., Maes, P.: Invisibilia: revealing invisible data using augmented reality and internet connected devices. Presented at the UBICOMP/ISWC 2015 ADJUNCT, Osaka, Japan (2015)
6. Restivo, M.T., Cardoso, A.: Experiment@Portugal 2012 – ongoing activities. In: 2013 International Conference on Interactive Collaborative Learning (ICL), Kazan, Russia (2013)
7. Marques, J.C., Rodrigues, J., Restivo, M.T.: Augmented reality in groundwater flow. In: Presented at the REV 2014: 11th International Conference on Remote Engineering and Virtual Instrumentation, Porto, Portugal (2014)
8. Cardoso, A., Restivo, M.T., Quintas, M.R., Chouzal, M., de Fatima, Rasteiro, M., Marques, J.C., Menezes, P.: Online Experimentation: Experiment@ Portugal 2012. In: 11th International Conference on Remote Engineering and Virtual Instrumentation, REV 2014, Porto, Portugal, pp. 303–308 (2014)
9. Unity 3D, 15 September 2015. http://unity3d.com
10. Rogers, M.P.: Bringing unity to the classroom. J. Comput. Sci. Coll. **27**, 2012 (2012)
11. Maiti, A., Kist, A.A., Smith, M.: Key aspects of integrating augmented reality tools into peer-to-peer remote laboratory user interfaces. In: Presented at the REV 2016: 13th International Conference on Remote Engineering and Virtual Instrumentation, Madrid, Spain (2016)
12. Mono Project, 30 October 2017. http://mono-project.com

A Literature Review on Collaboration in Mixed Reality

Philipp Ladwig[(✉)] and Christian Geiger[(✉)]

University of Applied Sciences, 40476 Düsseldorf, Germany
{philipp.ladwig,geiger}@hs-duesseldorf.de

Abstract. Mixed Reality is defined as a combination of Reality, Augmented Reality, Augmented Virtuality and Virtual Reality. This innovative technology can aid with the transition between these stages. The enhancement of reality with synthetic images allows us to perform tasks more easily, such as the collaboration between people who are at different locations. Collaborative manufacturing, assembly tasks or education can be conducted remotely, even if the collaborators do not physically meet. This paper reviews both past and recent research, identifies benefits and limitations, and extracts design guidelines for the creation of collaborative Mixed Reality applications in technical settings.

1 Introduction

With the advent of affordable tracking and display technologies, *Mixed Reality* (MR) has recently gained increased media attention and has ignited the imaginations of many prospective users. Considering the progress of research and enhancement of electronics over recent years, we inevitably will move closer to the ultimate device which will make it difficult to distinguish between the virtual world and reality. *Star Trek's Holodeck* can be considered as an ultimate display in which even death can take place. Such a system would provide realistic and complete embodied experiences incorporating human senses including haptic, sound or even smell and taste. If it were possible to send this information over a network and recreate it at another place, this would allow for collaboration as if the other person were physically at the place where the help is needed.

As of today, technology has not yet been developed to the level of Star Trek's Holodeck. Olson and Olson [24,25] summarized that our technology is not yet mature enough, and that "distance matters" for remote collaboration. But many institutes and companies have branches at different locations which implies that experts of different technical fields are often distributed around a country or even around the world. But the foundation of a company lies in the expertise of their employees and in order to be successful, it is critical that the company or institute shares and exchanges knowledge among colleagues and costumers. Remote collaboration is possible via tools such as *Skype, DropBox or Evernote*, but these forms of remote collaboration usually consists of "downgraded packets of communication" such as text, images or video. However, machines, assembly

© Springer International Publishing AG, part of Springer Nature 2019
M. E. Auer and R. Langmann (Eds.): REV 2018, LNNS 47, pp. 591–600, 2019.
https://doi.org/10.1007/978-3-319-95678-7_65

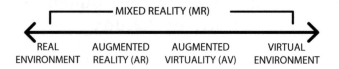

Fig. 1. Reality - Virtuality Continuum by Milgram and Kishino [19].

tasks and 3D CAD data are increasingly getting more complex. Exchanging 3D data is possible, but interacting remotely in real time on real or virtual spatial data is still difficult [24,25]. At this point MR comes into play, which have the potential to ease many of the problems of todays remote collaboration.

Milgram and Kishinio [19] defined the *Reality-Virtuality Continuum*, as depicted in Fig. 1, which distinguish between four different stages: Reality is the perception of the real environment without any technology. *Augmented Reality* (AR) overlays virtual objects and supplemental information into the real world. An example of an AR device is *Microsoft HoloLens*. *Augmented Virtuality* (AV) captures real objects and superimposes them into a virtual scene. A video of a real person, showed in a virtual environment, is an example for AV. *Virtual Reality* (VR) entirely eliminates the real world and shows only computer generated graphics. *Head Mounted Displays* (HMD) such as the *HTC Vive* or *Oculus Rift* are current examples of VR devices. This paper focuses on *Mixed Reality* which is defined by Milgram and Kishinio as a blend between AR and AV technology.

In the last three decades, research has shown a large amount of use cases for collaboration in MR: Supporting assembly tasks over the Internet [2,3,7,23,34], conducting design reviews of a car by experts who are distributed geographically [12,22] and the remote investigation of a crime scene [5,30] are only a few examples of collaborative applications in MR. Especially the domain of Remote Engineering and Virtual Instrumentation can benefit from remote guidance in MR. For example, many specialized, expensive and recent equipment or machines can only be maintained by highly qualified staff and are not often available at the location upon request if the machine happens to become inoperative. Furthermore, remote education could assist in the prevention of such emergency cases and help to spread specialized knowledge more easily.

The following sections chronologically describe the progress of research over recent decades. A predominant scenario can be observed in user studies: A remote user helps a local user to complete a task. Although, different authors use different terms for the participants of a remote session, we will use the abbreviations *RU* for remote user and *LU* for local user.

2   Research Until the Year 2012

A basis function of collaboration in every study examined in this paper is bidirectional transmission of speech. Every application uses speech as a foundation for communication. However, language can be ambiguous or vague if it describes spatial locations and actions in space. Collaborative task performance increases significantly when speech is combined with physically pointing as Heiser et al. [8] state. Some of the first collaborative systems, which uses MR, were video-mediated applications as presented by Ishii et al. [9,10]. A video camera, which was mounted above the participant's workplace, captured the work on the table and transmitted it to other meeting participants on a monitor. A similar system was developed by Kirk and Fraser [11]. They conducted a user study in which the participants had to perform a Lego assembly task. They found, that AR not only speeds up the collaboration task but it was also easier for the participants (in regards to time and errors) to recall the construction steps in a self-assembly task 24 h later when they were supported by MR technology instead of only listening to voice commands.

Baird and Barfield [2] and Tang et al. [34] prove that AR reduces the mental workload for assembly tasks. Billinghurst and Kato [4] reviewed the state of research on collaborative MR of the late 90's and concluded that there are promising applications and ideas, but that they scratch just only the surface of possibilities. It must be further determined, in which areas MR can be effectively used. Furthermore, Billinghurst and Kato mention that the traditional *WIMP-interface* (*Windows-Icons-Menus-Pointer*) is not appropriate for such a platform and must be reinvented for MR.

Klinker et al. [12] created the system *Fata Morgana* which allows for collaborative design reviews on cars and is capable to focus on details as well as compare different designs.

Monahan, McArdle and Bertolotto [20] emphasize the potential of *Gamification* for educational purposes: "Computer games have always been successful at capturing peoples imagination, the most popular of which utilize an immersive 3D environment where gamers take on the role of a character" [20]. Li, Yue and Jauregui [14] developed a VR e-Learning system and summarize that virtual "e-Learning environments can maintain students interest and keep them engaged and motivated in their learning" [14].

Gurevich, Lanir and Cohen [7] developed a remote-controlled robot with wheels, named *TeleAdvisor*, which carries a camera and projector on a movable arm. The RU sees the camera image, can remotely adjust the position of the robot and his arm with aid of a desktop PC and is able to project drawings and visual cues onto a surface by the projector. A robot, which carries a camera, has the advantage of delivering a steady image to the RU while a head-worn camera by the LU lead to jittery recordings, which can cause discomfort for the RU. Furthermore, a system controlled by the RU allows mobility, flexibility and eases the cognitive overhead for the LU, since the LU does not need to maintain the *Point-of-View* (PoV) for the RU.

To summarize this section, the transmission of information were often restricted until the year 2012 due to limited sensors, displays, network bandwidth and processing power. Many system rely on video transfer and were not capable of transmitting the sense of "being there" which restricts the mutual understanding of the problem and the awareness of spatial information.

3   New Technology Introduces a Sustainable Change

After the year 2012, more data became available for MR collaboration due to new technology. The acquisition and triangulation of 3D point clouds of the environment became affordable and feasible in real time. Better understanding of the environment results in more robust tracking of MR devices. Furthermore, display technology was enhanced and enabled the development of inexpensive HMDs. Tecchia, Alem and Huang [35] created one of the first systems which is able to record the workplace as well as arms and hands of the RU and LU with a 3D camera and allows the entrance of the triangulated and textured virtual scene by an HMD with head tracking. The system revealed improvements in performance over a 2D-based gesture system. Sodhi et al. [31] combines the *Mircosoft Kinect* and a short range depth sensor and achieved 3D reconstruction of a desktop-sized workplace and implemented a transmission of a hand avatar to the remote participant. Instead of a simple pointing ray, a hand avatar allows for the execution of more complex gestures, therefore delivering more information among the participants for creating a better mutual understanding.

Moreover, the system by Sodhi et al. [31] is capable of recognizing real surfaces. Understanding surfaces of the real environment allows for realistic physical interactions such as collision of the hand avatar with real objects such as a table. If the position of real surfaces are available within the virtual world, snapping of virtual objects to real surfaces is possible as well. This allows for decreased time in placing virtual object in the scene such as a furniture or assembly parts.

If the environment is available as a textured 3D geometry, it can be freely explored by the RU. Tait and Billinghurst [33] created a system which incorporates a textured 3D scan of a workplace. It allows the RU to explore the scene with keyboard and mouse on a monoscoping monitor and allows the selection of spatial annotations. It was found that increasing view independence (fully independent view vs. fixed or freeze views of the scene) leads to a faster completion of collaborative tasks and a decrease in time spent on communication during the task. Similar results are found by Lanir et al. [1] and explain: "A remote assistance task is not symmetrical. The helper (RU) usually has most of the knowledge on how to complete the task, while the worker (LU) has the physical hands and tools as well as a better overall view of the environment. Ownership of the PoV (*Point-of-View*), therefore, does not need to be symmetrical either. It seems that for helper-driven (RU-driven) construction tasks there is more benefit in providing control (of the PoV) to the helper (the RU)" [1].

Oda et al. [23] uses *Virtual Replicas* for assembly tasks. A Virtual Replica is a virtual copy of a real-existing, tracked assembly part. It exists in real life for the LU and it is rendered as a 3D model in VR for the RU. The position of the virtual model is constantly synchronized with the real environment. Many assembly parts of machines have complex forms and in some cases it is difficult for the LU to follow the instructions of the RU in order to achieve the correct rotation and placement of such complex objects. Therefore, virtual replicas, controlled by the RU, can be superimposed in AR for the LU which eases the mental workload for the task. Oda et al. found that the simple demonstration of how to physically align the virtual replica on another machine part is faster compared to making spatial annotations onto the Virtual Replicas as visual guidance for the LU which allows for an easier placement. Oda et al. employs physical constraints such as snapping of objects to speed up the task similar to Sodhi et al.

Poelman et al. [30] developed a system which is also capable of building a 3D map of the environment in real-time and was developed with the focus to tackle issues in remote-collaborative crime scene investigation. Datcu et al. [5] uses the system of Poelman et al. and proves that MR supports *Situational Awareness* of the RU. Situational Awareness is defined as the perception of a given situation, its comprehension and the prediction of its future state as Endsley described [6].

Pejsa et al. [26] created a life-size, AR-based, tele-presence projection system which employs the Microsoft Kinect 2 for capturing the remote scene and recreate it with the aid of a projector from the other participant's side. A benefit of such a system is that nonverbal communication cues, such as facial expressions, can be better perceived compared to systems where the participants wear HMDs which covers parts of the face.

Mueller et al. [21] state that the completion time of remote collaborative tasks, such as finding certain virtual objects in a virtual room, benefits by providing simple *Shared Virtual Landmarks*. Shared Virtual Landmarks are objects, such as virtual furniture, which helps to understand deictic expressions such as "under the ceiling lamp" or "behind the floating cube".

Piumsomboon et al. [28, 29] developed a system which combines AR and VR. The system scans and textures a real room with a Microsoft HoloLens and shares the copy of the real environment to a remote user who can enter this copy by a HTC Vive. The hands, fingers, head gaze, eye gaze and *Field-of-View* (FoV) were tracked and visualized among both users. Piumsomboon et al. reveal that rendering the eye gaze and FoV as additional awareness cues in collaborative tasks can decrease the physical load (as distance traveled by users) and make the task (subjectively rated by the users) easier. Furthermore, Piumsomboon et al. offers different scalings of the virtual environment. Shrinking the virtual copy of the real environment allows for a better orientation and path planning with help of a miniature model in the users hand similar as Stoakley, Conway and Pausch [32] show.

In summary, since technology has become advanced enough to scan and understand the surface of the environment in real time, important enhancements for collaboration tasks were achieved and attested as important for efficient remote work. 3D reconstruction of the participants' body parts and the environment allows for (1) better spatial understanding of the remote location (free PoV) (2) as well as better communication because of transmission of nonverbal cues (gaze, gestures) and (3) allows for incorporating the real surfaces with virtual objects (virtual collision, snapping). Furthermore, the 3D reconstruction of the environment implies better understanding of the environment which, in turn, leads to (4) more robust tracking of devices (phones, tablets, HMDs, Virtual Replicas) and (5) new display technologies enables more immersive experiences which lead to better spatial understanding and problem awareness for both users.

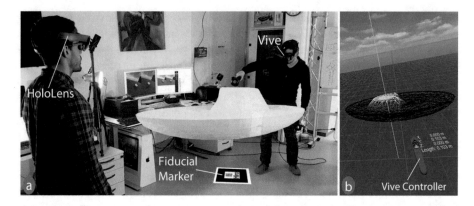

Fig. 2. (a) View of a third collaborator through his HoloLens: Users design a sail ship in a local collaboration scenario. One user is immersed by an VR HMD (HTC Vive) while his collaborators uses an AR device (HoloLens). (b) VR view of the Vive user: The sail ship in the middle and the Vive controller at the bottom can be seen.

4   Insights from a Development of a Collaborative Mixed Reality Application

We have developed an application in order to apply recent research outcomes and we want to share our lessons learned of combining two tracking systems. Our application is an immersive 3D mesh modeling tool which we have developed and evaluated previously [13]. Our tool allows creating 3D meshes with the aid of an HMD and two 6 *Degree-of-Freedom* controllers and is inspired by common desktop modeling applications such as Blender and Autodesk Maya. We have extended our system with a server-client communication which enables users with different MR devices to join a modeling session. Our tool can simulate how colleagues can collaboratively develop, review and discuss ideas, machine parts or designs.

It is created with the intent to be as flexible as possible. This includes: First, the users are free to choose an AR or VR device such as HTC Vive or Microsoft HoloLens. Second, the user can work with real objects, virtual replicas or entirely virtual items. Third, the system is capable to work locally in the same room, depicted in Fig. 2a, or remotely at different places.

A use case demonstrates how our system works and give insights of connecting and merging two different MR systems: A LU, using a HTC Vive, starts the modeling application and hosts a session. A RU scans a fiducial marker with his HoloLens in order to join the session. The marker has two purposes. First, it contains a QR code with connection details such as an IP address to the server. Second, it represents the origin of the tracking space of the remote Vive system. This allows the HoloLens user to place the virtual content of the server (content of the HTC Vive side) to any place in his real environment. Additionally, this approach also enables the user to synchronize the tracking spaces in the same room by placing the marker on the origin of the Vive tracking system, as shown in Fig. 2a.

Our first tests showed that we can successful merge two different tracking systems, such as the HTC Vive and the HoloLens, but we experienced some issues: The tracking system of the Vive interferes with the tracking system of the HoloLens as soon as the users approach closer than one meter to each other. It lead to tracking errors for the HTC Vive. Furthermore, we experienced that the HoloLens' processing power is limited due relative low technical specifications compared to a workstation which limits the complexity of the rendered scene. Moreover, we have identified that even the local network connection in our collaboration scenario in the same room reveals delays which are noticeable and could interfere with natural interaction, nonverbal cues and gestures.

5   Research Agenda, Technology Trends and Outlook

This paper has shown examples of remote collaboration which prove the performance and potential of MR. Although important enhancements and research results have been discovered in recent years, we still have a long way to go until we have achieved the ultimate display for collaboration - Star Trek's Holodeck.

A major concern of research, which up to this point has been scarcely investigated, is the collaboration between multiple teams. The focus in past research has been mainly conducted on collaboration between two persons, but how to exchange complex data and interact between multiple groups has yet to be researched further. Lukosch et al. [17] have taken the first steps in this direction but stated that further research is necessary. Piirainen, Kolfschoten and Lukosch [27] mention that a difficulty of collaborative remote work in teams is developing a consensus about the nature of the problem and specification. Situational Awareness cues and Team Awareness cues need to be outlined.

Another important point on the agenda is how to maintain focus of the users to certain events and parts in the environment. Awareness cues are in general an ongoing topic of research and must be investigated further. Müller, Rädle and Reiterer [21] ascertain that a technique is needed to put events, collaborators or objects into the users' focus, which are not in the field of view. Pejsa et al. [26] and Masai et al. [18] emphasize the importance of nonverbal communication cues such as facial expression, posture and proxemics which are important contributors to empathy but these cues are still difficult to transmit with today's hardware.

A relative rarely investigated field of research is comfort in MR, though it is an important area for the usage of an application over a long period of time. Up to this point, a real use case could look like this: A worker conducts a demanding assembly task for hours on an expensive machine by remote guidance. But the weight of the HMD, the usability of the application and the fatigue in his arms from making gestures for interacting with the device lead to a growing frustration by the worker which lead, in turn, to errors of the assembling. Piirainen et al. [27] advise not to underestimate user needs and human factors: "From a practical perspective the challenges show that the usability of systems is a key." Today, a general problem and consideration for every MR application is comfort for the user. Only a few years ago, VR and AR hardware was used to be bulky and heavy and research in regards of comfort was theoretically in vain. Research in MR is mainly focused on technical feasibility and compares productivity between non-MR and MR application. However, comfort and usability is important, if long-term applications are required, but research of comfort is scarce. Ladwig, Herder and Geiger [13] consider and evaluate comfort for MR application. Lubos et al. [15] revealed important outcomes for comfortable interaction and did first steps into this direction.

Moreover, perceiving virtual haptic is widely an unresolved problem in MR and researcher tries to substitute it with the aid of constraints such as virtual collisions and snapping, as Oda et al. shows [23]. Furthermore, Lukosch et al. [16] and Billinghurst [4] mention that further research is needed which particular tasks can be effectively solved and managed with MR.

Better tracking technologies, faster networks, enhanced sensors and faster processing will move us to the Holodeck and maybe even beyond. Further areas of research will arise with the advent of new technologies such as machine learning for object detection and recognition. MR devices of the future will not only recognize surfaces of the environment, but also detect objects such as machine parts, tools and humans.

6 Design Guidelines

Past research and our lessons learned revealed many issues which can be concluded into design guidelines for the development of MR applications:

Provide as Much Information About the Remote Environment as Possible. Video is a minimum requirement. A 3D mesh of the environment is better [5, 23, 28–31]. An updated 3D mesh in real-time seems to be the best case [35].

Provide an Independent PoV for Investigating the Remote Scenery. It allows better spatial perception and problem understanding [1, 28, 29, 33, 35].

Provide as Much Awareness Cues as Possible. Transmitting speech is fundamental. Information of posture of collaborators such as head position, head gaze, eye gaze, FoV [28, 29] is beneficial. For pointing by hand is a virtual ray sufficient but a static hand model [31] or even a full tracked hand model is better and conveys more information such as natural gestures [28, 29]. Provide cues for events happen outside the FoV of the users and provide Shared Local Landmarks [21]. To avoid cluttering the view of the users, awareness cues can be turned on and off [21].

Consider Usability and Comfort. If a long-term usage is desired, take a comfortable interface for the user into account and consider human factors [13, 15, 27].

References

1. Lanir, J., Stone, R., Cohen, B., Gurevich, P.: Ownership and Control of Point of View in Remote Assistance, p. 2243. ACM Press, New York (2013)
2. Baird, K.M., Barfield, W.: Evaluating the effectiveness of augmented reality displays for a manual assembly task. Virtual Reality **4**(4), 250–259 (1999)
3. Billinghurst, M., Clark, A., Lee, G.: A survey of augmented reality augmented reality. Found. Trends Hum. Comput. Interact. **8**(2–3), 73–272 (2015)
4. Billinghurst, M., Kato, H.: Collaborative mixed reality. In: Mixed Reality, pp. 261–284. Springer, Heidelberg (1999)
5. Datcu, D., Cidota, M., Lukosch, H., Lukosch, S.: On the usability of augmented reality for information exchange in teams from the security domain. In: Proceedings - 2014 IEEE Joint Intelligence and Security Informatics Conference, JISIC 2014, pp. 160–167. IEEE (2014)
6. Endsley, M.R.: Toward a theory of situation awareness in dynamic systems. Hum. Factors J. Hum. Factors Ergon. Soc. **37**(1), 32–64 (1995)
7. Gurevich, P., Lanir, J., Cohen, B.: Design and implementation of TeleAdvisor: a projection-based augmented reality system for remote collaboration. Comput. Support. Coop. Work (CSCW) **24**(6), 527–562 (2015)
8. Heiser, J., Tversky, B., Silverman, M.I.A.: Sketches for and from collaboration. Vis. Spat. Reason. Des. **III**, 69–78 (2004)
9. Ishii, H., Kobayashi, M., Grudin, J.: Integration of inter-personal space and shared workspace. In: Proceedings of the 1992 ACM Conference on Computer-Supported Cooperative Work - CSCW 1992, pp. 33–42. ACM Press (1992)
10. Ishii, H., Miyake, N.: Toward an open shared workspace: computer and video fusion approach of TeamWorkStation. ACM **34**(12), 37–50 (1991)
11. Kirk, D., Fraser, D.: The Effects of Remote Gesturing on Distance Instruction. Lawrence Erlbaum Associates, Mahwah (2005)
12. Klinker, G., Dutoit, A.H., Bauer, M., Bayer, J., Novak, V., Matzke, D.: Fata Morgana - a presentation system for product design. In: Proceedings - International Symposium on Mixed and Augmented Reality, ISMAR 2002, pp. 76–85. IEEE Computer Society (2002)
13. Ladwig, P., Herder, J., Geiger, C.: Towards precise, fast and comfortable immersive polygon mesh modelling. In: ICAT-EGVE 2017 - International Conference on Artificial Reality and Telexistence and Eurographics Symposium on Virtual Environments. The Eurographics Association (2017)
14. Li, Z., Yue, J., Jauregui, D.A.G.: A new virtual reality environment used for e-Learning. In: 2009 IEEE International Symposium on IT in Medicine & Education, pp. 445–449. IEEE (2009)
15. Lubos, P., Bruder, G., Ariza, O., Steinicke, F.: Touching the sphere: leveraging joint-centered kinespheres for spatial user interaction. In: Proceedings of the 2016 Symposium on Spatial User Interaction, SUI 2016, pp. 13–22. ACM (2016)
16. Lukosch, S., Billinghurst, M., Alem, L., Kiyokawa, K.: Collaboration in augmented reality. Comput. Support. Coop. Work CSCW Int. J. **24**(6), 515–525 (2015)
17. Lukosch, S., Lukosch, H., Datcu, D., Cidota, M.: Providing information on the spot: using augmented reality for situational awareness in the security domain. Comput. Support. Coop. Work (CSCW) **24**(6), 613–664 (2015)
18. Masai, K., Kunze, K., Sugimoto, M., Billinghurst, M.: Empathy glasses. In: Proceedings of the 2016 CHI Conference Extended Abstracts on Human Factors in Computing Systems, CHI EA 2016, pp. 1257–1263. ACM Press, New York (2016)

19. Milgram, P., Kishino, F.: A taxonomy of mixed reality visual displays. IEICE Trans. Inf. Syst. **12**(12), 1321–1329 (1994)
20. Monahan, T., McArdle, G., Bertolotto, M.: Virtual reality for collaborative e-learning. Comput. Educ. **50**(4), 1339–1353 (2008)
21. Müller, J., Rädle, R., Reiterer, H.: Remote collaboration with mixed reality displays. In: Proceedings of the 2017 CHI Conference on Human Factors in Computing Systems, CHI 2017, pp. 6481–6486. ACM Press (2017)
22. Nvidia: Nvidia Holodeck. https://www.nvidia.com/en-us/design-visualization/technologies/holodeck/. Accessed 25 Jan 2018
23. Oda, O., Elvezio, C., Sukan, M., Feiner, S., Tversky, B.: Virtual replicas for remote assistance in virtual and augmented reality. In: Proceedings of the 28th Annual ACM Symposium on User Interface Software & Technology, UIST 2015, pp. 405–415. ACM Press (2015)
24. Olson, G.M., Olson, J.S.: Distance matters. Hum. Comput. Interact. **15**(2–3), 139–178 (2000)
25. Olson, J.S., Olson, G.M.: How to make distance work work. Interaction **21**(2), 28–35 (2014)
26. Pejsa, T., Kantor, J., Benko, H., Ofek, E., Wilson, A.D.: Room2Room: enabling life-size telepresence in a projected augmented reality environment. In: Proceedings of the 19th ACM Conference on Computer-Supported Cooperative Work & Social Computing, CSCW 2016, pp. 1714–1723. ACM Press (2016)
27. Piirainen, K.A., Kolfschoten, G.L., Lukosch, S.: The joint struggle of complex engineering: a study of the challenges of collaborative design. Int. J. Inf. Technol. Decis. Mak. **11**(6), 1087–1125 (2012)
28. Piumsomboon, T., Day, A., Ens, B., Lee, Y., Lee, G., Billinghurst, M.: Exploring enhancements for remote mixed reality collaboration, pp. 1–5 (2017)
29. Piumsomboon, T., Lee, Y., Lee, G., Billinghurst, M.: CoVAR: a collaborative virtual and augmented reality system for remote collaboration. In: SIGGRAPH Asia 2017 Emerging Technologies on SA 2017, pp. 1–2. ACM Press (2017)
30. Poelman, R., Akman, O., Lukosch, S., Jonker, P.: As if being there. In: Proceedings of the ACM 2012 Conference on Computer Supported Cooperative Work, CSCW 2012, no. 5, p. 1267 (2012)
31. Sodhi, R.S., Jones, B.R., Forsyth, D., Bailey, B.P., Maciocci, G.: BeThere: 3D mobile collaboration with spatial input. In: Proceedings of the SIGCHI Conference on Human Factors in Computing Systems, CHI 2013, pp. 179–188 (2013)
32. Stoakley, R., Conway, M.J., Pausch, R.: Virtual reality on a WIM. In: Proceedings of the SIGCHI Conference on Human Factors in Computing Systems, CHI 1995, pp. 265–272. ACM Press, New York (1995)
33. Tait, M., Billinghurst, M.: The effect of view independence in a collaborative AR system. Comput. Supp. Coop. Work CSCW Int. J. **24**(6), 563–589 (2015)
34. Tang, A., Owen, C., Biocca, F., Mou, W.: Comparative effectiveness of augmented reality in object assembly. In: Proceedings of the Conference on Human Factors in Computing Systems, CHI 2003, p. 73 (2003)
35. Tecchia, F., Alem, L., Huang, W.: 3D helping hands: a gesture based MR system for remote collaboration. VRCAI Virt. Real. Continuum Appl. Ind. **1**(212), 323–328 (2012)

REMLABNET and Virtual Reality

Tomas Komenda[✉] and Franz Schauer

Faculty of Applied Informatics, Tomas Bata University, Zlin, Czech Republic
komendatomas@seznam.cz, fschauer@fai.utb.cz

Abstract. Our recent research in remote laboratory management systems (REMLABNET—www.remlabnet.eu) deals with questions such as how to make the user experience stronger and how to help users understand complex phenomena behind remote experiments and the laws of physics governing the experiment. At our current stage of technological development, we have both sufficiently powerful hardware and software to create an impressive virtual user interface which could be a help to this mission. An extended mixed reality taxonomy for remote physical experiments was proposed to identify goals of the future REMLABNET research and development. The first part of this paper describes classes of taxonomy and reasons why they were set up in this way. The second part mentions the chosen method of our research and our current progress.

Keywords: Remote physical experiment · Mixed reality · Virtual reality
User experience

1   Introduction

REMLABNET (version 2017) is a Remote Laboratory Management System (RLMS) which allows combining the research world of physics with modern elements of information technologies. The system provides remote control of both real experiments and simulation of experiments via a web interface and offers possibilities to process, observe, record and study results [1, 2]. These remote experiments include a server with a set of physical objects of study (rigs). Distance education offers a possibility to mix real objects with simulated ones and to create hybrid rigs (or to replace the whole process with simulation) [3]. The disadvantage is the absence of users in the laboratory. It is necessary to broadcast current events from the laboratory to users. Observation of both the rig and the measuring process is an essential part that in many cases leads to a real understanding of the task. The quality of such interaction has a major impact on the user's learning curve and their satisfaction with RMLSs. REMLABNET now provides a web interface with control elements and 2-D video streaming. We decided to extend REMLABNET and start a research project towards user interface and user experience. Rapid development and technological progress in virtual reality (VR) devices (e.g. smart glasses and VR helmets) bring new possibilities that RLMSs can offer to their users. VR has been linked to remote experiments for three decades [4]. One of the primary goals of virtual reality is to provide immersive environments that transfer participants from real life to a virtual one. This idea has the potential to improve remote physical experiments by providing a rich sense of interactive control and immersion in the environment. VR

© Springer International Publishing AG, part of Springer Nature 2019
M. E. Auer and R. Langmann (Eds.): REV 2018, LNNS 47, pp. 601–609, 2019.
https://doi.org/10.1007/978-3-319-95678-7_66

became a mature discipline combining computer graphics, computer vision, object recognition, simulation, artificial intelligence, and data mining. It also contains a lot of subcategories, e.g. augmented reality (AR), augmented virtuality (AV), mediated reality (MR), simulated reality (SR), and others. But to date, VR and RLMSs have not been well integrated because of a lot of technical and logical limitations [5]. Due to this complexity, it is an imperative to evaluate the risks and benefits of such a development. Our first steps were to identify REMLABNET in a mixed reality continuum and classify possible methods of research and subsequent development. For the purposes of REMLABNET and common development of remote physical experiments, we extend standard mixed reality taxonomy [8]. The first part of this paper describes classes of taxonomy and reasons why they were set up in this way. The second part mentions the chosen way of development and our current progress.

2 Mixed Reality Taxonomy

The taxonomy counts in important aspects and trends of modern RLMSs and remote-driven physical experiments, which REMLABNET is [2, 6]. On the other hand, we do not aim to cover the needs of more complex and extensive remote laboratory systems that combine multiple disciplinary approaches [7]. We proposed four classes based on user's devices (Table 1 and Fig. 1). Current technical maturity, accessibility, price, and backward compatibility with current systems were taken into account.

Table 1. Proposed classes

Class	Device	Short description
0.		A laboratory experiment or local computer-driven experiment. There is no remote access here. The user has to be present in the laboratory. This class is defined by the generally mixed reality taxonomy and is not a subject of this paper
1.	Monitor, display	A laboratory experiment is remotely controlled. Events and results are observed on a standard monitor or display (our current stage in REMLABNET)
2.	Monitor, display + augmented reality glasses (transparent and lightweight VR headsets)	This class is a superset of Class 1. Smart glasses allow seeing data displayed on a standard monitor through the glasses. Glasses bring an AR extension that can be related to events shown on the screen (a virtual RLMSs guide), events in the laboratory, and other users (e.g. 3-D mirroring of the rig, video chat with other users, collaborative learning, sharing of the results). This class guarantees a backward compatibility with Class 1 (it is possible to take the glasses off and use RLMS like Class 1)
3.	Virtual reality headset (helmet)	This class is a superset of Class 2; however, this class is not backward compatible with Class 1 or Class 2. There is no possibility of seeing the monitor through the headset. It is possible to take the headset off and use the monitor, but not both at the same time. A wide-angle stereoscopic camera can stream real events from the laboratory. This streamed scene can be extended by a hybrid or fully virtual object. In extreme cases, the whole scene can be entirely virtual

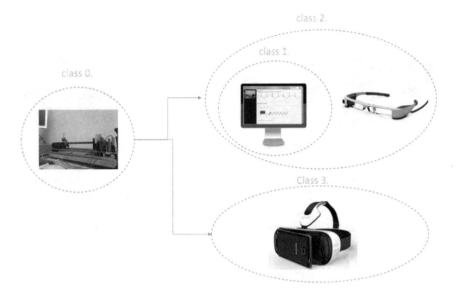

Fig. 1. Proposed class diagram

These classes do not cover all devices and their combinations (e.g. smartphones, tablets, smart lenses, holograms, and various types of VR helmets). It is obvious that a monitor can be replaced with a tablet and smart glasses with a smartphone. For REMLABNET purposes, we chose EPSON Moverio BT-300[1](smart glasses) as a representative of Class 2. As a representative of Class 3, we decided to use a combination of Samsung Galaxy S7 and Samsung Galaxy Gear[2] together with 180° wide-angle 3-D LucidCam[3] (Fig. 2).

To improve user's experience, it is important to identify the level of utilization of the system into relevant classes. For these objectives to be achieved, we can use S. Mann's mixed reality continuum [8]. Figure 3 (on the left) shows a two-dimensional continuum whose horizontal axis describes the degree of virtuality and vertical axis completes the degree of mediality. However, the user experience classification related to remote physical experiments requires more. RLMSs users remotely control and observe some events processed in the laboratory. Let's assume that a user is carrying out measurement of a simple pendulum. The rig is physically present in the laboratory.

---

[1] EPSON Moverio BT-300 Smart Glasses (AR/Developer Edition). Si-OLED display (0.43″ wide panel, 1280 × 720 × RGB, virtual screen size 80′), a high-resolution camera (5 million pixels), Intel Atom chipset, Android 5.1, price about $800 (April 2017). https://tech.moverio.epson.com/en/bt-300/.

[2] Samsung Galaxy Gear VR and Samsung Galaxy S7 Edge. 2560 x 1440 pixel Super AMOLED, 96° field of view, 60 Hz, price about $1000 ($900 + $100, April 2017). http://www.samsung.com/uk/smartphones/galaxy-s7/overview/, http://www.samsung.com/global/galaxy/gear-vr/.

[3] LucidCam 3-D VR camera (prototype). 4 K-30 fps/eye videos, wifi supported for the live stream, the range of view 180° × 180°, price about $500 (April 2017). https://lucidcam.com.

Fig. 2. Samsung Galaxy Gear (left), EPSON Moverio BT-300 (middle) and LucidCam (right)

A simple pendulum is moving. The results are taken via sensors. The user can watch a video stream of events. In this case, the situation is categorized as real reality (RR). If the video stream is complemented with the information about results (charts, numbers, tables), the RR is mediated. However, it is possible not to stream video and show virtual models of the rig (2-D or 3-D) instead, and the events in the laboratory can be demonstrated on these models (there can be many reasons for this attitude, e.g. high network latency, fault of video devices, etc.). In this case, the user observes an entirely virtual model which mirrors real events of the rig. The classification is mediated virtual reality. It does not matter whether the user uses a common monitor or display. The recent trend in remote physical experiments is to combine or completely replace physical rigs (or their parts) with simulation [9]. This innovation is based on the idea of Simulation-based Learning (SBL). Due to this, mixed continuum has been extended into a three-dimensional space (Fig. 3 right). This space is not binary. Instead, it is more of a continuum, with many variations that are possible between the two extremes. For the sake of simplicity and clarity, however, the space is divided into discrete quadrants. Any quadrant can be perceived as a discrete point (it is possible to reach this stage—for purposes

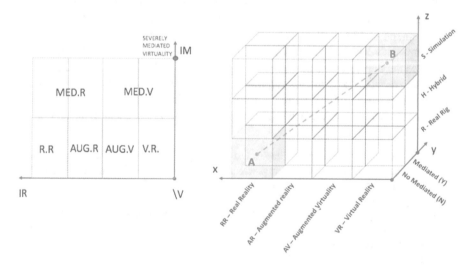

Fig. 3. Mixed reality continuum (left) and its extended version for remote physical experiments (right)

of future REMLABNET development we chose this approach) or as a continuum with a potentially more detailed classification.

We proposed a classification based on the class number and the relevant quadrant (*CLASS.X.Y.Z*). We can see two extremes in this taxonomy. *0.RR.N.R* (Class 0 and a quadrant marked as *A* in Fig. 3) would be a local experiment (observed by user's own eyes) running at the real rig. *1.RR.N.R* (Class 1) would be a remote experiment observed via 2-D video stream (watched on a common monitor or display). On the other hand, *3.VR.Y.S* (Class 3 and a quadrant marked as *B* in Fig. 3) would be a completely simulated experiment (observed via a VR headset consisting only from virtual elements), which is so realistic that users are not able to decide whether they are carrying out the measurement via simulation or they are observing a mirroring of the real rig. In the most extreme point of this stage, this remote experiment would be very close to the idea of simulated reality [10].

3   REMLABNET and the Extended Taxonomy

For a long time, REMLABNET was classified as *1.RR.Y.R*. It was possible to control and observe only real rigs via a web interface (HTML, JavaScript) [1]. However, students sometimes have problems understanding the complex phenomena behind remote experiments and the laws of physics governing these experiments. To provide insight into the physics of a certain phenomenon, a mixed multiparameter simulation, working with analytical formulations of the respective law, can be of great help. For this purpose, we designed a new module and implemented it into the remote experiment software, enabling simulations embedding into remote experiments outputs [1] (Fig. 4).

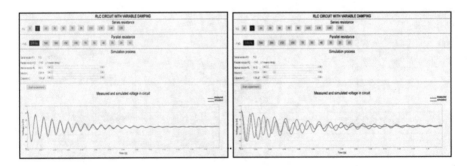

Fig. 4. The time domain response of RLC circuit to step Remote experiment (red) with embedded simulation (blue). Remote experiment data not fitted by simulation data (left) and remote experiment data 100% fitted by simulation data (right)

According to the proposed taxonomy, the embedded simulation moved REMLABNET into *1.RR.Y.H* (Fig. 5).

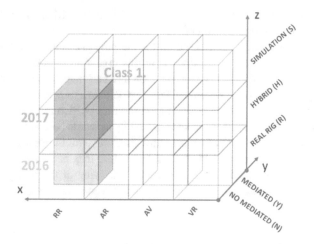

Fig. 5. REMLABNET (version 2016 and 2017) classification

4 REMLABNET and Migration from Class 1 to Class 2

To enhance user experience with the potential for better understanding of complex phenomena behind remote experiments, one of our current REMLABNET projects focuses on the migration of user interface from Class 1 to Class 2 (with backward compatibility). The complete list of steps requiring this research and the simultaneous development process is shown in Fig. 6.

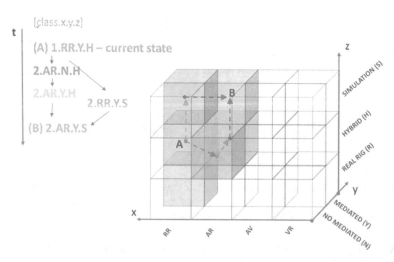

Fig. 6. REMLABNET (version 2017) and the diagram of migration from Class 1 to Class 2

The transition from stage *1.RR.Y.H* to *2.AR.N.H* reveals a tremendous potential of AR. We experiment with the mirroring of the rig and related events in the laboratory.

A virtual 3-D copy of the real rig can be watched via smart glasses and the rig can be remotely controlled via smart glass sensors. For these objectives, EPSON Moverio BT-300 smart glasses were elected as a mature and affordable device. In this step, a prototype is being developed in Unity3D[4]. Unity supports stereoscopic rendering for different VR devices, including the Moverio BT-300 or Samsung Gear VR. In this stage, the control of the experiment is in cooperation with computer – keyboard and mouse (same as in Class 1). However, we are currently researching reasonably advanced handling (e.g. a hand gesture could set a virtual pendulum in motion and this action could be transmitted into the real remote rig). The transition to stage *2.AR.Y.H* will bring the mediality of AR. Smart glasses extend the exploitable space where various measurement-related metrics and charts can be displayed. Then a way for remote cooperative learning is opened. Moverio BT-300 (like other similar smart glasses) has a high-resolution camera. Because users have their smart glasses on while measuring, everything that is seen (including the monitor, keyboard, mouse, their hands and virtual elements of the scene) can be streamed to the remote teacher or other students. The following picture (Fig. 7) shows the proposed architecture of REMLABNET in this stage.

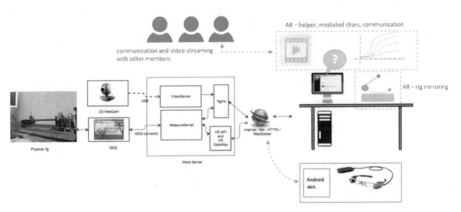

Fig. 7. Proposed schematic representation of a federated Remote Laboratory Management System (RLMS) – REMLABNET (www.remlabnet.eu), extended into *2.AR.Y.H* via EPSON Moverio BT-300 and a cooperative education module

Full replacement of the real rig via simulation is not possible yet, but mixed embedded simulation is the first step that has to be done before *1.RR.M.S* and *2.AR.M.S* will be reached. These stages offer a possibility to run the measurement either via the real rig or a fully simulated environment. High acquisition and maintenance costs of redundant rigs (to provide high availability and scaling of experiments) can be significantly reduced. Simulation can make this remote experiment accessible in case the real rig is out of order or being used by another user. This stage will be discussed in more details in future.

---

[4] http://www.unity3d.com.

5   Potential of Class 3 and Future Work

Class 2 can "take parts of the laboratory to users", but this ability together with the transparency of the smart glass display allows only AR stages to be reached. Class 3, unlike Class 2, provides the capacity to assemble user interface through the whole spectrum of a mixed reality continuum. Due to 3-D 180° wide-angle cameras, which are able to stream three-dimensional video, it is possible to build a prototype that can "take users into the laboratory". On the other hand, it is possible to offer a severely virtualized and simulated environment. Unfortunately, a few technical limitations do not allow the production of remote experiments for this class on a large scale. The biggest flaws are usually the insufficient resolution of the camera or VR headsets or network limitation between users and the laboratory [5]. Theoretical research on this topic is being conducted. We are developing a prototype which will be able to stream 3-D wide-angle video into VR headset. The combination of Samsung Galaxy S7 and Samsung Galaxy Gear together with 180° wide-angle 3-D LucidCam is used for prototyping. Our effort is to achieve *3.AR.Y.H*, although currently we are in *3.RR.Y.R* (Fig. 8). Nevertheless, our priority is to make significant progress with Class 2 because the combination of the traditional approach (displays, monitors) and lightweight and transparent smart glasses is more natural for users (in spite of the fact that VR headsets will soon provide much more possibilities).

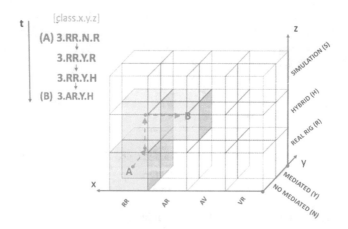

Fig. 8. REMLABNET (version 2017) and the diagram of migration into Class 3

Acknowledgment. The support of the project of the Swiss National Science Foundation (SNSF) - "SCOPES", No. IZ74Z0_160454 is highly appreciated. The support of the Internal Agency Grant of the Tomas Bata University in Zlin No. IGA/FAI/2017 for PhD students is acknowledged.

References

1. Schauer, F., Krbecek, M., Beno, M., Gerza. M., Palka, L., Spilakova, P., Komenda, T., Ozvoldova, M., Gerhatova, Z., Tkac, L.: REMLABNET IV - LTI federated remote laboratory management system with embedded multiparameter simulations. In: Proceedings of REV 2017, New York, USA, pp. 340–349 (2017)
2. The Consortium of Remlabnet (www.remlabnet.eu) is formed by three laboratories in Faculty of Education, Trnava University in Trnava, next, Faculty of Informatics, Tomas Bata University in Zlin and Faculty of Mathematics and Physics, Charles University in Prague
3. Poliakov, M., Henke, K., Wuttke, H.D.: The augmented functionality of the physical models of objects of study for remote laboratories. In: Proceedings of REV 2017, New York, USA, pp. 148–157 (2017)
4. Hradio, R., Torre, L., Galan, D., Cabrerizo, F.J., Herrera-Viedma, E., Dormido, S.: Virtual and remote labs in education: a bibliometric analysis. Comput. Educ. **98**, 14–38 (2016)
5. Smith, M., Maiti, A., Maxwell, A.D., Kist, A.A.: Object detection resource usage within a remote real time video stream. In: Proceedings of REV 2017, New York, USA, pp. 792–803 (2017)
6. Schauer, F., Krbecek, M., Beno, P., Gerza, M., Palka, L., Spilaková, P., Tkac, L.: REMLABNET III - federated remote laboratory management system for university and secondary schools. In: Proceedings of 13th International Conference on Remote Engineering and Virtual Instrumentation, REV 2016, Madrid, pp. 232–235 (2016)
7. Orduña, P., Zutin, D.G., Govaerts, S., Zorrozua, I.L., Bailey, P.H., Sancristobal, E., Salzmann, C., Rodriguez-Gil, L., DeLong, K., Gillet, D., Castro, M., López-de-Ipiña, D., García-Zubia, J.: An extensible architecture for the integration of remote and virtual laboratories in public learning tools. IEEE Revista Iberoamericana De Tecnologias Del Aprendizaje **10**(4), 223–233 (2015)
8. Mann, S.: Mediated Reality with implementations for everyday life. In: Presence: Teleoper. Virtual Environ. (the online companion to the MIT Press journal) (2002). presenceconnect.com
9. Schauer, F., Gerza, M., Krbecek, M., Ozvoldova, M.L.: Remote Wave Laboratory - real environment for waves mastering. In: Proceedings of REV 2017, New York, USA, pp. 350–356 (2017)
10. Aggarwal, V., Kuhlman, R.: Simulated reality. Electronics **5**(5), 18–21 (2007)

Exposing Robot Learning to Students in Augmented Reality Experience

Igor Verner[1(✉)], Michael Reitman[2], Dan Cuperman[1], Toria Yan[3], Eldad Finkelstein[2], and Tal Romm[1,2]

[1] Technion – Israel Institute of Technology, Haifa, Israel
ttrigor@technion.ac.il, dancup@ed.technion.ac.il,
ty.romm@gmail.com
[2] PTC Corp., Haifa, Israel
{reit,efinklestein}@ptc.com
[3] Massachusetts Institute of Technology, Boston, MA, USA
toria@mit.edu

Abstract. This paper considers a learning process in which the student teaches the robot new tasks, such as lifting unknown weights, via reinforcement learning procedure. Using CAD software, we ran virtual trials using the robot's digital twin in place of physical robot trials. When performing the task, the robot measures and sends the value of the weight to an IoT controller implemented on the ThingWorx platform and receives parameters of the optimal posture found through the virtual trials. When we presented the robot learning process to high school students they had difficulty fully understanding the robot's dynamics and selection of posture parameters. To address this difficulty, we developed an augmented reality interface which allows students to visualize robot postures on the digital twin and monitor the change in parameters (such as the center of gravity) measured by virtual sensors. The student can select a weightlifting posture and control the robot to implement it.

Keywords: Robot learning · Weightlifting · Virtual twin · Internet of Things
Augmented reality

1 Introduction

The industry is evolving towards development and acquisition of distributed smart systems integrated through the Internet of Things. New organizational concepts like Industry 4.0 define a "smart factory" as one where machinery and equipment are able to improve processes through automation and self-optimization. These functions are supported by data flow among and within manufacturing cells, allowing fast response to changes in the production process. The term "Industry 4.0" symbolizes new forms of technology and artificial intelligence within production technologies. Innovative use of big data and turning analytical knowledge into competitive advantage are two key competencies needed to successfully orchestrate Industry 4.0 networks [1]. The components are "smart connected products", which encompass three core elements: physical,

M. E. Auer and R. Langmann (Eds.): REV 2018, LNNS 47, pp. 610–619, 2019.
https://doi.org/10.1007/978-3-319-95678-7_67

smart, and connectivity. Physical elements refer to the product's mechanical and electrical parts. Smart elements refer to sensors, controllers and enhanced user interface. Connectivity elements refer to communication through internet connection [2]. Their development cannot rely solely on mechanical engineering. It requires true interdisciplinary systems engineering, because operating a smart, connected product requires a supporting cloud-based communicating system. Manufacturers need to integrate staff with varied work styles and from diverse backgrounds and cultures [3].

To satisfy the needs of the changing technology world, STEM education in school should upgrade towards teaching technology and design of smart connected products and complex engineering systems. This requires the introduction of new science and engineering concepts as well as new technologies that are not yet taught in schools. Two quickly developing technology areas are Internet of Things (IoT) and augmented reality (AR). In complex engineering systems, it is often hard to understand the relations and constraints between the system's components. To analyze such systems, it is necessary to apply new technologies to make the systems more transparent. There is a growing understanding that the technologies of IoT and AR could be effective in answering this need. Our study has been conducted since 2015 at the Technion Center for Robotics and Digital Technology Education through collaboration with the PTC Israel R&D office. It aims to explore the use of IoT and AR to support students' learning with learning robots.

2   Learning Through Communication with Robots

Learning through interaction with robots has attracted growing interest of researchers [4, 5]. Conventionally, human communication with robots was made through a computer. Modern intelligent robots can be communicated with directly and have learning capabilities [6], which allow both the human and the robot can learn from the interaction.

Our study considers a situation in which the student sets up a robot learning scenario, and the robot learns to execute the desired task. The student evaluates the change in the robot behavior and improves the learning setting. In this situation, the student is also a teacher of the learning robot.

Educational researchers pointed out that learning-by-teaching can be most beneficial [7]. Okita and Schwartz [8] relate the effectiveness of learning-by-teaching to recursive feedback that a teacher gets when evaluating performance and understanding of the learner, and improves teaching based on this feedback. The researchers argued that the theory can be applied to learning with intelligent systems, which they named "teachable agents". This theory provides a methodological basis for educational studies of learning with learning robots.

IoT and AR are discussed in the literature as educational technologies mainly with regard to computer networking and visualizations, while their use for engineering education in schools is still young. An educational methodology for teaching these technologies and their use for learning engineering systems still needs to be developed.

We seek to contribute to this development through a case study of the learning process in which the student teaches the robot to perform a weightlifting task.

3   Robot Weightlifting Task

Planning basic handling tasks to be executed by humanoid robots, such as weightlifting, is an evolving research topic [9, 10]. In the weightlifting task, if the mass and size of the weight are known to the robot, then its posture can be controlled in the open loop. The control policy is to prevent the robot from falling down by maintaining its static and dynamic stability [10]. If the mass and size of the weight are unknown, the closed loop control is needed. Here, the control policy can be determined analytically, but analytic solutions for humanoid robot weightlifting can be complex [11]. The authors proposed an alternative approach based on reinforcement learning through trial-and-error.

Our motivation for exploring humanoid robot weightlifting came from developing a fetch-and-carry humanoid robot for the RoboWaiter contest [12, 13]. In the contest assignment, the mass, size, and location of the weight were predetermined. Following the contest, we tackled the challenge of lifting a weight when its mass is unknown to the humanoid robot. Below is a short reflection of the past stages of our project.

In the first phase, we constructed a humanoid robot using a ROBOTIS Bioloid Premium kit and programmed it using RoboPlus software to implement the reinforcement learning procedure described below [14]. The robot grasped an unknown weight while sitting down and estimated its mass. Then, the robot performed trials to stand up and lift the weight for different values of body tilt angles. The robot evaluated whether it succeeded or failed the task by determining whether it remains standing or has toppled over. Because of the limited capacity of the robot controller, the data of the trials were communicated via a Bluetooth interface with a local computer which provided the robot with the tilt angle value suitable for successfully lifting the specific weight. Such straightforward trial-and-error procedure can only be applied in the case of a trivial robot posture defined by one parameter (tilt angle).

At the second phase of our research, in order to extend the robot learning procedure for postures defined by three parameters (tilt, knee, and ankle angles), we developed the connected system presented in Fig. 1.

The robot constructed at the first stage of the project was upgraded by adding grippers to suit barbell lifting. The digital twin was constructed and calibrated using the Creo Parametric CAD software as a virtual counterpart intended to test robot functioning in the simulation mode instead of testing the physical robot. The ThingWorx server receives and analyzes data of virtual trials from the local computer, which runs the simulator, and sends recommendations for weightlifting posture to the robot through the local computer used as a routing point [15].

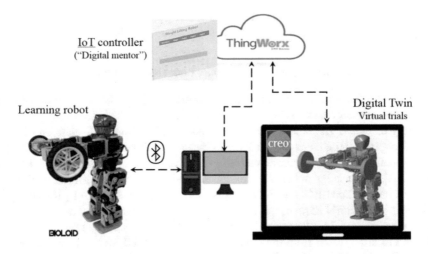

Fig. 1. The implemented robot learning scheme.

4   Pilot Courses

4.1   School Outreach Course on Robot Learning

This course was given to a group of twelve 11th graders majoring in systems engineering in 2016. The students constructed animal-like robots and programmed different robot learning scenarios using BIOLOID and RoboPlus. For example, two students developed a Smart Puppy robot that can be trained to acquire new behaviors by reinforcement (Fig. 2). They built a 15 degree of freedom puppy robot that tested the environment by a color sensor, received positive reinforcements through a proximity sensor and negative via a touch sensor.

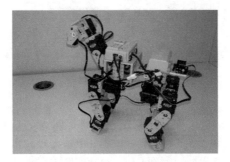

Fig. 2. Smart Puppy robot developed by the students

Then, the students designed and programmed robot motion responses to different sensory inputs and used operant conditioning learning, based on positive/negative reinforcement feedback, to train Puppy for basic obedience skills. The survey conducted at the end of the course indicated that all the students highly evaluated the course expressed strong interest to take part in more activities with learning robots.

4.2   Teacher Education Course on Smart Connected Products

A strategy for learning with a smart connected product (SCP) and its virtual twin was piloted in a teacher education course. In the course, they studied the design and manufacturing of SCPs. The course participants developed instructional units on different aspects of product life management and participated in laboratory sessions. The lab task was to create a model of smart connected gate valve for an irrigation system. The task included CAD modeling of the valve in Creo Parametric and constructing a LEGO-based physical model. The LEGO model consisted of a DC motor activated by Raspberry Pi. The operation control sequence of the valve was programmed in Python and used two input parameters: (i) air humidity, measured by a local sensor; (ii) rain forecast for a selected location, taken from the internet. An interactive UI for the valve was implemented in ThingWorx. The UI allowed the user to monitor the values of the input parameters and the valve state (Open/Closed), and to manually control the valve operation, overriding the automatic control (Fig. 3).

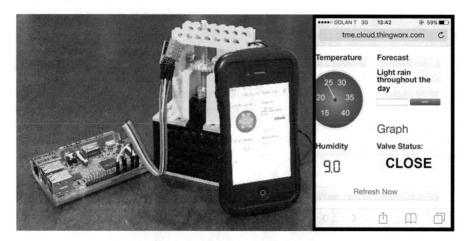

Fig. 3. A smart connected gate valve.

Eleven students majoring in mechanical engineering education participated in the course, among them nine prospective and two in-service teachers. All the students completed the course. In the post-course questionnaire, they noted the contribution of the course to their awareness of IoT and the importance of teaching the subject in school. At the same time, the students mentioned learning difficulties due to the complexity of the working environment that included such complex systems as Creo, ThingWorx, and Raspberry Pi, within the one-semester course.

4.3   School Outreach Course on Robot Learning with Virtual Twin

This course is delivered in the current academic year to a group of fifteen 10th and 12th graders majoring in computer science. The use of simpler robots for student experimentation allows more room to learn the subject of IoT.

The students started from constructing models of an excavator robot as a wheeled platform equipped with a 4 degrees of freedom excavating mechanism consisting of a boom, arm, and bucket, mounted on a turret (Fig. 4A). The robot weightlifting task was to locate and pick up an unknown weight that rested on the ground without tipping over.

A.   B.

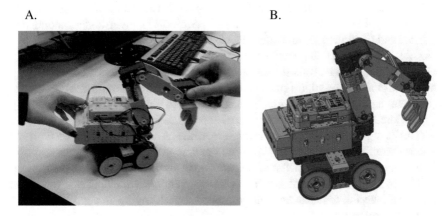

Fig. 4. Excavator: A. The physical model; B. The digital twin

To make the kinematic problem manageable for high school students, we simplified it, assuming that the turret and bucket angles were fixed. Thus, the excavator's position is uniquely determined by one parameter, which could be either picking distance, boom angle, or arm angle. Because the kinematic analysis of weightlifting by mathematical modeling could be a complex assignment for school students, we determined a suitable picking distance for each weight through reinforcement learning.

The robot trials are performed as follows: the robot moves to the object, picks it up and evaluates its weight in a retracted position. Then, the robot gradually extends the picking distance until tipping over, as indicated by its IR sensors.

The robot is connected to the IoT controller in the same way as presented in Fig. 1. During the learning session, the robot sends the weight value and the boom angle value for each trial to the controller. During the working session the controller serves as a digital mentor, sending the maximal picking distance for stable lifting to the robot upon request.

A digital twin created using Creo Parametric (Fig. 4B) will be used later in the course to substitute physical trials by virtual ones.

Our challenge in this course was to facilitate an understanding of the robot learning process by the students who have limited engineering background. To address the challenge and make the learning process transparent, we developed an augmented reality solution. This solution will be based on the AR environment that we developed for the humanoid robot weightlifting case, as described in details in the next section.

5   Using AR to Transparentize Robot Learning

Results of the pilot courses indicated that they efficiently exposed the participating students to the principles of robot learning. However, we observed that they often had difficulty to fully understand the robot dynamics and the logic for selecting motion parameters during weightlifting. This motivated us to start research towards an augmented reality tool that would enrich the robotic environment and facilitate transparency and deeper understanding of the mechanisms of robot learning. There is a large amount of literature on the effectiveness of different AR-based visualization tools in STEM education, but very little research on using AR for student-robot interaction. The main question of our research has been: how can AR tools facilitate understanding robot dynamics and the principles and concepts of robot learning?

Looking at industrial applications of Augmented Reality for engineering processes, we saw that the main reasons for adopting AR technology are related to its ability to visualize digital information (enhance user's view of physical world by overlaying digital data) and interact both with physical and digital objects via AR interface [16].

Therefore, we set a goal to develop an AR experience that would connect the physical robot and its digital twin via IoT infrastructure, while providing the user with the ability to control them both in real time. When manipulating the physical robot or emulating its behavior on the digital twin, the AR experience would also visualize essential parameters like position of center of gravity. A mixed reality environment like this should greatly facilitate the transparency and improve the learning by students who interact with the learning robot.

We used ThingWorx Studio (integrated authoring environment developed by PTC Corp) to implement AR experience connected to ThingWorx platform. User controls allow the student to operate physical robot directly from AR experience while visualizing physical parameters (weight) and manipulating the digital twin to simulate expected movements (see Fig. 5).

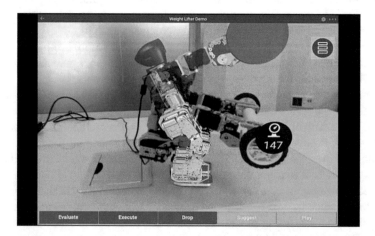

Fig. 5. The AR experience viewed with the ThingWorx View application, available for iOS, Android, and Windows.

Connectivity with physical robot is implemented through a python-based proxy due to software/hardware limitations on the robot controller (a python script on routing computer is used to exchange information to/from the robot). Once the user hits a button to trigger an action on the physical instance, this action is recorded in a Thingworx property which the python script is listening to. Once the script identifies that the property has changed, it communicates a request to the robot using a reserved message via Bluetooth to perform the needed action. When the robot receives the request, it performs the action and responds back via Bluetooth connection.

The implemented controls for the physical robot are Evaluate, Execute and Drop. Evaluate sends a request to the physical robot to initiate the weight evaluation sequence. Once the weight is evaluated, its value is sent back and stored in a ThingWorx property and is automatically updated on AR view. Execute sends the request to set the robot into a posture selected by the user and initiate the weightlifting task. Drop sends the request to the physical robot to initiate the motion sequence for dropping the weight it is holding.

Two additional controls, Suggest and Play, allow the user to manipulate the digital twin of the robot. The geometry and main animation sequences were transferred from Creo Parametric (where the digital twin is created), to Creo Illustrate, and were used in ThingWorx Studio to be added as integral part of AR experience. The communication with the digital twin is based on event binding within the Thingworx environment. The Suggest button triggers a method to show candidate postures for the given weight (received through Evaluate) – it will show three possible postures that the robot can set to successfully lift the given weight. These candidate postures are generated based on a massive virtual experiment with the digital twin, and correspond to different locations of the center of gravity (forward-most, backward-most and precisely in the middle). The user can review these candidate postures and select one for the simulation or for the execution. The Play button invokes a JavaScript function to show the kinematic movement of the digital twin executing the task according to the selected posture.

This AR experience supports two main views that the user can switch between them using a toggle button in AR interface. The overlay view positions the digital twin just over the physical twin with the purpose to show exact alignment between the two, or observe any discrepancy (see Fig. 5). The side-by-side view has the digital twin positioned next to the physical one, not occluding it and allows using it as a virtual teacher visualizing and validating the posture selected by the user – prior to executing it on the physical robot (Fig. 6).

One interesting side effect discovered when testing the new AR environment is the fact that it eliminates the need to have the student collocated with the robot. Since all user controls are included as part of the AR interface, there is no difference whether the student is located in the lab next to the robot or many miles away looking at the robot via web camera. Such ability to fully transcend the distance factor makes this approach uniquely positioned for the evolving area of remote teaching.

Fig. 6. Side-by-side view of the robot and the digital twin in the candidate posture. The center of gravity is visualized on the digital twin – both in static positions and during the movement.

6   Conclusion

The approach proposed in our study extends the scope of educational robotics, which traditionally focuses on practices with preprogrammed robots. The research results indicate that the challenge of developing learning robots can engage novice engineering students in experiential learning of innovative concepts and technologies, such as parametric design, digital prototyping and simulation, digital twin modeling, connectivity and internet of things. We found that those concepts and technologies are within the grasp of understanding of freshmen and high school students. When presented as part of holistic experience for design and operation of learning robots, these advanced concepts and technologies were quickly understood and accepted by the students, in some cases faster than expected.

The lack of solid engineering background may become an obstacle for understanding and practical application of these concepts. Here the use of augmented reality can provide breakthrough opportunities for learning with learning robots. An AR-based interface allows for the visualization of deep engineering data received and generated by the connected system while observing in real time as it operates. It also allows the student to control the engineering system and observe its response to various states and conditions, both driven by external interactions and defined by the human operator. Implementing the learning experience with AR-based controls opens unlimited opportunities for customizing user interface and for full remote operating.

In the next phase of the research, we will continue to practically explore our approach in high school outreach and teacher education courses towards development of strategies for learning with learning robots in a mixed reality experience.

References

1. Richert, A., Shehadeh, M., Plumanns, L., Grob, K., Schuster, K., and Sabina, J.: Educating engineers for Industry 4.0: virtual worlds and human-robot-teams. Empirical studies towards a new educational age. In: Proceedings of EDUCON, pp. 142–149 (2016)
2. Porter, M.E., Heppelmann, J.E.: How smart, connected products are transforming competition. Harvard Bus. Rev. **92**(11), 64–88 (2014)
3. Porter, M.E., Heppelmann, J.E.: How smart, connected products are transforming companies. Harvard Bus. Rev. **93**(10), 96–114 (2015)
4. Kanda, T., Hirano, T., Eaton, D., Ishiguro, H.: Interactive robots as social partners and peer tutors for children: a field trial. Hum. Comput. Interact. **19**(1), 61–84 (2004)
5. Verner, I., Polishuk, A., Klein, Y., Cuperman, D., Mir, R.: A learning excellence program in a science museum as a pathway into robotics. Int. J. Eng. Educ. **28**(3), 523–533 (2012)
6. Arumugam, S., Kalle, R.K., Prasad, A.R.: Wireless robotics: opportunities and challenges. Wirel. Pers. Commun. **70**(3), 1033–1058 (2013)
7. Fiorella, L., Mayer, R.: The relative benefits of learning by teaching and teaching expectancy. Contemp. Educ. Psychol. **38**, 281–288 (2013)
8. Okita, S.Y., Schwartz, D.L.: Learning by teaching human pupils and teachable agents: the importance of recursive feedback. J. Learn. Sci. **22**(3), 375–412 (2013)
9. Harada, K., Kajita, S., Saito, H., Morisawa, M., Kanehiro, F., Fujiwara, K., Kaneko, K., Hirukawa, H.: A humanoid robot carrying a heavy object. In: Proceedings of the IEEE International Conference on Robotics and Automation, pp. 1712–1717 (2005)
10. Arisumi, H., Miossec, S., Chardonnet, J.R., Yokoi, K.: Dynamic lifting by whole body motion of humanoid robots. In: IEEE/RSJ International Conference on Intelligent Robots and Systems, pp. 668–675 (2008)
11. Rosenstein, M.T., Barto, A.G., Van Emmerik, R.E.: Learning at the level of synergies for a robot weightlifter. Robot. Auton. Syst. **54**(8), 706–717 (2006)
12. Ahlgren, D.J., Verner, I.M.: Socially responsible engineering education through assistive robotics projects: the RoboWaiter competition. Int. J. Social Robot. **5**(1), 127–138 (2013)
13. Verner, I., Cuperman, D., Cuperman, A., Ahlgren, D., Petkovsek, S., Burca, V.: Humanoids at the Assistive Robot Competition RoboWaiter 2012. In: Robot Intelligence Technology and Applications 2012, pp. 763–774, Springer, Heidelberg (2013)
14. Verner, I., Cuperman, D., Krishnamachar, A., Green, S.: Learning with Learning Robots: A weight-lifting project. Robot Intelligence Technology and Applications, vol. 4, pp. 319–327. Springer, Berlin (2017)
15. Verner, I., Cuperman, D., Reitman, M.: Robot online learning to lift weights: a way to expose students to robotics and intelligent technologies. Int. J. Online Eng. **13**(8), 174–182 (2017)
16. Porter, M.E., Heppelmann, J.E.: Why every organization needs an augmented reality strategy. Harvard Bus. Rev. **95**(6), 46–57 (2017)

Framework for Augmented Reality Scenarios in Engineering Education

Matthias Neges[1(✉)], Mario Wolf[1], Robert Kuska[2], and Sulamith Frerich[2]

[1] Chair for IT in Mechanical Engineering, RUB, Bochum, Germany
{matthias.neges,mario.wolf}@itm.rub.de
[2] Virtualisation of Process Technology, RUB, Bochum, Germany
{kuska,frerich}@vvp.ruhr-uni-bochum.de

Abstract. The goal of the presented approach is to show a method suitable for better integration of real-time sensor data into practical education, without leaving the students to sort out the digital content by themselves. The authors want to empower teachers on-site to show their students relevant sensory data, effectively controlling the content the students can use and explore themselves. The students are enabled to find individual approaches towards the learning scenario, take different perspectives of the plant into account and try several virtual steps before the experiment is undertaken by themselves. The two main functions of the presented framework are the authoring of augmented reality content and controlling the augmented reality content of the student's smart devices via the teacher's master view. The authors created a simple setup phase, which is usable on-site, utilizing only one device in the master view mode. For students, the usage is even simpler, as their content is controlled via the master view. The framework technically supports an unlimited number of student clients to be controlled by one teacher view. The functionality has been established and validated with two experimental setups, both situated within the context of chemical engineering education.

Keywords: Augmented reality · Remote lab · Practical engineering education

1 Introduction

"Industry 4.0" aims to achieve high flexibility to face the growing competitive pressure between companies. Throughout all industrial branches the increasing automated production processes result in a rising complexity of machinery [1]. The "early adopters" of industry 4.0 were mostly found in automotive and mechanical engineering, but the process industry starts adapting the new technologies, too [2]. The national platform for industry 4.0 released the agenda for a research project "Plug and Produce" [3], which aims at maximized flexibility in the process industry and is part of the reference model architecture RAMI 4.0 [4]. Additionally to the changes to the machinery in industry 4.0, strengthening competences of employees is the major challenge in the ongoing process of digitalization [5].

© Springer International Publishing AG, part of Springer Nature 2019
M. E. Auer and R. Langmann (Eds.): REV 2018, LNNS 47, pp. 620–626, 2019.
https://doi.org/10.1007/978-3-319-95678-7_68

Engineering education includes lectures teaching theoretical knowledge and seminars practicing its content. Laboratories are part of the practical education where students are applying their background knowledge onto real life scenarios. However, the way of learning the background knowledge is mainly based on manuscripts or textbooks. Thus, information about experimental machines has to be linked by the student himself.

Augmented reality is defined as overlaying reality with additional virtual content and offering a higher degree of intuitional use compared to conventional documentation [6]. Schlick et al. [7] found that graphical instructions are superior to textual ones in MRO (maintenance, repair and overhaul) processes whenever the repetition rate is low. Furthermore, Dini and Mura conducted a study concerning the usage patterns and implementation of augmented reality applications in the industry [8]. They have shown that only 11% of the augmented reality usage happens for training purposes, while nearly two thirds are used directly to support maintenance oriented tasks. Jeřábek et al. state, that augmented reality's ability as a technical resource is to learn data interpretation and management when used to achieve didactic goals [9].

Using hands-on laboratories to educate students requires a specific case and a simplified apparatus, respectively. Usually, executing a case endures several hours. Within that time, various process conditions initially set and should be able to achieve a stationary state. Although the presence of a supervisor is a necessity for safety, in addition to checking the preparation of the students beforehand, the laboratory itself should always be safe even when misused.

The accessibility of laboratories for students is limited if growing numbers of students need further assistance [10]. One way to handle this is widening the possibilities given to the students by using remote, virtual or hybrid laboratories [11]. While virtual laboratories emulate real life apparatuses and their handling, remote laboratories produce real experimental data for students to analyse. Furthermore, they get into contact with modern technology as the remote control, and they learn how to work with machines not to be seen directly. Still, they need to be able to work with this machine as they would do it in real life hands-on-scenarios. This challenge has various approaches [12]. While an increased accessibility is achieved, the lack of assistance while performing an experiment needs an upgrade of the preparation beforehand. Students are on their own, so they need individual and more detailed explanations of the apparatus and what they will experience [13]. Lots of experiments are still of low complexity, although the installation itself can be very complex.

Thus, more complex systems such as extrusion processes which are dealing with lots of measured data and require fast and controlled interaction, are not very common in hands-on education. The teaching is based mainly on theoretical background or within hypotheses. In general, the use of augmented reality for instance enhances learning achievements, helps to understand, promotes self-learning and is able to visualize invisible processes within an apparatus [14]. Once this technology is used for complex scenarios, students are enabled to learn at real life laboratories while the control and safety observation remains in the hands of the teacher.

2   Aims and Requirements

The vision behind the concept is shown in Fig. 1. Given that both the teacher and the student handle smart devices, which are connected to the same network, the teacher makes use of his knowledge and preparation to give the student insight into the laboratory machine's workings. With the ability to control the focus of the student's augmented reality overlay, the teacher creates step by step instructions and explanations, which are based on sensory data the students can explore themselves. The approach is not aimed at replacing all interaction in the laboratory with app based communication, but to augment the argumentation of teachers with interactive elements. The students are therefore enabled to find individual approaches towards the learning scenario, take different perspectives of the apparatus into account and try several virtual steps before the experiment is undertaken by themselves. They follow a real experiment, which is performed by a teacher, to identify/suggest the next steps while the teacher can control the ideas and instantly show the experiments' behavior. This way, learning gets more individual by using digital technologies.

Fig. 1. Vision of the AR application's use

To validate the presented approach, the authors aim for a prototypical implementation for a preliminary evaluation in practical engineering education. The findings of this evaluation will be used for further improvement of the approach. The first step is to create a software which offers a basic functionality, but can be extended in future to whatever aim a teacher wants to reach, equipped with didactical support. Furthermore, it should be independent of the apparatus itself, so it can be transferred to other laboratories and installations.

The concept for the framework for augmented reality scenarios in engineering education should meet the following requirements:

1.   Offer intuitive visualization in augmented reality (AR).
2.   Offer a central platform for data exchange.
3.   Support distinctive roles for teachers and students in the AR application.
4.   Offer a generic interface for operating data in the central platform.

5. Offer teachers the ability to create data panels from operating data.
6. Offer teachers the ability to create education scenarios on machinery fast and easy.
7. Offer students the ability to view operating data panels placed by the teacher.
8. Offer teachers the ability to guide the students' visualization options.

3   Concept

The approach at hand synchronizes the views of both the teachers and his students, while displaying sensor data. Petrolo et al. state that Internet of Things platforms provide ubiquitous connectivity between smart devices [15], which makes them ideal for hosting data for distributed applications like in this case.

The usage of augmented reality in higher education is still relatively rare, but several approaches exist that make use of either motivational aspects of AR [16] or to support of the cooperative aspect of diagnosis [17] with digital information.

The concept consists of the following components:

- The apparatus (a hydraulic system) as the center of the experiment
- The instructor who prepares his lessons by creating and positioning AR panels
- Multiple learning students
- The smart device applications for teachers and students
- An Internet of Things (IoT) platform that manages the sensor data and the synchronization of all connected clients

The connections and dependencies between these components is shown in Fig. 2. In an initial phase, the apparatus needs to be connected to the IoT platform, so that the available (and relevant) sensory data is present, up to date and defined by type before the teacher plans the experiment. When planning, the teacher creates type-specific AR

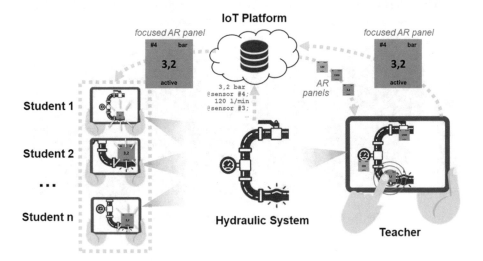

Fig. 2. Concept overview

panels (orange panels represent pressure, blue panels represent flow rate etc.) and places them in the learning environment via touch gestures.

In the lecture itself, the teacher activates his currently relevant panel by gesture. There are two options given in this moment. The first is to reveal the hidden panel to the students additionally to the ones that have been unlocked before. The second option is to show only the last activated panel.

4   Prototype Implementation and Validation

The authors designed a generic interface for sensor data to support a wide variety of machines, as the nature of machinery in engineering education is experimental. The two main functions of the presented framework are the authoring of augmented reality content and controlling the augmented reality content of the student's smart devices via the teacher's master view.

The authors created a simple setup phase, which is usable on-site, utilizing only one smart device in the master view mode. For students, the usage is even simpler, as their content is controlled via the master view. However, the students' pose (meaning both their position and their orientation) towards the apparatus is individual. The framework technically supports an unlimited number of student clients to be controlled by one master view (see Fig. 2). The functionality has been established and validated with two experimental setups, both situated within the context of chemical engineering education. Figure 3 shows both the teacher's (left) and the student's views with a focused element set (element p0). This element is the only one seen by the students. By using this focussing feature, the teacher can lead the students' attention to a specific measuring point. The students have to analyse and interpret the pending measurements to identify the behaviour of the apparatus and recommend necessary changes of the execution. This is a first attempt to implement different functions in the interaction of teacher and student. It can be extended by implementing further objects, enabling feedback and individual setups to enhance the learning experience of the students.

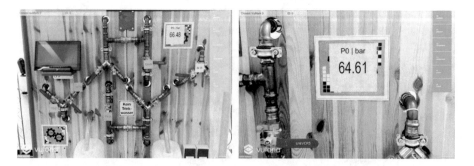

Fig. 3. Teacher's view to the left, student's view to the right

5   Conclusions

The first implementation showed great abilities for its future usage with classes. Interviewed teachers emphasized its intuitive approach. The next step will be the evaluation of innovation, integrity and usage in a small group of students to gather feedback from both students and researchers for further optimization by observation of the intuitiveness in usage and interviewing their experiences. In addition, it is planned to adapt the approach for several other experimental plants and implement greater functionality and therefore enable its better integration in a course and evaluation. Functionalities on the students' view can be graphically visualised by using time-dependent measurement data, visualized and animated insights of an apparatus, video pop-ups to show failures or misuse parameters, questionnaires, overlapping heat maps and so on. The usage depends on the case, the apparatus and what exactly should be taught. The teacher's view can gain further control of the students' view by building up a full scenario of several aspects beforehand, enabling individual choices of the instructor while performing the experiment.

This contribution shows the great potential of digital technologies for creating augmented reality scenarios within engineering education. The authors firmly believe that there is great potential for the future, and our findings are enabling further achievements.

Acknowledgements. We would like to thank Simon Schröder for his contributions regarding the implementation of the augmented reality application shown in the validation.

References

1. Vogel-Heuser, B., Bauernhansl, T., Hompel, M.T. (eds.): Handbuch Industrie 4.0 Bd.2: Automatisierung, 2. Aufl. 2017. Springer Reference Technik. Springer, Berlin, Heidelberg, s.l. (2017)
2. Vogel-Heuser, B., Bauernhansl, T., Hompel, M. (eds.): Handbuch Industrie 4.0 Bd.4: Allgemeine Grundlagen, 2nd edn. Springer Reference Technik. Springer, Berlin, Heidelberg, s.l (2017)
3. Plattform Industrie 4.0, ZVEI - Zentralverband Elektrotechnik- und Elektronikindustrie e.V. (2017). Industrie 4.0 Plug-and-Produce for Adaptable Factories: Example Use Case Definition, Models, and Implementation
4. DIN Deutsches Institut für Normung e.V. Referenzarchitekturmodell Industrie 4.0 (RAMI4.0) (DIN SPEC 91345:2016-04) (2016). https://www.beuth.de/de/technische-regel/din-spec-91345/250940128. Accessed 09 Oct 2017
5. Acatech – Deutsche Akademie der technikwissenschaften (ed.) Kompetenzen für Industrie 4.0: Qualifizierungsbedarfe und Lösungsansätze. acatech POSITION. Herbert Utz Verlag GmbH, München (2016)
6. Azuma, R.T.: A survey of augmented reality. Presence Teleoperators Virtual Environ. **6**(4), 355–385 (1997). https://doi.org/10.1162/pres.1997.6.4.355
7. Schlick, C.M., Jeske, T., Mütze-Niewöhner, S.: Unterstützung von Lernprozessen bei Montageaufgaben. In: Schlick, C.M., Moser, K., Schenk, M. (eds.) Flexible Produktionskapazität Innovativ Managen: Handlungsempfehlungen für die Flexible Gestaltung von Produktionssystemen in Kleinen und Mittleren Unternehmen. Springer Vieweg, Berlin (2014)

8. Dini, G., Mura, M.D.: Application of augmented reality techniques in through-life engineering services. Procedia CIRP **38**, 14–23 (2015). https://doi.org/10.1016/j.procir.2015.07.044

9. Jeřábek, T., Rambousek, V., Wildová, R.: Specifics of visual perception of the augmented reality in the context of education. Procedia Soc. Behav. Sci. **159**, 598–604 (2014). https://doi.org/10.1016/j.sbspro.2014.12.432

10. Frerich, S., Heinz, E., Müller, K.: RUB-Ingenieurwissenschaften expandieren in die virtuelle Lernwelt. In: Frerich, S., Meisen, T., Richert, A., et al. (eds.) Engineering Education 4.0, pp. 25–29. Springer International Publishing, Cham (2016)

11. Frerich, S., Kruse, D., Petermann, M., et al.: Virtual labs and remote labs: practical experience for everyone. In: 2014 IEEE Global Engineering Education Conference (EDUCON). IEEE, pp. 312–314 (2014)

12. Kruse, D., Frerich, S., Petermann, M., et al.: Remote labs in ELLI: lab experience for every student with two different approaches. In: 2016 IEEE Global Engineering Education Conference (EDUCON), pp. 469–475. IEEE (2016)

13. Kruse, D., Kuska, R., Frerich, S., et al.: More than "did you read the script?". In: Auer, M.E., Zutin, D.G. (eds.) Online Engineering & Internet of Things, vol. 22, pp. 160–169. Springer International Publishing, Cham (2018)

14. Akçayır, M., Akçayır, G.: Advantages and challenges associated with augmented reality for education: a systematic review of the literature. Educ. Res. Rev. **20**, 1–11 (2017). https://doi.org/10.1016/j.edurev.2016.11.002

15. Petrolo, R., Loscri, V., Mitton, N.: Cyber-physical objects as key elements for a smart cyber-city. In: Guerrieri, A., Loscri, V., Rovella, A., et al. (eds.) Management of Cyber Physical Objects in the Future Internet of Things Methods Architectures and Applications, 1st edn., pp. 31–50. Springer International Publishing, Cham (2016)

16. Souza-Concilio, I.A., Pacheco, B.A.: The development of augmented reality systems in informatics higher education. Procedia Comput. Sci. **25**, 179–188 (2013). https://doi.org/10.1016/j.procs.2013.11.022

17. Martín-Gutiérrez, J., Fabiani, P., Benesova, W., et al.: Augmented reality to promote collaborative and autonomous learning in higher education. Comput. Hum. Behav. **51**, 752–761 (2015). https://doi.org/10.1016/j.chb.2014.11.093

Poster: SIMNET: Simulation-Based Exercises for Computer Network Curriculum Through Gamification and Augmented Reality

Alvaro Luis Fraga[1]([✉]), María Guadalupe Gramajo[1], Federico Trejo[2], Selena Garcia[2], Gustavo Juarez[2,3], and Leonardo Franco[3]

[1] Universidad Tecnológica Nacional, Facultad Regional Santa Fe, Santa Fe, Argentina
alvarofraga@ieee.org, gramajoguadalupe@gmail.com
[2] Universidad Tecnológica Nacional, Facultad Regional Tucumán,
San Miguel de Tucumán, Argentina
{federicotrejo,selenagarcialobo,juarez.gustavo}@ieee.org
[3] Depto. de Lenguajes y Ciencias de la Computación, Universidad de Málaga, Málaga, Spain
lfranco@lcc.uma.es

Abstract. Gamification and Augmented Reality techniques, in recent years, have tackled many subjects and environments. Its implementation can, in particular, strengthen teaching and learning processes in schools and universities. Therefore, new forms of knowledge, based on interactions with objects, contributing game, experimentation and collaborative work. Through the technologies mentioned above, we intend to develop an application that serves as a didactic tool, giving support in the area of Computer Networks. This application aims to stand out in simulated controlled environments to create computer networks, taking into account the necessary physical devices and the different physical and logical topologies. The main goal is to enrich the students' learning experiences and contribute to teacher-student interaction, through collaborative learning provided by the tool, minimizing the need for expensive equipment in learning environments.

Keywords: Gamification · Augmented reality · Simulation-based exercises

1 Introduction

Economic inequality is a reality, it affects every sector of our society, and is mainly present in Third World countries. The immersion of graduates from public universities in third world countries becomes difficult, as they do not have the latest technologies or the necessary equipment to acquire the needed skills to perform effectively in real work problems. In the education sector, there are public institutions that do not have state-of-the-art technological elements or equipment that allow simulations or practices to be carried out in controlled environments for computer networks related careers, in contrast to private institutions or First World Countries. This scenario contributes negatively to students and graduates' employability.

Gamification and Augmented Reality (AR) techniques have positioned as attractive and interesting strategies to improve and innovate on many domains being education one of

© Springer International Publishing AG, part of Springer Nature 2019
M. E. Auer and R. Langmann (Eds.): REV 2018, LNNS 47, pp. 627–635, 2019.
https://doi.org/10.1007/978-3-319-95678-7_69

them. Applying them has strengthened the teaching and learning processes in schools and universities. Hence, it is possible to construct new forms of learning based on interactions with objects, providing games, experiments, challenges and collaborative work.

Gamification is defined as the use of game design elements and techniques, in non-game related contexts, to engage people and solve problems [1]. Implementing such techniques facilitates changes in the classic classroom activities, providing incentives for students through reward and scoring systems [2]. Using game mechanics in the classroom allows students to make mistakes and restart play, giving them the freedom to fail and experiment without fear [3].

In this sense, several researches that suggest capturing characteristics, capabilities, rules, and strategies provided by video game environments to promote creative thinking and lead to new ways of addressing real-world problems. Morschheuser et al. [4] describes as the concept of collaboration provided by the game dynamics through gamification techniques, influences the dynamics of the group, generating commitment, shared emotions, sense of belonging and definition of group norms. Game-based learning has been shown to have potential to modern students who are growing up in an era of interactive media and video games, making gambling in the classroom attractive and motivating [5]. Most studies highlight positive aspects of gamification, because they generate greater engagement and enjoyment, but these results often depend on the context of the gamified system and the characteristics of the player.

AR application shows potential in the education fields [6]. It can help to improve critical thinking skills, problem solving, and teamwork through different activities [7]. Students easily engage in the learning process and can enjoy the experience through game mechanics embedded in the AR activity. Because of AR many complex scenarios can be replicated and approached from a different perspective, such as physical and biological phenomena as well as astronomical phenomena.

Some of the advantages of this technology are the following: (i) the equipment that students need is a cell phone and a pattern card; (ii) the application can be used anywhere and anytime; (iii) and can be developing to be used as a standalone application [8]. Therefore, AR has become a popular technology used in educational settings. In this role, AR has also turned into an important focus of research in recent years. One of the most important reasons AR technology is widely accepted because it no longer requires expensive hardware and sophisticated equipment. AR can be defined as a technology that superimposes virtual objects (increased components) in the real world. These virtual objects seem to coexist in the same space as real-world objects [9]. Several studies show that AR can improve the teaching and learning experience [10, 11].

The main objective of this research is to enrich students' learning experiences and improve the teacher-student interaction through collaborative learning. Therefore, a tool that allows the virtualization of controlled environments to create computer network infrastructure simulations challenges by combining gamification and AR techniques is presented in this contribution to achieve this objective.

This research also intends to be a support application to institution that do not have the budget to equip their laboratories with certain and necessary equipment to carry out the computer network practices.

The rest of the paper is structured as follows. Section 2 describes some related works with the application of gamification and AR techniques in the academic field. Section 3 is divided in two subsections. The first one describes the guidelines needed to use SIMNET in classroom. The second part includes its architecture and description of the components used in the development of the application. Finally, Sect. 4 shows the conclusions and future work.

2 Related Works

This section will mention some related works in the field. These studies have made use of the techniques we have mentioned in the introduction to this paper.

Ibáñez et al. [12] present a learning application that make use of AR technology to aboard the basic concepts of electromagnetism. This study also compared an AR-based application with an equivalent web-based application to study its learning effectiveness and the level of enjoyment of high school students. The analysis showed that the AR application led participants to achieve higher levels concentration than those achieved by web-based application users.

Quintero et al. [13] presented an AR application to promote spatial visualization in calculus courses for engineering students. The aim of this research was presenting to students a cognitive experience of the surface as the intersection of planes. This experience was driven for its first pilot test. The results concluded that the traditional approach of learning mathematics cannot remain unchanged and it is necessary new techniques that contribute to the students learning experience.

PhysicsPlayground [14] is an AR application that allows simulate physical phenomena. This application can be used by special equipment and was developed using PhysX API from NVIDIA. PhysicsPlayground helps to teach and learn mechanics. This tool allows students to construct virtual models to study physical properties, verify formulas, develop theories and actively take part in physical education, encouraging experimentation and understanding. Authors highlight the nearly haptic interaction when building and running experiments in the application. Also, authors remark the possibility to change parameters in real time to take new perspectives of the results.

AR facilitates the observation of events that cannot to be easily observed with the naked eye [11]. Therefore, this can be exploited to increase students' motivation and help them to acquire better research skills [15].

According to [16], AR most significant advantage is its "unique ability to create hybrid learning environments that combine digital and physical objects, facilitating the development of processing skills such as critical thinking, problem-solving and communication through interdependent collaborative exercises".

In summary, it is clear how AR and gamification have been useful for improving learning experiences, both to provide student another perspective from which to approach the comprehension of certain complex issues such as visualization of phenomena invisible to our eyes. It can also be useful to provide virtual scenarios in which the equipment is not available to carry them out in reality or to simulate a practice.

That is why our proposal makes use of these technologies to develop an application that can provide the student with a new experience in computer networks.

In the following section, we will describe the methodology to apply the application in the classroom and the components used to develop it.

3   SIMNET: Methodology and Implementation

This section will present the scenario adopted to make a pilot test of the application developed. It will describe the characteristics it provides, and the design chosen for its development.

3.1   Usage Methodology

Each one of these course assignments has a scenario description, an aim related to the specific exercise and a QR code. This code attached to each exercise allows interaction between the application and the assignment. Once every student will have been supplied with the application on their mobile devices, they will access the environment developed for the activity.

The Fig. 1 shows the activities flow to consider when using SIMNET. This workflow consists of 7 activities: *Definition of student teams* (1), *Selection of virtual scenarios* (2), *Challenges* (3), *Optional goals* (optional activity) (4), *Assignment execution* (5), *Highest score team selection* (6), *Winning team* (7).

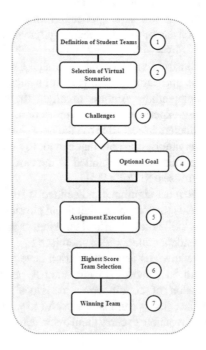

Fig. 1.   Methodology workflow

Initially, teachers will separate students into groups of up to 4 members to start the assignment as indicated in step (1). As soon as students have formed the teams and entered the application. The students will register the group and full name as it is shown in Fig. 2a, to record the group to which each member belongs. This information is added to a spreadsheet in a cloud service. Teachers are the only ones that will be able to access to this spreadsheet. Then, they will scan the QR codes to begin the activity.

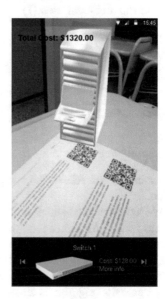

Fig. 2. (a) Screenshot of the application main screen; (b) Screenshot of the application in use.

The activities will be related to complete a certain virtual implementation of a computer network as indicated in step (2). Using components pre-assigned for each exercise. The components will appear in the application's toolbox when the activity begins. Students will be able to aboard the problems which represents the challenges indicated in step (3), with different virtual components. Having multiple ways to solve each assignment, so optional objectives have been added. Optional objectives range from the use of a specific component to accomplish the task to the lowest possible use of economic resources, step (4). The teams that decide to solve them will get extra points in the final score.

An example of an activity is shown in Fig. 2b, where students are encouraging to build a rack for certain number of personal computer. Next students must proceed with the assignment execution defined by the teacher using the application, step (5). The students have many switch options and type of connectors. Each activity has their component, information and cost expressed in dollars. After the student decides, which components will be use, they must select them and the components will be added to the rack and the cost of his solution will be display at the left corner of the screen.

Once the student achieves a possible solution to the problem. The application notifies them that a possible solution was found.

These activities encourage teamwork and cooperative work bringing together the above-mentioned benefits of using gamification. Every member of a team also can submit their own answer, but at the end of the assignment, only the best score will be taken as the final mark.

Next, the teacher arranges the results using the spreadsheet from the cloud service, a portion of the spreadsheet is exemplified in Table 1, as indicated in step (6) and finally lists the winning team, step (7).

Table 1. Portion of student registration sheet

Group name	Student name	Exercise 1	Exercise 2	Exercise 3
Team 1	Student 1	$3250	$2050	$1099
Team 1	Student 2	$3250	$2050	$1099
Team 2	Student 4	$4000	$2200	$1300

3.2   Implementation

This subsection describes the components used to develop the prototype of AR application.

We have designed and developed the first version of this proposal targeting mobile devices with an Android operating system up to version 4.4. This application makes use of the Qualcomm Vuforia SDK framework, jMonkeyEngine() and the Android platform.

Qualcomm Vuforia SDK software requires a version 3 or higher of the operative system to ensure that the application will work. The Vuforia SDK includes cutting-edge computer vision algorithms aimed at recognizing and tracking an increased variety of objects, including markers, target images, 3D objects, among others for any user developing the software. Although they are unnecessary for the development of sensor-based AR applications, they allow for easy deployment of computer vision-based AR applications.

jMonkeyEngine is a free gaming engine that brings the 3D graphics of programs to life. It provides intermediate software for 3D graphics and games that frees you from the exclusive coding in low-level OpenGL, for example, by providing an asset system for importing models, pre-defined lighting, physics and special effects modules that are not embedded in 3D objects.

The Android development toolkit includes the Eclipse Integrated Development Environment (IDE), the Android development toolkit for Eclipse (ADT) and many versions of the Android operative system. Specifically, to effectively use these technologies, it is necessary for the device to have a rear camera, GPS module, and gyroscope, and accelerometers and compass.

An AR application can be structured in three layers:

- Upper layer: Contains the domain logic, also known as the application layer.
- Intermediate layer: Modules or components that enable the use of AR, such as visualization, tracking, recording, and interaction.
- Lower layer: Conformed by the operative system, this layer integrates the components providing interfaces to manage all resource of the device.

After determining the architecture and its components, it is necessary to define how information flows in a typical AR application. In recent years, developers have converged on the well-used method of combining these tools using a similar order of execution, the AR control flow or rather the AR control loop. This diagram, Fig. 4, should be read from the left to right. Displays the main sequence of tasks of an AR application, which is repeated in a cycle.

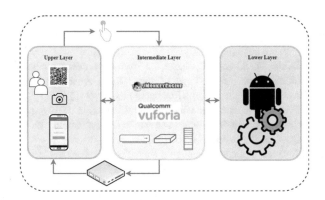

Fig. 4. SIMNET Architecture Schema.

First, the camera is accessed, and the current frame is retrieved and displayed. Then, depending on the AR, sensor data is retrieved, or the camera frame is processed with Computer Vision algorithms to search and track objects. Afterward, virtual objects are transformed and moved into the virtual world to match the real world. Finally, user interactions and gestures are processed to further change virtual objects or perform domain-specific tasks. At the same time, it is necessary to add domain-specific logic and other activities related to the operating system.

4   Conclusion

Augmented reality allows the students interaction between the real and virtual worlds. The capabilities AR of merging virtual and real worlds together have given birth to new approaches that allow for improving the quality of learning. The effectiveness of AR can be further extended when it combines with other types of techniques such as gamification. Gamification improves the student's engagement, motivation, and performance when carrying out a certain task; it does so by incorporating game mechanics and elements, thus making that task more attractive. Combining both techniques can generate great benefits in the educational area, allowing to experience new ways of learning, generate interest in the students and provide new skills that facilitate their insertion into the workplace.

For this reason, we presented this tool to simulate computer networks challenges in a collaborative environment that allows interaction with virtual devices, through an application that combines the advantages of AR and gamification techniques.

In future works, authors will present a pilot experience using the application in the computer network course starting next year.

Acknowledgments. LF and GEJ acknowledge support from grant TIN2014-8516-C2-1-R from MICINN-SPAIN which includes FEDER funds, and from the Universidad de Málaga (España) through "Ayudas Plan Propio".

References

1. Deterding, S., Dixon, D., Khaled, R., Nacke, L.: From game design elements to gamefulness: defining gamification. In: Proceedings of the 15th International Academic MindTrek Conference, Envisioning Future Media Environments, pp. 9–15 (2011)
2. Hanus, M.D., Fox, J.: Assessing the effects of gamification in the classroom: a longitudinal study on intrinsic motivation, social comparison, satisfaction, effort, and academic performance. Comput. Educ. **80**, 152–161 (2015)
3. De-Marcos, L., Domínguez, A., Saenz-de-Navarrete, J., Pagés, C.: An empirical study comparing gamification and social networking on e-learning. Comput. Educ. **75**, 82–91 (2014)
4. Morschheuser, B., Riar, M., Hamari, J., Maedche, A.: How games induce cooperation? A study on the relationship between game features and we-intentions in an augmented reality game. Comput. Hum. Behav. **77**, 169–183 (2017)
5. Glover, I.: Play as you learn: gamification as a technique for motivating learners. In: Proceedings of the 2013 World Conference on Educational Multimedia, Hypermedia and Telecommunications (Edmedia 2013) (2013)
6. Saidin, N.F., Halim, N.D.A., Yahaya, N.: The potential of augmented reality technology in education: a review of previous research. In: Proceedings of the International Graduate Conference on Engineering, Science and Humanities (IGCESGH). UTM (2014)
7. Phon, D.N.E., Ali, M.B., Halim, N.D.A.: Collaborative augmented reality in education: a review. In: 2014 International Conference on Teaching and Learning in Computing and Engineering (LaTiCE), pp. 78–83 (2014)
8. Kiat, L.B., Ali, M.B., Halim, N.D.A., Ibrahim, H.B.: Augmented reality, virtual learning environment and mobile learning in education: a comparison. In: 2016 IEEE Conference on e-Learning, e-Management and e-Services (IC3e), pp. 23–28 (2016)
9. Azuma, R., Baillot, Y., Behringer, R., Feiner, S., Julier, S., MacIntyre, B.: Recent advances in augmented reality. IEEE Comput. Graph. Appl. **21**(6), 34–47 (2001)
10. Medicherla, P.S., Chang, G., Morreale, P.: Visualization for increased understanding and learning using augmented reality. In: Proceedings of the International Conference on Multimedia Information Retrieval, pp. 441–444 (2010)
11. Wu, H.-K., Lee, S.W.-Y., Chang, H.-Y., Liang, J.-C.: Current status, opportunities and challenges of augmented reality in education. Comput. Educ. **62**, 41–49 (2013)
12. Ibáñez, M.B., Di Serio, Á., Villarán, D., Kloos, C.D.: Experimenting with electromagnetism using augmented reality: impact on flow student experience and educational effectiveness. Comput. Educ. **71**, 1–13 (2014)
13. Quintero, E., Salinas, P., González-Mendivil, E., Ramirez, H.: Augmented reality app for calculus: a proposal for the development of spatial visualization. Procedia Comput. Sci. **75**, 301–305 (2015)
14. Kaufmann, H., Meyer, B.: Simulating Educational Physical Experiments in Augmented Reality. ACM (2008)

15. Sotiriou, S., Bogner, F.X.: Visualizing the invisible: augmented reality as an innovative science education scheme. Adv. Sci. Lett. **1**(1), 114–122 (2008)
16. Dunleavy, M., Dede, C., Mitchell, R.: Affordances and limitations of immersive participatory augmented reality simulations for teaching and learning. J. Sci. Educ. Technol. **18**(1), 7–22 (2009)

Applications and Experiences

Using Learning Theory for Assessing Effectiveness of Laboratory Education Delivered via a Web-Based Platform

Shyam Diwakar[1(✉)], Rakhi Radhamani[1], Nijin Nizar[1], Dhanush Kumar[1], Bipin Nair[1], and Krishnashree Achuthan[2]

[1] School of Biotechnology, Amrita Vishwa Vidyapeetham, Amritapuri, Clappana P.O., Kollam 690525, Kerala, India
shyam@amrita.edu
[2] School of Engineering (VALUE Centre), Amrita Vishwa Vidyapeetham, Amritapuri, Clappana P.O., Kollam 690525, Kerala, India

Abstract. Learning styles are defined as a characteristic feature that determines cognitive and psychosocial behavior of learners, perceiving of knowledge, interaction and processing of information in different learning environments. Applying learning style theories in pedagogic concept has brought multiple dimensions in categorizing learning strategies, although such studies on laboratory skill education are limited. Kolb's Experiential Learning Model, has been widely accepted as an efficient pedagogical model of learning. In this paper, we explore the pedagogical basis for designing life sciences laboratory education and applying Kolb's learning style inventory for classifying learners into different categories. In the context of bioscience laboratory education, most students learners were reflective observers or assimilators (60%) or divergers (20%) and hence seemed most apt for virtual laboratory based education due to acclivity to standard demonstrations and lectures. Only 20% were convergers or accommodators. Unlike in some engineering students, this classification suggests bioscience laboratory education may be complemented using web-based tools and will need better assessments and virtualization methods.

Keywords: Learning style · Kolb model · Virtual labs · Laboratory education

1 Introduction

The concept of individual learning patterns among diverse student groups and different ways of information processing in a classroom setting have led to investigating different learning styles and its pedagogical influences on learning process [1]. Several existing theoretical models and learning style instruments have been explored for categorizing learning styles [2]. Basis for experiential learning styles are described by Lewin models of the experiential learning process, Dewis Model of learning and Piaget's model of learning and Cognitive Development [3]. Experiential learning model by Kolb is a well-accepted pedagogical learning model for effective conceptualization of individual learning process [4]. Kolb theory focuses on learning and teaching design with

© Springer International Publishing AG, part of Springer Nature 2019
M. E. Auer and R. Langmann (Eds.): REV 2018, LNNS 47, pp. 639–648, 2019.
https://doi.org/10.1007/978-3-319-95678-7_70

constructivist approach on understanding how people construct their knowledge [5, 6]. Theory emphasizes importance of experience in learning with link between theoretical knowledge and laboratory practice with integration of experience, cognition, perception and behavior in the learning scenario [7]. Learning Style Inventory instrument (LSI) associated with Kolb model classifies individual preferences for concrete experiences, abstract conceptualization, reflective observation and active experimentation and classified the learners into divergers, convergers, assimilators and accommodators [8, 9]. Other learning models included Peter Honey and Alan Mumford's model, which classified learning stages as Activist, Reflector, Theorist and Pragmatist [10], VARK model, which classified learning styles as unimodal, bimodal, trimodal and quadrimodal depending on learner's preference on perceiving or receiving information [11]. Gregorc's Mind Styles Model, which is a modified form of Kolb's learning focused on how learners process the information [12, 13].

Laboratory learning is an integral part of science and engineering education for applying theoretical knowledge with practical experiences [14]. Learning laboratory skills require hands-on training for improving analytical thinking and problem-solving skills [15]. Recently, university education has involved from classroom-based face-to-face interaction-rich lectures, with limited focus on laboratory skill-based methods. Laboratory skill pedagogy has shifted towards using information and communication technologies (ICT) for enhancing laboratory skills among STEM students. Use of such ICT-enabled visual information has been reported to facilitate cognitive learning, improves memory retention [16] and strengthens student motivation towards active learning process [17]. Studies reported the importance of including this pedagogical method in educating a group of students with minimum or no involvement of an instructor [18]. It has been shown that virtual laboratories address challenges related to providing hands-on laboratory experiences to science, engineering and information security students [19]. Pedagogical aspects of bioscience virtual laboratories in the main stream of education sector have been discussed elsewhere [20]. Some studies have also correlated the experiential learning process with student learning outcomes in the context of life science and in general bioscience undergraduate and postgraduate education [21, 22].

Unlike engineering or mathematics, bioscience laboratory education involve concept and skill-specific methods that challenges educators to employ multiple methods to assess learning outcomes. With that in focus, this paper analyses stages of Kolb's cycle among a group of university students who used virtual laboratory tools in addition to standard laboratory education within the context of University-level bioscience courses. The study has implications in evaluating preferred learning styles among cohorts of bioscience students by integrating Kolb's learning cycle and could be used to assess learning outcomes in current and future Massive Open Online Courses (MOOCs) platform.

1.1   Kolb's Experiential Learning

Kolb's Experiential Learning theory focuses on the importance of experience in learning by promoting link between classroom education and the real lab practices. Kolb introduced learning as an integration of experience, perception, behavior and cognition and suggested that effective learners possess Concrete Experience Ability (CE), Reflective Observation Ability (RO), Abstract Conceptualization Ability (AC), and Active Experimentation Ability (AE). In our study for concrete experience, learners participate effectively in a laboratory session by following the step-by-step instructions for demonstrating a theoretical concept. In, reflective observation (RO), learners reflect experiences from the laboratory activity which included discussions and reflective questions based on hands-on experiences. Abstract conceptualization (AC) refers to creation or conceptualization of a theoretical model by utilizing the learning experiences. Active experimentation (AE) emphasis on doing the experiment and learners apply theories for decision making and problem-solving skills (see Fig. 1). Corresponding to these stages, different learning styles were identified and learners were classified depending on their learning style preferences. Convergers preferred to learn using practical application of experimental concepts, theories and logistics. Divergers learned from observation and emphasizes more on innovative and imaginative ways of performing tasks. Accommodators used hands-on experiences for learning, and prefers trial and error methods for problem solving and discovery learning. Assimilators learned using reflective observation and uses logical theories for understanding the concepts [7, 23, 24].

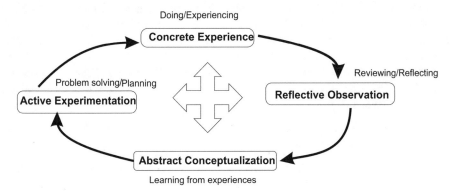

Fig. 1. Stages in Kolb's learning cycle for bioscience laboratory education (Adapted from Kolb, 1984)

2 Methodology

2.1 Sample Collection

This study was performed as a use-case with a student cohort of age group 21–25 in an Indian University during February–November 2017. The target population was postgraduate students pursuing courses on Biotechnology, Microbiology and Bioinformatics. The students were trained on 320+ virtual laboratory courses developed by the authors and are available freely online (http://vlab.amrita.edu) [25]. From the target population of 100 samples, 84 participants successfully completed the process. 16 data points were not taken into consideration due to incompleteness in the questionnaire responses. Gender consisted of 70 females (83%) and 14 males (17%), 36% pursuing post-graduation in Biotechnology, 40% post-graduation in Microbiology and 24% pursuing post-graduation in Bioinformatics.

2.2 Experimental Design

The students were trained to use virtual laboratories and included the use of these ICT-enabled based labs in their curriculum during a semester. The experiment group (CVL) in the study followed traditional classroom teaching with hands-on training using virtual labs. Control group (CBL) followed classroom based teaching without any lab experiences. Learning style of students were analyzed with a Learning Style Inventory questionnaire based on Kolb experiential learning model.

2.3 Data Collection

Feedback data was collected as feedback-based questionnaire and was divided into two sections. First part included the basic questions on age, gender, educational information and knowledge on computers and use of virtual laboratories in education system. The second part included Kolb's Learning Style based questionnaire which was used for understanding the learning style of students. The second questionnaire included 16 questions which corresponds to assessment of learning style among the group of study participants. Students showed their responses on a Likert Scale (Strongly Agree - 5, Agree - 4, Neither Agree nor Disagree - 3, Disagree - 2, Strongly Disagree - 1) to the respective questions of analysis. Following sections describes the feedback questions for learning style analysis.

Concrete Experience

1. I found virtual labs as reference material for my classroom studies.
2. I found virtual labs as a learning platform to test my laboratory skills before performing the experiment in a physical laboratory.
3. Equipment based training in virtual labs were useful and engaged me in learning.
4. ICT-enabled virtual labs offered an interactive learning material to meet the education goals in schools/universities (including rural and urban localities).

Reflective Observation

1. I remember the step-by-step procedure of the experiment when I see what is happening rather than hearing experimental protocol.
2. I prefer to use virtual labs prior to actual experimentation rather than reading a laboratory manual.
3. Observing the experimental processes (such as changing the color of the media while performing specific test for microorganisms in Microbiology lab) helped me to remember results/concepts of the experiment.
4. While using virtual labs, I followed the work flow (Theory-procedure-self-evaluation-virtualization-assignment-reference-feedback) of each experiment.

Active Experimentation

1. Learning manuals provided in virtual labs helped to achieve a clear understanding of the related topics.
2. After using virtual lab experiments, are you confident to perform it in a real lab? (Without instructor support and without making any critical errors).
3. Virtual labs helped student communities (below average and/or above average students) to score better in examinations.
4. Virtualization of experiments (as in virtual labs - simulated experiments) resembling physical lab scenarios supported learning by doing.

Abstract Conceptualization

1. The concepts of the experiments were explained clearly in the virtual labs website.
2. Virtual labs train the students in improving both laboratory skills (equipment and reagent usage) and technical skills (such as PCR, Electro blotting, Chromatography and so on).
3. The experiments designed in a simpler way so that you could finish the lab session with minimal errors.
4. Usage of virtual labs helped reduce time spent in learning concepts when compared to traditional classroom learning.

2.4 Data Analysis

Data was reported as percentages to determine the preferred learning style of the study groups. Mean scores were also tabulated for understanding the relevance of individual elements in the Kolb's learning cycle. Correlation analysis was not included in this study at this stage.

3 Results

3.1 Virtual Labs as a Preferred Tool for Laboratory Education

In the general feedback section, all participants reported that they were familiar with computer usage with more than 10 years of usage experience. Students also reported the use of animation and simulation labs in their laboratory learning with much reduced time as compared to completing a lab exercise in a physical laboratory (see Fig. 2).

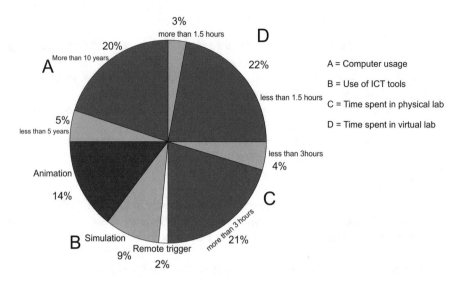

Fig. 2. General feedback report from students indicating the usage of computers, ICT tools, time-spent in physical laboratory and virtual laboratories

3.2 Virtual Labs Suggest Bioscience Students Were Mostly "Assimilators" or "Divergers"

Through direct feedback, participants reported virtual lab-based education enhanced their experiential learning process for life science education. The mean scores given by the participants (CVL) for each stage of the Kolb learning process is shown in Table 1. Results indicated that optimal learning in students happened when the four phases of learning cycle is activated. Comparatively higher mean scores for CE and RO indicated that student's preference of learning from experiences, watching and listening. Comparatively low mean score for AE indicated student's preference of including hands-on lab practices for improving their knowledge level (see Table 1).

Table 1. Mean scores estimated from study groups for Kolb's Experiential Learning Cycle

Elements of Kolb's Learning Cycle	Mean score
Concrete experience activity	4.35
Reflective observation activity	4.29
Abstract conceptualization activity	4.03
Active experimentation activity	3.93

The students were classified into different groups depending on their learning style preferences. The scores given for each feedback question was tabulated and the percentage analysis was performed to classify the study groups into accommodators, assimilators, convergers and divergers (see Fig. 3).

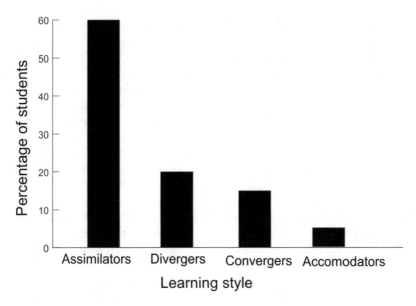

Fig. 3. Classification of students based on learning style preferences using Kolb's Experiential Learning

Most bioscience students were assimilators (60%) in the study. Students showed preference to using virtual labs as demonstrations. Control group students (CBL) who opted traditional learning approach preferred blending of theoretical model and laboratory experience towards knowledge retention. Another percentage of the students favored to be divergers, focusing on details and the "why" of the experimental process rather than the "how". Only a few students (20%) were convergers or accommodators.

4   Discussion and Conclusion

This paper employs the framework of Kolb's model, a known pedagogical learning model, for assessing laboratory education using ICT-enabled virtual labs in university-level bioscience education delivered through a web-based platform. In this preliminary study assessing different phases of Kolb's learning cycle, direct feedback data on the basis of Kolb's Learning Style Inventory were collected from the students.

Most students were assimilators or divergers suggesting some technical education styles may not be suitable for assessing bioscience students. Although the impact of knowledge retention and transfer of knowledge was shown previously, this study based on Kolb's experiential learning on virtual lab based education showed that students have been adapted to the ICT-based learning as new experiences indicating their tangible experience allowed more assimilators than accommodator and converger-style learners. Data suggests students using ICT-enabled laboratory education, learning could happen through reflective experiences and ability to conceptualize from observations and experiences for theoretical applications. Some results were not within the scope of this study and may need the assessment of such ICT tools in stand-alone laboratory skill education. With bioscience experiments, virtual labs lead to better knowledge transformation into mental models. For example, students reported quantitative plots of population ecology virtual labs using data observation generated using the virtual laboratory experiments allowing teachers to test of understanding among students. Also, students who performed virtual lab experiments prior to physical laboratory, scored higher in all our tests [15] indicating the transformation of experience into mental models was more explicit in students with those who used virtual lab experiments.

The student exercise in blended learning mode suggested virtual labs augmented their ability in problem solving and decision making for improved learning outcomes and allowing a novel outlook into virtual laboratories beyond concrete learning tools for stand-alone bioscience laboratory skill education. This assessment with virtual labs within the Kolb's framework suggested implementing blended learning approach in laboratory skill education integrating knowledge depiction to self-driven student inquiry.

In this paper, virtual lab based education was explored as stand-alone and blended education models in the context of Kolb's experiential learning theory. Activation of four phases of the learning cycle could be crucial for better laboratory skill-learning outcomes. While usage of mean scores suggests some of Kolb's experiential learning theory in identifying assimilator, accommodator, converger and diverger students and their learning skills, learning styles, it will further aid teachers in reflective methods. Online implementations for STEM education open up a novel dimension and sustainable development in teaching and learning, further extending its application in a research for humanitarian outreach. The study is being extended with multiple student cohorts consisting of engineering and science students from different geographically distinct continents and varying backgrounds, to derive a more generalized understanding of how Kolb's learning cycle implicates student learning outcomes with the usage of virtual and remote experimentations.

Acknowledgement. This work derives direction and ideas from the Chancellor of Amrita Vishwa Vidyapeetham (University), Sri Mata Amritanandamayi Devi. The work is supported by the Sakshat project of National Mission on Education through ICT, Department of Higher Education, Ministry of Human Resource Department and Government of India, and by Embracing the World.

References

1. Miller, P.: Learning styles: the multimedia of the mind. Research report. Learn. Styles Multimed. Mind. Res. Rep. 11 (2001)
2. Cassidy, S.: Learning styles: an overview of theories, models, and measures. Educ. Psychol. **24**, 419–444 (2004)
3. Miettinen, R.: The concept of experiential learning and John Dewey's theory of reflective thought and action. Int. J. Lifelong Educ. **19**, 54–72 (2000)
4. Kolb, D.A.: Experiential Learning: Experience as the Source of Learning and Development. Prentice-Hall, Englewood Cliffs, NJ (1984)
5. Abdulwahed, M., Nagy, Z.K.: Towards constructivist laboratory education: case study for process control laboratory. In: 2008 38th Annual Frontiers in Education Conference, pp. S1B–9–S1B–14 (2008)

6. Baker, M., Robinson, S., Kolb, D.: Aligning Kolb's experiential learning theory with a comprehensive agricultural education model. J. Agric. Educ. **53**, 1–16 (2012)
7. Kolb, A.Y., Kolb, D.A.: Experiential learning theory: a dynamic, holistic approach to management learning, education and development. Presented at the SAGE handbook of management learning, education and development (2009)
8. Abdulwahed, M., Nagy, Z.K.: Applying Kolb's experiential learning cycle for laboratory education. J. Eng. Educ. **98**, 283–294 (2009)
9. Gooden, D.J., Preziosi, R.C., Barnes, F.B.: An examination of Kolb's learning style inventory. Am. J. Bus. Educ. **2**, 57–62 (2009)
10. Lesmes-Anel, J., Robinson, G., Moody, S.: Learning preferences and learning styles: a study of Wessex general practice registrars. Br. J. Gen. Pract. **51**, 559–564 (2001)
11. Peyman, H., Sadeghifar, J., Khajavikhan, J., Yasemi, M., Rasool, M., Yaghoubi, M.Y., Mohammad Hassan Nahal, M., Karim, H.: Using VARK approach for assessing preferred learning styles of first year medical sciences students: a survey from Iran. J. Clin. Diagnostic Res. **8**, 1–4 (2014)
12. Seidel, L.E., England, E.M.: Gregorc's cognitive styles: preferences for instructional and assessment techniques in college students. Annu. Conv. Am. Psychol. Soc. 1997 (1997)
13. Uzun, A., Goktalay, S.B., Öncü, S., Şentürk, A.: Analyzing learning styles of students to improve educational practices for computer literacy course. Procedia Soc. Behav. Sci. **46**, 4125–4129 (2012)
14. Nedungadi, P., Haridas, M., Raman, R.: Blending concept maps with online labs (OLabs): case study with biological science. In: ACM International Conference Proceeding Series, 10–13 August, pp. 186–190 (2015)
15. Achuthan, K., Francis, P., Diwakar, S.: Augmented reflective learning and knowledge retention perceived among students in classrooms involving virtual laboratories. Educ. Inf. Technol. **22**(6), 2825–2855 (2017)
16. El-Sabagh, H.: The impact of a web-based virtual lab on the development of students' conceptual understanding and science process skills (2011)
17. Narciss, S., Proske, A., Koerndle, H.: Promoting self-regulated learning in web-based learning environments. Comput. Human Behav. **23**, 1126–1144 (2007)
18. Mitra, S., Dangwal, R.: Limits to self-organising systems of learning—the Kalikuppam experiment. Br. J. Educ. Technol. **41**, 672–688 (2010)
19. Konak, A., Clark, T.K., Nasereddin, M.: Using Kolb's experiential learning cycle to improve student learning in virtual computer laboratories. Comput. Educ. **72**, 11–22 (2014)
20. Diwakar, S., Radhamani, R., Sujatha, G., Sasidharakurup, H., Shekhar, A., Achuthan, K., Nedungadi, P., Raman, R., Nair, B.: Usage and diffusion of biotechnology virtual labs for enhancing university education in India's urban and rural areas. In: E-Learning as a Socio-Cultural System: A Multidimensional Analysis, pp. 63–83 (2014)
21. Radhamani, R., Sasidharakurup, H., Kumar, D., Nizar, N., Nair, B., Achuthan, K., Diwakar, S.: Explicit interactions by users form a critical element in virtual labs aiding enhanced education—a case study from biotechnology virtual labs. In: 2014 IEEE Sixth International Conference on Technology for Education, pp. 110–115. IEEE (2014)
22. Diwakar, S., Radhamani, R., Sasidharakurup, H., Kumar, D., Nizar, N., Achuthan, K., Nair, B.: Assessing students and teachers experience on simulation and remote biotechnology virtual labs: a case study with a light microscopy experiment. In: Vincenti, G., Bucciero, A., Vaz de Carvalho, C. (eds.) 2nd International Conference on e-Learning e-Education and Online Training (eLEOT 2015), pp. 44–51. Springer International Publishing, Novedrate, Italy (2016)
23. Sharlanova, V.: Experiential learning. Trakia J. Sci. **2**, 36–39 (2004)

24. Lin, S.-C., Lin, Y.-Y., Lin, J.-Y., Cheng, C.-J.: A study of Kolb learning style on experiential learning. In: Third International Conference on Education Technology and Training, Wuhan, China, pp. 299–302 (2012)
25. Raman, R., Nedungadi, P., Achuthan, K., Diwakar, S.: Integrating collaboration and accessibility for deploying virtual labs using VLCAP. Int. Trans. J. Eng. Manag. Appl. Sci. Technol. 2, 547–560 (2011)

Vocational Education for the Industrial Revolution

Enrique Blanco[✉], Fernando Schirmbeck, and Claiton Costa

Serviço Nacional de Aprendizagem Industrial, Rio de Janeiro, Brazil
{enrique.blanco, fernando.schirmbeck,
claiton.costa}@senairs.org.br

Abstract. Our deal is to present a VET Methodology to Industry 4.0. A VET Methodology that privileges the pedagogical strategies focused on the development of students' competences. We emphasize the required competences for an Industrial Revolution. This is implies not only understanding how the workers can to interact with the machines, equipment and systems, but also identifying which are the technical, social and methodological competences necessary to training people who will interact in these innovative environments. But it is not enough knowing which skills should be taught, but mainly having clear how such skills should be taught and acquired by the students to be applied to the Industry 4.0.

Keywords: Vocational education · Skills · Industry 4.0

1 What Is Industry 4.0 Context?

We are living the fourth Industrial Revolution, known as Industry 4.0. The effects of this revolution already appear at economy and at production of goods and services sectors, but its consequences can influence our vision on science and technology. Are the current and future workers prepared to face this revolution as Industry 4.0 workers? In fact, the ability to make decisions, the virtual connection of processes, the communication between machines (IOT), the product customization to supply customer's demands and Big Data Analytics will not be available for all industrial processes. But the current students and the future workers will be in contact with this new reality in some specific areas of industry and must be trained by a coherent VET to Fourth Industrial Revolution. We understand the urgency of a VET to a Revolution that is going on at the present. It won't occur immediately, but this is the moment to prepare ourselves for this new reality.

1.1 What Professional Is Expected by the Industry 4.0?

The expected professional for this new scenario is the one who is able to articulate the knowledge and technical skills they have learned to overcome situations never faced at VET schools. This is very important, because as the world of work is extremely dynamic and the reality of companies is different, there are no conditions to predict all kind of situations that might occur in every day scenario of an industry. So a definition

© Springer International Publishing AG, part of Springer Nature 2019
M. E. Auer and R. Langmann (Eds.): REV 2018, LNNS 47, pp. 649–658, 2019.
https://doi.org/10.1007/978-3-319-95678-7_71

of list of skill for a 4.0 worker is not enough. There must be a methodology that allows the student to transcend learning and to develop a capacity for self-learning. In this sense, it is not enough to teach technical skills, but it is fundamental to develop social and methodological skills based on a consolidated method in active learning processes. How to develop these skills? Humans will coexist with collaborative robots in the factory environment, robots will be moved by Artificial Intelligence (AI), communicating and deciding without the participation of people through IOT. These machines and equipment mainly work in routine activities, so the professional will not be busy with repetitive activities. Therefore, the Fourth Industrial Revolution companies will need professionals who are not dedicated to routine tasks but, on the other hand, they will need to master specific technologies of Industry 4.0 and will need to control process by having a more strategic and multidisciplinary vision. In addition to this, decision-making, data interpretation and information collected by the automated process will be an important skill to be developed. This causes changes in VET process and it needs specific methodology for this. Workers must be flexible to change rapidly between technologies. Therefore, the VET process will become more strategic than today. Workers must develop and be able to manage their own learning process because they need to be connected with the constant changes of technology, procedures in the current and future world of work. This reality is totally different from the current vocational training program being conducted and therefore we need a VET Methodology suitable for this new reality.

1.2   Who Is the Center of Industry 4.0?

Workers must be able to analyze scenarios, configurations and interpretation of data obtained by automated processes for their strategic decisions. So, the center of the process won't be the machine, but humans. The machines, the artificial intelligent and the intelligent systems will be associated with human beings. We need to have the Artificial Intelligence working to serve the Natural Intelligence. The worker of the Industry 4.0 is the one who develops dynamically his/her competences according the interpretation of the settings of the world of work that is in constant transformation, such as scientific and technological knowledge. In this way, a VET that is based on developing abilities is more appropriate than the teaching and the learning based on contents and subjects. The VET 4.0 Project, an "International Erasmus + KA2 project 2016–2018" presents the challenges for VET to Industry 4.0 reality. They said: "the future development is rather unpredictable and that teachers, trainers, workers and VET students are challenged to learn and to acquire new competences for a new world of work [1]. And more: "So far, this development is not yet a topic in curricula although in some companies working world 4.0 has started to be already reality" [1]. So, a question is proposed: what are the abilities that the professionals need to develop the professional competences for this industry? Are we prepared to develop these abilities for these future professionals? We are currently faced with difficulties in finding strategic and multidisciplinary professionals capable of dominating emerging technologies. We'll present how we develop present and future workers to Industry 4.0 world of work.

2   Vocational Education for the Industrial Revolution

The curricula of the Mechatronics and Industrial Automation courses of SENAI (National Industrial Apprenticeship Service) is linked to the profile of the future employees who will work in the Industry 4.0. The technical and pedagogical staff analyze skills needed to train these workers and establish technical, social and methodological skills that will be involved in the teacher's practice and strategies of our VET Methodology [2, 3]. We develop course plans according to the National Vocational Education Itineraries. These Itineraries presents the VET curricula, which are connected with the transformation and reality of the world of work – these transformations incorporate the reality of the fourth Industry Revolution.

A study about the necessary skills for Industry 4.0 analyzes the BRICS context related the change in demand for core work-related skills, 2015–2020, for all industries and conclude that cognitive abilities, system skills and complex problem solving skills are the three top skills expected to be high in demand and will continue to remain important [4]. In that context the VET occupies a prominent role. To Industry 4.0, the employees will have to gain new skills, but "the core qualification and skills imparted in the current technical and vocational education will still remain important and will have to be updated with the evolution of industry technology" [4]. SENAIs' VET Methodology brings the basic qualification as well as the new skills of the industrial environment to the classroom and to its laboratories where students have access to the current world of work. These strategies focus on active learning: Problem Situation Learning (PBL), Research, Project and Case Studies. These strategies materialize in Learning Situations (Challenges) from the industry work context that destabilize the students by degrees of complexity. The development of technical, social and methodological skills aims at building professional competences to the Industry 4.0 workers. The study mentioned emphasizes that "the majority of (VET) teachers is not aware of the dimension of change and not prepared to face the innovations. There hardly exists learning material for VET students nor does it for the further education of teachers and trainers" [1].

We recognize this scenario. That is why SENAI's current VET Methodology have been training future workers to take decision making and get strategic vision for personal and professional challenges and to solve real problems in the Industry 4.0 Mainly, workers develops the capacity for self-learning and solving problems and challenges because they have been trained through the teaching and learning processes that are based on the Skills' Logic [2, 3, 6–9] and not on the contents' logic. The VET for the industrial revolution is the one that integrates the necessaries areas of knowledge for the development of the required skills to the world of work.

2.1   The Methodology Practice for the Industry 4.0

Teachers of SENAI's VET Schools apply this Methodology that takes several phases performed throughout the teach and learning process. It involves technical analysts in education, pedagogical coordinators, technical coordinators and teachers at schools. We will explain each phase of the application of SENAI Vocational Education Methodology, that aims future professionals to work in Industry 4.0.

Phase 1 – Curriculum Analysis and Skills Identification: we analyze the course curriculum, in this case, the Mechatronics and Industrial Automation and the skills will be taught in each curricular unit. The skills are listed in the internal organization of curricular units, the formative contents, the necessary pedagogical environments (laboratories, classrooms, libraries), the softwares, the remote laboratories, 3D and 2D simulators, the educational kits, machines and necessary equipments. Each curricular unit sets the specific workload to develop their skills throughout the VET course. Each skill that the future professional needs results from the competencies that industry will need. For example, we have an ability that requires "defining the fault analysis procedure that is used to maintain the equipment/control and automation device". Therefore, the logic of subjects based on the simple transmission of content is not coherent with pedagogical strategies. The SENAI VET Methodology structures the pedagogical units denominated curricular units that goes beyond the subjects' logic. Teachers apply the content of the training with coherent and meaningful pedagogical mediation for the student to solve the problems proposed. We expect that the worker will perform satisfactorily applying skills in practice. After this phase, we need to identify how the student will be able to develop necessary skills, as we'll demonstrate in the next phase.

Phase 2 – Selection of Evaluation Criteria: to each skill we define the parameters for the judgment of the quality of the performance expressed by the student, according the performance expected. These criteria allow us to observe if the student will be able to develop the satisfactory skill. Thus, the relationship between what the student expresses and what we expect is fundamental for observable and evaluable performances of each student. To develop the skills that support professional competences, teachers' pedagogical strategies must go beyond the contents and develop technical activities. For this, the teacher should: mediate learning, change the focus of teaching to learn, develop challenging Learning Situations, encourage autonomy, initiative and proactivity, and problem solving skills. Mobilizing knowledges, skills and attitudes to overcome the different challenges of the world of work: this is the aim to develop skills in VET for Industry 4.0. For this, our teachers who come from industry, breaking with the practice of passing on knowledge and repeating routine professional practice and encourage students to overcome challenges. Students become active and main subjects of the teaching and learning process. Tables 1 and 2 define the types of evaluation criteria.

Table 1. Types of evaluation according to quantitative and qualitative criteria

Quantitative evaluation criteria	The focus is on the quantitative explanation for numerical indicators. For the definition of quantitative evaluation criteria, the teacher must be sure about the skills to be developed, aiming to achieve the defined professional profile, and allowing the student to practice them (the skills) that will be faced in several different situations
Qualitative evaluation criteria	The Qualitative Evaluation has as its parameters of judgement, quality criteria such as visual aspect and several other aspects of the product, and also the autonomy, initiative, creativity and student's participation in a given challenge

Table 2. Types of evaluation according to critical and desirable criteria

Critics evaluation criteria	The critics evaluation criteria are essential, that is, they are the criteria that the student must necessarily achieve during the development of some Learning Situation, in order to prove that he/she is able to continue without difficulties. If a certain skill is not achieved by a student in Learning Situations, the teacher must offer new Learning Situations trying different pedagogical approaches allowing the student to developed the necessary skills
Desirable evaluation criteria	The desirable evaluation criteria is also very important. But they are not essential, because its desirable and not critics. This criteria should be developed during the process, but when not achieved, doesn't affect the learning process as a whole. New learning situation should be proposed again throughout the course

Phase 3 – Formative Content Selection: in order to the student develop the skills of curricular unit, the teachers must select the specific content directly linked to a certain skill. That's why we invert the logic of the content-based learning and teaching process because the main purpose of this VET methodology is the development of the skill, so that the content is only a way for its construction, but not the purpose of the VET. The purpose of VET is the development of skills. Our classes are not based on simple presentation of curricular units contents disconnected from a specific goal. For example, when we teach a subject such as the Ohm Law, contents are specifically and pedagogically integrated in order to work as a support to help solving real and practical problems that involves the application of this Law, so the student can make connections with the formative contents in specific practices brought by Learning Situations. The teachers' pedagogical strategies construct specifics and contextualized situations in order to develop the necessary skills. That's the main reason why Skills' Strategy is quite different from contents strategies in VET learning and teaching process.

Phase 4 – Evaluation Process: we have used an Evaluation Process that occurs in three moments: diagnostic evaluation, formative evaluation, summative evaluation. This evaluation process is linked to the development of the skills guaranteeing performance and improvement of the students. Thus, it is possible to connect theory and practice to solve problems based on the interpretation of the settings of the world of the work, which is constantly changing. The student becomes the protagonist of the teaching process and mediator of his/her own learning process. The Fig. 1 presents the workflow of evaluation process.

(a) **Diagnostic Evaluation:** allows the teacher deepening your understanding about students' previous knowledge. This evaluation is done by teachers based on specific assessment tools and techniques, by means of experiential activities and tests that demonstrate the mastery and linked between theory and practice of the students. When students demonstrate that they can not link new knowledge to prior knowledge, we do not use content storage and repetition practices. We have checked the previous knowledge of the students and consider their real experience in order to elaborate the Learning Situations. We contextualize the teaching

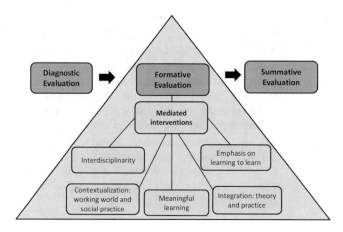

Fig. 1. Workflow of the whole Evaluation Process

strategies according to the students' reality so that the learning process becomes meaningful for them. Diagnostic evaluation is the first phase of the global evaluation process. At this moment, the teacher needs to know the trajectory of the students until they came to school, cause from there, students begin to interact with the teachers and with the contents and the course practices. Table 3 defines what to identify e how to make a diagnostic evaluation.

Table 3. Diagnostic Evaluation Process

What to identify?	Diagnostic evaluation is a way to avoid future difficulties, allowing the teacher to act on the causes of these possible events. Teacher can work in different levels: Group: allows teacher identifies current understanding of the skills, educational level, among other aspects; Routes of teaching and learning: identifies group and individual strategies and the kind of assessment that will be applied during the process; Teachers: identifies available time and the students motivation to define routes for teaching and learning with this group
How to do?	Teachers get information about the students in the school secretary and with other teachers. Pedagogical coordination also provides informations, describing some of the general characteristics of the students, socioeconomic situation, age, information about the progress of the class, difficulties and previous achievement. The diagnostic evaluation exceeds the contents and the classroom, it covers institutional, social and culture issues

(b) **Formative Evaluation:** this evaluation regulates the process of teaching and learning during the pedagogic practice in the classroom. This evaluation is formed by the mediation of learning, emphasizing how to learn to learn, contextualization of the world of work, social practices, meaningful learning and the integration between theory and practice. During the teaching and learning route, teachers should know what their students are learning and how students learn in order to reorient the route if its necessary. This is a continuous evaluation aimed to develop necessary skills. Table 4 defines what to identify e how to make a formative evaluation.

Table 4. Formative Evaluation Process

What to identify?	Formative evaluation analyzes the process during activities, such as the performance of each student. Therefore, teachers analyze: mediate learning: Have I mediated learning process of my students? Meaningful learning: do students understand the meaning of my lessons? Skills development: do Learning Situations help students to develop their skills that support the competences? Students feedback: have I analyzed the students' mistakes and sent them feedback to reorient the teaching and learning process?
How to do?	The formative evaluation reorient the learning process. We work with the guiding principles of teachers practice, as learning mediation. We asked about their difficulties frequently in order to commit them with the teaching and learning process. The student is the center of the process. Motivation strategies are used so that students overcome the challenges of their own professional construction

The process of Formative Evaluation involves pedagogical practices. Teachers develop strategies, teaching techniques and/or dynamics for students to build skills that will be verified in Summative Evaluation, through Learning Situations. For example, social skills are easily developing through experiences. If the goal is to develop teamwork or relationship skills, group activities are ideal. When teachers choose your strategies to teach, they must adapt each strategy to what they want to develop. Teachers consider the quality of the performance they want from the student, in addition to the time available to develop the complete Learning Situation. An important point in this VET Methodology is that these strategies developed by the teachers themselves represent exactly the skills needed for Industry 4.0 workers, as shown in Table 5.

(c) **Summative Evaluation:** Our VET system is based on Certifications. That's why we need qualitative and quantitative results in order to get a final performance. These are The Summative Evaluation, the last step of the overall Evaluation Process. According SENAI VET Methodology, Learning Situations organize educational activities to developing skills during the Formative Evaluation Process. Thus, as the teacher's practice is significant, reading text, long oral expositions and passive learning are replaced by activities that allow the student to apply skills and knowledge to solve real challenges. In Summative Evaluation, teachers work on challenging learning strategies and active learning. This strategy prepares future workers for the reality of Industry 4.0. It is the most appropriate way to evaluate and regulate skill development. We relate formative contexts and evaluation criteria for each challenge. The teacher can choose several challenging learning strategies, as shown in Table 6.

Table 5. Teaching strategies

Practical activity/resolving challenges	Performed in laboratories or workshops, this strategic proposes learn by doing, through theory and practice. Involves cognitive abilities (planning) and psychomotor in execution of process and products (goods or service)
Demonstration	Technics exhibitions, proceeds, machines operation, activities execution related to a job are registered in the teaching situation plan
Technological practice	In specific environment or labs, this activity analyze de quality standard, according to specific standards, rules, composition, viability, functionality of the prototype or products, by means of a specific methodology. Strategies of practice technological are the laboratory tests, among others
Group dynamic	People's motivation is the goal. Teachers should choose creative dynamics that foster student cooperation and mutual acceptance and involvement in a relaxed environment
Group work	Mobilization of students for the collective construction of knowledge and joint activities on the subject. At the same time, group work has important skills such as discussing, choosing, sharing and listening
Dialogued exhibition	Exposition of topics and topics that arouse interest, curiosity, and active student participation using appropriate educational resources. Teachers encourage dialogues, questions, reflections and criticisms using the questions during class
Technical visit	Participation and analysis of actual work processes. The technical visit in companies or industries is a good strategy to break the routine in the classroom. Teachers request reports on specific topics from the technical visit

Table 6. Challenging learning strategies

Case studies	Teachers develop problematic work situations from their own experience in companies and ask students to critically analyze the problem and find different solutions using technical arguments
Research	From a study of a challenging problem a research is performed and data and information are collected to investigate the responses. The search allows the student learn and deepen his knowledge
Problematic situations	Students make decisions for themselves to solve proposed problems and learn to deal with problematic situations that occur in industrial settings. Students do not wait for answers or help from the teacher
Projects	A project starts from a plan of action and produces goods or services. It is flexible, it can involve variables and problems that have not been identified previously, but that emerge during the process

The Table 7 defines what to identify and how to do the Summative Evaluation.

Table 7. Summative Evaluation Process

What to identify?	The whole Evaluation Process (Diagnostic + Formative + Summative) is intentionally based on integral formation of the student and the teaching and learning process. Thus, the result of the Summative Evaluation is the conclusion of the teaching and learning process focused on skill development
How to do?	Summative assessment is also a process, it allows to evaluate student learning at the end of the teaching and learning process by Learning Situations. It allows to decide on the promotion or not of the student, considering the performance achieved

In this last step of Evaluation Process, the Summative Evaluation is applied. The Learning Situation must be contextualized, has sociocultural value, evokes knowledges, problem solving, test hypothesis. Students solve real and practical problems making connections with the formative contents in specific practices brought by Learning Situations. The elements of a Learning Situation are: technical, social and methodological skills to be developed by each curricular unit; the evaluation criteria through which students demonstrate their skills; the selection of the formative contents that allows students to reach their skill development. The Learning Situation results the skill development, based on the evidence of the evaluations criteria.

3   Actual Results

The evaluation criteria focus on skill development rather than simple test evaluation. The class plan uses educational technologies and other pedagogical didactic resources focused on the development of technical, social and methodological skills. The focus is the development of the student and the process of teaching and learning based on the Skills' Logic [2, 3, 6–9]. Teachers from all SENAI VET schools oversee student performance in the industry along with industry supervisors who confirm the effectiveness of our methodology through student performance reports in their work activities. We believe that these results are based on our methodological decision that emphasizes the logic of skills rather than content logic. Learning Situations are not simple tests, but highlight the entire teaching and learning process based on the Evaluation Process we present.

Thus, this Methodology is adequate and consistent to develop futures professionals for the fourth industrial revolution. In "Effects of Industry 4.0 on vocational education and training" study, Sabine Pfeiffer says: "Vocational schools must be modernized, and their teaching staff must be offered continuing education and training – not only in the area of IT and data security, and in the new technologies, but, more importantly, in new and participation-based learning methods" [5]. That's why, is not enough to know what skills should be taught, but mainly how these skills should be taught so that they can materialize in the world of work by the workers of Industry 4.0. This is the focus of our VET Methodology.

4   Conclusions

We've been working with the method in all SENAI VET Schools, in Rio Grande do Sul/Brasil, training teachers to develop Learning Situations and class planning based on the development of technical, social and methodological skills to training workers for the Industrial Revolution. We understand this method can be used for other VET institutions, but it's necessary to train teachers in small groups and develop continuing education courses so that the appropriate project will be in the classroom. Thus, teachers and students can develop technical, social and methodological skills, based on the VET method for industry 4.0.

References

1. Vocational Education and Training in the Working World 4.0: Challenges for VET (2018). http://www.vet-4-0.eu/vet-4-0-53.html. Accessed 1 Dec 2017
2. SENAI: Itinerário Nacional de Capacitação Docente, Brasil (2017)
3. SENAI: Metodologia SENAI de Educação Profissional, Brasil (2013)
4. Aulbur, W., Arvind, C.J., Bigghe, R.: Skill development for Industry 4.0: BRICS skill development working group. Roland Berger GMBH (2016)
5. Pfeifer, S.: Effects of Industry 4.0 on vocational education and training. Institute of Technology Assessment, Austrian Academy of Sciences (2015)
6. Zabala, A., Arnau, L.: Como aprender e ensinar competências, Artmed (2010)
7. Le Boterf, G.: Desenvolvendo a competência dos profissionais, Artmed (2007)
8. Dolz, J., Ollangier, E. (orgs): O enigma da competência em educação, Artmed (2004)
9. Perrenoud, P., Thurler, M.G. (orgs): As competências para ensinar no século XXI, Artmed (2002)

Students' Perception of E-library System at Fujairah University

Ahmad Qasim Mohammad AlHamad[(✉)]
and Roqayah Abdulraheim AlHammadi

IT College, Fujairah University, Fujairah, UAE
aqdl4@yahoo.com, world-senes@hotmail.com

Abstract. The move from conventional to digital libraries isn't just a technological development, it requires an adjustment in the worldview by which people access and connect with data. The main aim of this paper is to study the students' perception of e-library system at Fujairah University, UAE. A questionnaire was conducted of close-ended questions that was distributed to random sample students at Fujairah University, UAE. A simple random sample of 75 students were chosen randomly sample from a larger population to study the students' perception of the e-library system at Fujairah University.

Keywords: Advanced libraries · Conventional libraries · Information overload
e-Resources

1 Introduction

A conventional library is portrayed by the accompanying accentuation on capacity and safeguarding of physical things, especially books and periodicals, listing at an abnormal state instead of one of detail, e.g., author and subject files rather than full content, perusing in view of physical nearness of related materials, e.g., books on sociology are close to each other on the racks and latency; data is physically gathered in one place; readers must travel to the library to realize what is there and make utilization of it.

To differentiate, an advanced library varies from the above in accentuation on access to digitized materials wherever they might be located, with digitization killing the need to possess or store a physical thing, classifying down to singular words or glyphs, perusing in view of hyperlinks, catchphrase, or any characterized measure of relatedness. Materials on a similar subject, should not be almost near one another in any physical sense, communicate innovation; readers require not visiting a computerized library except electronically; for them the library exists at wherever they can get to it, e.g., home, school, office, extra [1].

Advanced libraries are rapidly turning into the standard at schools and colleges here in UAE and abroad as approaches to grow the materials accessible to understudies and to enable them to sharpen their exploration abilities. Yet, this generation age whose thought of research is a speedy troll through Google-should be urged to investigate its school's advanced libraries. According to [2], advanced libraries work at three levels: First is any online data. Second comes specific accumulations, for example, ones for

M. E. Auer and R. Langmann (Eds.): REV 2018, LNNS 47, pp. 659–670, 2019.
https://doi.org/10.1007/978-3-319-95678-7_72

scientists and researchers. Third is a much tighter gathering of papers, inquire about undertakings, or different materials that might be doled out by educators. The last would be like conventional held materials, however with the computerized library, they are accessible all day, every day to more than one individual at any given moment [3].

One difficulty to the utilization of a library's assets, and specifically its electronic assets, is that they are not seen as being direct. As opposed to an Internet web search tool, where a solitary catchphrase inquiry will normally bring about a great many hits, regardless of what the point, in the library, students need to pick a specific database and be more particular in the pursuit words they utilize. Besides, database subjects regularly cover, with contrasts in dates, diary and subjects secured, and whether the material is full-message or not. Also, the library may have a print membership to a specific title that isn't full-message electronically, or the title might be open full-message through another database than the one initially looked. Along these lines, not exclusively do understudies need to locate the applicable references, however they likewise need to know how to find the article after that. This implies juggling many screens, numerous advancements, multi-entrusting electronic occupations, and obviously, knowing where to search for this essential data. Finally, there is the extra perplexity that more library databases utilize Web-based innovations. Since the interface is consistent there does not appear to be an unmistakable, on the screen, distinction between Web-based library assets and general Web-based assets. The greater part of the above likewise expect the understudy is capable in the utilization of PCs. It is very certain that scanning for data has progressed toward becoming "inflexibly connected to PC innovation" [4].

The researchers of this paper studied the students' perception of e-library system at Fujairah University, UAE. The researchers tried to address the following main questions. For what use do students use E-library? How E-library helps students improve competencies? Why we should increase library access and development to student? How we can conserve the usage of e-library and help others use it also? To answer these questions, a questionnaire was conducted of close-ended questions that was distributed to random sample students at Fujairah University, UAE. A simple random sample of 75 students were chosen randomly sample from a larger population. The individuals are chosen at random and not more than once to prevent the bias that would negatively affect the validity of the result. The results of the questionnaire were analysed and will be introduced on the final version of this paper.

2 Discussion

Seeing how students explore this web of assets is vital in helps teachers and web developers to create and evaluate teaching method intended to educate students in library utilization. Students are increasingly Web-smart, huge numbers of them having been raised around PCs and the Internet. Students register with an assorted variety of PC and Web-seeking abilities and experience. Students might not have been presented to library assets, or not know about which assets a library may have, or how to make utilization of them. It is in this way important to us to attempt and comprehend what attributes will influence one understudy to stretch out and investigate library assets, while another might not.

Students' library use is another variable influencing the use of electronic resources. It is reasonable to assume that the more an undergraduate uses the library, the more familiar the student will be with its resources, including its electronic resources. However, if students use the library primarily as a quiet and convenient place to study, they may not be aware of its resources at all, as compared to the student who never puts a foot in the library. Several studies have shown that undergraduates use the library mostly as a place to study and make photocopies, but do not make great use of some of the available library services, such as interlibrary loan and the reference desk [5].

A great part of the writing concerning students' utilization of advanced assets is focused in inquire about related with the data proficiency development. Data education alludes to the way toward distinguishing data sources, getting to the sources, assessing them and utilizing them suitably. Much late research concentrates on student's capacity to look and assess data and how they lead seeks, understudies see themselves as exceptionally gifted or specialists at data proficiency abilities, for example, Internet looking, assessing the dependability and believability of online sources or understanding the moral issues encompassing advanced data.

Appropriate to this research, the Project Information Literacy PIL inquire about takes a gander at students look into capacities in two settings: examine related with a class task and research that students direct for regular day to day existence reasons. According to [6] students had a troublesome time beginning exploration ventures for their classes, checking them and deciding the nature of their endeavors. They were not master clients of the innovation nor the assets accessible to them on their grounds, especially defenders. They likewise found that students utilized library assets (on and disconnected) for scholarly purposes however swung to web crawlers and their family and companions organize for regular day to day existence search. Students revealed that they utilized an assortment of look techniques for course related versus regular day to day existence inquire about, finding that Google, online journals, and Wikipedia were sources to swing to for regular daily existence explore.

Regularly, this was led in open finished organization with students seeking more just wondering and enthusiasm than with an unmistakable heading, as they had with course related research. Course related research, be that as it may, ended up being more disappointing as understudies revealed trouble discovering materials or finding assets. They additionally found that regardless of student's notoriety for being enthusiastic PC clients who are familiar with new advancements, that few of them utilized Web 2.0 application for teaming up on assignments or research. While the PIL examine brought about a depiction of undergrad practices, it neglected to sufficiently control for student's orders, which was intensified by the idea of the exploration venture itself. It is vital for instance, to consider the distinctions of research directed by students in science in correlation with those in the humanities.

3   Methodology

I have chosen the primary data research method, which is collecting original and fresh data [7] for the first time for understanding how students use the digital library and for what it's used. The primary research will be including a questionnaire conducted of questions printed and distributed to students at UOF. The reason questionnaire will be chosen as one of primary research method conducted to employees is because people do not have time to do interviews or focus groups to share their opinions or experience about how they frequently use digital library and for what. According to [8] the best way to get the feedbacks in a minimum time with low cost is through questionnaires.

In this research, the questionnaire is conducted of Close-ended questions that would be distributed to students of UOF. I have used the simple random sampling method which is choosing a randomly sample from a larger sample or population [9]. The individuals are chosen at random and not more than once to prevent a bias that would negatively affect the validity of the result.

4   Findings and Discussions

To collect the questionnaire answers and analyse it, the data should be converted into numerical data and gets transferred in tables, and then transformed to charts by using data analysis tools, which are Pie charts, bar charts, histograms, etc. In this research, I used the pie chart and bar charts to make the data clear and understandable by the human eye.

4.1   Personal Information

According to Fig. 1, the data shows that among the 76 students available at UOF, there were 25% of students age ranges between 18–25, 45% of students age ranges between 26–32, 25% of students age ranges between 33–40, 5% of students are above 40 years old (Tables 1 and 2).

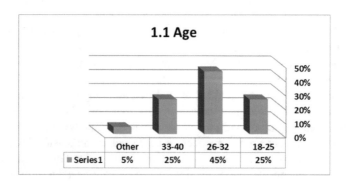

Fig. 1. Age

Figure 2 shows that 36% of students at UOF are males and 64% of the rest are females (Table 3).

Figure 3 shows that 3% of students at UOF hold diploma degree, 83% of students at UOF hold bachelor degree, 13% of students at UOF hold master degree and 1% hold other educational degrees (Table 4).

Table 1. Age

Age	
18–25	19
26–32	34
33–40	18
Others	4

Table 2. Gender

Gender	
Male	27
Female	48

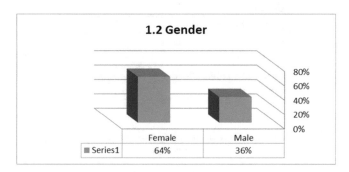

Fig. 2. Gender

Table 3. Level of education

Level of education	
Diploma	2
Bachelor	62
Master	10
Other	1

Figure 4 shows that 40% of students at UOF are enrolled at the IT college, 45% of students at UOF are enrolled at the BAS college, 6% of students at UOF are enrolled at the Arabic college, 8% of student at UOF are enrolled at the PR & media college and 1% of students are at other colleges within UOF (Table 5).

Figure 5 shows that 4% of access the E library from home, 42% of access the E library from the university, 5% of access the E library from their phones and 49% use all access.

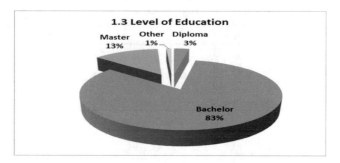

Fig. 3. Gender

Table 4. College

College	
Information technology	30
Business administration	34
Arabic	4
Media and public relation	6
Other	1

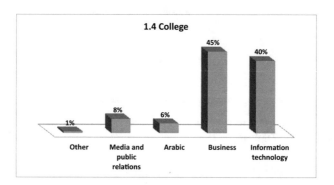

Fig. 4. College

Table 5. Access to e-library

Access to e-library	
Home	3
University	31
Phone	4
All	37

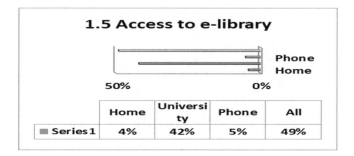

Fig. 5. Access to e-library

4.2   General Questions About E-library

Figure 6 shows that 5% of students visit the E library on daily basis, 35% of students visit the E library once a week, 57% of students visit the E library once a month and 3% didn't visit the library yet (Tables 6 and 7).

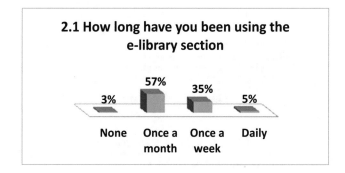

Fig. 6. Library access way

Table 6. Library access way

How long have you been using the e-library section	
Daily	4
Once a week	27
Once a month	43
None	1

Table 7. Hour per week

How many hours you spend weekly using the e-library	
Less than 1 h a week	35
2–3 h a week	36
7–9 h a week	4
Over 10 h a week	0

Figure 7 shows that 47% of students visit the E library for less than on hour in a week, 48% of students visit the E library for 2–3 h a week, 5% of students visit the E library for 7–9 h a week (Table 8).

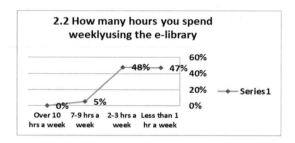

Fig. 7. Hours per week

Table 8. E-library usage purpose

The purpose(s) you mainly use the e-library for	
Research	44
Education	29
Other usage	2

Figure 8 shows that 58% of students visit the E library for research, 38% of students visit the E library for studying (Table 9).

Figure 9 shows that 7% of students face slow internet speed while visiting the library, 10% face difficulties in finding relevant information, 30% see so much load on information in the e-library, 30% face difficulties in loading and downloading files and 23% face difficulties accessing the e library for privacy issues (Table 10).

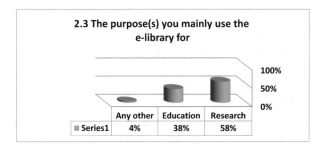

Fig. 8. E-library usage purpose

Table 9. E-library problems

What trouble do you face most frequently in using the e-library	
Slow access speed	5
Difficulty in finding relevant information	6
Overload of information on the internet	26
It takes too long to view/download pages	26
Privacy problem	14

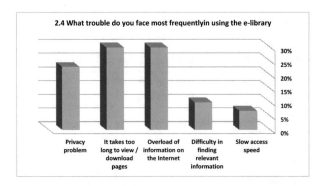

Fig. 9. E-library problems

Figure 10 shows that 4% of student are fully satisfied from e resources while visiting the library, 37% are partially satisfied, 38% are least satisfied and 21% had no comments (Table 11).

Figure 11 shows that 27% of student saw that the e library document are saving time, 11% saw that its more informative, 9% commented that is much expensive,

Table 10. E-library system user satisfaction

Up to what extent, are you satisfied with the e-Resource facilities provided by the university library	
Fully	3
Partially	27
Lease satisfied	28
No comments	15

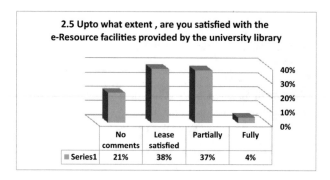

Fig. 10. E-library system user satisfaction

Table 11. Comparison between e-library and traditional library

In your opinion, using e-library as compared to use of conventional document is	
Time saving	26
More informative	11
More expensive	9
Easy to use	15
More preferred	14
More flexible	14
More effective	10

15% commented that it's easy to use, 14% commented that is preferable to them, 14% commented that its much flexible and 10% commented that its more effective (Table 12).

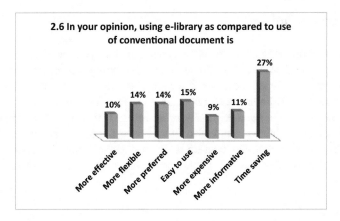

Fig. 11. Comparison between e-library and traditional library

Table 12. Academic efficiency

How the use of e-library has influenced your academic efficiency	
Use of conventional documents has increased	5
Dependency on the e-Resources has increased	28
Expedited the research process	26
Improved professional competence	16

Figure 12 shows that 6% of student saw that e library helped them use conventional documents, 38% saw that its more independent, 35% commented that they are more into research now, 21% improved their competencies.

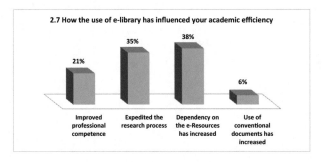

Fig. 12. Academic efficiency

5 Recommendations

Aligning the recommendations to the research question which is why do students use E-library and how E-library helps students improve competencies and why we should increase usage of E-library access and development to student and how we can conserve the Usage of E-library and help others use it also. The following are keys of recommendation that have emerged from the research findings: Student require high speed internet in the library. They also like to have the ability to load and download files quickly. Another recommendation is the implementations of research methods to find relevant information. Also, making each college with different library.

References

1. Ready, R.: Digital vs. Traditional Libraries. Online material retrieved from WTEC, February 1999. http://www.wtec.org/loyola/digilibs/02_03.htm
2. Fosmire, M.: Information literacy and engineering design: developing an integrated conceptual model. IFLA J. **38**(1), 47–52 (2012)
3. Sharp, J.E.: Digital libraries, how do you get students to use them? Retrieved from Stanford (2005). https://tomprof.stanford.edu/posting/673
4. Whitmire, E.: A longitudinal study of undergraduates' academic library experiences. J. Acad. Librariansh. **27**(5), 379–385 (2001)
5. Jacobson, F.: Gender differences in attitudes toward using computers in libraries: an exploratory study. Libr. Inf. Stud. Res. **13**(3), 267–279 (1991)
6. Head, A.J., Eisenberg, M.B.: Lessons learned: how college students seek information in the digital age. Project Information Literacy First Year Report with Student Survey Findings (2009). http://projectinfolit.org/pdfs/PIL_Fall2009_finalv_YR1_12_2009v2.pdf
7. Parab, P.: Slide share. Retrieved from Data Collection-Primary & Secondary (2013). http://www.slideshare.net/parabprathamesh/primary-sec
8. Saunders, D.C., et al.: Collecting primary data. Retrieved from CIPD (2005). http://www.cipd.co.uk/NR/rdonlyres/E4D6775E-07B6-4BCF-A912-C3DE563C3F74/0/1843980649SC.pdf
9. Jewel, M.H.: Sampling. Retrieved from Slide Share (2012). http://www.slideshare.net/jeweliiuc/sampling-13638951

Virtual Working Environment Scheduling of the Cloud System for Collective Access to Educational Resources

Irina Bolodurina, Leonid Legashev, Petr Polezhaev,
Alexander Shukhman$^{(\boxtimes)}$, and Yuri Ushakov

Orenburg State University, Orenburg, Russia
ipbolodurina@yandex.ru, silentgir@gmail.com,
newblackkpit@mail.ru, shukhman@gmail.com,
unpk@mail.ru

Abstract. This paper describes the cloud system for collective access (CSCA) to virtual working environments as a means of providing an economically profitable remote access to paid and free software for educational institutions of secondary education. The problem of efficient CSCA scheduling to optimize the usage of virtual working environments and software licenses has been studied in details. The mathematical model of cloud system resources control is presented. The statistical analysis of fitness function value distribution is performed.

Keywords: DaaS · Scheduling · Cloud system · Statistical analysis

1 Cloud System for Collective Access to Virtual Working Environments

1.1 Introduction

Nowadays many educational institutions of secondary education are facing problems with regular renewal of the computer park and software purchase, installation and configuration, that mostly caused by insufficient financing and lack of qualified stuff. One of the solutions is to create cloud system for the collective access (CSCA) to virtual working environments [1], as it shows on Fig. 1. This system is based on DaaS (Desktop-as-a-Service) scheme [2]. The proposed approach is to create intelligent methods for efficient administration of CSCA resources within constraints.

Users get access to virtual desktops with installed software via Internet browser. Each user forms one request from the organization. The user sets the required number of VM instances, picks needed software from the list and sets a work schedule. The DSS (Data Storage System) contains the VM images and license servers for paid software. The resources orchestrator module is used for optimal scheduling. Administrators and moderators provide the technical support. The users only needed low-cost computers to use CSCA functionality. Collective using of CSCA leads to reduced costs of the software and computer components purchase and opportunity to use modern software in the educational process.

© Springer International Publishing AG, part of Springer Nature 2019
M. E. Auer and R. Langmann (Eds.): REV 2018, LNNS 47, pp. 671–677, 2019.
https://doi.org/10.1007/978-3-319-95678-7_73

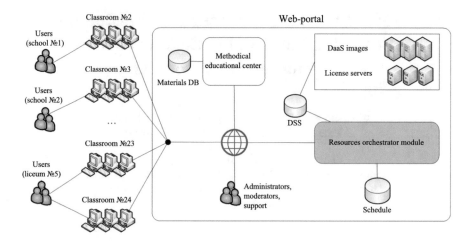

Fig. 1. Cloud system of collective access to virtual working environments

There is a lot of research on scheduling problem, virtual machines placement and cloud technologies using in educational process. One of the most famous books about scheduling algorithms is [3]. Different classifications of scheduling problem are described. NP-completeness of scheduling problem is shown. Exponential-time scheduling is unacceptable for CSCA. Due to this using of heuristic algorithms is needed. Paper [4] described load balancing of virtual machines resources in cloud environments. Authors proposed scheduling strategy of virtual machines placement based on genetic algorithms. SaaS (Software as a Service) model in China's education system is described in [5]. Students get access to the office suite of applications, libraries and cloud applications (GoogleApps, Zoho Office). The main disadvantage is the use of open-source software only. South Ural State University has developed a cloud educational platform "Personal virtual computer" [6], based on EaaS (Education as a Service) model. Access to the virtual desktops is performed by paid software Citrix XenDesktop. Analysis of the existing literature shows the lack of efficient solutions for the CSCA resources administration within the constraints of corresponding timeslots and software license counts.

1.2 Mathematical Model of Cloud System Resources Control

The mathematical model of the cloud system is developed to describe the basic stages of operation of the CSCA and to provide DaaS experience.

Cloud system is defined as tuple $C_{cloud} = \{Z_{temp}, S_{soft}, R_{req}, U_{users}, F_{flav}, D_{data}, \tilde{S}\}$, where: $Z_{temp} = \{Z_{temp}^1, Z_{temp}^2 \ldots, Z_{temp}^i\}$ – set of scheduling templates, $S_{soft} = \{S_{soft}^1, S_{soft}^2, \ldots, S_{soft}^j\}$ – accessible software set, $R_{req} = \{R_{req}^1, R_{req}^2, \ldots, R_{req}^n\}$ – requests, U_{users} – number of users, $F_{flav} = \{F_{flav}^1, F_{flav}^2, \ldots, F_{flav}^m\}$ – set of VM flavors, $D_{data} = \{D_{data}^1, D_{data}^2, \ldots, D_{data}^p\}$ – set of datacenters, $\tilde{S}$ – set of schedules.

Schedule template is defined as tuple $Z_{temp}^i = \{Z_{title}^i, Z_{length}^i, Z_{intervals}^i\}$, where: Z_{title}^i – template title, $Z_{length}^i \in N$ – session duration, $Z_{intervals}^i = \{[z_r, z_r + Z_{length}^i] | z_r \in N\}$ – set of schedule timeslots.

Free and paid software is defined as tuple $S_{soft}^j = \{S_{OS}, S_{title}, S_{license}, S_{install}, S_{delete}\}$, where: $S_{OS} \in \{OS_1, OS_2, \ldots, OS_o\}$ – supported operating system, S_{title} – software title, $S_{license}$ – number of licenses, $S_{install}$ – install time, S_{delete} – deletion time.

VM flavor is defined as tuple $F_{flav} = \{C_m, M_m, D_m\}$, where: C_m – number of cores, M_m – RAM in Gb, D_m – disk space in Gb.

Each request is defined as tuple $R_{req}^k = \{\hat{Z}_{temp}^i, R_{count}^k, \hat{S}_{soft}^k, t_{arrive}^k, T_k, \hat{T}_k, w_k, R_{status}^k\}$, where: $\hat{Z}_{temp}^i \in Z_{temp}, i = 1, .., |Z_{temp}|$ – schedule template, R_{count}^k – number of virtual machines, $\hat{S}_{soft}^k \subseteq S_{soft}$ – subset of software, t_{arrive}^k – request arrival time, $T_k \subseteq Z_{intervals}$ – needed timeslots, $\hat{T}_k \subseteq Z_{intervals}$ – picked timeslots, w_k – weight coefficient, $R_{status}^k = \{-1, 0, 1\}$ – request status. R_{status}^k is equal 1 in case when user's request sets in the schedule at the demanded time, $\hat{T}_k \equiv T_k$. Otherwise, R_{status}^k is equal 0 and user gets access to the CSCA resources at the different time, $(\hat{T}_k \neq T_k) \wedge (|\hat{T}_k| = |T_k|)$. If there is no possibility to satisfy user's request within current constraints of CSCA, then R_{status}^k is equal -1, $\hat{T}_k = \emptyset$.

A cloud system schedule S is represented as follow:

$$S = \begin{vmatrix} R_{status}^1 & \hat{T}_1 & \hat{S}_{soft}^1 & R_{count}^1 & T_1 & \hat{Z}_{temp}^1 \\ R_{status}^2 & \hat{T}_2 & \hat{S}_{soft}^2 & R_{count}^2 & T_2 & \hat{Z}_{temp}^2 \\ R_{status}^3 & \hat{T}_3 & \hat{S}_{soft}^3 & R_{count}^3 & T_3 & \hat{Z}_{temp}^3 \\ \ldots & \ldots & \ldots & \ldots & \ldots & \ldots \\ R_{status}^n & \hat{T}_n & \hat{S}_{soft}^n & R_{count}^n & T_n & \hat{Z}_{temp}^n \end{vmatrix}. \tag{1}$$

Generated requests are added to the queue with FCFS (First-Come, First-Served) service disciplines. The scheduler maximizes the value of fitness function, which is described by Eq. (2):

$$F(S) = \sum_{k=1}^n w_k x_k \rightarrow \max_{S \in \tilde{S}}, \tag{2}$$

$$\forall S \in \tilde{S}, \forall k = \overline{1, n} : x_k(S) = \begin{cases} \alpha \cdot R_{count}^k, & if \ \hat{T}_k \equiv T_k, \\ \beta \cdot R_{count}^k, & if \ (\hat{T}_k \neq T_k) \wedge (|\hat{T}_k| = |T_k|). \\ -\gamma \cdot R_{count}^k, & if \ \hat{T}_k = \emptyset. \end{cases}$$

Software licenses constraints are described by the following Eq. (3):

$$\forall j = 1..|S_{soft}|, \forall l = [t_{start}, t_{end}] : G_{S,j,l} \leq S_{license}^j, \tag{3}$$

where $G_{S,j,l}$ – number of virtual machines running the j-th software at the l-th time interval according to the schedule S.

The number of placed VMs constraints are described by Eq. (4):

$$\forall k = \overline{1,n}: H(R^k_{req}) = R^k_{count},$$ (4)

where $H(R^k_{req})$ – number of placed VMs.

The scheduling problem is NP-complete. We have implemented two evolutionary algorithms: simulated annealing (SA) and genetic algorithm (GA). Initial schedule was creating by using the Round-Robin (RR) method of cyclic load distribution.

Simulated annealing is a stochastic algorithm, which allows finding optimal solution of scheduling problem and avoiding a local maximum of the fitness function value. The current schedule mutates randomly and the fitness function value is calculated in each iteration of SA algorithm. The current schedule is accepted as the best solution if the fitness function value is greater than the previous one. In case when the fitness function value is less than the previous one, the current schedule is also accepted as the best solution with probability (5):

$$P = e^{-\left|\frac{F(s_{current})-F(S_{best})}{t_{curr}}\right|}.$$ (5)

Each individual in the genetic algorithm represents one of the schedule options. The initial population size corresponds to the requests number. To ensure the diversity of the initial schedule options, we apply the Round-Robin method to the random permutations of requests in the queue. The crossover operator provides the inheritance of child the optimal VM assignments from the both parents within the previous population. The mutation operator changes a schedule, which is randomly chosen from the current population. During the selection operation the individuals with the maximum fitness function value (1) are saved as the base for the next population. The detailed implementation of both algorithms is described in [7].

2   Experimental Research

2.1   Cloud System Simulator

To carry out the experiments we implemented the CSCA simulator, using C++ language and Visual Studio 2012 IDE. Later we performed statistical analysis of fitness function value distribution. 150 experiments of requests generation and scheduling were performed within CSCA simulator for a large number of users. The simulated annealing algorithm satisfies 89.7% of initial requests in average. It's 1.88 times more effective than Round-Robin method. The genetic algorithm satisfies 86.5% of initial requests in average. It's 1.48 times more effective than Round-Robin method. The simulated annealing average execution time is 35 ms. The genetic algorithm average execution time is 210 ms.

2.2 Statistical Analysis

Fitness function values are varies in the interval from −40 to 80. We divide this interval into 24 segments, each of length 5, and calculate the hit frequency of the fitness function value (2) in each segment for the proposed heuristic algorithms, as it shown in Fig. 2.

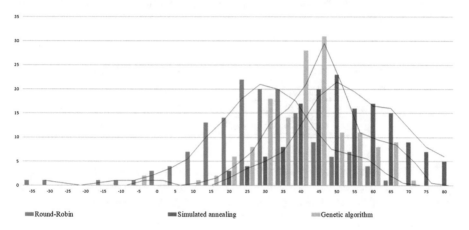

Fig. 2. Interval distribution of fitness function value

After analyzing the graphs, we put forward a hypothesis H_0, that the observed frequency distribution of the fitness function value is a normal distribution. Pearson's chi-squared statistical test [8] (χ^2) is used to accept or reject this hypothesis at significance level $\alpha_P = 0.05$ and the number of degrees of freedom $K = 24 - 2 - 1 = 21$. Test values are compared to one-sided critical value, which is a $K_{crit} = 32.7$ for the corresponding α_P level and the K number. Table 1 shows the statistical test results. Hypothesis H_0 is accepted for SA algorithm data set since the test statistic is less than the critical value of χ^2. Hypothesis H_0 is rejected for GA and RR algorithms data sets since the test statistic exceeds the critical value of χ^2.

Table 1. Pearson's chi-squared statistical test (χ^2)

Algorithm	$\overline{X}$	σ	K_{obs}
Round-Robin	26.26	16.52	124.39
Simulated annealing	49.36	13.91	6.82
Genetic algorithm	38.73	13.26	87.33

In both cases there are outliers in univariate data sets. Grubbs' test [9] is used to detect such outliers and reject them. To test whether the minimum value x_{min_obs} is an outlier, the Grubbs' test statistic is:

$$\tau_1 = \frac{\overline{X} - x_{min\_obs}}{\sigma}. \qquad (6)$$

Critical value of the τ-distribution with significance level of α_P is $\tau_{crit} = 3.81$. For the both GA and RR algorithms the x_{min_obs} and x_{min_obs+1} values were rejected. Table 2 shows the χ^2-test results for the corrected data sets. Hypothesis H_0 is finally accepted for GA and RR algorithms data sets since the test statistic is less than the critical value of χ^2.

Table 2. Grubbs' test to detect outliers in a univariate data sets

Algorithm	$\overline{X}$	σ	K_{obs}
Round-Robin	27.09	15.00	8.11
Genetic algorithm	39.35	12.20	17.96

The hypothesis H_0 is accepted for the all three data samples of fitness function value distribution, hence we can use Student's t-test [10] for independent samples with unequal variances (also known as Welch's t-test). Statistic t_e defined by the following formula:

$$t_e = \frac{\overline{X_1} - \overline{X_2}}{\sqrt{\frac{\sigma_1}{n_1} + \frac{\sigma_2}{n_2}}}. \qquad (7)$$

Table 3 shows the significant differences between the data samples of proposed algorithms. It can be concluded that the simulated annealing algorithm with the highest expected value is the most efficient solution for implementation.

Table 3. Student's t-test for independent samples with unequal variances

Sample 1	Sample 2	t_e	t_{crit}
RR	SA	13.28	1.96
RR	GA	7.71	
SA	GA	6.60	

3   Conclusion

CSCA as the means of providing an economically profitable remote collective access to paid and free software was described. The problem of efficient CSCA scheduling has been studied in details. A new mathematical model which is formalizing the process of cloud system operation was developed. The statistical analysis of fitness function value distribution is performed. Simulated annealing algorithm shows the best results by all criteria.

Acknowledgment. The research has been supported by the Russian Foundation for Basic Research (projects 16-07-01004, 16-29-09639, 16-47-560335), and by the President of the Russian Federation within the grant for young scientists and Ph.D. students (SP-2179.2015.5).

References

1. Shukhman, A.E., Polezhaev, P.N., Legashev, L.V., Ushakov, Y.A., Bolodurina, I.P.: Creation of regional center for shared access to educational software based on cloud technology. In: Global Engineering Education Conference (EDUCON), pp. 916–919 (2017)
2. Eaves, A., Stockman, M.: Desktop as a service proof of concept. In: 13th Annual Conference on Information Technology Education (SIGITE 2012), pp. 85–86 (2012)
3. Brucker, P.: Scheduling Algorithms. Springer, Heidelberg (2007)
4. Hu, J., Gu, J., Sun, G., Zhao, T.: A scheduling strategy on load balancing of virtual machine resources in cloud computing environment. In: Third International Symposium on Parallel Architectures, Algorithms and Programming (PAAP), pp. 89–96 (2010)
5. Wang, B., Xing, H.: The application of cloud computing in education informatization. In: IEEE International Conference on Computer Science and Service System (CSSS), pp. 2673–2676 (2011)
6. Kostenetskiy, P.S., Semenov, A.I., Sokolinsky, A.I.: Creation of educational platform "Personal Virtual Computer" based on cloud computations. In: International Supercomputer Conference, pp. 374–377 (2011)
7. Bolodurina, I.P., Legashev, L.V., Polezhaev, P.N., Shukhman, A.E., Ushakov, Y.A.: Request generation and intelligent scheduling for cloud educational resource datacenter. In: 8th International Conference on Intelligent Systems, pp. 747–752 (2016)
8. Pearson, K.: On the criterion that a given system of deviations from the probable in the case of a correlated system of variables is such that it can be reasonably supposed to have arisen from random sampling. London Edinburgh Dublin Philos. Mag. J. Sci. **50**(302), 157–175 (1900)
9. Grubbs, F.: Procedures for detecting outlying observations in samples. Technometrics **11**, 1–21 (1969)
10. Ruxton, G.D.: The unequal variance t-test is an underused alternative to student's t-test and the Mann-Whitney U test. Behav. Ecol. **17**, 688–690 (2006)

Activities of Euro-CASE Engineering Education Platform

Petar Bogoljub Petrovic[1,2] and Milos Srecko Nedeljkovic[1,2(✉)]

[1] Faculty of Mechanical Engineering, University of Belgrade, Belgrade, Serbia
{pbpetrovic,mnedeljkovic}@mas.bg.ac.rs
[2] Academy of Engineering Sciences, Belgrade, Serbia

Abstract. A platform dedicated to the problems of engineering education (EngEdu) has been established by European Council of Academies of Applied Sciences, Technologies and Engineering (Euro-CASE). The renewed working group (WG) of this EngEdu Platform started to work in Sept. 2017 and now tries to establish a direct link with REV conference participants and organizers, and disseminate and exchange ideas and results. Current topics of interest are presented as well as one of the projects going on.

Keywords: Engineering education · Euro-CASE

1 Introduction

Recognizing the importance of engineering education and the influence of new e-technologies, the European Council of Academies of Applied Sciences, Technologies and Engineering (Euro-CASE) established a platform dedicated to the problems of engineering education [1]. Nine national academies of engineering sciences are actively contributing to the renewed Working Group (WG) with the following representatives: Prof. Albert Albers, acatech, Germany; Prof. Hanna Bogucka, PAN, Poland; Prof. Gerard Creuzet, NATF, France; Prof. Janez Možina, IAS, Slovenia; Prof, Kurt Richter, ÖAW, Austria; Prof. David Timoney, IAE, Ireland; Prof, Petr Zuna, EACR, Czech Republic; Prof. Nick Tyler, RAEng, United Kingdom; and Prof. Petar B. Petrovic, Chairman, AESS, Serbia (with valuable contributions of Prof. Vlastimir Matejic, AESS, Serbia).

The first kick-off meeting of the renewed working group (WG) of this Engineering Education (EngEdu) Platform was held in September 2017, in Belgrade, Serbia, so its work is in full progress now.

2 Basic Outline

When it comes to higher education, the European Commission, and above all, EC Directorate General for Education and Culture (DG EAC), almost invariably link their strategic goals, initiatives, activities to the integral corpus of higher education, with the general view that: "National governments are responsible for their education and

M. E. Auer and R. Langmann (Eds.): REV 2018, LNNS 47, pp. 678–685, 2019.
https://doi.org/10.1007/978-3-319-95678-7_74

training systems and individual universities organize their own curricula. However, the challenges facing higher education are similar across the EU and there are clear advantages in working together" and in this regard further states: "The European Commission works closely with policy-makers to support the development of higher education policies in EU countries in line with the Education and Training 2020 strategy (ET2020). The modernization agenda for higher education fixes five key priorities for higher education in the EU."

In contrast to this attitude and despite the fact that engineering is the key driver of innovation (in terms of technological innovation as a crucial factor for job creation, economic growth, and prosperity, which is explicitly stated in the above quote), there is no explicit focus of DG EAC on engineering education.

The above mentioned, but also a broader analysis of the EC position in policy-making for higher education in Europe, can lead to several basic findings that are highly relevant for defining the framework of Euro-CASE undertaking to build the Engineering Education Platform:

1. **There is no explicit visibility of engineering education** in the broad corpus of EC policies that are related to higher education in Europe;
2. **The role of the EC in the field of higher education is predominantly focused on the national level** of the EU Member States, and limited to the support for reforms of national education systems and actions through a cooperation framework known as Strategic Framework for European Cooperation in Education and Training, ET 2020, which comprises the development of complementary EU-level tools, mutual learning and the exchange of good practice via the open method of coordination; in fact, EU higher educational system is entirely the responsibility of national governments;
3. **EC higher education is directly linked to research and innovation**, which forms a generic basis for fulfilling the vital need of Europe to create jobs, economic growth, and prosperity – basically, there is quite a strong foundation to establish a correlation between the strategic objectives in the field of higher education and European Innovation Policy which is organized within the context of the Innovation Union.

However, in implicit terms, the EC associates many of its strategic orientations and policies most directly with engineering and then consequently with engineering education. For example: the strategic orientation to the re-industrialization of the European economy (European Industrial Renaissance, and the related Industry 4.0, as a reflection of the need for mass digitization of manufacturing processes, i.e., transformation of the existing production base towards Cyber-Physical Production Systems in order to ensure the highest possible performance of competitiveness, sustainable leadership of the European industry through the framework in global terms), corpus of policies and instruments related to the strategic necessity to direct science to economy and market (European Technology Platforms, Key Enabling Technologies, Public-Private Partnership for Research, Horizon 2020 – Thematic priority 2: Industrial Leadership, just to mention a few of them).

2.1 The Value-Added Contribution of the Platform

Prospective users of the Platform findings: EU Commission, Euro-CASE, National Academies of Engineering, higher education institutions across Europe, national governments and other stakeholders.

Value to European Commission - There is an obvious gap between the explicitly expressed opinion of the EC on engineering education and opinion on the implicitly expected role of engineers in society, i.e. the contribution the EC expects from engineering in the implementation of the strategic objectives of Europe and its key development policies and policies related to the Grand Societal Challenges. In terms of the above observation, (if it is sustainable?), the key value added contribution to the EC should, on the one hand, highlight the discrepancy, or a kind of contradiction, and provide evidence to support this claim, and, on the other hand, propose specific recommendations, measures and activities to successfully respond to this kind of challenge, with a wider consensus of stakeholders which necessarily involves the level of Member States (understanding of EU diversity and specificity of engineering education and attempt to develop an initiative for creating something like Smart Specialization Strategy for Engineering Education – S3 for EngEdu in Europe). The point is to provide the EC with engineers' perspective and understanding of the current state and the future of engineering education in Europe.

In addition to the foregoing, the benefits the EC can potentially gain from the findings and other content presented in EngEdu Platform are as follows:

1. Insight into the current state of engineering education within the EU;
2. Understanding the non-compliance of the role of engineering and engineering education outcomes – at European, regional and national level;
3. Identification of desirable changes in the engineering education system;
4. Identification of the actual and desirable "common European core" in the educational profile of engineers;
5. Modernization of the engineering education system;
6. Synthesis of measures and policies for increasing interest of the upcoming generation in engineering;
7. Synthesis of measures and policies for adequate diffusion of "good practice" in engineering education;
8. Other (the list is not exhausted here – the subject of further discussion and agreement).

2.2 Approach

Engineering education is a subject that Euro-CASE has been engaged in for many years through the formal framework – Engineering Education Platform – EngEdu Platform. EngEdu Platform I (the renewed one is II) came to an end in 2011, producing three Final Reports summarizing results of the series of questionnaires filled in by the member academies: (1) Inspiring the young generation – good practice case studies in Europe, (2) Ranking in Engineering Sciences, and (3) Bologna process 2010.

In May 2013, Euro-CASE Board re-established this Platform, i.e., EngEdu Platform II, chaired by Professor Petr Zuna (Engineering Academy of the Czech Republic – EACR). In the further work, the continuity will be achieved through valuable contributions generated in the previous period, whereby it is suggested to significantly expand the scope so that a holistic approach would be used: (1) to cover the total area of engineering education, (2) to make Euro-CASE activities in this area more visible to key external stakeholders, both at EU and at national level of Member States, and partly at the global level, and (3) to make the entire process of achieving the objectives of the Platform more convergent and efficient through a new methodological framework.

When selecting objectives, topics that will be covered and methodology of the work, one should be aware of the fact that engineering education is exceptionally complex and broad area. Due to the interconnectedness of different social interests and different perspectives, this is also an extremely sensitive subject, with many unresolved issues for which it is quite certain that it is almost impossible to find a compromise solution acceptable to all. In this context, it is extremely important to identify the unresolved issues for which it is quite clear that compromise solutions, acceptable to all or majority, can be found, and make a clear distinction between them and "difficult" topics. The following is proposed to be taken as an initial set of highly relevant topics of general character:

1. The current status of engineering education (from the perspective of the stated or assumed role that engineering should have) as seen by: (a) employers and their associations, (b) scientific and higher education communities, (c) Government authorities, and (d) other institutions engaged in the development at the European, regional and national level.
2. Desirable and possible changes (improvements) in the current state of engineering education in terms of the expected role of engineering for future technological, economic, social, environmental and other development.
3. National, regional and European system policies, measures and key actors for performing desirable changes in engineering education.
4. Outline of the system for monitoring and evaluating the impact of changes on systems of educating engineers.
5. Other (list of proposed topics is not exhausted by this – the subject of further discussion and harmonization);

whereby, within the proposed, or possibly extended list of general topics, 3 to 5 subtopics would be chosen, i.e. particular thematic priorities with a high degree of relevance for the Platform objectives, which would be a subject of further analysis. The selection of these particular thematic priorities is the subject of EngEduWG future discussions and agreement.

In terms of determining the appropriate methodological approach, the following set of questions should be considered/answered:

1. For which level the role of engineering is expressed or assumed in the coming overall development: the European (European development strategies), regional (in case there are regional development strategies), national (unavoidable)?

2. Data collection, identification and analysis of the current situation, proposed changes – redefinition of: (a) social aspects of the role of engineering and (b) engineering education system, carried out at the national level (i) exclusively according to the unique methodology prescribed in advance by EngEduWG, or (ii) according to the methodology freely defined by each national Academy, or (iii) using a kind of "mixture" of (i) and (ii)?
3. The whole work on this Platform is: (a) fully defined, carefully planned, reliably monitored and carefully redefined whenever necessary, or (b) carried out in a way that follows the Soft System Methodology (SSM)?
4. The financial plan of revenues and operating costs of the Platform is: (a) made in advance, followed and redefined in justified cases, or (b) there is no ex ante approach so financial issues are resolved on a case-by-case basis, from the ongoing until the follow-up meeting of EngEduWG or other planned activity of the Platform? This topic is discussed below;
5. Other (list of unresolved issues is not exhausted by this - the subject of further discussion and harmonization);

The main deliverable of the Platform is Euro-CASE Position Paper on Engineering Education (in Europe, or in general?) – Euro-CASE PP on EngEdu, which uses arguments to express the opinion of Euro-CASE through findings, beliefs, attitudes and recommendations on the main issues of engineering education from the perspective and understanding of engineers.

3   Working Resources

The main working resource of Engineering Education Platform is a voluntary ad-hoc working group – EngEduWG, consisting of representatives of the Academies, members of Euro-CASE, which have expressed interest and willingness to actively engage in the Platform. Strength, agility and dedication to EngEduWG, as well as the commitment of member Academies to actively support this undertaking, are a critical factor in the success of the Platform. EngEduWG will be formed upon the initiative of AESS and the invitation extended by Euro-CASE Secretariat in Paris to all member Academies.

In addition to engaging Euro-CASE internal working resources, **a significant factor in the success of the Platform is the external interaction with institutions and experts relevant to engineering education at the EU level** (in particular EC DGs) and equally important at national and regional level. It is also advisable to establish interaction with relevant institutions in non-European regions, in particular the USA and developed and fast-growing Asia economy (the potential of Euro-CASE partner engineering academies should be used). Besides being a very respectable resource for collecting valuable information, EduEngWG can make its work more visible and recognizable through external interactions, build mutual understanding and dialogue, and establish some kind of (potentially permanent) informal partnerships.

4   Financial Resources

The work on Engineering Education Platform is a complex undertaking that requires substantial financial resources. The issue of financing is unresolved at the moment and requires further consideration as a matter of great significance for the successful implementation of the entire undertaking. Generally speaking, there are different possibilities for financing the work of the Platform, starting from the national, then regional and finally EU level, where potential sources of support for such initiatives can be found.

5   Current Activities

EngEdu WG established key objectives of the Platform; methodology; organization of work and a plan of overall activities on the Platform until its completion, in order to work on; Evidence-based scientific policy advice to EC, external visibility (general and professional public, media); interaction with the stakeholders outside Euro-CASE, in particular EC and EU institutions that are directly or indirectly interested in EngEdu; financing the work of the Platform; etc.

Current focus of the EngEdu WG activities is in two priority topics:

1. **Knowing–Doing Gap in Engineering Education**—Understanding and bridging the gap between theory and practice in the European engineering education, with particular focus on the following aspects: 1. Excellence through INTERDISCIPLI-NARITY and adaptability to RAPID TECHNOLOGY INNOVATION cycles; 2. Integration of Industry 4.0 in Engineering Education—Challenges for the human factor in the next generation manufacturing (Survival in Industry 4.0 requires Engineering Education 4.0!); 3. New BUSINESS THINKING, Taking RISKS and Dealing with UNCERTAINTY—The importance of soft skills and how to collaborate in the "global village"/business thinking and entrepreneurial components in the curricula of engineering education, and
2. **Big Data and Learning Analytics**—Technology-enhanced teaching and learning environments, and decision making in engineering education through educational data mining and learning analytics, with focus on the following aspects: 1. New Understanding of Learning Processes based on educational data mining and learning analytics—identifying constructs of student cognition to promote in technology-enhanced learning environments; 2. Creating capacities for effective and meaningful collecting, analyzing and visualizing information from educational Big Data; 3. Open Pedagogy/Education and related opportunities, potentials, and challenges for building a European model of engineering education—Engineering Learning 4.0 (human-machine collaboration/interaction/networking in the educational immersive digital environment).

6 Enabling Web-Based Remote Laboratory Community and Infrastructure - SCOPES Project of Swiss National Science Foundation [2]

Academy of Engineering Sciences of Serbia (AESS), through the University of Belgrade Faculty of Mechanical Engineering (UB-FME) is actively participating in the above mentioned project dedicated to the sophisticated engineering education in the field of complex fluid measurements. The project leader is Dr. Denis Gillet on behalf of the EPFL - Ecole Polytechnique Federale de Lausanne.

The project is in accordance with the two priority topics of Euro-CASE, thus making concrete contribution, namely: understanding and bridging the gap between theory and practice in engineering education; technology-enhanced teaching and learning environments, creating capacities for effective and meaningful collecting, analyzing and visualizing information; and open pedagogy/education opportunities, potentials, and challenges for building a European model of engineering educa-tion—Engineering Learning 4.0.

Figures 1, 2 and 3 briefly show some illustrative examples which are described in detail in [3, 4] and could be better studied there.

Fig. 1. Remote educational set-up.

Fig. 2. LabView screen for the set-up control and measurements.

Namely, the setup for remote experimental measurements for the students studying fluid machinery has been originally designed and manufactured. Measurements can be made locally, as well as remotely. Distance learning-by-doing is under implementation and the final results will be reported until the end of project by mid-2018.

The research and results of this project are in direct correspondence with the aims described in two priority topics of Sect. 5 in this paper. The outline of the project has been presented during the kick-off meeting of the EduEng WG in Belgrade.

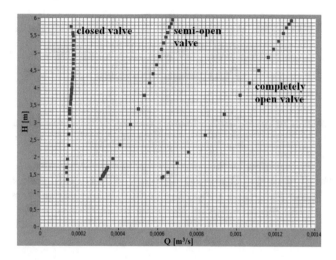

Fig. 3. Screenshot of the pipe system characteristics measurements.

7   Anticipated Outcomes

Anticipated outcome is a stronger connection and cooperation between REV conference and EngEdu WG, and dissemination of activities of both institutions. Also, the incorporation of common directives into a teaching practice.

Actual current outcome is presented through a special topic more elaborated in the paper as Creating and Using Remote Labs for Engineering Education - Enlarging capacities, open educational access and learning by doing remotely, as a topic elaborated during the Kick-off meeting, and already connected to the well-established GoLab project (EPFL - Ecole Polytechnique Federale de Lausanne).

References

1. http://www.euro-case.org/index.php/activites/item/21-engineering-education.html
2. http://www.go-lab-project.eu/partner/%C3%A9cole-polytechnique-f%C3%A9d%C3%A9rale-de-lausanne
3. Nedeljkovic, M.S., Cantrak, Dj.S., Jankovic, N.Z., Ilic, D.B., Matijevic, M.S.: Virtual instrumentation used in engineering education set-up of hydraulic pump and system. In: Proceedings of the 15th International Conference on Remote Engineering and Virtual Instrumentation (REV2018), 21–23 March 2018, pp. 341–348. University of Applied Sciences Duesseldorf, Germany (2018). Paper-ID: 1166
4. Nedeljkovic, M.S., Jankovic, N.Z., Cantrak, Dj.S., Ilic, D.B., Matijevic, M.S.: Engineering education lab setup ready for remote operation - pump system hydraulic performance. In: Proceedings of the 2018 IEEE Global Engineering Education Conference (EDUCON), 17–20 April 2018, pp. 1175–1182. Santa Cruz de Tenerife, Canary Islands, Spain (2018)

Virtual Instrumentation Used in Engineering Education Set-Up of Hydraulic Pump and System

Milos Srecko Nedeljkovic[1], Djordje Cantrak[1(✉)], Novica Jankovic[1],
Dejan Ilic[1], and Milan Matijevic[2]

[1] Hydraulic Machinery and Energy Systems Department,
Faculty of Mechanical Engineering, University of Belgrade, Belgrade, Serbia
{mnedeljkovic,djcantrak,njankovic,dilic}@mas.bg.ac.rs
[2] Department for Applied Mechanics and Automatic Control,
Faculty of Engineering, University of Kragujevac, Kragujevac, Serbia
matijevic@kg.ac.rs

Abstract. Numerous students attend courses in hydraulic turbo pumps at the Faculty of Mechanical Engineering University of Belgrade, as well as from other faculties. They have oral lectures and exercises with numerical tasks, but also laboratory exercises. This easily driven upgraded installation is very convenient for laboratory exercises for various groups of students. In this paper is demonstrated upgrade of the present installation with transducers instead of U-tube analogue manometers, acquisition system and LabVIEW application for hydraulic pump diagnostics and flow meter calibration. Pressure transducers are also previously calibrated. This installation has three regulating valves and two for tanks emptying. In this way students could form two simple and one complex pipeline. Serious and parallel combinations of pipe lines could be demonstrated, as well as pump by-pass regulation. Hydraulic pipe curve could be formed for one position of the valve on the pump pressure side and various pump frequencies. This could be repeated for other valve positions. In addition pump head characteristic for one rotation number is formed by changing the valve position. Good overlapping with manufacturer's curve is demonstrated. Cavitation and turbulent swirling flow phenomena could be also visualized and demonstrated on this installation with transparent pump housing, as well as partly suction and pressure pipes.

Keywords: Hydraulic pump · Laboratory work · Virtual instrumentation

1 Introduction

Bernoulli and continuity equations, as well as Euler equation for turbomachinery are theoretical background for the lessons presented on the designed installation at University of Belgrade, Faculty of Mechanical Engineering, Laboratory for Hydraulic Machinery and Energy Systems (UB FME HMES) [1–4]. Following lessons of applied fluid mechanics and specific topics in hydraulic turbomachinery could be presented here: calibration of the non-standard Venturi flow meter, simple and complex pipe systems,

© Springer International Publishing AG, part of Springer Nature 2019
M. E. Auer and R. Langmann (Eds.): REV 2018, LNNS 47, pp. 686–693, 2019.
https://doi.org/10.1007/978-3-319-95678-7_75

determination of the centrifugal pump head and flow for various regimes, i.e. pump characteristic curve after standard ISO 9906 [5], pump regulation with valves and frequency, as well as demonstration of cavitation and turbulent swirling flow.

Numerous students attend courses in hydraulic turbo pumps at UB FME HMES, as well as from other faculties and sometimes in courses in industry. They have oral lectures and exercises with numerical tasks, but also laboratory exercises. This demonstration-educational installation is very convenient for laboratory exercises for various groups of attendees. In this paper is demonstrated upgrade of the existing installation with calibrated transducers instead of U-tube analogue manometers, acquisition system and LabVIEW application for hydraulic pump diagnostics and flow meter calibration [3, 4, 6].

Data in e-form and consequent automatic recalculation of measured values and generation of e-report is possible.

2 Demonstrational Installation

Test rig, its functionality and possible experiments are described in the following sections.

2.1 Test Rig

The demonstrational-educational test rig, designed and manufactured at HMES, is presented in Fig. 1.

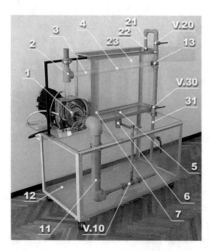

Fig. 1. Demonstrational-educational installation [2].

The main elements of the installation are: 1–pump with the transparent casing, 2–elbow, 3–transparent pipe, 4–transparent Venturi flow meter, 5 and 6–valve for emptying the upper tank, 7–scale on the calibration reservoir, V.10–suction valve,

11–suction pipe, 12–lower reservoir (volume is 250 l), 13–T-joint, V.20–valve, 21–pipe for filling the upper reservoir, 22–nozzle, 23–upper reservoir (volume is 55 l), V.30–valve, 31–pipe to lower reservoir. Used pump is Grundfos, model UPE 50-120 F, in-line type. Its original casing is replaced with a new transparent classical one, with axial inlet and radial outlet. Pump speed could be controlled manually on the pump housing or by infrared remote controller, type R-100.

Connected differential pressure transmitters at positions 8 and 9 in Fig. 2, are TPd-101, manufactured by Institute of Chemistry, Technology and Metallurgy, Department of Microelectronic Technologies, University of Belgrade, Belgrade, Serbia. They have measurement range 0–3 bar, power supply 14–26 VDC, output 4–20 mA. They are calibrated on the oil deadweight tester in LHM and calibration curve is imported in the LabVIEW application.

2.2   Test Rig Functionality

Scheme of the test rig is presented in Fig. 2.

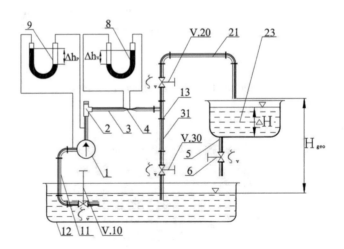

Fig. 2. Demonstrational-educational installation.

On this test rig following laboratory exercises could be demonstrated:

- Venturi flow meter calibration (valve V.30 is closed);
- Pump testing procedure after standard ISO 9906 [5];
- Pump cavitation test;
- Pump circulates water in the main, lower reservoir $H_{geo} = 0$ (valve V.20 is closed);
- Pump transports water on the specific geodesic height $H_{geo} = $ const (valve V.30 is closed);
- Pump regulation with by-pass. In this case pipes on the pump pressure side are in parallel and regulation is performed by valve V.30;
- Energy efficiency issues: Comparison of pump regulation with throttling valve, pump speed and by-pass.

Numbers of the test rig elements in Fig. 2 correspond to those in Fig. 1. A pressure transmitter on the pump suction side could be added for cavitation experiments. All pressure transmitters are presented in Fig. 2 with U-tubes for better physical interpretation and understanding of the pressure values and differences.

Students first learn how to start and stop the centrifugal pump by following the next procedures:

1. Check if the pump is filled with water.
2. If not, open the cap on the T-joint 2.
3. Close valves V.10, V.20 and V.30.
4. Fill up with approximately 3.5 l of water.
5. Close the cap on the T-joint 2 and start the pump.
6. Open the valve on the pump suction side V.10.
7. Slowly open the valve(s) V.20 and/or V.30.

   Closing procedure is:

1. Close the valve(s) V.20 and/or V.30. Remark: All valves on the pump pressure side should be closed.
2. Close the valve V.10.
3. Turn off the pump.

Pump pumps water from the lower reservoir (12) through suction pipe (11) with valve V.10, pipe (3) and Venturi flow meter (4) to the T-joint (2) to both reservoirs, if valves V.20 and V.30 are open.

2.3 Experimental Determination of the Pump Characteristic Curve

In this case valve V.20 is closed. Pump speed is adjusted to one value and valve V.30 is also closed. The flow rate is zero now. Pump head (H) is defined as follows:

$$H = \frac{p_{II} - p_I}{\rho g} + \frac{c_{II}^2 - c_I^2}{2g} + (z_{II} - z_I), \tag{1}$$

where p is an average pressure, c is an average velocity, z is a geodesic height of the cross-section center, ρ is a fluid density, g is an acceleration due to gravity, indexes I and II denote pump inlet and outlet measuring cross-sections, respectively. Measuring cross-sections are defined after ISO 9906 [5]. Pressure difference is measured by use of the differential pressure transmitter 9. Average velocity (c) is calculated in the following manner:

$$c_i = Q/A_i, \tag{2}$$

where A_i is a surface of the inner cross-section ($A_i = D_i^2 \pi/4$, where D_i is an inner diameter in the measuring cross-section, $D_I = 56.2$ mm and $D_{II} = 30$ mm) and i = I, II. Difference in geodesic heights is constant and is $\Delta z = z_{II} - z_I$. Flow rate is measured by

Venturi flow meter, i.e. by differential pressure transmitter 8. Electric motor consumption (P) is measured with electric power meter. After all these measurements pump unit efficiency (η) could be determined as follows:

$$\eta = \rho QHg/P. \tag{3}$$

Now, valve V.30 is open slowly, in steps, to the fully open position. Pump characteristic curve could be determined now for the specified speed. This could be repeated for other pump speeds.

2.4   System Characteristic Curve Experimental Determination

If it is our intention to determine the system characteristic curve in the $H_{geo} = 0$ mode, valve V.20 should be closed. and valve V.30 is, for the beginning, in the completely open position. Its curve could be determined by varying pump rotational speed (Fig. 6). Closing the valve V.30 and varying the pump rotational speed would generate another pipe hydraulic curve. Students could also observe a complex hydraulic system (pipes working in parallel) by opening both valves V.20 and V.30. Pump duty point is now changed. Changes in flow direction in pipe 21 (to upper tank, no flow and to the T-joint (13)) can be demonstrated by changing the pump speed and consequently, generating different energy in the T-joint (13). In this installation could be performed also experiments with cavitation - initial and 3%, but they are not presented here due to the scope of paper.

3   LabVIEW Application

LabVIEW application is developed for this hydraulic system [3, 4, 6]. Students can follow the procedure described in the previous section and record signals from various differential pressure transmitters. In this case these are two transmitters connected, one for pump head measurements, while the second one for pump flow rate. Front panel is shown in Fig. 3, where: 1 - stop, 2 - start, 3 - pause, 4 - record, 5 - chart for experimentally obtained pump characteristic curve (Q-H curve), 6 - pump current efficiency, 7 - pump current head in [m], 8 - pump current flow in [m³/s], 9 - pump current head in [kPa], 10 - display of the current pressure transmitter connected on Venturi flow meter in [kPa], 11 - current pump duty point (red square). This red square denotes real time pump measurements, This is useful due to the fact that each new valve V.30 position needs first flow stabilizing and afterwards measurements. Pump characteristic curve is formed by closing the valve V.30 (Fig. 3).

Pressure transmitters' calibration curves are imported into the generated software and connected to the eight channel input module National Instruments NI-9203. USB CompactDAQ chassis cDAQ-9174 is connected to a laptop. In the developed, LabVIEW application exist four screens. In Fig. 4 is presented the first "welcome" or input data screen.

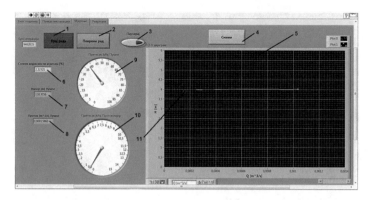

Fig. 3. LabVIEW application front panel [3, 4].

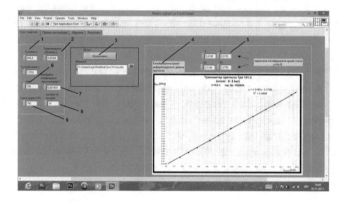

Fig. 4. LabVIEW application first (welcome or input) screen [3, 4].

In input screen are presented following necessary data: 1 - water density, 2 - acceleration due to gravity, 3 - possibility to export data, 4 - calibration curve of the differential pressure transmitter, 5 - calibration curve coefficients, 6 - acquisition card measuring frequency, 7 - calibration coefficient of the Venturi flow meter, 8 - number of samples, 9 - pump unit power. Most of these data are necessary for further measurements and data processing. Students generate, by themselves, in groups, differential pressure transmitters calibration curves and import them into the presented LabVIEW application.

In Fig. 5 are presented experimentally obtained results for three pump speeds. They are important for students' electronic reports: 1 - experimentally obtained pump duty point, 2 - date of measurements, 3 - pump flow rate, and 4 - pump head. Sampling rate could be adjusted in the recording mode and all presented measured data are averaged. Pump speed n_1 is the highest, while n_3 is the lowest of these presented in Fig. 5.

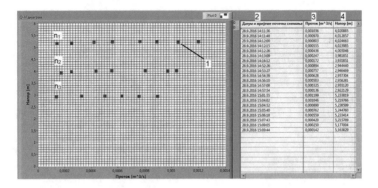

Fig. 5. Experimentally obtained and presented results [3, 4].

Pump characteristic curves are obtained for various positions of the valve V.30. Various parameters could be followed on the front panel, such as pump head and flow rate, water density, pump speed and etc. Some new controls are planned to be installed and connected such as all three valves control. This test rig upgrade is in progress. Diagrams, as well as results in table format, can be exported for students' electronic reports (Fig. 5). In this LabVIEW application could be determined what, in addition, to be exported for students' data processing necessary for reports and oral exam. In addition to pump characteristic curve determination, students can determine system characteristic curves. This can be done, for example, for the installation without geodesic height for various positions of the valve V.30, as described in Sect. 2.4. Three pump and system characteristic curves are presented in Fig. 6.

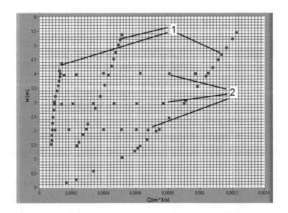

Fig. 6. Three pump and system characteristic curves [3, 4].

All these points in Fig. 6 are, in fact, pump duty points measured with two differential pressure transmitters. Students demonstrate their knowledge by conducting the experiments, discussing and presenting the results in an appropriate form.

4   Conclusions

Pump head characteristics and duty points determination, pump starting process, different ways of pump regulation and fluid flow phenomena such as cavitation and turbulent swirling flow could be demonstrated in the presented demonstrational-educational test rig, designed and manufactured at the University of Belgrade Faculty of Mechanical Engineering Hydraulic Machinery and Energy Systems Department. Pump can be tested after international standard ISO 9906. This is a great step in presentation of the main lessons to students in practical manner. They could also easily drive pump and determine its duty points by themselves, what is a great benefit for future engineers. As Venturi flow meter can be calibrated, pump duty points in various regimes could be determined. Complex hydraulic system of three pipes in parallel and serial operation can be demonstrated. Energy efficiency is demonstrated for various regimes. Energy saving could be demonstrated by comparison of the flow regulation by throttling valve and pump speed variation., as well as by use of by-pass. Experiments are monitored with application developed in LabVIEW software. Students perform experiments, collect data and export them for their e-reports. Enrichment of the fluid flow phenomena and hydraulic pump lessons with visualization and measurements that could be performed on the presented installation is very important in education. This enables easier learning process.

Acknowledgment. The work presented in this paper was partly funded by the SCOPES project IZ74Z0_160454/1 "Enabling Web-based Remote Laboratory Community and Infrastructure" of Swiss National Science Foundation and by Ministry of Science and Technological Development Republic of Serbia, Project TR 35046 what is gratefully acknowledged.

References

1. Protić, Z., Nedeljković, M.: Pumps and Fans - Problems, Solutions, Theory, 5th edn. Faculty of Mechanical Engineering, University of Belgrade, Belgrade (2006)
2. Lapadatović, M.: Demonstrational-educational setup for testing pumps. Dipl. Thesis, Faculty of Mechanical Engineering, University of Belgrade, Belgrade (2000)
3. Petrović, M.: Project of the pump installation for serial mode and software for data acquisition. M.Sc. Thesis, Faculty of Mechanical Engineering, University of Belgrade, Belgrade (2016)
4. Lasica, N.: Project of the pump installation for parallel mode and software for data acquisition. M.Sc. Thesis, Faculty of Mechanical Engineering, University of Belgrade, Belgrade (2016)
5. ISO 9906:2012 Rotodynamic Pumps – Hydraulic Performance Acceptance Tests – Grades 1, 2 and 3
6. National Instruments: Basics II: Development, Course Manual, Course Software Version 8.0, May 2006 Edition

Study of Remote Lab Growth to Facilitate Smart Education in Indian Academia

Challenges and Perks in Bringing Laboratory Education to a Billion

Venkata Vivek Gowripeddi[1]([✉]), Kalyan Ram Bhimavaram[2],
J. Pavan[1], Nithin Janardhan[1], Amrutha Desai[1], Shubham Mohapatra[1],
Apurva Shrikhar[1], and C. R. Yamuna Devi[1]

[1] Dr. Ambedkar Institute of Technology, Bangalore 560056, KA, India
vivek.vg@hotmail.com, pavanj278@gmail.com,
nitnik313@gmail.com, amruthadesai1997@gmail.com,
shubhm2496@gmail.com, appu028@gmail.com,
yamuna_devicr@yahoo.com
[2] BITS-Pilani KK Birla Goa Campus, Sancoale 403726, Goa, India
kalyanram.b@gmail.com

Abstract. In a country with a growing population of over 1.2 billion and over 1.7 million engineers graduating every year, it becomes more imperative to have new methods of smart education and which are scalable across the masses. This paper discusses about the need for remote labs in Indian context and identifies the key parameters that determine their growth.

Preliminary reports of different remote labs in India, across all disciplines are tabulated. Their impact is measured both qualitatively and quantitatively through described parameters and their relative effectiveness and is measured. The cost of remote labs, their technology and cost are compared and contrasted with one another thus establishing the key factors for its success and impact.

Outcomes of the preliminary tests show that remote labs in colleges with high student to machine equipment ratio has the highest impact. From normal per student average lab time of 3 h/week, the touch time has increased to over 10.5 h/week. The concept has resounded very well in rural areas where there are learning centers being established to connect with labs thousands of kilometers away.

It is clearly evident that remote labs are both a necessity and highly impactful in the Indian context with outcomes exceeding the expectations. Authors' feel with right approach and execution; smart education can be provided to millions of students in Indian Academia.

Keywords: Smart education · Remote labs · Indian Academia
Learning tools · Outcome based education

© Springer International Publishing AG, part of Springer Nature 2019
M. E. Auer and R. Langmann (Eds.): REV 2018, LNNS 47, pp. 694–700, 2019.
https://doi.org/10.1007/978-3-319-95678-7_76

1 Introduction

The idea for the paper comes from the authors' involvement in the Indian academia as researchers, educators and facilitators of remote labs. The group is one of the few teams across the country working on remote labs - spreading ideas, motivating teams and developing remote labs suitable for the Indian Environment.

1.1 Importance of Education in India

In modern day India, after its independence in 1947, there has been a tremendous change in the attitude towards education. Education which was only limited to the elite and the rich in pre-independent India was made open and to be available to all [1]. Leaders, Government, Reformers, Parents all agreed education was the most important aspect for development of the nation and educating children became the first priority. This has led to increase in the number and spread of colleges, universities and schools all across the country and enrollment in them. From Fig. 1, it can be seen that in last six decades, India's literacy has grown from 12% in 1947 to 74% in 2011 and over the last 5 years' enrolment in colleges for higher education has grown by over 5 million to 34.6 million enrolled in 2015–2016 as depicted in Fig. 2.

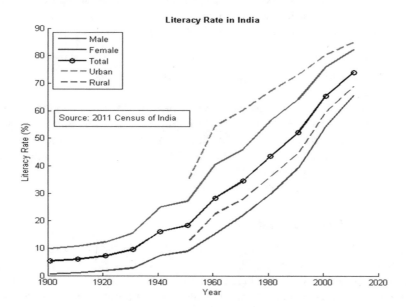

Fig. 1. Literacy growth in India

1.2 Growth of Technical Education

India's technical program was started initially started to build the telecommunication, space and oil and gas machinery for domestic use. With western world growing at an accelerated pace due to electronics and Internet revolution, it became more required

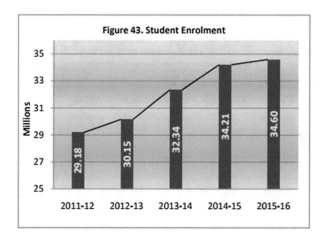

Fig. 2. Enrollment in higher education over last 5 years [2]

than ever to have more engineers and researchers. This created the need in India to generate English speaking technical graduates and this created a technical college boom in India. So, over the last two decades, the number of engineering and technical colleges and universities in India multiplies ten folds. The number technical colleges went up from 892 in 1998 to over 10, 000 today [3].

2   Current Laboratories and Their Shortcomings

As with every boom, comes a recession. With exponentially increase in the number of colleges, quality of education offered by these institutions suffered, Since, most of the institutions focused on the numbers rather than quality. Today's surveys reveal – "Lack of skills in engineers makes then unemployable". Practical Laboratory education plays key part in skill development [4]. Although we can agree, remote labs are future of labs, let us look at problems of current laboratories particular to Indian Scenario [5].

2.1   Low Machine to Student Ratio

Most of the lab equipment is expensive and requires space and room which is costly affair as real estate is expensive. So, most engineering colleges have one experiment apparatus for every 15 students, students are split in groups of 3–5 students to conduct an experiment as shown in Fig. 3. In India, machine to student ratio is 1 to 15 or more compared to western world where it is 1:3 or less [6]. This causes the students to get a touch time of just 3 h/time compared to at least 9 h/week in more developed countries.

Fig. 3. How experiments are performed in colleges

2.2 Outdated Machinery

Most colleges and universities have experimental apparatus which dates back to 90s and early 2000s. Most of the funds are utilized for building new classrooms and residential complexes such as hostels, dining halls for increasing intake of students. Colleges are increasingly focused on getting more admission which generates more revenue rather improving the quality by upgrading current machinery [7].

2.3 Lack of Connect Between Institutions

Sharing of resources between institutions facilitates collaborative research and learning [8]. That kind of connect doesn't exist formally between different colleges and universities in India. This hinders research and also results in wastage of funds as redundant apparatus are procured by two universities, who otherwise with proper communication could have shared apparatus and procured different apparatus.

3 Challenges Faced to Establish and Maintain Remote Labs in Indian Academia and Potential Solutions

Remote Labs numbers are increasing in India at a very slow pace due to following:

3.1 Cost and Affordability

Remote lab establishment requires a significant initial investment in terms of setting up remotely accessible experiments, server, design and installation [9]. However, the irony is large and well-funded institutions who have sufficient funds to setup a full-fledged remote lab, are generally not interested and invest in fairly outdated technologies. On the other hand, small and self-funded institutes who interested to setup a remote lab cannot afford such a huge investment.

Cost and features of the remote lab of those found in the US, Australia or Europe must be scaled down to Indian Environment such that they are made more affordable and cost friendly. Providing financial assistance such as periodic repayments rather one-time investment might boost remote lab setups.

3.2 Lack of Trained Personnel

Although there is fairly minimal maintenance when it comes to remote labs, most of the attenders and lab-in-charge personnel do not possess the required skills to maintain the remote lab such as maintaining servers, identifying problems when lab stops functioning, necessary, maintaining the laboratory conditions etc. [10].

Hence, it is very much required to conduct training programs and workshops for lab-in-charges and periodically update their knowledge with such events. Necessary precautions should be taken and possible failures identified and personnel should be trained to tackle these.

3.3 Lack of Institute-Industry Interaction and Indigenous Companies

There are a few companies who are working on remote labs in India. India being geographically vast and culturally different, there is often no connect between the manufacturer and potential customer for technologies such as remote labs. Due to this, institutes are often approached by foreign companies whose equipment is too costly or trust issues and support prevent the sale from happening.

Increasing indigenous companies and establishing a good network of Institute-Industry Interactions is highly necessary for growth in remote lab sales.

4   Benefits and Impact of Remote Labs in Indian Scenario

In the previous two sections, the current state of laboratories and challenges faced were discussed. In this section, light is thrown on positives such as the benefits, advantages and impact of remote labs on students in Institutes across India.

4.1   Increased Accessed Time

The major goal and perk of remote lab is the exponential increase in the touch time per student [11]. Table 1 clearly shows how touch time/week increased from 3 h in case of normal lab to 10.5 h in case of remote lab. It also identifies how tradeoff between outreach to students and touch time creates interesting scenarios – (1) Doubling the student outreach will only decrease the touch time by a third. (2) Keeping touch time same as a normal, remote lab can serve nearly 3.5 times the students as normal lab.

4.2   Access to Rural Students

More than 60% of Indian students live in Rural areas and more than 80% of Indian population lives in Rural areas. As most of the educational institution in these areas cannot afford advanced state-of-the-art equipment, as shown in rows 3 and 4 of Table 1, rural students can greatly benefit from remote labs as all they need at their end is PC/Laptop with internet connection [12]. Increasingly these days, with Government policies and social outreach programs, villages where electricity was once non-existent are brimming with high speed broadband and 4G towers to provide connectivity. It can only get better for rural students with the growth of remote labs [13].

Table 1. Figures illustrating student access time

Type of lab	No. of students (1)	No. of groups (5/group) (2)	No. of accessible hours/day (3)	Number of accessible days/week (4)	Total hours (3) * (4) = (5)	Touch time (hours/week) (5)/(2) = (6)
Normal lab	80	16	8	6	48	3
Remote lab for self use	80	16	24	7	168	10.5
Remote lab for self use & access to rural students	120	24	24	7	168	7
Remote lab for self use, rural students & other university students	160	32	24	7	168	5.25
Remote lab for self use & sharing access to maximum students	280	56	24	7	168	3

4.3 New Alliances and Connected Labs

With increasing number of installations. new alliances and agreements to share labs can be formed, thus resulting in a more connected network of institutions and their laboratories [14]. This will promote learning, research and sharing of ideas [15, 16].

5 Conclusion

In this paper, we have discussed (1) Importance of education in India, (2) Current state of laboratories, (3) Different challenges to establishment of remote labs such as cost, lack of trained personnel, alliances etc. (4) Perks of remote labs such as increased access time, increased outreach. The goal of this paper is to throw light on the education system, laboratory education of the largest democracy in the world and share the impact remote labs can have on this system and challenges that stand in the way. It is our sincere request to every educator, researcher, industrialist to see and solve the problems, opportunities and rewards in this land and contribute for a better tomorrow. We hope our little effort goes a long way in making meaningful transformation to education system in India and across the world.

Acknowledgment. Authors wish to thank World Bank's TEQIP fund for research and development of various institutes across the country. Many institutes and people associated with them have provided with valuable insights, opinions and data. Authors express their sincere gratitude towards all of them.

References

1. Murthi, M., Guio, A.C., Dreze, J.: Mortality, fertility, and gender bias in India: a district-level analysis. Popul. Dev. Rev. **21**(4), 745–782 (1995)
2. Premawardhena, N.C.: http://mhrd.gov.in/sites/upload_files/mhrd/files/statistics/AISHE2015-16.pdf (2012)
3. Cheney, G.R., Ruzzi, B.B., Muralidharan, K.: A profile of the Indian education system. Prepared for the New Commission on the Skills of the American Workforce (2005)
4. Dunnette, M.D., Campbell, J.P.: Laboratory education: impact on people and organizations. Ind. Relat. J. Econ. Soc. **8**(1), 1–27 (1968)
5. Argyris, C.: On the future of laboratory education. J. Appl. Behav. Sci. **3**(2), 153–183 (1967)
6. Pruthvi, P., Jackson, D., Hegde, S.R., Hiremath, P.S., Kumar, S.A.: A distinctive approach to enhance the utility of laboratories in Indian academia. In: 2015 12th International Conference on Remote Engineering and Virtual Instrumentation (REV), pp. 238–241, IEEE, Feb 2015
7. Natarajan, R.: Emerging trends in engineering education-Indian perspectives. In: Proceedings of 16th Australian International Education Conference, vol. 30, Sept 2002
8. Kozma, R.B.: National policies that connect ICT-based education reform to economic and social development. Hum. Technol. Interdisc. J. Hum. ICT Environ. **1**(2), 117–156 (2005)
9. Ferrero, A., Salicone, S., Bonora, C., Parmigiani, M.: ReMLab: a Java-based remote, didactic measurement laboratory. IEEE Trans. Instrum. Meas. **52**(3), 710–715 (2003)
10. Nordhaug, O.: Human Capital in Organizations: Competence, Training, and Learning. Scandinavian University Press, Oslo (1993)
11. Nickerson, J.V., Corter, J.E., Esche, S.K., Chassapis, C.: A model for evaluating the effectiveness of remote engineering laboratories and simulations in education. Comput. Educ. **49**(3), 708–725 (2007)
12. Diwakar, S., Kumar, D., Radhamani, R., Sasidharakurup, H., Nizar, N., Achuthan, K., et al.: Complementing education via virtual labs: implementation and deployment of remote laboratories and usage analysis in South Indian villages. Int. J. Online Eng. **12**(3), 8–13 (2016)
13. Achuthan, K., Sreelatha, K.S., Surendran, S., Diwakar, S., Nedungadi, P., Humphreys, S., Sreekala S., et al.: The VALUE@ Amrita virtual labs project: using web technology to provide virtual laboratory access to students. In: 2011 IEEE Global Humanitarian Technology Conference (GHTC), pp. 117–121. IEEE (2011)
14. van Joolingen, W.R., de Jong, T., Lazonder, A.W., Savelsbergh, E.R., Manlove, S.: Co-Lab: research and development of an online learning environment for collaborative scientific discovery learning. Comput. Hum. Behav. **21**(4), 671–688 (2005)
15. Kraut, R.E., Fussell, S.R., Brennan, S.E., Siegel, J.: Understanding effects of proximity on collaboration: implications for technologies to support remote collaborative work. In: Distributed Work, pp. 137–162 (2002)
16. Lowe, D., Berry, C., Murray, S., Lindsay, E.: Adapting a remote laboratory architecture to support collaboration and supervision. In: Proceedings of the Sixth International Conference on Remote Engineering and Virtual Instrumentation, International Association of Online Engineering, pp. 103–108 (2009)

Ant Colony Algorithm for Building of Virtual Machine Disk Images Within Cloud Systems

Irina Bolodurina, Leonid Legashev, Petr Polezhaev,
Alexander Shukhman$^{(\boxtimes)}$, and Yuri Ushakov

Orenburg State University, Orenburg, Russia
ipbolodurina@yandex.ru,
silentgir@gmail.com, shukhman@gmail.com,
newblackkpit@mail.ru, unpk@mail.ru

Abstract. Nowadays, the development of the cloud systems for collective access to virtual working environments, software, virtual stands and laboratories is very actual. Typically, such systems are created based on virtual machines (IaaS – Infrastructure as a Service), which provide remote access to their resources (including access to desktops). In the case of a large number of cloud system users it is necessary to store a large number of virtual machine disk images with various installed software. But it requires significant financial costs, when public cloud providers are used. In this study it is proposed to build disk images automatically and store them for a short time.

Keywords: Steiner tree · Cloud system · Virtual desktop
Ant colony optimization

1 The Virtual Machine Disk Images Building Problem

1.1 Introduction

Cloud computing is changing the way how students work, and it helps to deliver modern software and other cloud resources through the Internet [1].

Students can get access to virtual desktops with installed software for educational purposes via Internet browser. One of the cases of cloud system usage in education are programming and informatics contests. There are many preferences of students in choosing development environment, compilers, office applications and graphic editors. Typically each user can form one on demand request and select needed software from the list of available resources. In the case of a large number of cloud system users it is necessary to store a large number of virtual machine disk images with various operating systems and installed software. But it requires significant financial costs to store them, when public cloud providers are used.

1.2 The Formalization Problem of Building Virtual Machine Disk Images

The optimal algorithm for building disk images should be used in the configurator of cloud system. It analyzes the current schedule and builds a number of virtual machine

© Springer International Publishing AG, part of Springer Nature 2019
M. E. Auer and R. Langmann (Eds.): REV 2018, LNNS 47, pp. 701–706, 2019.
https://doi.org/10.1007/978-3-319-95678-7_77

images within a short time interval. The main functions of this configurator are building VM images, storing the most often used images, installing software packages to the stored images. To solve this problem we implement five algorithms. They are variations of the ant colony algorithm and the Prim's algorithm.

Firstly, we should formalize our problem.

Let F_{flav} be a set of VM disk images to build. They are terminal images. Each image F_{flav}^i is represented by a binary vector, which describes a subset of a global set of all available software packages S_{soft}. Let us describe a directed acyclic graph $G = (V, E)$, its vertices V are all possible disk images. Each arc $(u, v) \in E$ corresponds to transition from disk image u to image v by installation of a single software package. Its weight is the installation time of the corresponding software package. Let F_{flav}^0 be a start image. It has only operating system and no educational software. Let $Path(F_{flav}^i)$ be a path from the start image to the terminal image F_{flav}^i. It depicts the software installation sequence. Also, let us denote the installation tree as $I = \bigcup_{i=1}^{m} Path(F_{flav}^i)$. It is the union of all the paths from the start to terminal images. $S_{install}(I)$ is the installation time of all terminal images. So, we have an optimization problem (1). We should minimize $S_{install}(I)$ by selecting installation tree I.

The optimization problem reduces to the search of the minimal Steiner tree for a directed graph with an exponential number of vertices. It is known to be NP complete [2]. Various types of application of the Steiner trees in computer networks are discussed in [3–6].

$$S_{install}(I) \rightarrow \min_{I} \tag{1}$$

To solve the problem of building virtual machine images (1) we propose to use various versions of the ant colony algorithm and the Prim's algorithm. We need to initialize the following parameters of the modifications of ant colony algorithms: the number of ants, the pheromone coefficient α' and the heuristic coefficient β', the initial pheromone τ_0, the pheromone evaporation coefficient ρ and the installation time for corresponding software packages.

We propose the Original Ant Colony Optimization (OACO) algorithm for building disk images. We place ants to terminal vertices (images), one ant per terminal vertex. At each iteration we move them randomly in the reverse direction of graph arcs, until they reach the start image or the vertex visited by another ant at the current iteration of the algorithm. When no ant can move, the paths traversed by them form a Steiner tree. After this ants go back along their traversed paths and update the pheromone for their arcs.

Let us formalize the steps of OACO algorithm:

Step 0. Initialize the number of iteration $t = 1$ and the best result $S^* = \infty$, $I^* = \varnothing$.

Step 1. Ants are distributed to the terminal vertices (one ant per terminal vertex). Let us denote them as $Ants = \{ant_i\}_{i=\overline{1,m}}$.

Step 2. While $t \leq T_{max}$ do the following:

Step 2.1. Let $M = Ants$ be a set of ants, which can move;

Step 2.2. For each $ant_i \in M$ set up its path of traversed arcs as an empty set $P_i = \varnothing$;

Step 2.3. Let $V' = \varnothing$ be a set of visited vertices;

Step 2.4. While $M \neq \varnothing$ (there are ants, which do not reach the start vertex or the path of another ant) randomly select one $ant_i \in M$ and do the following:

Step 2.4.1. Let v_i be the current vertex of ant_i, if $v_i \in V'$ (v_i was previously visited by some ant), then $M = M \backslash \{ant_i\}$ (ant_i cannot move, remove it) and go to the step 2.4;

Step 2.4.2. $V' = V' \cup \{v_i\}$ - add v_i to the set of visited vertices;

Step 2.4.3. Randomly select the next vertex u_i from the vertices of ingress arcs of v_i according to the probabilities (2):

$$P_{u_i v_i}(t) = \frac{[\tau_{u_i v_i}(t)]^{\alpha'} * [\eta_{u_i v_i}(t)]^{\beta'}}{\sum\limits_{k} [\tau_{u_k v_i}(t)]^{\alpha'} * [\eta_{u_k v_i}(t)]^{\beta'}}, \tag{2}$$

where $\tau_{u_i v_i}(t)$ - the pheromone of the arc (u_i, v_i) (it equals to the initial value τ_0, if this arc was not previously visited by any ant), $\eta_{u_i v_i} = 1/S_{install}((u_i, v_i))$ - the inverse value of the software package installation time, which corresponds to the arc (u_i, v_i);

Step 2.4.4. $P_i = P_i \cup \{(u_i, v_i)\}$ - add the selected arc to the path of ant_i;

Step 2.4.5. Go to the step 2.4;

Step 2.5. $I = \bigcup\limits_{i=1}^{m} P_i$ - union the paths of all ants to get the tree and calculate its cost S_t (installation time), if $S_t < S^*$ then $S^* = S_t$ and $I^* = I$

Step 2.6. For each $ant_i \in Ants$ move it back to its terminal vertex along its path P_i and update the pheromone for each traversed arc (u_i, v_i) according to the formula:

$$\tau_{u_i v_i}(t+1) = \tau_{u_i v_i}(t)(1 - \rho) + \frac{1}{S_t} \rho. \tag{3}$$

Step 2.7. $t = t + 1$ and go to the step 2.

Step 3. Return the best solution - the values of S^* and I^*.

In addition, we propose several modifications of this algorithm.

Software Coefficient Usage Ant Colony Optimization (SCUACO) algorithm is the same as the OACO algorithm except formula (2), which is replaced with the following:

$$P_{u_i v_i}(t) = \frac{[\tau_{u_i v_i}(t) + U_{u_i v_i}]^{\alpha'} * [\eta_{u_i v_i}(t)]^{\beta'}}{\sum\limits_{k} [\tau_{u_k v_i}(t) + U_{u_i v_i}]^{\alpha'} * [\eta_{u_k v_i}(t)]^{\beta'}}, \tag{4}$$

where $U_{u_i v_i}$ - the software usage coefficient, which corresponds to software package installed for the arc (u_i, v_i).

SubTree Ant Colony Optimization (STACO) algorithm is also the same as the OACO algorithm except the formula for the greedy decision factor of the next vertex selection:

$$\eta_{u_i v_i} = \frac{1}{\sum_i S_{install}(P_{0 \to F^i_{flav}}) + S_{install}(P_{0 \to u_i})},$$ (5)

where $P_{0 \to F^i_{flav}}$ is the path from the start image to the terminal image F^i_{flav}, $S_{install}(P_{0 \to F^i_{flav}})$ - its installation time for the worst case, $P_{0 \to u_i}$ is the path form the start image to the image u_i, $S_{install}(P_{0 \to u_i})$ - its installation time for the worst case. Summation is carried out only for F^i_{flav}, which can be reached from the image v_i.

We implement a modification of the classic Prim's algorithm for undirected graphs to process our directed graph G. It begins the Steiner tree from the start vertex and in the cycle at each iteration it adds to the tree a vertex with a minimum distance (installation time) from one the tree vertices. The algorithm stops, when it reaches all the terminal vertices. Also, for optimization it does not consider the vertices, from which there are no paths to any terminal vertex.

In addition we implement two hybrids of the ant colony optimization and the Prim's algorithms.

The first of them is the Prim's Ant Colony Optimization (PACO) algorithm. It is the modified variant of the OACO algorithm. In PACO algorithm at each iteration all the ants move from their terminal vertices until they reach the start one. Then the subgraph of the original graph G with the visited vertices V' and the arcs from G is created. After this, the modified Prim's algorithm is executed for this subgraph. The weight of the result tree is used to update the pheromone only for the tree arcs, not for the arcs traversed by ants. This heuristic gives the best result (see part 3).

The Second Prim's Ant Colony Optimization (SPACO) algorithm is the same as the PACO algorithm except the pheromone update stage. SPACO updates the pheromone only for the arcs traversed by ants.

2 Experimental Research

2.1 Cloud System Simulator

To carry out the experiments we implemented the cloud system simulator, using C# language and Visual Studio 2012 IDE. Parameters when generating images are as follow: available software count is 25; software installation time is randomly distributed in [2, 20]; terminal images count is randomly distributed in [15, 20]; picked software subset size is randomly distributed in [3, 8]; iterations count is 2500; initial pheromone value is 0.02. We varied the pheromone weight α', heuristic coefficient and pheromone evaporation coefficient ρ and we calculated mean function values (1) for 25 experiments of each variation. Results for each of six algorithms are shown in Fig. 1.

2.2 Experimental Results

Now we can specify the most optimal ant colony optimization parameters for each of six algorithms, as it's shown in Table 1. We can also specify minimum value of (1),

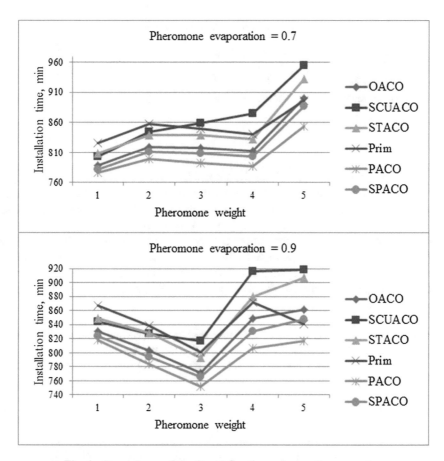

Fig. 1. Dependence of the fitness function value on the parameters

Table 1. Optimal parameters of the proposed algorithms

	OACO	SCUACO	STACO	Prim's	PACO	SPACO
α'	1	1	1	5	3	1
β'	5	5	5	1	3	5
ρ	0.9	0.9	0.3	0.9	0.9	0.9

average iteration on which the best solution was found and average operating time in ms, as it's shown in Table 2.

Analyzing the graphs (Fig. 1), we can say that SCUACO algorithm finds its best solutions when pheromone weight α' is small. If pheromone weight α' is high, Prim's algorithm works better than OACO and is approaching SPACO results. Both SCUACO and STACO algorithms modify probability (2), which negatively affected the calculation results compared to OACO. PACO algorithm shows the best results across all experiments of finding a minimal software installation sequence (1).

Table 2. Metrics of the proposed algorithms with optimal parameters

	OACO	SCUACO	STACO	Prim's	PACO	SPACO
Min value	770.84	802.84	792.84	800.72	751.44	766
Best iteration	1024	1341	890	1325	736	1284
Operating time	52874,64	66895,88	208934,24	65,08	28961,24	51114,36

3   Conclusion

This paper proposes the approach for dynamic building of virtual machine disk images for virtual working environments in the cloud systems. The optimization problem to determine the optimal sequence of software installation for obtaining a given set of disk images was formalized. It is a Steiner problem for constructing a minimal tree for directed graph. The five variations of ant colony algorithm and Prim's algorithm were proposed to solve it. It demonstrated efficiency of PACO algorithm during experimental studies on the cloud system simulator. Proposed solution can be used in educational institutions to provide faster access to virtual environments with installed software.

Acknowledgment. The research has been supported by the Russian Foundation for Basic Research (projects 16-07-01004, 16-29-09639, 16-47-560335), and by the President of the Russian Federation within the grant for young scientists and PhD students (SP-2179.2015.5).

References

1. Shukhman, A.E., Polezhaev, P.N., Legashev, L.V., Ushakov, Y.A., Bolodurina, I.P.: Creation of regional center for shared access to educational software based on cloud technology. In: Global Engineering Education Conference (EDUCON), pp. 916–919 (2017)
2. Foulds, L., Graham, R.: The Steiner problem in phylogeny is NP-complete. Adv. Appl. Math. **3**(1), 43–49 (1982)
3. Liu, L., Song, Y., Zhang, H., Ma, H., Vasilakos, A.V.: Physarum optimization: a biology-inspired algorithm for the Steiner tree problem in networks. IEEE Trans. Comput. J. **64**, 818–831 (2013)
4. Song, Y., Liu, L., Ma, H., Vasilakos, A.V.: A biology-based algorithm to minimal exposure problem of wireless sensor networks. IEEE Trans. Netw. Serv. Manag. J. **11**, 417–430 (2014)
5. Liu, B.H., Nguyen, N.T., Pham, V.T., Wang, W.S.: Constrained node-weighted Steiner tree based algorithms for constructing a wireless sensor network to cover maximum weighted critical square grids. Comput. Commun. J. **81**, 52–60 (2016)
6. Han, X., Liu, J., Liu, D., Liao, Q., Hu, J., Yang, Y.: Distribution network planning study with distributed generation based on Steiner tree model. In: IEEE PES Asia-Pacific Power and Energy Engineering Conference (APPEEC-2014) (2014). https://doi.org/10.1109/appeec.2014.7066185

Work in Progress: Pocket Labs in IoT Education

Christian Madritsch$^{(\boxtimes)}$, Thomas Klinger, and Andreas Pester

Engineering and IT, Carinthia University of Applied Sciences - CUAS,
Villach, Austria
`c.madritsch@fh-kaernten.at`

Abstract. This work in progress describes the ongoing initiative at CUAS to use Internet of Things (IoT) technologies for educational purposes. First, it describes the relevance of Internet of Things technologies in education, next it focuses on different hardware platforms used by CUAS. The main part describes the implementation of one specific example developed by students, the MIA-project. Finally, it gives an outlook on how future student projects will benefit from the lessons learned.

Keywords: Internet of Things (IoT) · Bachelor degree program
Raspberry Pi · Arduino · Teensy · Blend micro
Project-based learning · Pocket labs

1   Introduction

Internet of Things (IoT) technologies tend to be one possible future of engineering in general. In the last 2–3 years in Higher Education the trend to a Maker University is one of the mainstream trends. For engineering study programmes, which traditionally had both, a theoretical and a lab part, this means at first instance the close (thematically and from a timeline point of view) connection of the experimental and theoretical part of teaching.

IoT technologies incorporate many different aspects of technical systems. They usually consist of a microprocessor/microcontroller which can be used stand alone or with an operating system, e.g. Linux. The application software can be programmed using the traditional C/C++ approach or by applying emerging programming languages like Python and Processing. IoT systems are equipped with interfaces to the outside world, both analogue and digital. Sensors, attached to these interfaces can measure a wide range of physical entities with astonishing precision and sampling rate. Last, but not least, IoT systems are equipped with communication means to interact with other IoT devices. Bluetooth, Zigbee, WiFi, and LoRaWAN are some of the possible choices.

© Springer International Publishing AG, part of Springer Nature 2019
M. E. Auer and R. Langmann (Eds.): REV 2018, LNNS 47, pp. 707–713, 2019.
https://doi.org/10.1007/978-3-319-95678-7_78

During the last years, CUAS has successfully introduced various kinds of IoT systems and technologies into a growing group of bachelor degree courses (e.g. Real-Time Systems, Bus systems and Protocols, Algorithms and Data-structures,). This work in progress description provides an overview about different kinds of IoT systems currently used at CUAS. It also explains one specific IoT student-project in detail.

2   IoT at CUAS

At CUAS, we are using different IoT architectures and systems. Since different applications demand different levels of complexity, speed, composability, and interoperability, the one fits all solution cannot and will not be applied.

Arduino is one major IoT device used by CUAS. Due to its ease of use and high flexibility, it is the ideal choice for beginners as well as for experienced students. The integrated development environment (IDE) with its built-in Processing programming language and the high number of extension boards (shields) with corresponding software libraries provide a limitless playground for various kinds of projects.

We use Arduino as a base platform which means that depending upon the current requirements of our applications, students need to choose a clone-version. Two clones we are using are the Blend micro and the Teensy board.

The Blend Micro board (see Fig. 1) is at its heart an Arduino blended with a Bluetooth Low Energy (BLE) interface into a single board. It is targeted for makers to develop low power Internet-Of-Things (IoT) projects quickly and easily. The micro-controller unit (MCU) is Atmel ATmega32U4 and the BLE chip is a Nordic nRF8001. Blend Micro runs as BLE peripheral role only, it allows BLE central role devices to establish connections with it [1].

The Teensy 3.2 USB development board (see Fig. 2) is a complete USB-based microcontroller development system in a very small footprint capable of implementing many different types of projects. The Teensy is a breadboard-friendly development board. Each Teensy 3.2 comes pre-flashed with a bootloader and can be configured directly using USB. It can be programmed in C or by using Arduino sketches for Teensy. The processor on the Teensy also has full access to the USB and can emulate any kind of USB device [2].

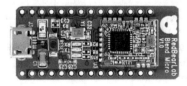

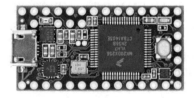

Fig. 1. Blend Micro board.   **Fig. 2.** Teensy 3.2 board.

The Raspberry Pi 3 Model B (see Fig. 3) has a 1.2 GHz 64-bit quad-core ARMv8 CPU, 802.11n Wireless LAN, Bluetooth 4.1, Bluetooth Low Energy (BLE). It is used in combination with extension-boards, breadboards, cameras, sensor/actor kits, etc. The Raspberry Pi is setup with the appropriate OS (Raspbian, Windows 10 IoT, Ubuntu,). Programming takes place using Python, C or C++ [3].

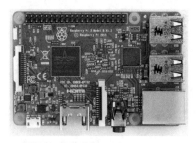

Fig. 3. Raspberry Pi 3.

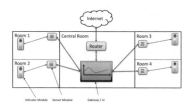

Fig. 4. MIA Overview.

3 Managed Indoor Air (MIA) Student Project

3.1 General Overview

The goal of the MIA student project was to develop an air measurement system using IoT technologies. The three main aspects of this project were: the choice and configuration of the hardware, software development for different target platforms, and the configuration of the appropriate communication interfaces.

The whole application was split into three IoT modules (see Fig. 4):

- the Sensor Module, capable of measuring the CO_2 concentration of the air, the air temperature and the humidity,
- the Indicator Module, providing a simple and intuitive way to show humans when to open a window to ventilate, and
- the Gateway Module, which is used to graphically represent the acquired data from the sensor module and it also acts as the gateway into the internet and portal for a cloud-based storage system.

To interconnect the three different modules, a communication architecture was developed (see Fig. 5). The final choice was to use BLE to connect the Sensor module to both, the Indicator module and the Gateway module and to use WiFi to connect the Gateway module via a router to the internet.

3.2   Sensor Module

The purpose of the Sensor module (see Fig. 6) is to measure the CO2 concentration of the air, the air temperature and the humidity and to transmit this data using BLE to the Gateway Module.

The Sensor module consists of the following components:

– Teensy 3.2 (Arduino compatible IoT device),
– Adafruit BME280 (temperature and humidity sensor),
– Telaire T6615 (CO2 Sensor), and
– BLE2 Click (BLE Module).

The CO2 Sensor and the BLE Module are connected to the Teensy using UART, the BME280 uses I2C as its interconnect (see Fig. 7).

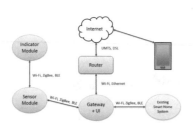

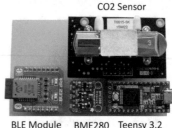

Fig. 5. MIA Communications Overview.

Fig. 6. Sensor module.

3.3   Indicator Module

The purpose of the Indicator module (see Fig. 8) is to provide a simple and intuitive way to show humans when to open a window to ventilate by using a

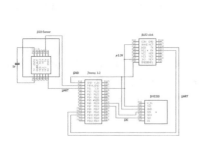

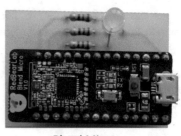

Fig. 7. Sensor module internal interconnects.

Fig. 8. Indicator module.

RGB LED. If the LED is turned off, the air quality is good. If the LED turns on to yellow, the air quality is medium, and if the air quality is bad, the LED turns to red. After the air quality has reached a good level again, the LED turns off.

The Indicator module consists of the following components:

- Blend Micro Board,
- RGB Led, and
- BLE module (integrated).

3.4 Gateway Module

The purpose of the Gateway module (see Fig. 9) is to graphically represent the acquired data from the sensor module and to act as the gateway into the internet and portal for a cloud-based storage system.

The Gateway module consists of the following hardware components:

- Raspberry Pi 3,
- Raspberry Pi 7" LCD Touchscreen,
- BLE Module (integrated), and
- WiFi 802.11n (integrated).

The Raspberry Pi 3 is using Raspbian a Debian Linux clone as operating system. The following software components are also installed on this IoT-device (see Fig. 10):

- Apache 2.4.10 Webserver,
- SQLite 3,
- Python 2.7,
- PHP/HTML,
- Google Charts API, and
- Chromium.

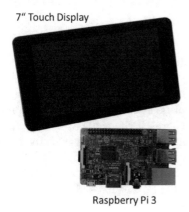

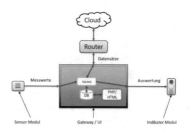

Fig. 9. Gateway module.   **Fig. 10.** Software components.

SQLite is managing the database (DB), Apache Webserver acts as the external interface to the DB and PHP/HTML are the transport protocols used. The Google Charts API is used to draw diagrams and Python is used as the programming language.

3.5   Overall Results

The IoT student project MIA was very successful. Three students have been working over a period of seven months and completed the project not just in time but also three weeks earlier. The budget was not overspent and two independent MIA systems have been build. They are being used at CUAS and they certainly are references for future projects in this field.

In Figure 11, the graphical representation of the measurement results on the Gateway module can be seen. The same representation can be accessed via the Internet using a browser or a Smartphone.

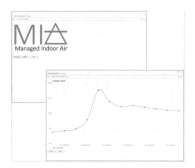   

Fig. 11. GUI Measurement results.   **Fig. 12.** Final MIA system.

The project group also designed and constructed housings for the individual modules, which can be seen in Fig. 12. The in-house 3D-printer of CUAS has been used.

4   Outlook

Due to the large and obvious relevance of IoT technologies, students are very much interested in similar and future IoT projects. Not all projects require a complete hardware and software design from scratch the challenge in the future will be to identify candidates for the IoT approach and the others intended for custom embedded systems design.

As the developed modules can be used for mobile measurements as well as for laboratory exercises, they satisfy the definition of Pocket Labs [6], being portable electronic devices for mobile measurements.

References

1. http://redbearlab.com/blendmicro/
2. https://www.pjrc.com/teensy/
3. https://www.raspberrypi.org/
4. Laplante, P.A.: Real-Time Systems Design and Analysis: Tools for the Practitioner. Wiley, Hoboken (2011)
5. Analog Devices, ADALM1000 Evaluation Board. http://www.analog.com/en/design-center/evaluation-hardware-and-software/evaluation-boards-kits/eval-adalm1000.html
6. Klinger, T., Madritsch, C.: Collaborative learning using pocket labs. In: International Conference on Interactive Mobile Communication, Technologies and Learning, Greece, November 2015

Demonstration: Using IPython to Demonstrate the Usage of Remote Labs in Engineering Courses – A Case Study Using a Remote Rain Gauge

Alberto Cardoso[1(✉)], Joaquim Leitão[1], Paulo Gil[1,2], Alfeu S. Marques[3], and Nuno E. Simões[3]

[1] CISUC, Department of Informatics Engineering, University of Coimbra, Coimbra, Portugal
{alberto,jpleitao}@dei.uc.pt
[2] Department of Electrical Engineering, FCT, NOVA University of Lisbon, Lisbon, Portugal
psg@fct.unl.pt
[3] INESC-Coimbra, Department of Civil Engineering, University of Coimbra, Coimbra, Portugal
{jasm,nunocs}@dec.uc.pt

Abstract. The use of collaborative tools that can contribute to share and demonstrate the usage of remote experiments, to support teaching and to enhance the learning process, is of great importance in several educational contexts and particularly in engineering courses. Jupyter/IPython notebooks are one of these tools that provide a programming environment to develop and share scientific contents and that can promote the access to remote and virtual labs. Teaching and learning activities in different high education courses, especially in engineering subjects, can benefit of using this type of resources. This paper presents an IPython-based approach to show how to interact with a remote rain gauge to obtain data about the rainfall in a given location, which may be useful in different learning contexts, namely in programming or environmental science subjects.

Keywords: Jupyter · IPython · Remote laboratories · Rainfall measurement
Teaching · Collaborative tools

1 Introduction

The development of Information and Communication Technologies (ICT) and the evolution of the concept of Internet of Everything, in which the interconnectivity between things and people is promoted and supported, represent an opportunity to develop and provide resources online with specialized contents for different uses, but with particular significance for several subjects of engineering areas.

In this context, laboratory classes using experimental environments play a crucial role and represent a challenge for teachers to improve the quality of materials and develop new approaches for teaching, taking advantage of collaborative tools and of the use of remote and virtual laboratories.

The use of online experimentation represents a great opportunity to support teaching and learning activities complementing the laboratory activities and motivating students

© Springer International Publishing AG, part of Springer Nature 2019
M. E. Auer and R. Langmann (Eds.): REV 2018, LNNS 47, pp. 714–720, 2019.
https://doi.org/10.1007/978-3-319-95678-7_79

to perform practical works, acquiring knowledge, understanding the concepts and achieving experimental skills in a flexible way.

Although there are several platforms to provide resources online, Jupyter[1] is an open-source project that can be used to support interactive data science and scientific computing across several programming languages as, for example, Python. This project grew out of the IPython project [1], which initially provided an interface only for the Python language and continues to make available the canonical Python kernel for Jupyter.

Thus, Jupyter notebooks (or IPython notebooks, particularly when the Python programming language is used) give the necessary support to implement the concept of "Literate Computing" and "Reproducible Research", providing tools to develop and make available narratives anchored in a live computation, which offer the possibility of communicating knowledge and research based on data and results in a readable and replicated way. Notebooks are accessed through a web browser and are designed to support the workflow of scientific computing, from an initial interactive exploration phase to publishing a comprehensive record of computation [2]. The contents of a notebook are organized into cells, which can include text, images, code or math operations, that weave together to produce an interactive document. This approach corresponds to an evolution of the interactive shell or REPL (Read-Evaluate-Print Loop) that has long been the basis of interactive programming [3, 4]. Therefore, instructors can provide students with self-contained IPython notebooks, representing excellent resources for teaching and learning purposes.

In this context, this paper aims to exemplify and demonstrate the use of IPython notebooks to interact with remote and virtual laboratories, in particular showing the usage of a remote rain gauge to obtain data about the rainfall in a given location, which can be useful in different learning contexts, namely in programming or environmental science subjects of different engineering courses.

2   Notebooks

Considering the Jupyter project, notebooks are open with a web browser, making practical the use of the same interface running locally like a desktop application, or running on a remote server. Thus, notebooks provide a programming environment that can be shared and offers many advantages for students and instructors as these are free and open source software, as well as it can be used to promote teaching/learning based on innovative approaches and reproducible research [5]. For example, a teacher can provide the notebooks on a web server and easily give students access. Each notebook file is documented in the JSON format with the extension '.ipynb', making easier the process to write, manipulate and share these files. So, notebooks can be used to record a computational piece of code in order to explain it or a given subject in detail to others, and a variety of tools help users to conveniently share notebooks [2].

In association with Jupyter, there are several services available, such as "*nbconvert*", "*nbviewer*", "*binder*" or "*nbgrader*", which complement the project tools and

---

[1] http://jupyter.org/ (last accessed: January 27, 2018).

make it more powerful. *Nbconvert* converts notebook files into different file formats, including HTML, LaTeX and PDF, making them accessible without needing any Jupyter software installed. *Nbviewer* (https://nbviewer.jupyter.org/) is a hosted web service that provides an HTML view of notebook files published anywhere on the web. These HTML views have a major advantage over publishing converted HTML directly because they link back to the notebook file, given the possibility to interested readers to download, run and modify it themselves. *Binder* (http://mybinder.org/) allows sharing of live notebooks including a computational environment in which users can execute the code. With this, authors can publish notebooks, for example on GitHub, along with an environment specification in one of a few common formats in an interactive and immediately verifiable form [2]. *nbgrader*, is a tool designed to facilitate the grade process of notebooks, by providing an interface that blends the autograding of notebook-based assignments with manual human grading. Additionally, it streamlines and simplifies the process of assignment creation, distribution, collection, grading, and feedback [6].

Considering this set of tools, among others, the scientific code, its computational environment, data and execution conditions can be maintained in a *git* repository, enabling its maintenance and reusability. Consequently, numerous papers have been published supported by notebooks to reproduce the analysis or the creation of key results. Moreover, there are already available various examples of books as a collection of IPython notebooks, such as "Python for Signal Processing" [7] or "QuTiP Lectures as IPython Notebooks" [8].

This work aims to demonstrate that notebooks can also be used to simplify the interaction with remote and virtual experiments, taking advantage of their services and functionalities.

3   The Remote Experiment

Precipitation can be defined as any form of water (snow, rain, sleet, freezing rain, and hail) that falls to the earth's surface. The amount and duration of precipitation during rainfall events is of great importance for environmental monitoring, natural resources mapping and meteorological data acquisition purposes.

Precipitation can be measured using different sensors. The precipitation measuring instrument (udometer) considered in this work is a "tipping-bucket rain gauge", which uses a small cup that fills with precipitation and when is full, it will tip and empty. Using a wired or wireless communication channel to connect the sensor to a Raspberry Pi computer, a counter tracks how many times the cup tips, and after a given time interval, this number is converted to a precipitation measurement in millimeters. Figure 1 illustrates the udometer used in this work.

Fig. 1. Example of a tipping-bucket rain gauge (http://pronamic.com (last accessed: January 27, 2018)).

In order to remotely monitor the precipitation, a web server is running in the Raspberry Pi computer, which sends data to a remote computer that acts as station manager, using an Internet connection. This computer is used to manage the system and to store and deliver data through web services. For this, an API with various endpoints was developed offering different possibilities of interaction using RESTful web services, configured as URL commands. These web services include commands to fetch data in real time, in a given period of time or between two specific dates.

The following commands exemplify the way the manager station (located at http://hydra.dei.uc.pt/station/manager/...) can be inquired to get data about the amount of precipitation, acquired by the udometer connected with a given web server (identified by <node>):

- Fetch data registered in a given hour:
    - .../api/data/<node>/<year>/<month>/<day>/<hour >/
- Fetch data registered in a given day:
    - .../api/data/<node>/<year>/<month>/<day>
- Fetch data registered in a given month:
    - .../api/data/<node>/<year>/<month>/
- Fetch data registered between two specific dates, specifying the sampling period (<search_type> as hour or minute):
    - .../api/data/search/<search_type>/<node>/<date1>/<date2>/

To use the available web services, it is necessary to send requests (commands) to web service endpoint URL and receive responses from it. The services receive and send data in JavaScript Object Notation (JSON). The requests are sent in a web browser and the responses are displayed within a webpage.

The following command exemplifies the request to fetch the amount of precipitation recorded each hour on November 23, 2017:

- Command: http://hydra.dei.uc.pt/station/manager/api/data/DEC1/2017/11/23/:
  Response in JSON: {"status": 200, "data": {"events": [0.6, 0.4, 4.0, 5.8, 0.4, 6.2, 0.6, 0.0, 0.6, 0.0, 0.2, 0.0, 0.0, 0.2, 0.2, 0.2, 2.2, 1.4, 1.0, 0.0, 0.6, 0.0, 0.0, 0.2]}, "success": true}

The obtained response is a dictionary object with 3 elements with the keys: "status", "data" and "success". The value associated with "status" specifies that the operation was successful or not (200 signifies that it was successful). The second pair with the key "data" has another dictionary as value, which key identifies the meaning of the collected data (in this case "events" represents the events transmitted to the station manager) and the corresponding value is a list of the amount of precipitation obtained in each of the 24 h of the day, in millimeters (mm). The value of "success" is Boolean and determines if the request was completed successfully or not.

4   Example of an IPython Notebook

Considering the environment provided by the remote lab described in the previous section, the use of notebooks to interact with remote udometer arises as a natural and very interesting option, where it is possible to show as the web services can be used to obtain data about the precipitation in the location where the udometer is positioned.

Moreover, the Jupyter notebooks offer the ideal setting to describe the concepts related with precipitation, using contents as text or images, and program different approaches and methodologies to analyze, process and visualize data fetched from the remote system, using one of the several programming languages available.

Therefore, the notebooks represent a very useful tool that can be used in different teaching and learning contexts, namely in programming or environmental science subjects of different engineering courses. Given the large flexibility to structure the content of the notebooks, as well as their characteristics that favors collaboration and sharing, they can support different teaching/learning approaches and strategies, taking advantage of the interaction with the remote/virtual laboratories, which can motivate students and improve their performance.

To give an example of how to get and visualize data (precipitation in mm) fetched from the remote udometer presented in the previous section, Fig. 2 presents a cell of an IPython notebook (available at GitHub[2]) programmed in Python 2.7:

```
In [1]:  from requests import get
         import matplotlib.pyplot as plt
         r=get("http://hydra.dei.uc.pt/station/manager/api/data/DEC1/2017/11/")
         d=r.json()
         data=d.get('data')
         e=data.get('events')
         days=range(1,31)
         plt.plot(days,e[:-1],'-o', label="November")
         plt.xlabel('Days')
         plt.ylabel(u'Precipitation [mm]')
         plt.title(u'Diary precipitation in Pólo II of UC')
         plt.legend()
         plt.show()
```

Fig. 2. Example of a cell of an IPython notebook.

---

[2] https://github.com/ (last accessed: January 27, 2018).

This cell is a simple but exemplary instance of code in Python that uses the "get" function from the "requests" module to get the JSON object returned by the request defined by the web service endpoint URL and the "matplotlib" module for data visualization purposes. Figure 3 shows the plot representing the diary precipitation, in mm, occurred during November, 2017, at "Pólo II" campus of the University of Coimbra.

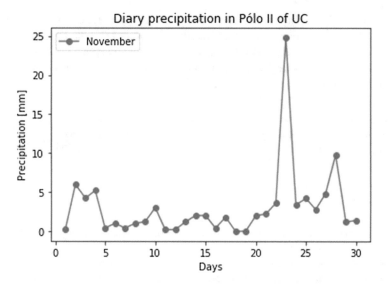

Fig. 3. Result of running the cell of the IPython notebook in Fig. 2.

Based on this illustrative example, several other notebooks can be developed with contents adjusted to the specific learning context, including text, images and code using data fetched from the remote system. For example, in a context of programming subjects based on the Python language, the notebook can consider cells to present and exemplify the use of different objects in Python and to develop various algorithms for data analysis and processing.

5   Conclusion

This paper presents and demonstrates the use of Jupyter/IPython notebooks as tools that provide a programming environment to develop and share scientific contents and that can promote the access to remote and virtual labs. The use of this type of resources, where text, images and code can be combined in a harmonious and comprehensible manner, can contribute to explore innovative approaches and improve teaching and learning activities in different high education courses, especially in engineering subjects.

In particular, this work presents the use of IPython notebooks to interact with a remote laboratory and get precipitation data at a given location, which can be useful in different learning contexts, namely in programming or environmental science subjects. The remote system includes a udometer connected to a Raspberry Pi computer where a web server is running to send data in JSON format through the Internet to a manager station that receive and response to web service endpoint URL requests.

The use of notebooks with access to remote labs can represent an important and very interesting tool in theoretical and practical classes, complementing the laboratory experiments and contributing for the enhancement of the students' learning process and their experimental skills.

This approach is being used in subjects of some engineering courses of the Faculty of Sciences and Technology of the University of Coimbra and the preliminary results are very encouraging to continue to use and to develop this type of online experimentation resources for teaching activities in different areas of engineering courses and to share research results.

Acknowledgement. This work was partially supported by Calouste Gulbenkian Foundation under U-Academy project [Project 2015/2016 FCG-138259]. This work was also partially supported by the Portuguese Foundation for Science and Technology (FCT), through the PhD scholarship SFRH/BD/122103/2016.

References

1. Pérez, F., Granger, B.E.: IPython: a system for interactive scientific computing. Comput. Sci. Eng. **9**(3), 21–29 (2007)
2. Kluyver, T., Ragan-Kelley, B., Pérez, F., Granger, B., Bussonnier, M., Frederic, J., Kelley, K., Hamrick, J., Grout, J., Corlay, S., Ivanov, P., Avila, D., Abdalla, S., Willing, C.: Jupyter Development Team: Jupyter Notebooks—a publishing format for reproducible computational workflows. In: ebook Positioning and Power in Academic Publishing: Players, Agents and Agendas, pp. 87–90 (2016). https://doi.org/10.3233/978-1-61499-649-1-87
3. Iverson, K.E.: A Programming Language. Wiley, New York (1962)
4. Spence, R.: APL demonstration, Imperial College London (1975). https://www.youtube.com/watch?v=_DTpQ4Kk2wA. Accessed 27 Jan 2018
5. Raju, A.B.: IPython notebook for teaching and learning. In: Natarajan, R. (ed.) Proceedings of the International Conference on Transformations in Engineering Education. Springer, New Delhi (2015). https://doi.org/10.1007/978-81-322-1931-6_91
6. Hamrick, J.B.: Creating and grading IPython/Jupyter notebook assignments with NbGrader. In: Proceedings of the 47th ACM Technical Symposium on Computing Science Education – SIGCSE 2016, pp. 242–242 (2016). https://doi.org/10.1145/2839509.2850507
7. Unpingco, J.: Python for Signal Processing, Springer (2014). https://github.com/unpingco/Python-for-Signal-Processing. Accessed 27 Jan 2018
8. Johansson, R.: QuTiP Lectures as IPython notebooks. https://github.com/jrjohansson/qutip-lectures. Accessed 27 Jan 2018

Poster: LabSocket-E, LabVIEW and myRIO in Real-Time/Embedded Systems Student Teaching and Training

Doru Ursutiu[1(✉)], Andrei Neagu[2], and Cornel Samoila[3]

[1] AOSR Academy, University « Transilvania » Brasov, Brasov, Romania
udoru@unitbv.ro
[2] University « Transilvania » Brasov, Brasov, Romania
andrei.neagu@unitbv.ro
[3] ASTR Academy, University « Transilvania » Brasov, Brasov, Romania
csam@unitbv.ro

Abstract. For many years now we used in our Creativity Laboratory and Cypress – National Instruments club Virtual Instrumentation technologies. With all the new hardware's and software's appeared technologies we try to build one easy scenario to use them starting from Highs Schools Children's, extend to the students and maybe to suggest industrials how fast they can use it and implement in industrial application. Thanks to our collaboration with "Bergmans Mechatronics LLC" we implemented step by step remote controlled laboratories using their LabSocket solution and now we do the same with their LabSocket-E.

Keywords: LabVIEW · LabSocket-E · Real-Time · Embedded systems
myRIO

1 Introduction

The technology is rapidly changing every year and new technologies are evolving faster than ever. The emergence of **Embedded Systems** had not only reduced the cost and risk but also improved the quality systems. As we know Embedded systems can be referred to as computer systems that is embedded in electro/mechanical systems. In the same time the **Real-Time** embedded systems are useful for monitor and controlling external environments. The usage of the real-time embedded system have practically improved in the recent years. It has widely adopted from bigger industries like military, aeronautics, robotics to commercial industries like telecom automobile and kitchen appliances.

In this context one must consider **myRIO** one embedded hardware/software device designed by National Instruments specifically to help students in the design of real, complex engineering systems more quickly and affordably than ever before. But in the same time together with **Raspberry Pi** and **Arduino** are becoming the most used platforms for teaching and learning embedded systems in the **DIY (Do it Yourself)**.

© Springer International Publishing AG, part of Springer Nature 2019
M. E. Auer and R. Langmann (Eds.): REV 2018, LNNS 47, pp. 721–727, 2019.
https://doi.org/10.1007/978-3-319-95678-7_80

LabVIEW (Laboratory Virtual Instrumentation Engineering Workbench) is a graphical programming environment that we intensively use in Creativity Laboratory and Cypress – National Instruments CLUB. High school children's and students can use to quickly and easy develop applications using multiple platforms and operating systems. The power of LabVIEW software consist in its ability to fast interface with many devices and instruments using hundreds of ready available libraries who help creative people to accelerate their understanding, training and development time to quickly acquire, analyse, and present data [1].

LabVIEW Virtual Instruments (VI's) are graphical, driven by dataflow and event-based programming, and are multitarget and multiplatform capable. Students have object-oriented flexibility and multithreading and parallelism features. LabVIEW VI's, using myRIO devices, can be deployed to real-time and FPGA targets.

The **myRIO Student Embedded Device** features Input-Output (I/O) on both sides of the device in the form of MXP and MSP connectors - including Analog Inputs AI, Analog Outputs AO, digital I/O lines DIO, LEDs, one push button, on-board accelerometer, a Xilinx FPGA, and a dual-core ARM Cortex-A9 processor (our model also include WiFi support) [2]. Students can program the myRIO device with LabVIEW or C and now using the new **LabSocket** (and LabSocket-E for embedded systems) they can extend the functionality of LabVIEW applications to the browser, without the use of browser plug-ins or requiring developers to write a single line of HTML or JavaScript code. With its on-board devices, seamless software experience, and library of courseware and tutorials, the myRIO Student Embedded Device provides an affordable tool for students and educators.

LabSocket-Embedded (or "LabSocket-E") is a new version of the LabSocket add-on tool for LabVIEW that allows developers to automatically create browser-based user interfaces for LabVIEW applications operating on National Instruments real-time platforms. These platforms include myRIO, sbRIO, FlexRIO, cRIO, and PXI. Support for Raspberry PI platforms is planned for 2017 Q4 [3].

(or ribbon).

2   Actual Problems in High School Children's and Students Education

The present paper try to answer to the actual needs of the children's and students: fast learning cycles in relation with the many new technologies and changes in hardware and software development, close connected with the exponential use of Internet and Smart Phones (thinking at the future industrial applications).

With our activities and developments we try to teach our students what is in fact Real-Time:

1. Real-time does not always mean fast
2. Real-time means absolute reliability
3. Real-time systems have timing constraints that must be met to avoid failure
4. Determinism is the ability to complete a task within a fixed amount of time

and how can be controlled one Real-Time Embedded system from a normal browser and/or Smart Phone without necessity to install any supplementary software tools on student devices.

Our answer to these actual important educational questions can be done using LabVIEW and LabSocket-E in combination with myRIO:

1. Compatible with National Instruments' real-time platforms using: Real-Time Linux, VxWorks, or Phar Lap operating systems
2. LabSocket-E is sold on a per-developer-seat basis. Developers may deploy their RT target software to an unlimited number of myRIO (or other RT) platforms
3. Remote access to VIs executing on desktop platforms is not supported by LabSocket-E. The current desktop version of LabSocket is recommended for this kind of applications.

In Fig. 1 we present one examples of facilities introduced by LabSocket-E in automated building Web controlled application for PC Web browser and iPhone browser.

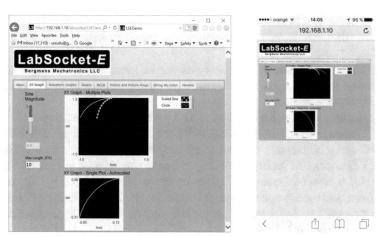

PC: IE browser control myRIO   iPhone: Safari browser control myRIO

Fig. 1. Control Real-Time application on myRIO using LabSocket-E

3 Applications and Developments

In the first stage of training children and students (trainee) will open the DEMO project downloaded from the site [3] and will run the LSEDemo.vi and with the provided Web link they will be able to control this demonstration from the web page (see Fig. 2 with the indicators to Web page, Demo project and LabVIEW application LSEDemo.vi).

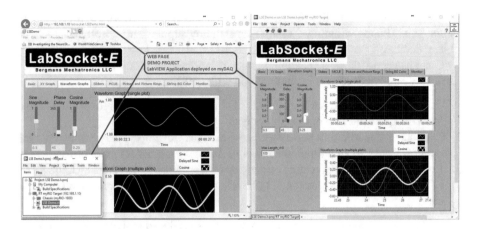

Fig. 2. Example how to use the facilities of LabSocket-E

This LSEDemo.vi was build using one TAB template in the LabVIEW PANEL and this TAB offer facility to fast understand the active function supported by the LabSocket-E.

After this first step Creativity Laboratory trainee will start to build similar applications and control the on-board available sensors and input-output myRIO devices or to add necessary outside sensors.

At this level we train students how they can easy adapt the Demo example to be used with the XYZ accelerometer from inside myRIO and control them from LabVIEW and from any PC or mobile device browser (after they build and deploy on myRIO the necessary applications). In the first TAB (Fig. 3a) page was inserted information about myRIO and his integrated 3-axes accelerometer; in the second TAB (Fig. 3b) we can see the measurement time, number of iterations and the WFG (wave form graphic) with the Z and YXZ signal evolution in time; in the last TAB (in Fig. 3c) we can see Primary Processing Loop Period and State.

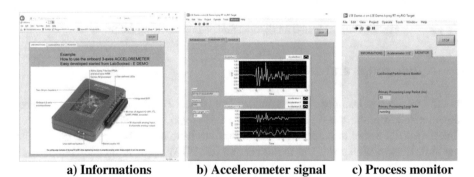

a) Informations   **b) Accelerometer signal**   **c) Process monitor**

Fig. 3. Interface for myRIO 3-axes accelerometer Web control on PC

The same interfaces can be seen on any smart phone (Fig. 4) and we can control one real time process programmed on the myRIO device using this simple to use LabSocket application combined with LabVIEW and myRIO development tools from National Instruments.

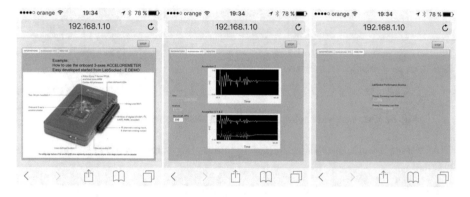

Fig. 4. Interface for myRIO 3-axes accelerometer Web control on iPhone

Now we like to present some step by step developments how to use myRIO with different laboratory tools and using the LabSocket-E to do web control of the system. First we like to us Fourier Education sound sensor DT320 (40–110 dB) connected to myRIO with PCB Adaptor.

Many of the described systems can be acquired directly by children's and students, at discounted prices, from the STUDICA (the Education Source for Academic Student Software & Technology Products) [4].

In the Fig. 5 we present one easy and simple system, fast to build using these devices to continuous monitoring of acoustical pollution and sound levels in our laboratories. We use this stem also in Meloterapy Master laboratory where we need to know every moment what was the precise level of the exposure to the selected musical audition and synchronize and correlate with the evolution level of brain waves measured with the MindWave from NeuroSky and all integrated in one LabVIEW common application [5, 6] (Fig. 6).

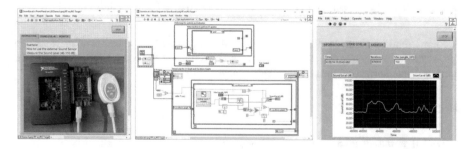

Fig. 5. Acoustical level measurement system and LabVIEW application

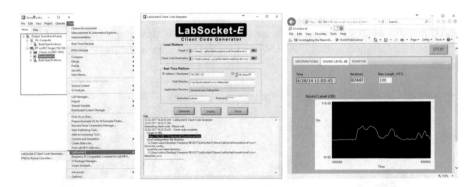

Fig. 6. Generate and deploy the sound level measurement and WEB interface

As we see using LabVIEW and Reconfigurable I/O using myRIO [7] children's and students:

1. Rapidly design custom hardware circuitry using LabVIEW like graphical development tools
2. Learn how to execute parallel tasks simultaneously
3. Do not need knowledge of VHDL or FPGA programming tools to build application and web control
4. Even in the case of the Master of Meloterapy, students learn LabVIEW quickly and can develop simple applications of acoustics, signal analysis, etc.

4   Conclusions

Students can easy use LabVIEW Real-Time, LabSocket-E and National Instruments hardware to create reliable, deterministic systems. Now they work at the university with myRIO and in future at their home using Raspberry PI.

They use myRIO LabVIEW Real-Time features built in tools for developing measurement, control and signal analysis applications and using LabSocket-E to deploy on myRIO and after to be able to control them from any PC or Smart Phone browser.

Select from a variety of scalable hardware targets to fit their application needs and combine this experiments with one fast and easy to use software solution.

Acknowledgment. I like to thanks to Bergmans Mechatronics LLC for their permanent support for Creativity Laboratory with all the new versions of software LabSocket, LabSocket-E and for creative discussions about the software development.

References

1. LabVIEW Introduction. http://www.ni.com/getting-started/labview-basics/
2. Learn myRIO. http://www.ni.com/academic/students/learn-rio
3. LabSocket-E. http://labsocket.com/Download/LabSocket-E_User_Guide.pdf

4. myRIO on STUDICA. http://www.studica.com/product/search?search=myRIO
5. Brain Wave Monitor. https://store.neurosky.com
6. Ursuţiu, D., Samoilă, C., Drăgulin, S., Constantin, F.A.: Investigation of music and colours influences on the levels of emotion and concentration. In: Auer, M., Zutin, D. (eds.) Online Engineering and Internet of Things. Lecture Notes in Networks and Systems, vol. 22. Springer, Cham (2018)
7. myRIO, Project Essentials Guide, Ed Doering. ftp://ni.com/evaluation/academic/ myRIO_project_essentials_guide__Feb_09_2016___optimized.pdf

Poster: Smart Applications for Raising Awareness of Young Citizens Towards Using Renewable Energy Sources and Increasing Energy Efficiency in the Local Community

Radojka Krneta[1(✉)], Snežana Dragićević[1], Andreas Pester[2], and Andreja Rojko[3]

[1] Faculty of Technical Sciences Čačak, University of Kragujevac, Kragujevac, Serbia
{radojka.krneta,snezana.dragicevic}@ftn.kg.ac.rs
[2] Carinthia University of Applied Sciences, Villach, Austria
pester@fh-kaernten.at
[3] ECPE European Center for Power Electronics e.V., Nuremberg, Germany
andreja.rojko@ecpe.org

Abstract. Sustainable use of energy and raising public awareness related to this issue is one of the most important priorities of the local communities all over Europe. Cities are consuming 80% of energy in the world, that is why focus on promoting and investing in energy efficiency (EE) of the city, investing in renewable energy sources (RES) and smart management solutions are essential for the sustainable development of smart cities. There is often a low level of awareness about measures for rational use of energy and energy efficiency among citizens in local communities, given the low energy efficiency in the production, distribution and consumption of energy in all sectors. Education and raising awareness of young citizens, therefore, play a key role in understanding why it is necessary to act locally and what can be done by individuals. In this context, this paper presents planned research activities related to raising awareness and positive attitudes among young citizens towards using RES and increasing EE in the local community within the scientific and technological cooperation between the Faculty of Technical Sciences in Čačak, University of Kragujevac in Serbia and the School of Engineering and IT, Carinthia University of Applied Sciences in Austria. Smart applications that can be used for raising awareness of young citizens towards using RES and increasing EE in the local community, such as online experiments (remote and virtual) with learning outcomes related to RES and EE and online monitoring of energy consumption and RES potential in the local community, are described in this paper. Also, developing of an innovative web-based platform targeted for improving energy efficiency and possibilities of using RES through consumer understanding, engagement and behavioral changes, which would be of use to the local authority in the energy sector, is proposed.

Keywords: Smart city · Energy efficiency · Renewable energy sources
Energy awareness · Smart IT solutions

© Springer International Publishing AG, part of Springer Nature 2019
M. E. Auer and R. Langmann (Eds.): REV 2018, LNNS 47, pp. 728–735, 2019.
https://doi.org/10.1007/978-3-319-95678-7_81

1 Introduction

Sustainable use of energy and public awareness raising is one of the most important priorities of the local communities all over Europe. There is an ever increasing need to adapt the ways in which energy is used in order to reduce energy consumption and limit the pollution associated with the energy use. Cities are consuming 80% of energy in the world, that is why focusing on promoting and investing in EE, RES and smart management solutions are essential for the sustainable development of smart cities. Smart cities deploy information and communication technologies (ICT) to be more intelligent and efficient in the use of energy resources, resulting in cost and energy savings, improved service delivery and quality of life, and reduced environmental footprint. There are plenty research initiatives, projects and papers related to the topic of smart cities. The EU Smart Cities Information System (SCIS) [1] brings together project developers, cities, institutions, industry and experts from across Europe to exchange data, experience and know-how and to collaborate on the creation of smart cities and an energy-efficient urban environment. da Cruz Neto [2] reveals survey results where it was discovered that such projects were already conducted or are currently running in 1119 cities. With the initiative presented in this paper, also a city of Čačak, Serbia has joined this group of the cities.

The institutional framework for the use of RES and implementation of EE measurements is regulated by a number of legal acts of Republic of Serbia [3], including the Energy Law and the Law on Efficient Use of Energy as well as other sectoral laws, programs and strategies governing these areas. Serbia adopted the National Action Plan for RES as a framework for the promotion of energy generated from RES and set the mandatory national targets for the share of renewable energy in gross final consumption of energy (27%) in 2020. National Strategy for Sustainable Development defines the increase of EE as one of the priorities for achieving sustainable development of the country. The Strategy and industrial development policy of Serbia from 2011 to 2020 has identified EE as a significant condition for achieving the goal set - improving competitiveness. Energy Sector Development Strategy of the Republic of Serbia for the period until 2025 with projections by 2030 assumes informing and educating the public about the need for EE improvement and possibilities of the use of RES as very important prerequisite for the desired change. Furthermore, it also recognizes the importance of creating the society consciousness concerning the energy value and the need for its rational use.

The key activity to be undertaken for achieving the goals foreseen in the strategic documents is to ensure the leading role of the local communities in the implementation of efficient use of energy and RES. The city of Čačak has a lasting determination towards the use of RES and EE measurements within its territory. In 2005 the Local Ecological Action Plan document was adopted, in 2011 Strategy for Sustainable Development of the City of Čačak, while Environmental Protection Strategy for the City of Čačak is in the process of adoption.

There is a significant potential in the territory of the city of Čačak for the use of RES sources, especially solar energy, because the city belongs to a favorable geographical area. According to the research of the solar energy potential in the territory of the Republic of Serbia carried out by the Ministry of Science and Environmental Protection in 2004, the annual average of the daily energy of global solar radiation on the horizontal surface for the city of Čačak amounts to 3.8–4.0 kWh/m^2 and the total annual global solar radiation energy on a horizontal surface is 1390–1460 kWh/m^2. For proper use of RES it is necessary to precisely identify the potentials at the local level. For these purposes, the measurements from a meteorological station that was installed in the city during May 2017 has started. This station is the first meteorological station in the Čačak region, which is set up according to the World Meteorological Organization standards and is networked into the system of meteorological stations of the Republic of Serbia.

In the territory of the city of Čačak there is a low level of awareness about measures for rational use of energy and EE, given the low EE in the production, distribution and consumption of energy in all sectors. Education and awareness raising for young citizens, therefore play a key role in understanding why it is necessary to act locally and what can be done by individuals. The local community can take a leading role and with a communication strategy on sustainable energy use they can encourage future local actions.

In the context of education and raising awareness of young citizens about measures for rational energy use and improved energy efficiency, this paper presents planned research activities towards using RES and increasing EE in the local community, within the scientific and technological cooperation between the Faculty of Technical Sciences in Čačak, University of Kragujevac in Serbia and the School of Engineering and IT, Carinthia University of Applied Sciences in Austria.

2 Developing Smart Solutions for Raising Awareness of Young Citizens Towards Using RES and Increasing EE in the Local Community

The scientific and technological cooperation between the Faculty of Technical Sciences in Čačak, University of Kragujevac and the School of Engineering and IT, Carinthia University of Applied Sciences will be established within bilateral project. The project aim is to contribute to the change of young citizens' behavior towards using RES and increasing EE in the local community through their educational experience on this topic and their engagement in the monitoring of energy consumption and RES potential in local community, promoting in this way energy-aware lifestyle.

The project will be realized in the next two years through exchanging experiences and good practices and developing research activities related to raising awareness and positive attitudes among young citizens towards using RES and increasing EE in the local community, as well as planning continuation of cooperation in this area. For the purpose of exchanging experience and transfer knowledge, technologies and best practices, several study visits between Serbian researchers and Austrian researchers will be realized during two project year. The following smart solutions based on ICT will be developed within project research activities:

- online experiments with learning outcomes related to RES and EE
- online monitoring of energy consumption and RES potential in local community
- web-based platform targeted for improving EE and possibilities of using RES in local community.

The change of young citizens' behavior towards using RES and increasing EE in the local community through their educational experience and their engagement in the monitoring of energy consumption and RES potential in local community will ensure their transformation from the passive end-users into actively engaged smart citizens that are using available digital tools.

2.1   Research Activities Leading to Raising Awareness of Young Citizens Towards Using RES and Increasing EE in the City of Čačak

The importance of considering the social side of implementation of renewable energy is widely recognized. The success of such initiatives namely depends to a large extent on their social acceptance, thus it is important to have a clear insight into what would be the social implications of deploying and diffusing renewable energy technologies.

Regarding the current situation in the field of renewable energy applications in the Čačak city, it is important to emphasize the lack of quantitative indicators in the planned areas of renewable energy applications and the efforts to promote their use. Besides the traditional use of wood for heating the use of RES is almost negligible in the City of Čačak.

For the optimal implementation of solar energy sources, it is necessary to precisely identify the potentials and possibilities of its application. It is necessary to create an appropriate solar potential analysis based on the measurement data to show the real situation in the city. Although the units of local self-government are not obliged to have a legal obligation to draft a document that would deal with the use of RES in the territory of the Čačak city, the existence of this document is very significant. Such document would recognize all the potentials of RES and reveal the possibilities for their implementation. In order to use the technologies of renewable energy use on a larger scale, it is necessary to educate the citizens about the possibilities and advantages, but also to invest in the research and realization of projects in this field.

The available data of solar energy and wind energy potential were supplied by the automatic meteorological weather station installed at Fruit Research Institute in Čačak. The meteorological station is installed in the center of town, away of an industrial zone, near the city park. Installed weather station MicroStep-MIS provides on a 1-min basis measurements of wind speed and wind direction, air temperature and relative humidity, atmospheric pressure, solar radiation and sunshine duration, precipitation and snow height (see Fig. 1). The data were recorded in the Renewable Energy Laboratory at the Faculty of Technical Sciences in Čačak and at the same time they are transferred to Republic Hydro-meteorological Service of Serbia.

Raising the level of environmental awareness of citizens is necessary when it comes to the use of energy from RES. Planned research activities leading to raising awareness of young citizens towards using RES and increasing EE in the city of Čačak will include integration of online experiments in learning modules related to RES and EE as well as engagement of young citizens (students from university and secondary schools) in online monitoring of energy consumption and RES potential in local community. Similar practice was already successfully implemented in similar projects [4], where remote and virtual experiments were used to complement learning materials on sustainable energy.

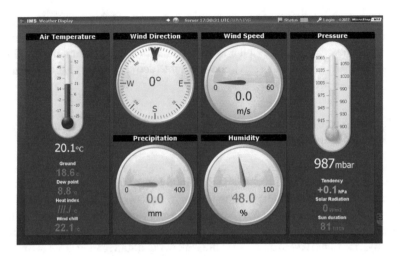

Fig. 1. Display of current measured values in MicroStep-MIS weather station application

Furthermore, it is planned to carry out additional actions for increasing energy awareness and improve sustainable behavior of young citizens in the city of Čačak, similar to the ones proposed in 10ACTION project [5]:

– Actions for children: solar drawing competitions and web game;
– Actions for teenagers: debates, design and photo competitions;
– Actions for university students: debates and ideas competition, exhibitions of renewable energy projects, and workshops.

Changing people's energy-use behaviors must go beyond one-way education. The activities must address barriers to change, as well as making the behaviors easy, convenient, relevant, and socially desirable. Research and case studies have revealed some factors that have proven effective and will be implemented in our activities [6]:

– Personal and interactive contacts: when young people are personally confronted with an opportunity to adopt more energy-efficient behavior, as opposed to having the opportunity presented through information materials or the media, their participation rises dramatically;
– Use vivid, relevant, personalized information: if the desired behaviors are pictured, young people are more likely to visualize themselves doing them;
– Emphasize a positive gain, not deprivation: emphasize what they will gain from adopting certain EE behaviors;
– Encourage active involvement and commitment: encourage public commitments by displaying or publishing the names of people who have pledged to undertake various EE actions;
– Provide incentives: the form of a monetary reward, award plaque or certificate of recognition;
– Promote energy-efficient behavior in conjunction with other environmental programs;

- Provide feedback: feedback showing how much energy the participants have saved over certain periods of time can be a strong motivation for further efforts in this direction.

Online experiments with learning outcomes related to RES and EE can successfully meet most of above mentioned factors leading to changing young citizens' energy-use behaviors. Development of these online experiments will be based on built infrastructure at the Faculty of Technical Sciences in Čačak and on a reach experience of researchers in this technology gained through the project NeReLa [7]. During realization of the project NeReLa, 796 students from four Serbian universities conducted and evaluated 21 different learning modules with remote experiments. Further, 391 student from secondary vocational schools had an opportunity to test NeReLa remote experiments within 27 exemplary classes with remote experiments realized in 16 secondary vocational schools in 11 cities of Serbia. High evaluation grades of university students (average grade of 4.52) and students from secondary vocational schools (average grade of 4.32) proved attractiveness of this educational resources for engineering students at both education level.

For the above listed reasons we decided to develop new set of online experiments (remote and virtual) where the measured data from MicroStep-MIS weather station will be used as input data. By conducting these experiments the university students and the students of secondary schools will solve different assignments with learning outcomes related to using RES and increasing EE. Measured values of environmental parameters will be used as input data for online experiments that students will conduct, thus providing young citizens with models that effectively depicture the environmental reality. By conducting these experiments the students will be enabled to learn how to change the world around them using their own surrounding environment as a living laboratory, transforming them in this way from passive end-users into actively engaged digital smart citizens [8].

Municipalities can encourage and involve their citizens directly in energy saving measures, thus contributing actively to the reduction of energy use and to climate change mitigation [9]. Engagement of young citizens in online monitoring of energy consumption and RES potential should be a part of public awareness raising strategy and actions on energy savings in their local community. Such engagement of digital smart citizens will contribute to a great extent to local community to develop towards smart city. This is in line with the following statement given by Shankar [10] "To ensure a greater share of online participation channels such as through smartphone applications and social media, municipalities needs to invest in smart people – not merely in smart technologies. Only then will tools like smartphones and mobile applications have the potential to revolutionize city governance and contribute to the making of people-centric smart cities. These cities need to have inclusive, innovative, and sustainable smart citizens". In a future step of this project the geographical distribution of EE in the municipalities and the percentage use of RES can be represented. Carinthia University of Applied Sciences (CUAS) has experience in such a geoinformation mapping of data and can provide the necessary guidance [11].

2.2   Web-Based Platform for Improving EE and Possibilities of Using RES in Local Community

The smart city concept assumes the inclusion of ICT solutions as an integral part of the concept [9]. ICT enables the establishment of online platforms for collective action, offering the potential of involving a large number of interest groups and thus enhancing interaction, democracy, and transparency of the planning process. Through such platforms, traditional—face-to-face—ways of establishing interaction have migrated to the Web, providing planners and decision-makers from energy sector in local community a variety of options for "e-engaging" citizens at the various planning stages [12]. In this sense, development of web-based platform dedicated to improving EE and possibilities of using RES in the city of Čačak through engagement of young citizens is planned in the final stage of the project. ICT applications such as web games, renewable energy projects, collections of measured data with the analysis of these data and visualization of result analysis as well as the set of online experiments that will be developed during the project by engaging students from university and secondary schools, will be available via the online platform to other citizens for further promotion of energy saving and using RES as well as to management of local community for decision making in energy sector, leading in this way to urban sustainability. Furthermore, this online platform will be used by local management for their online communication with citizens in the form of news, messages, success stories, recommendations, announcements of various awareness campaign and actions concerning increasing EE and use of RES in local community.

For a broader use and a management of users of online experiments a Remote Lab Management System (RLMS) can be used, to connect the experiments from Čačak with experiments from other providers or universities. CUAS is working on such kind of online lab infrastructures, which can solve questions of scalability, multi-user and multi-experiment access management [13].

3   Conclusions

Smart applications, technologies and services aimed at improving EE and possibilities of using RES through consumer understanding, engagement and behavioral changes are presented. Its innovation impact on raising awareness of young citizens towards using RES and increasing EE in the local community is discussed. Proposed online experiments, online monitoring and web-based platform can support the city and citizens' specific decision making, capable of dealing with objectives for urban sustainability. We are convinced that the results of this research will help the local authorities to use these IT tools for creating the base concept of smart city in order to improve quality of life of its inhabitants.

Acknowledgment. This paper describes bilateral project between Serbia and Austria "Smart solutions for raising awareness of young citizens towards using renewable energy sources and increasing energy efficiency in the local community", which will be supported by the Austrian

Federal Ministry of Science, Research and Economy and by the Ministry of Education, Science and Technological Cooperation of the Republic of Serbia.

References

1. The EU Smart Cities Information System (SCIS). http://smartcities-infosystem.eu/
2. da Cruz Neto, C.D.: Educating smart citizens: skills of the educators in smart cities. In: 20th International Conference on Interactive Collaborative Learning, ICL 2017, Budapest, Hungary, 27–29 September 2017, pp 1538–1544 (2017)
3. Ministry of Mining and Energy of the Republic of Serbia, Sector for Energy Efficiency and Renewable Energy Sources: Laws, Regulations, Regulations, Action Plan for Renewable Energy Sources, Action Plans for Energy Efficiency of the Republic of Serbia. http://www.mre.gov.rs/dokumenta-efikasnost-izvori.php
4. Rojko, A., Bauer, P., Prochazja, P., Pazdera, I., Vitek, O.: Development and experience with ICT based education in sustainably energy. In: 2015 IEEE International Conference on Industrial Technology (ICIT), pp. 3264–3269 (2015)
5. 10 ACTION Project. https://ec.europa.eu/energy/intelligent/projects/sites/iee-projects/files/projects/documents/10action_presentation_en.pdf
6. Creating an Energy Awareness Program - A handbook for federal energy managers, U.S. Department of Energy, Energy Efficiency and Renewable Energy, July 2007. http://www.inogate.org/documents/yhtp_ceap_hndbk.pdf
7. The project NeReLa website. http://nerela.kg.ac.rs/
8. Yacine, A., Stylianos, S., Demetrios, S., Gunnar, M.: A Cyberphysical learning approach for digital smart citizenship competence development. In: 26th International Conference on World Wide Web Companion, Perth, Australia, 3–7 April 2017, pp. 397–405 (2017)
9. Csobod, E., Grätz, M., Szuppinger, P.: Overview and analysis of public awareness raising strategies and actions on energy savings, Report on INTESE project. http://www.intense-energy.eu/uploads/tx_triedownloads/INTENSE_WP6_D61_final.pdf
10. Shankar A.: Why smart cities need smart citizens. The Hindu, 12 September 2016. http://www.thehindu.com/features/homes-and-gardens/why-smart-cities-need-smart-citizens/article8625075.ece
11. Kosar, B., Paulus, G., Erlacher, C.: Entwicklung eines Web-GIS-Portals für Energiekenndaten auf Gemeindeebene. In: Symposium und Fachmesse Angewandte Geoinformatik, Angewandte Geoinformatik 2013, AGIT 2013, Salzburg, Österreich, 3–7 July 2013, pp. 519–528 (2013)
12. Stratigea, A., Papadopoulou, C.A., Panagiotopoulou, M.: Tools and technologies for planning the development of smart cities. J. Urban Technol. 22(2), 43–62 (2015)
13. Garbi-Zutin, D., Auer M., Orduña, P., Kreiter, C.: Online lab infrastructure as a service: a new paradigm to simplify the development and deployment of online labs. In: 13th International Conference on Remote Engineering and Virtual Instrumentation (REV), pp. 208–214 (2016)

Author Index

A

Abreu, Paulo, 507, 536
AbuShanab, Shatha, 175
Achuthan, Krishnashree, 639
Ahlawat, Manish, 234
AlHamad, Ahmad Qasim Mohammad, 659
AlHammadi, Roqayah Abdulraheim, 659
Alves Portela Santos, Eduardo, 425
Al-Zoubi, Abdallah, 244
Anand Kumar, G., 261
Andrieu, Guillaume, 244
Angulo, Ignacio, 290
Antov, Dago, 32
Aranda, Ernesto, 100
Archibong, Mbuotidem Ime, 226
Arras, Peter, 197
Arun Kumar, S., 234
Atzori, Franco, 451
Azad, Abul K. M., 363

B

Babarada, Florin, 489
Baizán-Álvarez, Pablo, 69
Bajči, Brajan, 144
Barañano, Xabier Osorio, 128
Barragán, Pedro Paredes, 298
Batarseh, Majd, 244
Benachenhou, Abdelhalim, 136
Benattia, Abderrahmane Adda, 136
Bengloan, Gildas, 570
Bhimavaram, Kalyan Ram, 694
Biradar, Preeti S., 234
Blanco, Enrique, 649
Blazquez-Merino, Manuel, 69
Bochicchio, Mario A., 517

Bolodurina, Irina, 386, 671, 701
Boukachour, Hadhoum, 57
Brom, Pavel, 118
Brück, Rainer, 175
Brunetti, Giuliana, 451

C

Cai, Su, 549
Callaghan, Michael James, 570
Canals, Txetxu Arzuaga, 128
Cano-Ortiz, Sergio Daniel, 269, 528
Cantrak, Djordje, 304, 686
Cardoso, Alberto, 714
Castro, Manuel, 69, 298
Centea, Dan, 354
Chandan, H. R., 153
Chaos, Dictino, 100
Cherel, Léo, 570
Claesson, Lena, 22
Coppenrath, Matthias, 3
Costa, Claiton, 649
Cristobal, Elio San, 298
Cruz, Eneko, 290
Cuperman, Dan, 610

D

Darshan, H. O., 261
Das, Sayan, 312
Das, Vishnu, 234
de la Torre, Luis, 100
de la Vega Moreno, David, 128
Degreef, Phaedra, 458
Demir, Veysel, 363
Desai, Amrutha, 694
Devang, Nithin, 363

© Springer International Publishing AG, part of Springer Nature 2019
M. E. Auer and R. Langmann (Eds.): REV 2018, LNNS 47, pp. 737–740, 2019.
https://doi.org/10.1007/978-3-319-95678-7

Diaz-Labiste, Pedro Efrain, 269
Díaz-Zayas, Almudena, 110
Diez, Gabriel, 69
Diwakar, Shyam, 639
Dragićević, Snežana, 728
Dudić, Slobodan, 144
Dvorak, Jiri, 118
Dychka, Ivan, 559

E
Ece, Burak, 479
El Mostadi, Mohamed Ali, 570
Elbestawi, Mo, 354
Eng, Simon, 333

F
Fäth, Tobias, 80, 323
Ferreira, Paulo, 244
Ferrer, Julien, 570
Fidalgo, André, 244
Finkelstein, Eldad, 610
Fraga, Alvaro Luis, 627
Franco, Leonardo, 627
Frerich, Sulamith, 620

G
Galan, Daniel, 100
Galler, Kevin, 466
Garbi-Zutin, Danilo, 244
García Pascual, Gustavo, 110
Garcia, Selena, 627
Garcia-Loro, Félix, 69
Garcia-Zubia, Javier, 290
Gaurav, J., 261
Gebeshuber, Klaus, 378
Geiger, Christian, 591
Gericota, Manuel, 244
Gerza, Michal, 226
Gil, Paulo, 714
Golubeva, Tatyana, 165
Gomez Eguíluz, Augusto, 570
Gowripeddi, Venkata Vivek, 694
Gramajo, María Guadalupe, 627
Grieu, Jean, 57
Grout, Ian, 405

H
Haberstroh, Max, 207
Hees, Frank, 207
Henke, Karsten, 80, 323
Hernandez, Unai, 290
Hills, Catherine, 323
Hiremath, Panchaksharayya S., 186

Hutabarat, Windo, 42
Hutschenreuter, René, 80

I
Ilic, Dejan, 304, 686

J
Jacques, Harald, 269, 528
Janardhan, Nithin, 694
Jankovic, Novica, 304, 686
Jekic, Nikolina, 14
Juarez, Gustavo, 627

K
Kafuko, Martha, 90
Kapliienko, Tetiana, 197
Karthick, B., 186
Kist, Alexander A., 323, 415, 581
Klinger, Thomas, 707
Komenda, Tomas, 601
Komura, Ryotaro, 344
Konshin, Sergey, 165
Korkmaz, Hayriye, 479
Krbecek, Michal, 226, 312
Krneta, Radojka, 728
Kumar, Dhanush, 639
Kuriscak, Pavel, 118
Kuska, Robert, 620

L
Ladwig, Philipp, 591
Langmann, Reinhard, 3, 50, 269, 528
Larrondo-Petrie, Maria M., 252
Lecroq, Florence, 57
Legashev, Leonid, 671, 701
Leitão, Joaquim, 714
Leshchev, Sergey, 165
Li, Xiaowen, 549
Li, Yutan, 549
Liaqat, Amer, 42
Liu, Enrui, 549
Lohani, Vinod K., 443
Lombardía-Legrá, Lienys, 528
Lopez-Davalos, Laura, 42
Lundberg, Jenny, 22
Lustig, Frantisek, 118, 312

M
Macho-Aroca, Alejandro, 69
Madritsch, Christian, 707
Maiti, Ananda, 323, 415, 581
Makarova, Irina, 32
Manea, Elena, 489

Marques, Alfeu S., 714
Martinez-Cañete, Yadisbel, 269, 528
Matijevic, Milan, 304, 686
Matsuoka, Masato, 344
Maxwell, Andrew D., 323, 415, 581
McShane, Niall, 570
Meda, Prasanth Sai, 186
Mendes, Joaquim, 498, 507
Mendes, Luciano Antonio, 425
Merino Gómez, Pedro, 110
Mihaiescu, Dan, 489
Mikulowski, Dariusz, 435, 471
Milenković, Ivana, 144
Mironova, Natalia, 165
Mohapatra, Shubham, 694
Monni, Stefano Leone, 451
Moussa, Mohammed, 136
Murgia, Fabrizio, 451

N
Nair, Bipin, 639
Narasimhamurthy, K. C., 153
Neagu, Andrei, 721
Neagu, C. A., 217
Nedeljkovic, Milos Srecko, 304, 678, 686
Neges, Matthias, 620
Nikhil, J., 261
Nizar, Nijin, 639
Nogueira, Daniel, 536

O
Orduña, Pablo, 290
Oyekan, John, 333
Ozvoldova, Miroslava, 226, 312

P
Palacios, Asier Llano, 128
Paliwal, Priyanka, 186
Parfenov, Denis, 386
Parida, Sanjoy Kumar, 234
Parkhomenko, Andriy, 395
Parkhomenko, Anzhelika, 395
Pashkevich, Anton, 32
Pavan, J., 261, 694
Pester, Andreas, 14, 466, 707, 728
Petrovic, Petar Bogoljub, 678
Pilski, Marek, 471
Pita, Itziar Angulo, 128
Polezhaev, Petr, 671, 701
Popescu, Alina, 489
Prabhu, Vinayak, 333
Prajval, M. S., 186
Prathap, S., 234

R
Rábek, Matej, 283
Radhamani, Rakhi, 639
Ram, Kalyan B., 186, 234
Ravariu, Cristian, 489
Reitman, Michael, 610
Reljić, Vule, 144
Restivo, Maria Teresa, 498, 507, 536
Rodriguez-Artacho, Miguel, 298
Rodriguez-Gil, Luis, 290
Rojas-Peña, Leandro, 50
Rojko, Andreja, 728
Romm, Tal, 610

S
Saliah-Hassane, Hamadou, 298
Salis, Carole, 451
Salmerón Moreno, Alberto, 110
Samoila, Cornel, 217, 721
San Cristobal, Elio, 69
Sanabria-Macias, Frank, 269
Sato, Jun, 344
Sayan, Das, 226
Schaedler Uhlmann, Thiago, 425
Schauer, Franz, 226, 312, 601
Schirmbeck, Fernando, 649
Seidel, Felix, 80
Šešlija, Dragan, 144
Sharma, Ankit, 153
Shrikhar, Apurva, 694
Shubenkova, Ksenia, 32
Shubham, Shorya, 153
Shukhman, Alexander, 671, 701
Silva, Pedro, 498
Simões, Nuno E., 714
Singh, Ishwar, 354
Smith, Jeremy Dylan, 443
Smith, Mark, 581
Sokolyanskii, Aleksandr, 395
Sommer, Thorsten, 207
Stehling, Valerie, 207
Stepanenko, Aleksandr, 395
Subert-Semanat, Andrés, 159
Šulc, Jovan, 144
Sulema, Olga, 559
Sulema, Yevgeniya, 559

T
Tabunshchyk, Galyna, 197, 458
Tavares, Rafael, 536
Tiwari, Ashutosh, 42, 333
Tiwari, Divya, 42
Trejo, Federico, 627

Tshukin, Boris, 165
Tulenkov, Artem, 395

U
Uddin, Mohammed Misbah, 363
Ursutiu, Doru, 217, 489, 721
Ushakov, Yuri, 671, 701

V
Vaira, Lucia, 517
Van Merode, Dirk, 458
Vardasca, Ricardo, 498, 507
Verner, Igor, 610
Vijayalakshmi, M. B., 186
Vorhauer, Christoph, 378

W
Wanyama, Tom, 90, 354

Wilson, Marie Florence, 451
Winzker, Marco, 175
Woei, Lim Eng, 333
Wolf, Mario, 620
Wuttke, Heinz-Dietrich, 80, 323

Y
Yamuna Devi, C. R., 261, 694
Yan, Toria, 610
Yayla, Ayse, 479
Yonemura, Keiichi, 344

Z
Žáková, Katarína, 277, 283
Zalyubovskiy, Yaroslav, 395
Zapata-Rivera, Luis Felipe, 252

Printed in the United States
By Bookmasters